PRACTICAL GEOGRAPHY

PRACTICAL GEOGRAPHY

A Systematic Approach

Third Edition

ASHIS SARKAR

Orient BlackSwan

PRACTICAL GEOGRAPHY

ORIENT BLACKSWAN PRIVATE LIMITED

Registered Office
3-6-752 Himayatnagar, Hyderabad 500 029, Telangana, India
e-mail: centraloffice@orientblackswan.com

Other Offices
Bengaluru, Chennai, Guwahati, Hyderabad, Kolkata,
Mumbai, New Delhi, Noida, Patna, Visakhapatnam

First Published by Orient Blackswan Pvt. Ltd. 2015

Series cover and book design

Reprinted 2016, 2017, 2019, 2020, 2022

ISBN 978-81-250-5903-5

Typeset in Electra LT Std 10.5/12 by
Jojy Philip, New Delhi 110 015

Printed in India at
B.B. Press
Noida

Published by
Orient Blackswan Private Limited
3-6-752 Himayatnagar, Hyderabad 500 029, Telangana, India
e-mail: info@orientblackswan.com

For

My dear children

***Buntai** and **Wriju** who motivated me in writing this book with innovative ideas and are now pursuing new goals in life*

Contents

UNIT IV
Field Techniques

UNIT V
Advanced Tools of Analysis

Preface to the Third Edition

This edition is a further revision of *Practical Geography—A Systematic Approach*. The first edition appeared in 1997, and it was first revised in 2009. The earlier editions were so well-accepted by students, scholars, teachers and practitioners of geography, that I felt really honoured and obligated to revise the volume by way of updating the concepts, organising the content in a better and student-friendly manner, correcting the inadvertent printing errors and redrawing some of the figures. The major changes brought about in this edition are:

1. New text added with problems (chapter 1),
2. Revisions in formulae/equations (chapter 2);
3. New text added with revised matter (chapter 3);
4. More elaborate texts added for Correlation, Regression, Hypothesis Testing and Time Series Analysis (chapter 4);
5. Texts revised with further clarity for Hypsometric Curve, Ombrothermic Graph and Gini Coefficient (chapter 5);
6. Examples of Geological Map thoroughly revised and properly arranged (chapter 6);
7. Texts revised for Image Information and Processing; also Practice Stereopairs and Satellite Images added (chapter 9);
8. Text revised with keys to recognising specimens of rocks and minerals; new text added for anhydrite, barite, magnesite, turquose, coal and pumice (chapter 10).

Over the past 32 years, I have been fortunate enough to teach geography at both undergraduate and post-graduate levels in various government colleges across West Bengal and help my students and research scholars understand the mysteries of practical geography, especially the intricacies of field study, cartography, remote sensing and geographical information system. Their enthusiasm and commitment to research have helped me maintain my own level of interest in the discipline.

My students, scholars and colleagues have always been helpful, and discussions with experts from other disciplines have also enriched my knowledge and key concerns. Special thanks are due to Indrita Saha (Assistant Professor, Darjeeling Government College) and Priyank Pravin Patel (Assistant Professor, Presidency University), who were forthcoming with their help and support so that I could revise the book keeping quality in mind. It is impossible to name all those who have aided me in this project. But I remain grateful to them all.

The staff of Orient BlackSwan—Nilanjana Majumdar, Roopa Sharma and Sonali Sengupta—have been very supportive.

Finally, it is my beloved wife, Rupa Sarkar, who deserves special mention for being by my side always with her warm support, inspiration and encouragement.

Ashis Sarkar
Head, PG Department of Geography
Chandernagore College
Government of West Bengal

Chandernagore
January, 2015

Preface to the Second Edition

Since it first appeared in 1997, this edition is a more comprehensive revision of *Practical Geography—A Systematic Approach*, after taking on board suggestions made by my students, readers and colleagues. Seeing how well-accepted the first edition was, I prepared the present edition with students and scholars in mind to provide an introduction to the basic concepts and methods, and to develop a foundation that can be used as the basis for further study and research in geography.

Keeping the basic structure of the book unaltered, I have taken this opportunity to revise all the chapters to bring them up to date, as well as to add some new material, to delete the obsolescent and less interesting paragraphs, and to revise some infelicitous and unintelligible passages. The major changes are: *first*, chapters on statistical data presentation, geographical information system, remote sensing and field study have been thoroughly revised and updated; *second*, the remaining chapters have been reorganised with new additions and alterations; *third*, questions and problems have been appended at the end of each chapter; *fourth*, map samples, viz., weather map and geological map have been added for practice; *fifth*, maps, diagrams and graphs have been electronically produced to enhance cartographic quality; *sixth*, the most important matters have been inserted in box frames; *seventh*, the appendix has been enriched with the frequently asked topics (FAT); and *eighth*, references have been fully updated and reorganised, especially on the more advanced topics for the interested readers to follow.

Over the past 25 years, I have been fortunate enough to teach geography. I started this revision to help my students and research scholars understand the mysteries of practical geography, especially the intricacies of field study, cartography, remote sensing and geographical information system. Their enthusiasm and commitment to research have always been a factor in maintaining my own level of interest, and I take this opportunity to express my gratitude to all of them. I am especially indebted to Mr Priyank Pravin Patel and Ms Lila Mahato, both my research scholars for their thoughtful advice, cartographic and geo-informatic support, and encouragement.

My friends, colleagues and teachers have always been helpful, and the discussions with people from other disciplines have also added to my knowledge and awareness of the key issues. There are many people, in many places, who have assisted me with support, advice and ideas, both directly and indirectly; it is impossible to name all of them, but I am nevertheless grateful. Not to mention, I take full responsibility for all errors and omissions.

The staff of Orient BlackSwan has been extremely supportive, as always. Special mention may be made of Sreyashi Lahiri and Sonali Sengupta. I must thank Dibyendu Nandi for being so committed in applying his electronic skill in cartographic production, for his spirit and attitude, and his real passion for the field geography.

Finally, my wife, my daughter and my son deserve special mention for always being by my side with warm support, infinite inspiration and constant encouragement.

Ashis Sarkar
Head
Department of Geography
Presidency College, Kolkata

Kolkata
October, 2007

Preface to the First Edition

This book is an outcome of my own utter interest in the methodological issues of practical geography. As a student of geography and also as a practising teacher, I observed that there was a dearth of textbooks on practical geography as per the curriculum of the Indian universities. Consequently, I had to rely heavily on a large number of books and research papers which are being published abroad and are therefore not easily available. Even today there is no textbook on practical geography, systematic in treatise, precise, comprehensive and sound in concept and technique. This book, I presume, will surely represent my efforts to fill in this gap.

To serve the undergraduate and post-graduate students, this book is set out in the form of a course in map making (scale, map projection and surveying: chapter 1– 3), statistical analysis (data analysis and representation: chapter 4–6), map interpretation (geological map, weather map, topographical map, aerial photographs and satellite imageries: chapter 6–9), field techniques (identification of rocks and minerals and field study: chapter 10–11) and towards building a geographical information system (GIS: Chapter 12) so that it may be used both as a textbook as well as a reference book. All my efforts would be successful if the students find the book useful.

I would like to thank Bibekananda Roy for drawing up all the graphs, diagrams and maps so very meticulously. Truly speaking, it would be invidious to single out any individual who assisted me in several ways in preparing the manuscript; so I thank them all.

Finally, I am grateful to my wife Rupa Sarkar, who inspired me at every stage of writing this book and patiently tolerated disruptions forced upon her.

Ashis Sarkar
Professor and Head
Department of Geography
Presidency College

Kolkata
May, 1997

Unit I

Map Making

1

Presentation of Map Scales and Measurement on Maps

HIGHLIGHTS

- Statement Scale
- Numeric Ratio Scale
- Graphical Scale
- Construction of Graphical Scales
- Vernier Scale
- Direct or Positive Vernier
- Retrograde or Negative Vernier
- Scale and Map Reduction-Enlargement
- Finding Direction on a Map
- Finding Distance on a Map
- Finding Area on a Map

A map is a graphic representation of the features on the earth's surface. It is, therefore, a storehouse of spatial information, commonly used to evaluate the topologic and metric properties of the geographic features on a map, e.g., *distance, direction, connectivity and proximity*. These attributes enable us to identify the spatial patterns of association of the geographical features. Thus, a map conveys two fundamental properties—*locations and attributes at locations* and is therefore a very powerful tool for the spatial scientists, especially geographers.

Essentially maps bear a definite relationship between what has been represented on it and what exists on the curved surface of the earth. This relationship forms the fundamental basis for the concept of a map scale which is technically defined as the ratio between a distance measured on the map and the corresponding distance on the ground (Fig. 1.1). Hence, a map is simply the *scale-model* of the real world.

Map scales are represented in three basic and reversible forms—a *statement*, a *numeric ratio* and a *graph* and accordingly they are respectively called statement scale, ratio scale and graphical scale (Table 1.1).

On maps it is customary to show a map scale in all the three forms for the sake of the convenience of the user. It is normally placed within a box, being associated with the north arrow, positioned exactly in the middle, just above the scale bar (Fig. 1.2). The location of the scale box depends on the

Fig. 1.1 Concept of Scale

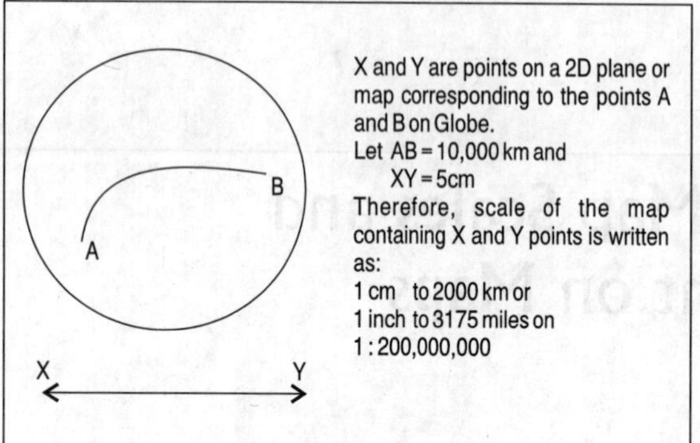

Table 1.1 Map Scales

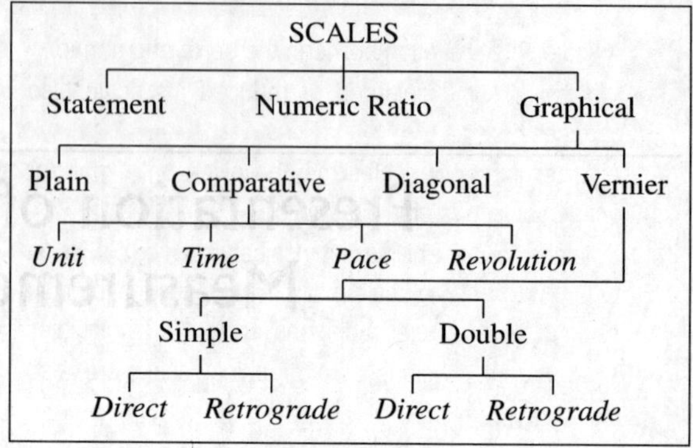

page layout, the best cartographic presentation being decided by—page border, location, size and shape of the outline map, available space, map index, key or legends.

STATEMENT SCALE

A map scale can well be represented in the form of a simple statement, in which the map distance is always expressed as a unit length. This is known as a *statement scale* and is written on a map as, 1 cm to 25 km or 1 inch to 50 miles. The value on the left-hand side of the statement indicates the map distance (cm/inch) and that on the right-hand side indicates the ground distance (km/mile).

Despite its simplicity, the statement scale has two demerits. First, a layman, ignorant of the various scales of measurements, will not be able to read the map. Second, a reproduction of the original map by reduction or enlargement necessitates a determination of the scale all over again.

NUMERIC RATIO SCALE

A map scale may also be expressed as a *numeric ratio*, in which the numerator is the map distance and the denominator is the corresponding ground distance, both being expressed in the same unit of measurement. Therefore, it is a dimensionless fraction. It is also known as a *ratio scale* or *Representative Fraction* (R.F.) (Fig. 1.2). For example, 1:100,000.

The greatest advantage of a numeric ratio scale is that a map with a given R.F. can be used universally. A statement scale or a graphical scale in any system of measurement can easily be computed from the R.F. and cartographically plotted as well.

Thus, 1:100,000 may be written as:

1 cm to 100,000 cm (CGS system)

$$\text{or,}\quad 1 \text{ cm to } \frac{100{,}000}{100{,}000} \text{ km}$$

or, **1 cm to 1 km**

1 inch to 100,000 inch (FPS system)

or, 1 inch to 100,000 ÷ 63,360 miles

or, **1 inch to 1.58 miles**

From a statement scale equation, R.F. can easily be determined. For example, 1 cm to 25 km may be

Fig. 1.2 Map Scales

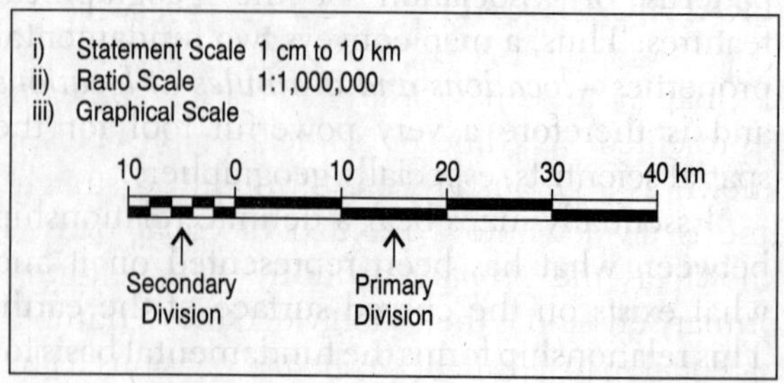

Properties of Scale

- A scale sets a relation between the earth and a map.
- When a map is enlarged or reduced, its scale also changes.
- All three forms of scales may be given on a map for optimum cartographic abstraction.
- Data about a straight line distance between two places on a map with a given scale may be conveniently acquired using a divider, a ruler, a diagonal scale and a calculator.
- The primary divisions of a graphical scale are also called cycles.
- The secondary divisions are very important as their magnitude defines the degree of precision of a map scale.
- The ground distance from a cadastral map/plan is precisely measured with a diagonal scale.
- To keep time, pace scales are largely used by the army in the field.
- A Vernier scale is always fitted with precision instruments like barometer, sextant, theodolite, planimeter, slide callipers, screw gauge, spherometer, etc.

Construction of Plain and Comparative Scales

Plain Scale

- Select and fix the magnitude of ground distance for one primary division and also for one secondary division.
- Based on the given problem or RF or statement, compute the map distance in cm or inch (x) for the corresponding value of one primary division.
- Draw a horizontal straight line and divide it with x for primary divisions (minimum number of division = 3).
- Graphically divide the first primary division (from the left side) into the desired number of secondary divisions.

Comparative Scale

- Draw two plain scales with identical numerical values of primary and secondary divisions in two different measurement units.
- Superimpose the two plain scales (one above the other) such that their 0-marked lines coincide.

written as 1 cm to 25×10^5 cm. Hence, R.F. becomes $1{:}25\times10^5$. Similarly, 1 inch to 50 miles may be written as 1 inch to 50×63,360 inch or 1 inch to 3,168,000 inch. Therefore R.F. becomes $1{:}32\times10^5$.

Note:

i. In R.F. conversion, both sides of the statement are expressed in the same unit of measurement.

ii. The denominator of the R.F. should be rounded off to the nearest hundred or thousand or million as applicable.

GRAPHICAL SCALE

A map scale may also be cartographically represented by a line or a linear graph. This is known as a *graphical scale.* Primarily, a straight line is divided into a number of equal parts—known as the *primary* divisions. The left-most primary division is then subdivided into a number of equal parts, called *secondary* divisions. The value of one secondary division defines the precision of the map scale (Fig. 1.1). A graphical scale can be constructed from either a statement scale or a numeric ratio scale. It can then be conveniently used by a layman to measure distances on a map. The graphical scale may take four different forms—a *plain scale*, a *comparative scale*, a *diagonal scale* and a *vernier scale*.

CONSTRUCTION OF GRAPHICAL SCALES

Plain Scale

This is the simplest form of a graphical scale and is shown as a linear graph (Fig. 1.2 and Fig. 1.3).

The primary or fundamental divisions (at least three) represent the ground distance in multiples of 1, 5 or 10. The primary division on the extreme left is subdivided into a suitable number of secondary divisions. The length of each secondary division depends on the

Table 1.2 Value of Primary and Secondary Divisions and R.F. Dimensions

R.F.	Value of a Primary Division (for 10 < x <100)		Number of Secondary Divisions
$1:x \times 10^6$	500 or 1000	(kilometres or miles)	5 or 10
$1:x \times 10^5$	50 or 100	(kilometres or miles)	5 or 10
$1:x \times 10^4$	5 or 10	(kilometres or miles)	5 or 10
$1:x \times 10^3$	1	(kilometres or miles)	5 or 10
	250 or 500	(metres or yards)	
$1:x \times 10^2$	25 or 50 or 100	(metres or yards)	5 or 10
$1:x \times 10^1$	5 or 10	(metres or yards)	5 or 10
$1:x \times 10^0$	1	(metres or yards)	3 or 5 or 10

Fig. 1.3 Plain Scale

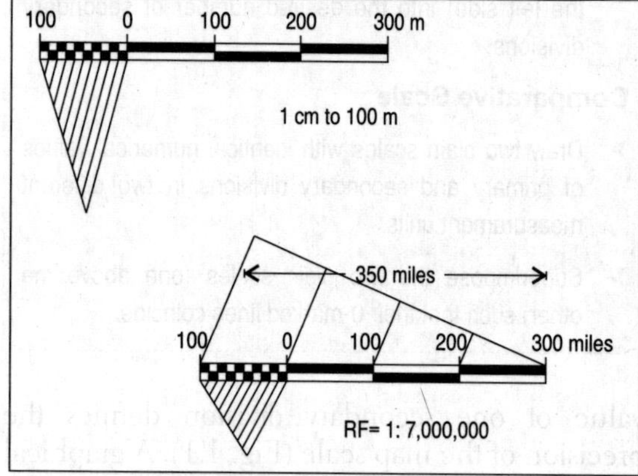

fractional length up to which precision in measurement is desired.

As a construction principle, the map distance for each primary division is first evaluated from either a statement scale or R.F. (Table 1.2). A straight line is drawn horizontally with a length suitable for the map (convenient class-room size being 6 inch or 15 cm). With the help of a divider and a diagonal scale, primary divisions are cut off and perpendicular straight lines (of ≤ 0.5 cm or 0.3 inch) are drawn at each division point. The first primary division from the left is then graphically subdivided into secondary divisions with the desired precision. The drawing is then properly labelled and bordered.

Example 1
Construct a scale to show 100 m on a primary division and 10 m on a secondary division from the statement, 1 cm to 50 m.

Calculation

A ground distance of 50 m ≅ 1 cm of map distance

or, a ground distance of 1 m ≅ $\frac{1}{50}$ cm of map distance

∴ a ground distance of 100 m ≅ $\frac{1}{50} \times 100$ cm of map ⇒ **2 cm**

Hence, the length of one primary division on a map for a ground distance of 100 m is 2 cm.

Example 2
Construct a scale to show 350 miles for R.F. 1:7,000,000.

Calculation
From the given R.F., the value of one primary division has been selected as 100 miles (Table 1.2).

Therefore, 7,000,000 inch on ground is represented by 1 inch on map

or, $\frac{7,000,000}{63,360}$ miles on ground ≅ 1 inch on map

or, 1 mile on ground ≅ $\frac{63,360}{7,000,000}$ inch on map

∴ 100 miles on ground

$$\cong \frac{63,360}{7,000,000} \times 100 \text{ inch on map} \Rightarrow 0.91 \text{ inch}$$

Hence, the map distance corresponding to 100 miles of ground distance is 0.91 inch. It is the length of one primary division on the map.

Comparative Scale

It is defined as a composite plain scale in which two linear scales representing different units of measurement, or time and distance, or pace and distance, or revolution and distance are superimposed for the sole purpose of comparison. They are respectively known as *Unit Scale, Time Scale, Pace Scale* and *Revolution Scale*.

Unit Scale

This type of comparative scale shows a comparison of distance measured in different but comparable units, e.g., kilometres-miles, metres-yards, etc. The two basic principles of construction are:

1) in both the scales, the ground distance equivalents of the primary and the secondary divisions are identical, and
2) the zeros of both the scales coincide during superimposition (Fig. 1.4).

Fig. 1.4 Unit Scale

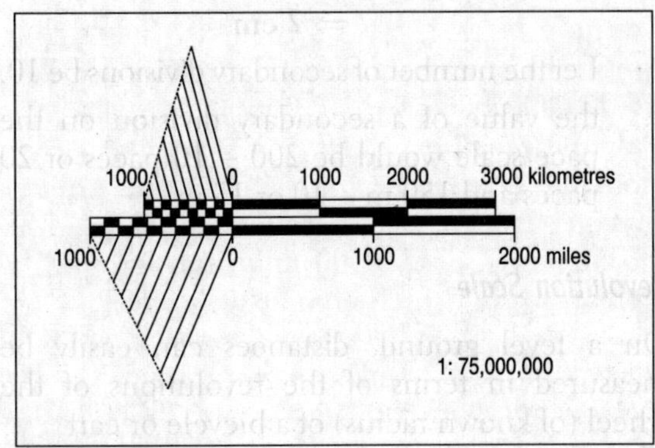

Example 3

Draw a comparative scale to show kilometres and miles on a map with R.F. 1:75,000,000.

Calculation

The value of a chosen primary division is 1,000 km (CGS system) and 1,000 miles (FPS system).

(i) to find a map distance corresponding to 1,000 km of ground distance:

75,000,000 cm on ground ≅ 1 cm on map

or, $\frac{75,000,000}{100,000}$ km on ground ≅ 1 cm on map

or, 1 km on ground ≅ $\frac{100,000}{75,000,000}$ cm on map

∴ 1,000 km on ground ≅

$\frac{100,000 \times 1000}{75,000,000}$ cm on map ⇒ **1.33 cm**

(ii) to find map distance corresponding to 1,000 miles of ground distance:

75,000,000 inch on ground ≅ 1 inch on map

or, $\frac{75,000,000}{63,360}$ miles on ground ≅ 1 inch on map

or, 1 mile on ground ≅ $\frac{63,360}{75,000,000}$ inch on map

∴ 1,000 miles on ground ≅

$\frac{63,360 \times 1,000}{75,000,000}$ inch on map ⇒ **0.84 inch**

Time Scale

This represents a relationship between time and distance. Usually, the speed of a moving body is transferred to a map scale in such a way that the ground distance describing the primary division in one is exactly divisible by the time interval describing the same primary division in the other (Fig. 1.5).

Fig. 1.5 Time Scale

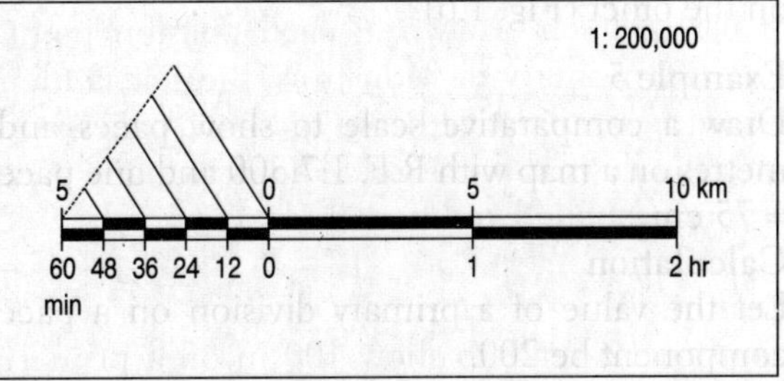

Example 4

Construct a comparative scale to show the march of a battalion advancing at the rate of 5 km per hr on a map with R.F. 1:200,000.

Calculation

The value of a chosen primary division for a distance scale is 5 km and the same on a time component is 1 hr or 60 min.

From the given R.F.,

$$200{,}000 \text{ cm on ground} \cong 1 \text{ cm on map}$$

$$\text{or, } \frac{200{,}000}{100{,}000} \text{ km on ground} \cong 1 \text{ cm on map}$$

$$\text{or, } 1 \text{ km on ground} \cong \frac{100{,}000}{200{,}000} \text{ cm on map}$$

$$\text{or, } 5 \text{ km on ground} \cong \frac{100{,}000 \times 5}{200{,}000} \text{ cm on map}$$

$$\Rightarrow \mathbf{2.5 \text{ cm}}$$

Let, the value of a secondary division on the distance scale be 1 km.

Therefore, the number of secondary divisions

$$= \frac{5 \text{ km}}{1 \text{ km}} \Rightarrow 5$$

Now, the value of one secondary division on the time scale $= \frac{1 \text{ hr}}{5} \Rightarrow \mathbf{12 \text{ min}}$

Pace Scale

It is a type of time scale especially used by the army during a quick reconnaissance survey. The length of a standard military pace is 30 inch or 75 centimetres. The comparative scale shows distance (yards or metres) on one side and pace on the other (Fig. 1.6).

Example 5

Draw a comparative scale to show paces and metres on a map with R.F. 1:7,500 and one pace = 75 cm.

Calculation

Let the value of a primary division on a pace component be 200.

Fig. 1.6 Pace Scale

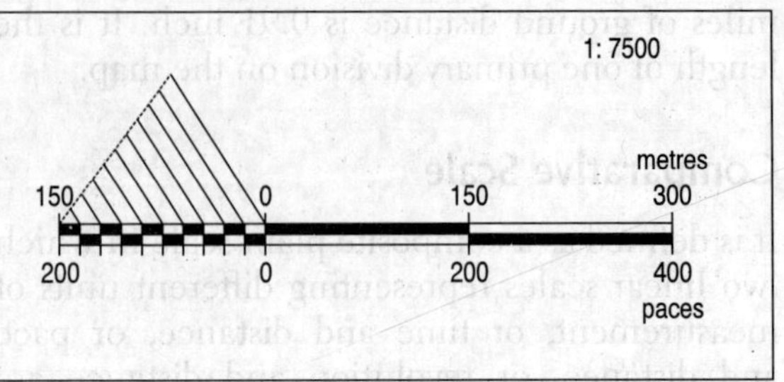

Therefore, 200 paces = 200 × 75 cm

$$= 15{,}000 \text{ cm}$$

$$= \frac{15{,}000}{100} \Rightarrow 150 \text{ m}$$

From the given R.F.,

$$7{,}500 \text{ cm on ground} \cong 1 \text{ cm on map}$$

$$\text{or, } \frac{7{,}500}{100} \text{ m on ground} \cong 1 \text{ cm on map}$$

$$\text{or, } 1 \text{ m on ground} \cong \frac{100}{7{,}500} \text{ cm on map}$$

$$\text{or, } 150 \text{ m on ground} \cong \frac{100 \times 150}{7{,}500} \text{ cm on map}$$

$$\Rightarrow \mathbf{2 \text{ cm}}$$

Let the number of secondary divisions be 10.

∴ the value of a secondary division on the pace scale would be 200 ÷ 10 paces or 20 paces and 150 m ÷ 10 or 15 m.

Revolution Scale

On a level ground, distances can easily be measured in terms of the revolutions of the wheel (of known radius) of a bicycle or cart. Therefore,

Ground Distance = Circumference of the wheel × No. of revolutions

The *revolution* scale shows distance on one side and the number of revolutions of the other. It is also a special type of time scale (Fig. 1.7).

Fig. 1.7 Revolution Scale

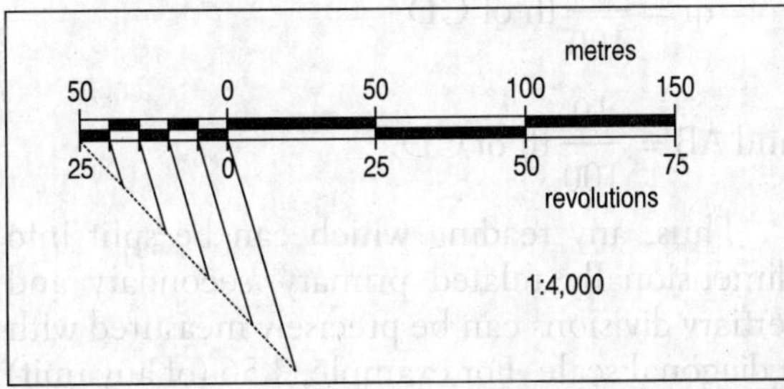

Example 6

Draw a comparative scale to show the distance and revolution of a bicycle wheel on a cadastral map with R.F. 1:4,000. The radius of the wheel is 31.83 cm.

Calculation

From the given R.F.,

4,000 cm on ground ≅ 1 cm on map

or, $\frac{4,000}{100}$ m on ground ≅ 1 cm on map

or, 1 m on ground ≅ $\frac{100}{4,000}$ cm on map

∴ 50 m on ground ≅ $\frac{100 \times 50}{4,000}$ cm on map

⇒ **1.25 cm**

Now,

Circumference of the wheel = 2.π.(radius)
= 2.π.31.83 cm
= 200 cm
= 2m.

∴ Number of revolutions in 50 m = $\frac{50\text{ m}}{2\text{ m}}$

⇒ **25**

Let, the number of secondary divisions be 5. Therefore, the values of the secondary divisions would be 50 ÷ 5 or 10 revolutions.

Diagonal Scale

In small and medium scale maps, accurate measurements can be obtained upto the smallest unit on the secondary division with a plain or comparative scale. However, on large scale or cadastral maps and plans (commonly used by surveyors, planners and geographers), measurements either in three dimensions, or upto 1/100th part of a primary division, are often desired, and here lies the importance and use of a diagonal scale.

A diagonal scale, unlike others, contains three types of divisions—*primary*, *secondary* and *tertiary*. Here, the total value of the tertiary divisions is exactly equal to the value of one secondary division. Similarly, the total value of the secondary divisions is equal to the value of one primary division. The principle of construction is based on the properties of similar triangles explained below:

In Fig.1.8, let CD represent a primary division on which a rectangle ACDE is drawn such that AE = DC and DE = CA and each angle is a right angle. DE and CA are graphically divided into ten equal parts (i.e., tertiary divisions) and through each division point horizontal lines are drawn parallel to DC. Similarly, DC and EA are graphically divided into ten equal parts (i.e., secondary divisions). B, the first division

Construction of a Diagonal Scale

- Break up the reading to fix the map distances for one primary, one secondary and one tertiary division.
- Find the magnitudes of ground distance for one primary, one secondary and one tertiary division.
- Find the number of primary, secondary and tertiary divisions for the scale reading to be shown.
- First, draw a plain scale with suitable numbers of primary and secondary divisions.
- Tertiary divisions are then drawn perpendicularly on the leftmost division of the main scale.
- Horizontal lines parallel to the base of the main scale are drawn through each tertiary division points.
- Secondary divisions are then marked off on the topmost horizontal line of the leftmost primary division. These are finally joined diagonally with those at the base of the left primary division.

Fig. 1.8 Graphical Construction of Secondary and Tertiary Divisions

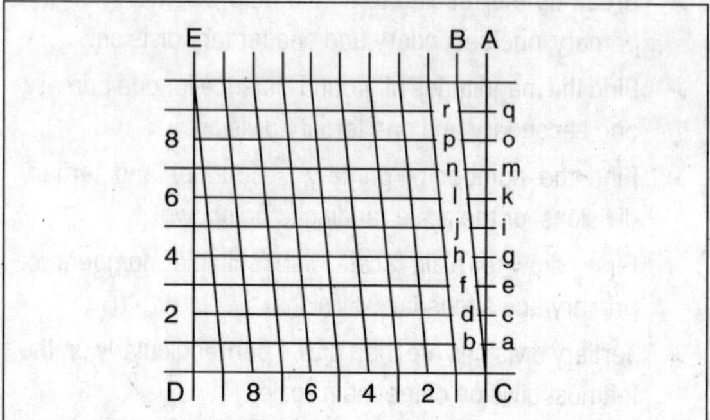

point to the left of A on AE, is joined diagonally with C and then lines parallel to BC are drawn through each secondary division point. The name diagonal scale is derived from the way in which the secondary divisions are joined, i.e., diagonally.

The mesh of lines form ten triangles (abC, cdC, efC, ghC, ijC, klC, mnC, opC, qrC and ABC) with identical angles and thus all are similar triangles.

Therefore, $\dfrac{ab}{AB} = \dfrac{Ca}{CA}$

or, $ab = AB.\dfrac{Ca}{CA}$

$= \dfrac{CA}{10} \cdot \dfrac{CD}{10} \cdot \dfrac{1}{CA}$

$= \dfrac{CD}{100}$

$= \dfrac{1}{100}$ th of CD (primary division)

Similarly,

$cd = \dfrac{2}{100}$ th of CD,

$ef = \dfrac{3}{100}$ th of CD,

$qr = \dfrac{9}{100}$ th of CD,

and $AB = \dfrac{10}{100}$ th of CD.

Thus, any reading which can be split into dimensionally related primary, secondary and tertiary divisions can be precisely measured with a diagonal scale. For example, 4.56 (of any unit) may be written as $\Rightarrow 4.0 + 0.5 + 0.06$

$\Rightarrow 4 \times 1.0 + 5 \times 0.1 + 6 \times 0.01$

$\Rightarrow 4 \times 1 + 5 \times \dfrac{1}{10} + 6 \times \dfrac{1}{100}$

Here, 4, 5 and 6 are the readings on the primary, secondary and tertiary divisions while their respective values are 1, $\dfrac{1}{10}$ th and $\dfrac{1}{100}$ th.

Similarly a reading 789 (of any unit) may be written as $\Rightarrow 700 + 80 + 9$

$\Rightarrow 7 \times 100 + 8 \times 10 + 9 \times 1$

Here, 100, 10 and 1 are respectively the values of the primary, secondary and tertiary divisions and 7, 8 and 9 are their corresponding readings on the diagonal scale. Readings with three related units can also be conveniently shown on a diagonal scale, the dimensional relation determining the number and value of the secondary and tertiary divisions respectively. For example, kilometre-metre-centimetre, mile-furlong-yard, furlong-yard-foot, yard-foot-inch, metre-decimetre-centimetre, decimetre-centimetre-millimetre, etc.

Example 7

Draw a diagonal scale to read 3.47 inches on R.F. 1:1 (Fig. 1.9).

Calculation

Reading to be shown: 3.47 inch

$\Rightarrow (3.00 + 0.40 + 0.07)$ inch

$\Rightarrow (3 \times 1.00 + 4 \times 0.10 + 7 \times 0.01)$ inch

Therefore, a primary division is to be divided into 10 equal parts $\left(\dfrac{1\,\text{inch}}{0.1\,\text{inch}}\right)$ and a secondary

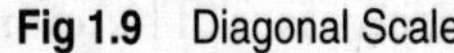

Fig 1.9 Diagonal Scale

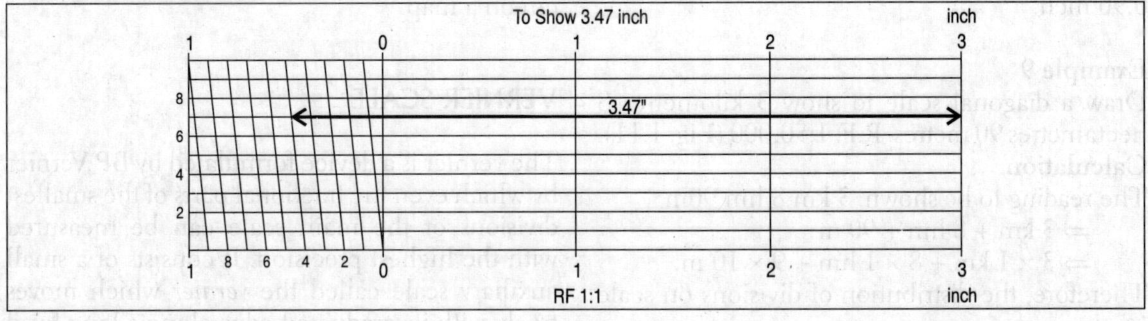

Fig 1.10 Diagonal Scale

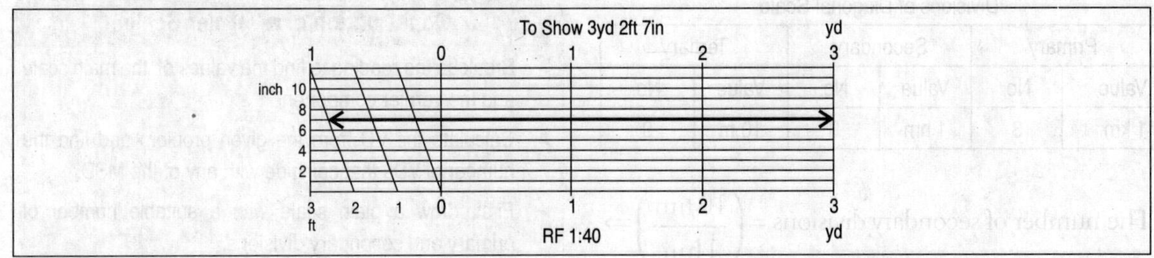

division also in 10 equal parts $\left(\dfrac{0.1\,\text{inch}}{0.01\,\text{inch}}\right)$ with the following readings:

Divisions of Diagonal Scale

Primary		Secondary		Tertiary	
Value	No.	Value	No.	Value	No.
1.00 inch	3	0.10 inch	4	0.01 inch	7

Since the R.F. is 1 : 1, the length of a primary division = **1.00 inch**

Example 8

Draw a diagonal scale to show 3 yards 2 feet 7 inch on R.F. = 1:40 (Fig.1.10).

Calculation

The reading to be shown: 3 yards 2 feet 7 inch

$\Rightarrow$ 3 yards + 2 feet + 7 inch

$\Rightarrow 3 \times 1$ yards + 2×1 ft + 7×1 inch

Therefore, the distribution of divisions on the diagonal scale is:

Divisions of Diagonal Scale

Primary		Secondary		Tertiary	
Value	No.	Value	No.	Value	No.
1 yard	3	1 ft	2	1 inch	7

The number of secondary divisions = $\left(\dfrac{3\,\text{ft}}{1\,\text{ft}}\right) \Rightarrow 3$, because 3 feet make a yard. Similarly, the number of tertiary divisions is $\left(\dfrac{12\,\text{inch}}{1\,\text{inch}}\right) \Rightarrow 12$, since 1 ft = 12 inch.

Now, from the given R.F.,

40 inch on ground $\cong$ 1 inch on map

or, $\left(\dfrac{40}{36}\right)$ yards on ground $\cong$ 1 inch on map

or, 1 yard on ground $\cong \left(\dfrac{36}{40}\right)$ inch on map

$\Rightarrow$ **0.90 inch**

Therefore, the length of a primary division is 0.90 inch.

Example 9

Draw a diagonal scale to show 3 kilometres 8 hectametres 90 metres. R.F. 1:50,000 (Fig. 1.11).

Calculation

The reading to be shown: 3 km 8 hm 90m

$\Rightarrow$ 3 km + 8 hm + 90 m

$\Rightarrow 3 \times 1$ km + 8×1 hm + 9×10 m.

Therefore, the distribution of divisions on scale is:

Divisions of Diagonal Scale

Primary		Secondary		Tertiary	
Value	No.	Value	No.	Value	No.
1 km	3	1 hm	8	10 m	9

The number of secondary divisions = $\left(\frac{10\text{ hm}}{1\text{ hm}}\right) \Rightarrow$ 10, because 1 km = 10 hm and the number of tertiary divisions = $\left(\frac{100\text{ m}}{10\text{ m}}\right) \Rightarrow$ 10, because 1 hm = 100 m.

Now from the given R.F.,

50,000 cm on ground $\cong$ 1 cm on map

or, $\left(\frac{50,000}{100,000}\right)$ km on ground $\cong$ 1 cm on map

$\therefore$ 1 km on ground $\cong \left(\frac{100,000}{50,000}\right)$ cm on map

$\Rightarrow$ **2.0 cm**

Therefore, the length of a primary division is 2 cm on a map.

VERNIER SCALE

The vernier is a device formulated by BP Vernier by which even the fractional parts of the smallest divisions of the main scale can be measured with the highest precision. It consists of a small auxiliary scale called the *vernier* which moves freely with its graduated edge along a long fixed

Construction of a Vernier Scale

- Break up the reading to find the values of the main scale and the vernier component
- Calculate the VC from the given problem and find the number of VDs that coincide with any of the MSD.
- First, draw a plain scale with a suitable number of primary and secondary divisions.
- Based on the relations of the vernier and main scale, divide the MSD-length (preferably on the left side) into the required number of VDs (x).
- Take, on a screw-fitted divider, the length corresponding to the distance covered by the coinciding VD.
- Place one end of the divider on the MSR-mark and slide it along the main scale until the other end coincides with a MSD.
- Now, mark the position of the left end of the divider on the main scale.
- Starting from this point, draw vernier divisions (x) using a paper strip.

Fig 1.11 Diagonal Scale

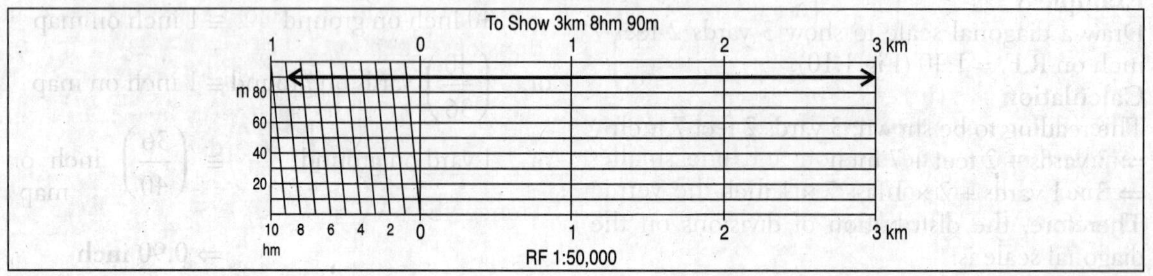

scale called the main or primary scale. The vernier carries an *index mark* that forms the zero of the vernier divisions and is denoted by an *arrow mark*. The scale may be drawn as a straight linear graph or as an arc. The device is based upon the fact that it is easier to see the exact coincidence of two lines than to judge the distance between the two lines and is used in precision instruments like theodolite, sextant, barometer, planimeter, abney's level, spherometer, screw gauge, slide callipers, etc.

Verniers may be of two types—*single* and *double*—depending on its construction respectively on one or both sides of the index mark. However, from the point of view of principles of construction, verniers may be divided into two classes—*direct* or *positive* and *retrograde* or *negative*.

DIRECT OR POSITIVE VERNIER

To read $1/n$ th part of the smallest division on the main scale (n being the total number of small divisions), (n – 1) primary divisions are taken and divided into n equal divisions on the vernier scale. The value, linear or angular, of one smallest main scale division is called the *least count* (d) of the main scale. Therefore, the value of (n – 1) division on the main scale is (n – 1)d ...(i)

Now, if the value of one smallest division on the vernier scale is v, the same reading on the vernier scale would be n.v ... (ii)

On principle, n.v = (n – 1)d

$$v = \left(\frac{n-1}{n}\right)d$$

$$= \left(1-\frac{1}{n}\right).d \qquad \text{... (iii)}$$

The difference between the values of one smallest main scale division and one smallest vernier scale division is defined as the vernier constant (VC). Therefore,

$$VC = d - \left(1-\frac{1}{n}\right)d$$

$$= d - d + \frac{d}{n}$$

$$= \frac{d}{n} \qquad \text{... (iv)}$$

While reading a vernier scale, the values of d and n are noted first and the vernier constant is evaluated. Next, it is carefully noted which of the vernier divisions exactly coincides with the main scale divisions. If this is the ith division, then the vernier scale reading showing the fractional part,

$$= i \times VC$$

$$= i \times \frac{d}{n} \qquad \text{... (v)}$$

Hence,

Total reading = Main scale reading + Vernier scale reading

= Main scale reading + (Vernier constant × No. of vernier divisions coinciding with any of the main scale divisions) ... (vi)

In direct verniers, the vernier divisions are shorter than the main scale divisions. Since both scales are graduated in the same direction, it is simpler to use and easy to construct.

RETROGRADE OR NEGATIVE VERNIER

In this type, (n + 1) smallest main scale divisions are taken and divided into n equal vernier scale divisions. Therefore,

$$n.v = (n+1)d$$

$$v = \left(1+\frac{1}{n}\right)d$$

Hence, the VC = (v – d)

Fig 1.12 Positive Vernier Scale

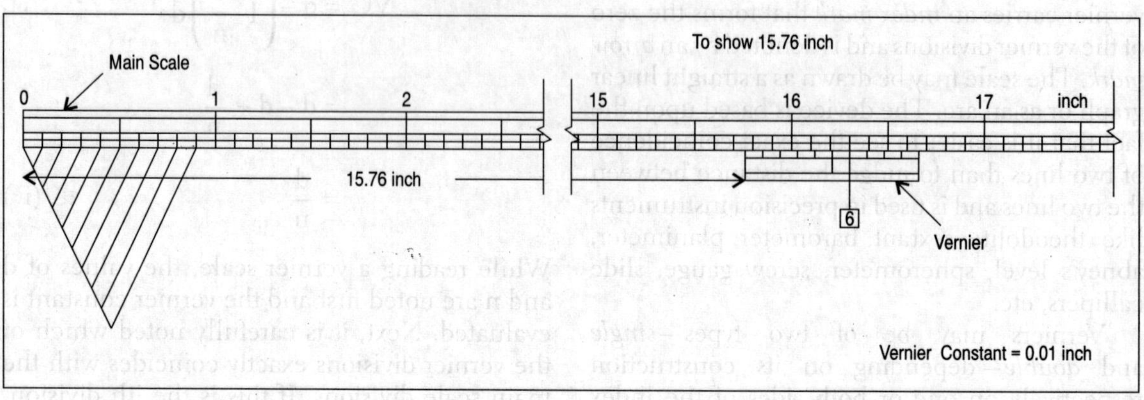

$$= \left(1+\frac{1}{n}\right)d - d$$

$$= \frac{d}{n}.$$

The principle remaining the same, the vernier divisions are longer than the main scale divisions. Therefore, the vernier scales are graduated in the direction opposite to that of the main scale.

Example 10

Draw a vernier scale to read 15.76 inch, given the least count of the main scale is 0.1 inch and 9 small main scale divisions are equal to 10 small vernier scale divisions (Fig.1.12).

Calculation

The least count of the main scale, d = 0.1 inch and the number of vernier scale divisions, n = 10

Therefore,

$$\text{Vernier Constant, VC} = \frac{d}{n}$$

$$= \frac{0.1\text{ inch}}{10} \Rightarrow \mathbf{0.01\ inch}$$

The reading to be shown

= 15.76 inch

= (15.70+0.06) inch

∴ the main scale reading = 15.70 inch

and the vernier scale reading = 0.06 inch

Hence, the index mark would lie between readings 15.7 inch and 15.8 inch on the main scale. The number of vernier divisions coinciding with any of the main scale divisions,

$$= \frac{\text{Vernier Scale Reading}}{\text{Vernier Constant}}$$

$$= \frac{0.06\text{ inch}}{0.01\text{ inch}} \Rightarrow 6$$

Example 11

Draw a vernier scale to read 4,000 miles, given R.F. 1:75 × 10^6. Precision of measurement must be 1/100th part of an inch.

Calculation

From the given R.F., the map reading for the ground distance of 4,000 miles,

$$= \frac{63{,}630}{75\times10^6} \times 4{,}000\text{ inch}$$

= 3.38 inch

= (3.30 + 0.08) inch

Therefore, the vernier scale reading is 0.08 inch and the vernier arrow would lie between 3.30 inch and 3.40 inch on the main scale.

$$\text{Vernier Constant} = \frac{1}{100}\text{ inch} \Rightarrow 0.01\text{ inch.}$$

Therefore, the least count of the main scale is 0.1 inch and there are 10 small vernier scale divisions to equal 9 small main scale divisions.

Fig 1.13 Positive Vernier Scale

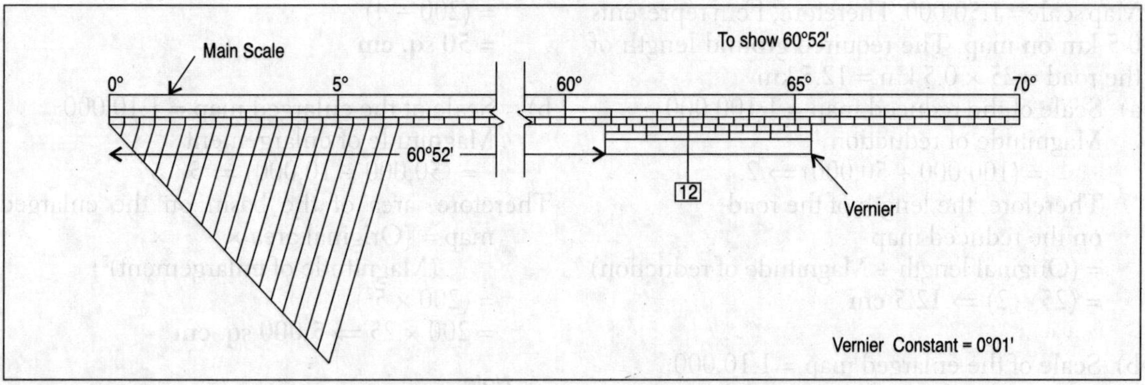

Now, the number of vernier divisions coinciding with any of the main scale divisions,

$$= \frac{0.08\,\text{inch}}{0.01\,\text{inch}} \Rightarrow 8$$

Example 12

Draw a vernier scale to read 60°52', given least count of the main scale is 20' and 19 small scale divisions are equal to 20 small vernier scale divisions (Fig. 1.13).

Calculation

The least count of the main scale,

$d = 0°20'$.

and the number of vernier scale divisions,

$n = 20$.

Therefore, the vernier constant,

$$VC = \frac{d}{n}$$

$$= \frac{0°20'}{20} \Rightarrow 0°1'$$

Now, the readings to be shown = 60°52'

$\Rightarrow 60°40' + 0°12'$

Therefore, the vernier scale reading is 0°12' and the vernier arrow mark would lie between 60°40' and 60°60' or 61°00' on the main scale. The number of vernier divisions coinciding with any of the main scale divisions

$$= \frac{0°12'}{1'} \Rightarrow \mathbf{12}$$

SCALE AND MAP REDUCTION-ENLARGEMENT

While working with maps it often becomes necessary to reproduce the original map in different desired sizes and layouts. Naturally the scale of the map changes with its reproduction as the original map is either enlarged or reduced. The following points are very important in this context:

- Large scale maps have smaller values of denominator in R.F. and vice versa.
- Measurements on maps, linear or areal, depends on map scale.
- Opisometer or Rotameter is used to measure distance between points on a map.
- Planimeter is used to measure area on a map
- Pantagraph is used for reduction and enlargement of maps.
- Reduction and enlargement of maps change its scale and thereby the linear and areal measurements.

Example 13

Length of a road on a map with scale 1:50,000 is measured as 25 cm. Find its ground length. Find also the map-lengths of the same: a) when the map is reduced to 1:100,000 and b) when the map is enlarged to 1:10,000.

Calculation

Map scale = 1:50,000. Therefore, 1 cm represents 0.5 km on map. The required ground length of the road = 25 × 0.5 km = 12.5 km

a) Scale of the reduced map = 1:100,000
Magnitude of reduction
= (100,000 ÷ 50,000) ⇒ 2.
Therefore, the length of the road on the reduced map
= (Original length ÷ Magnitude of reduction)
= (25 ÷ 2) ⇒ **12.5 cm**

b) Scale of the enlarged map = 1:10,000
Magnitude of enlargement
= (50,000 ÷ 10,000) ⇒ 5
Therefore, the length of the road on the enlarged map
= (Original length × Magnitude of enlargement)
= (25 × 5) ⇒ **125 cm**

Note:
Length on the new reduced map = original length ÷ N
Length on the new enlarged map = original length × N
(where, N = magnitude of enlargement or reduction)

Example 14

Area of a drainage basin on a map with scale 1:50,000 is measured as 200 sq.cm. Find its ground area. Find the map area: a) when the map is reduced to 1:100,000 and b) when the map is enlarged to 1:10,000.

Calculation

Map scale = 1:50,000
Therefore, 1 cm represents 0.5 km and, 1 sq. cm on map = 0.25 sq. km on ground. The required ground area of the drainage basin = 200 × 0.25 sq. km
= **50 sq. km**

a) Scale of the reduced map = 1:100,000
Magnitude of reduction
= (100,000 ÷ 50,000) ⇒ 2.
Therefore, area of the basin on the reduced map
= {Orginal area ÷ (Magnitude of reduction)^2}
= (200 ÷ 2^2)
= (200 ÷ 4)
= **50 sq. cm**

b) Scale of the enlarged map = 1:10,000
Magnitude of enlargement
= (50,000 ÷ 10,000) ⇒ 5
Therefore, area of the basin on the enlarged map = {Original area × (Magnitude of enlargement)^2}
= (200 × 5^2)
= 200 × 25 ⇒ **5,000 sq. cm**

Note:
Area on the new reduced map = original area ÷ N^2
Area on the new enlarged map = original length × N^2
(where, N = magnitude of enlargement or reduction)

Example 15

A map (1: 4000) is enlarged twice of its original size. Find the R.F. of the enlarged map.

Calculation

Magnitude of Enlargement = 2
Denominator of the R.F. of the original map = 4000
Therefore,
Denominator of the R.F. of the enlarged map = 4000 ÷ √2
= 2828
Therefore, R.F. of the Enlarged Map = 1 : 2828
≡ **1: 2800**

Example 16

A map of India (1: 50,000) is reduced one-third of its original size. Find the R.F. of the enlarged map.

Calculation

Magnitude of Reduction = 3
Denominator of the R.F. of the original map = 50,000
Therefore,
Denominator of the R.F. of the reduced map = 50,000 x √3
= 86,602
Therefore, R.F. of the Reduced Map = 1 : 86, 602
≡ **1 : 87,000**

Note:
Denominator of the R.F. of the new reduced map,
$D_R = D\sqrt{n}$
Therefore, R.F. of the Reduced Map = 1 : D_R
Denominator of the R.F. of the new enlarged map,
$D_E = D \div \sqrt{n}$
Therefore, R.F. of the Enlarged Map = 1 : D_E

FINDING DIRECTION ON A MAP

A map always has a *true north direction*, presented either by a *north arrow* or *meridians*. To find the direction of a point from another *point*, perform the following tasks:

- Locate the points (A and B) on the map, and join them by a straight line (fig 1.14).
- Draw lines parallel to the north arrow or meridians both at A and B. These are now north lines at A and B.

Fig 1.14 Finding Direction on Map: B from A = 39° 00' (N39°00'E) and C from B = 147° 00' (S33°00'E)

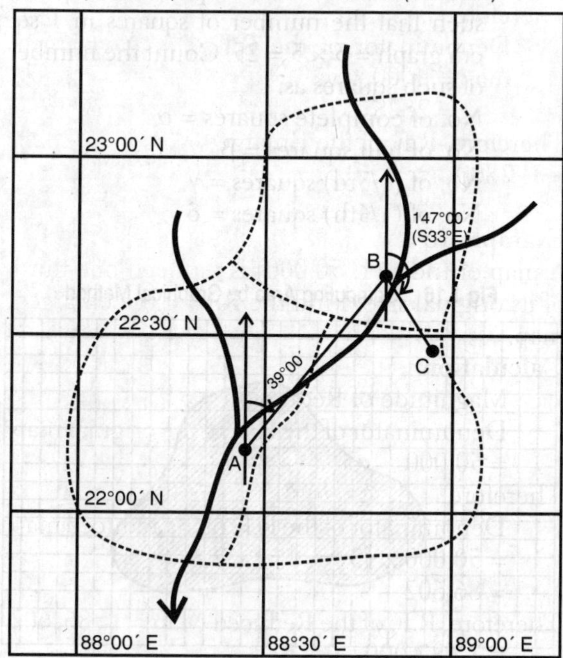

- With a protractor, measure the horizontal angle at A between the North line at A and the line AB. This gives the azimuth or true direction of BA, i.e., direction of A from B.
- Direction may be expressed either in whole circle system as whole circle bearing (360° > θ > 0°) or in quadrantal system as reduced bearing (90° > θ > 0°).
- When magnetic declination is known, magnetic north direction on map can be easily found and drawn with the help of a protractor by using the formula:

  Azimuth or True bearing = Magnetic Bearing + Magnetic Declination

  (+, when declination is east and –, when it is west)

Map Scale and Effective Resolution

Scale	*Effective Resolution (m)*	*Scale*	*Effective Resolution (m)*
1 : 2500	1.25	1 : 250,000	125
1 : 10,000	5	1 : 500,000	250
1 : 24,000	12	1 : 1,000,000	500
1 : 50,000	25	1 : 10,000,000	5000
1 : 100,000	50		

FINDING DISTANCE ON A MAP

A map is always drawn on a scale. Any linear measurement on a map is scale dependent. Therefore, accuracy of measurement of distance depends on both the cartographic accuracy and the type of scale used. For computing distances on a map, the statement scale is a must. If it is not given, it is to be computed first, either from the R.F. or the graph.

- *Measuring a Straight Line*

  Place a ruler over the distance to be measured on the map and note the map distance. Convert it into the ground distance by multiplying it with the scale factor or denominator or the magnitude in the right hand side of the statement.

Example: Map distance = d cm
Map scale = 1 cm to S km
Therefore, ground distance = d.S km

➤ *Measuring a Curved Line*

1. *Using a Divider:* Set a pair of dividers at a small unit distance apart, e.g., 0.20 cm. Step off the curved line from one end to the other. Repeat the process from the oposite direction. Counts will be the same.
   If this count is N,
   map distance = 0.20 N cm
   Therefore, ground distance
   = 0.20 N.S km, where S = scale factor

2. *Using a toned Thread:* Lay the thread over the curved line, carefully following each curve. Straighten it and then place against a ruler to find the map distance.
   If it is d cm, ground distance
   = d.S km, where S = scale factor

Fig 1.15 Opisometer

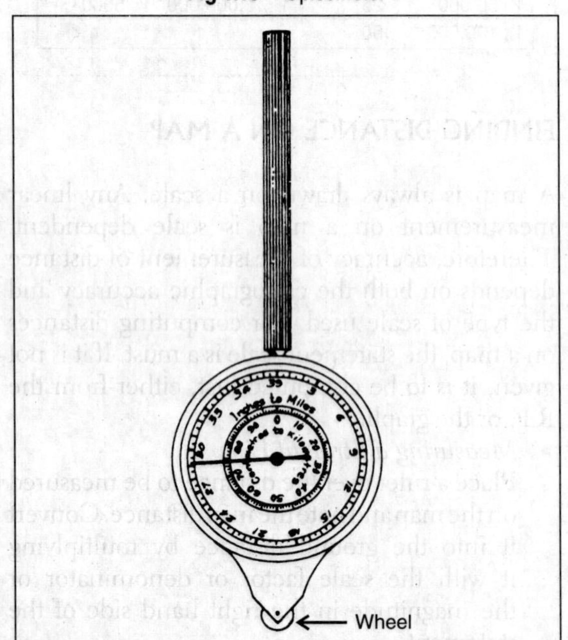

3. Using an Opisometer (Fig.1.15) or Rotameter: It is a small toothed wheel fitted with a recording dial. The wheel is rolled along the length of a line and the map distance is read off directly from the dial. If it is d cm,
   ground distance = d.S km

FINDING AREA ON A MAP

Measurement of area on a map is certainly related to linear measurements, which is scale dependent. Therefore, its accuracy depends on the type of scale used as well as the accuracy of the operator in manipulating the instrument.

➤ *Graphical Method*

1. *Square Procedure:* There are two simple ways to carry out this task—a. trace the area to be measured on to a piece of millimetre graph paper and b. superimpose a transparent millimetre graph paper on to the area (Fig. 1.16). Take 2 mm × 2 mm as the dimension of the operational square, such that the number of squares in 1 sq. cm graph = 5 × 5 = 25. Count the number of such squares as:
   No. of complete squares = α
   No. of half squares = β
   No. of (1/3rd) squares = γ
   No. of (1/4th) squares = δ

Fig 1.16 Computing Area by Graphical Method

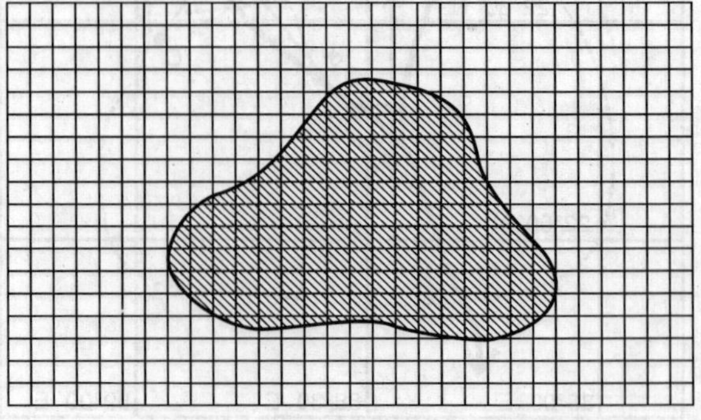

Therefore, total number of complete operational squares = $\left(\alpha + \frac{\beta}{2} + \frac{\gamma}{3} + \frac{\delta}{4}\right)$

If the map scale is 1 cm to 0.5 km, the map area of an operational square = (0.2 × 0.2) sq. cm or 0.04 sq. cm.

Therefore, the total map area

$$= \left(\alpha + \frac{\beta}{2} + \frac{\gamma}{3} + \frac{\delta}{4}\right). (0.04). (0.5 \times 0.5) \text{ sq. km}$$

$$= \left(\alpha + \frac{\beta}{2} + \frac{\gamma}{3} + \frac{\delta}{4}\right). (0.01) \text{ sq. km}$$

2. *Strip Procedure* (Fig. 1.17): Draw a series of parallel lines of a fixed unit distance apart either over the area or on a tracing paper placed on the map. Draw vertical lines at both ends of the strips. The vertical lines should be placed in such a way that the area to be excluded from the strip is as equal as possible to the area to be included in the strip.
Let, the total length of all the strips = 50 cm, the unit distance = 0.2 cm, and the map scale = 1:50,000.
Therefore, Map area = 50 × 0.2 = 10 sq. cm
Ground area = 10 × 0.5 × 0.5 sq. km = 2.5 sq. km

Fig 1.17 Computing Area by Strip Procedure

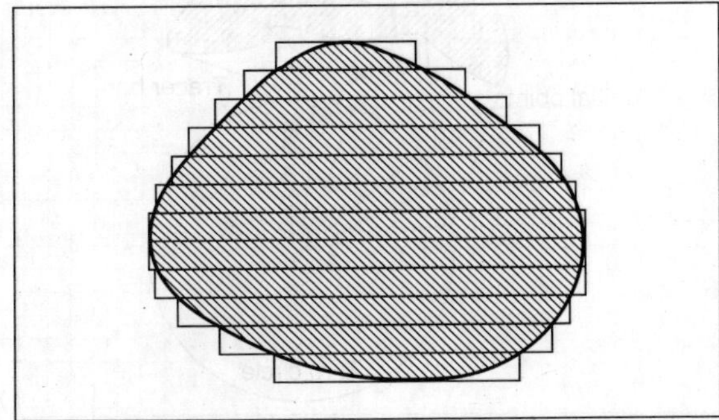

Fig 1.18 Computing Area by Geometrical Method

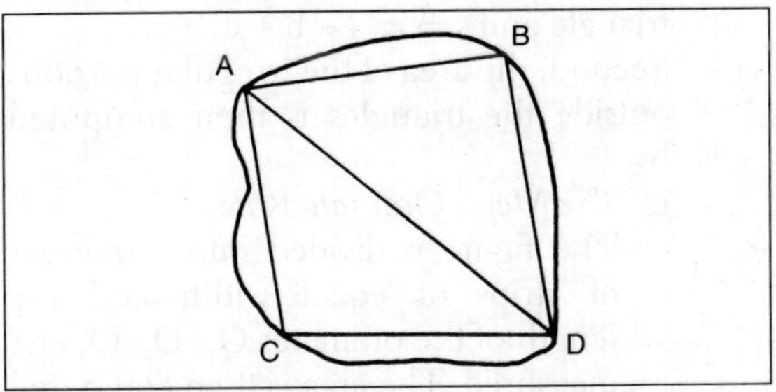

Fig 1.19 The Mean Ordinate Rule

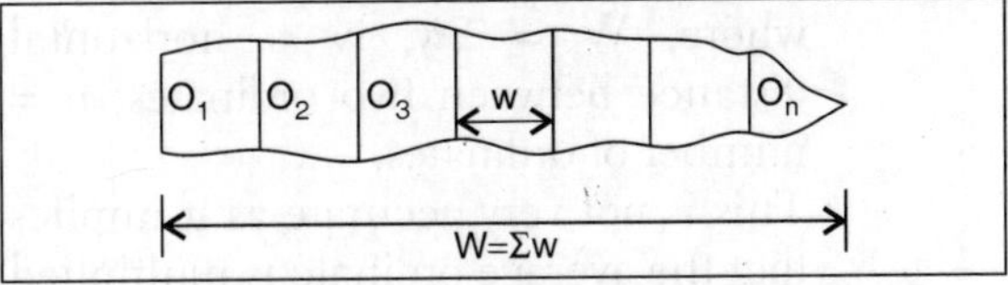

Fig 1.20 The Mid-Ordinate Rule

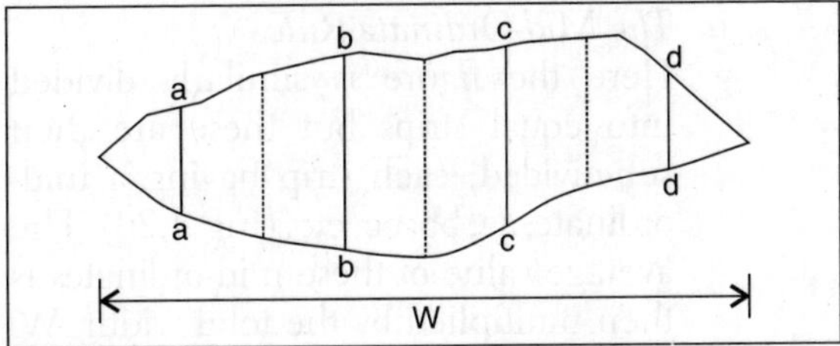

Fig 1.21 The Trapezoidal Rule

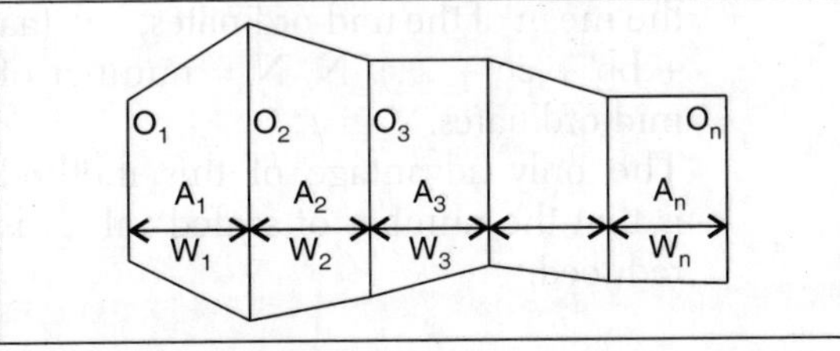

3. *Geometric Procedure* (Fig. 1.18–1.21): This method is best applied to a relatively simple outline.
First, divide the area into triangles and find the area covered by the traingles by adding up the area of each triangle using either of the following formulae (Fig. 1.18):
(a) Area = ½ × base × altitude or
(b) Area = $\sqrt{[s(s-a)(s-b)(s-c)]}$,

where a, b, c, are the three sides of a triangle and s = ½(a + b + c)

Second, the area of the irregular portions outside the triangles is then computed by:

i. *The Mean Ordinate Rule*

The figure is divided into a number of strips of equal width and the length of the ordinates O_1, O_2, O_3 etc. measured. The area is then computed as (Fig. 1.19)

$A = [\ \Sigma(\text{ordinates}) \times W\ ] \div n$

where, W = Σw, w = horizontal distance between two ordinates, n = number of ordinates.

This is not very accurate as it implies that the average ordinate is multiplied by the total width, W.

ii. *The Mid-Ordinate Rule*

Here the figure is similarly divided into equal strips but these are then sub-divided, each strip having a mid-ordinate, aa, bb, cc, etc. (Fig. 1.20). The average value of these mid-ordinates is then multiplied by the total width, W. The area is then computed as A = m.W where, W = width of the figure, m = the mean of the mid-ordinates, i.e., (aa + bb + cc + ...)/ N, N = number of mid-ordinates.

The only advantage of this method is that the number of scaled values is reduced.

iii. *The Trapezoidal Rule*

This is a more accurate method which assumes that the boundary between the extremities of the ordinates are straight lines and the figure is divided by evenly spaced ordinates into a number of trapeziums (Fig. 1.21). The area is then computed as.

$A = (O_1 + O_2 + O_3 + \ldots + O_n) \cdot (w/2)$

where, w = width of the trapeziums, and O_1, O_2, O_3 are the successive sides of the parallel trapeziums.

iv. *Instrumental Method*

A planimeter is the instrument used to measure the area of any irregular figure on maps. It is the best method as error can be minimised to a limit of ± 1% (Fig. 1.22). The procedures are as follows:

- Lay the map on a flat table and fix its 4 corners.
- Set the index mark on the tracing bar for the reading against the 1:1 scale, as given in the instrument's certification.
- Fix the anchor point firmly in the map outside or inside the figure.
- Mark a definite point on the outline of the figure, and set the tracing point exactly on it.
- Read the dial and the wheel and record it as initial reading (I.R.).
- Move the tracing point in a clockwise direction along the outline until it reaches the starting point.
- Read the dial and the wheel and record it as the final reading (F.R.).
- Repeat the procedure to take at least 5 such observations and find the area (Table 1.3). The area is then computed as,

Fig 1.22 Determination of Area by Planimeter

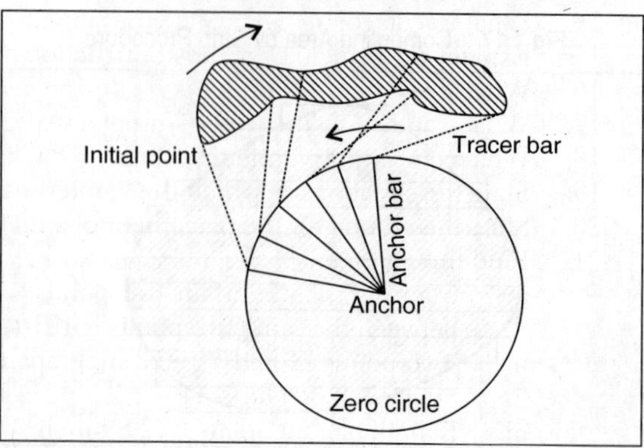

Table 1.3 Computation of Map Area by Planimeter

Observation	I.R.	F.R.	Difference	Mean No. of Revolution
1	1.234	2.136	0.902	
2	2.136	3.039	0.903	
3	3.039	3.942	0.901	0.9012
4	3.942	4.838	0.896	
5	4.838	5.742	0.904	

Map area = R × M + C,
where, R = mean number of revolutions, M = multiplier (given in the instrument's certification), C = constant

(***Note:*** C = 0, when anchor is placed outside the figure, and is equal to the area of zero circle, given in the instrument's certification, when anchor is placed inside the figure.)

Computations

Let the multiplier for 1 revolution
= 100 sq. cm (Note this from the instrument's certification).

Therefore, map area
= $0.9012 \times 100 \Rightarrow 90.12$ sq.cm.

If the map scale is 1: 50,000
1 cm = 0.5 km,
∴ Ground Area = $90.12 \times 0.5 \times 0.5$ sq. km
= 22.53 sq. km

EXERCISE

1. Define map scale and state its importance.
2. What are the different forms of the presentation of map scale?
3. Distinguish between the statement scale and ratio scale.
4. What are the disadvantages of a statement scale?
5. State why ratio scale has got the universal applicability.
6. What is meant by R.F.?
7. What are the advantages of a graphical scale?
8. Define the precision of a graphical scale.
9. What is a comparative scale? What are its different types?
10. What are the advantages of a comparative scale?
11. What is a time scale?
12. What is a pace scale?
13. What is a revolution scale?
14. What is a diagonal scale? State the situations in which it is mostly used.
15. Explain the principles of the construction of a diagonal scale.
16. What is a vernier scale?
17. What is meant by the least count of a scale?
18. What is meant by vernier constant?
19. State the situations in which the vernier scale is mostly used.
20. State the criteria for the placement of a map scale in a map layout.
21. Find the statement scale:
    - i. Map distance between two points A and B is 12.3 cm while the exact ground distance between the same two points is 1010 km
    - ii. Two points, X and Y (6.78 inch apart on a map), are actually 5682 miles away from each other
    - iii. Length of a meridian = 12.89 inch

iv. Length of the Equator = 30 cm
v. Length of the Tropic of Cancer = 13.56 inch
vi. Length of the Antarctic Circle = 18.72 cm
vii. A map with scale 1 cm to 8 km is enlarged twice its original size
viii. A map with scale 1 inch to 50 miles is reduced to one-fifth of its original size
ix. Distance between two successive parallels at 10° interval on the Central Meridian = 0.85 inch
x. Distance between two successive meridians at 8° interval of the Equator = 1.85 cm
xi. Map area within the Antarctic Circle = 12.64 sq. cm
xii. Map area of the northern hemisphere = 180 sq. cm
xiii. Map area of the north eastern hemisphere = 56 sq. inch
xiv. Map area between the Tropic of Cancer and the Arctic Circle = 42 sq. inch
xv. Diameter of a globe = 12 inch
xvi. Distance between two successive meridians at 5 degree interval on the 30th parallel = 1.00 cm

22. Find the Representative Fraction from the following information:
i. Map distance between two points P and Q is 14.6 cm while the exact ground distance between the same two points is 234 km
ii. Two points, M and N (8.80 inch apart on a Map), are actually 6666 miles away from each other
iii. Length of a meridian = 10.01 inch
iv. Length of the Equator = 33 cm
v. Length of the Tropic of Cancer = 9.99 inch
vi. Length of the Antarctic Circle = 12.21 cm
vii. A map with scale 1:2000 is enlarged ten times its original size
viii. A map with scale 1:50000 is reduced to one-fifth of its original size
ix. Distance between two successive parallels at 10° interval on the Central Meridian = 0.55 inch
x. Distance between two successive meridians at 10° interval on the Equator = 1.25 cm
xi. Map area within the Antarctic Circle = 11.11 sq. cm
xii. Map area of the northern hemisphere = 222 sq. cm
xiii. Map area of the northeastern hemisphere = 55 sq. inch
xiv. Map area between the Tropic of Cancer and the Arctic Circle = 33 sq. inch
xv. Diameter of a globe = 30 cm
xvi. Distance between two successive meridians at 10° interval on the 45th parallel = 1.00 cm

23. Draw a linear or plain scale with suitable primary and secondary divisions based on the following information:
i. Map distance between two points R and S is 4.44 inch while the exact ground distance between the same two points is 3333 miles
ii. Two points, A and B (22.22 cm apart on a Map), are actually 555 km away from each other
iii. Length of a meridian = 25 cm
iv. Length of the equator = 12 inch
v. Length of the Tropic of Cancer = 8.88 inch
vi. Length of the Antarctic Circle = 7.77 cm
vii. A map with scale 1:400 is enlarged thrice its original size
viii. A map with scale 1:5000 is reduced to one-tenth of its original size
ix. Distance between two successive parallels at 4° interval on the Central Meridian = 0.50 inch
x. Distance between two successive meridians at 6° interval of the Equator = 1.11 cm
xi. Map area within the Antarctic Circle = 3.45 sq. inch
xii. Map area of the northern hemisphere = 50 sq. inch
xii. Map area of the northeastern hemisphere = 100 sq. cm

xiv. Map area between the Tropic of Cancer and the Arctic Circle = 80 sq. cm
xv. Diameter of a globe = 20 cm
xvi. Distance between two successive meridians at 10° interval on the 60th parallel = 0.50 cm

24. Draw a unit comparative scale with suitable primary and secondary divisions based on the following information:
    i. Map distance between two points X and Y is 6.66 inch while the exact ground distance between the same two points is 1111 miles
    ii. Two points, P and Q (10.11 cm apart on a map), are actually 444 km away from each other
    iii. Length of a meridian = 20 cm
    iv. Length of the Equator = 12 inch
    v. Length of the Tropic of Cancer = 10 inch
    vi. Length of the Antarctic Circle = 10 cm
    vii. A map with scale 16 inch to a mile is enlarged twice its original size
    viii. Distance between two successive parallels at 1° interval on the Central Meridian = 0.12 inch
    ix. Distance between two successive meridians at 1° interval on the Equator = 0.22 cm
    x. Map area within the Antarctic Circle = 5 sq. inch
    xi. Map area of the northern hemisphere = 75 sq. inch
    xii. Map area of the northeastern hemisphere = 75 sq. cm
    xiii. Map area between the Tropic of Cancer and the Arctic Circle = 100 sq. cm
    xiv. Distance between two successive meridians at 1° interval on the 60th parallel = 0.3 cm
    xv. R.F. = 1:200,000,000
    xvi. R.F. = 1:25,000,000
25. Draw a Time scale with suitable divisions. Given, a column of army advances at the rate of 2 km per hour on a map with R.F. = 1:50,000.
26. Draw a pace scale with suitable divisions. Given, 1 pace = 60 cm for a map with R.F. = 1:6,000.
27. Draw a revolution scale with suitable divisions. Given, the diameter of the wheel of a bicycle = 30 inch for a mouza map with R.F. = 1:4,000.
28. Based on R.F. = 1: 75,000,000 draw a diagonal scale to read 5670 miles.
29. Based on R.F. = 1: 63,360 draw a diagonal scale to read 4.56 miles.
30. Based on R.F. = 1: 50 draw a diagonal scale to read 4 yards 1 ft 7 inch.
31. Based on R.F. = 1: 75,000 draw a diagonal scale to read 4 miles 6 furlongs 132 yards.
32. Based on R.F. = 1: 5,000,000 draw a diagonal scale to read 478 km.
33. Based on R.F. = 1: 5,000 draw a diagonal scale to read 666 yards.
34. Based on R.F. = 1: 4,000 draw a diagonal scale to read 555 metres.
35. Draw a vernier scale to read 4.84 inch. Given the least count of the main scale = 0.1 inch and 10 vernier divisions = 9 small main scale divisions.
36. Draw a vernier scale to read 5.34 inch. Given the least count of the main scale = 0.2 inch and 10 vernier divisions = 9 small main scale divisions.
37. Draw a vernier scale to read 4.84 inch. Given the least count of the main scale = 0.1 inch and least count of the vernier = 0.9 inch.
38. Draw a vernier scale to read 38°45'. Given the least count of the main scale = 30' and 30 vernier divisions = 29 small main scale divisions.
39. Draw a vernier scale to read 22°32'. Given the least count of the main scale = 20' and 60 vernier divisions = 59 small main scale divisions.
40. Draw a vernier scale to read 68°25'. Given the least count of the main scale = 15' and 15 vernier divisions = 14 small main scale divisions.
41. Find the R.F. of:

a. A large map drawn on scale, 1 cm to 10 km was reduced thrice of its original size.
b. A large map drawn on scale, 1 inch to 4 miles was reduced twice of its original size.
c. A small map drawn on scale, 1: 25,000,000 was enlarged 5 times its original size.
d. A small map drawn on scale, 1: 250,000 was enlarged 3 times its original size.

42. A large map drawn on scale, 1: 10,000 was reduced thrice of its original size. Draw a comparative scale based on the R.F. of the reduced map.
43. A large map drawn on scale, 1 cm to 50 km was reduced half of its original size. Draw a comparative scale based on the R.F. of the reduced map.
44. A small map drawn on scale, 1: 63,360 was enlarged 4 times its original size. Draw a comparative scale based on the R.F. of the enlarged map.
45. A small map drawn on scale, 1 inch to 16 miles was enlarged 7 times its original size. Draw a comparative scale based on the R.F. of the enlarged map.

2

Map Projection

HIGHLIGHTS

- Scale Factor
- Deformation
- Classification of Map Projections
- Terminologies in Map Projection
- Measurements on a Globe
- Polar Zenithal Gnomonic Projection
- Polar Zenithal Stereographic Projection
- Polar Zenithal Orthographic Projection
- Polar Zenithal Equidistant Projection
- Polar Zenithal Equal-Area Projection
- Simple Conical Projection with I Standard Parallel
- Simple Conical Projection with II Standard Parallels
- Conical Equal-Area Projection with I Standard Parallel
- Conical Equal-Area Projection with II Standard Parallels
- Conical Orthomorphic Projection with I Standard Parallel
- Conical Orthomorphic Projection with II Standard Parallels
- Bonne's Projection
- Sinusoidal Projection
- Polyconic Projection
- Cylindrical Equal-Area Projection
- Mercator's Projection
- Gall's Stereographic Projection
- Mollweide's Projection

The term *projection* is commonly used to designate the phenomena of image production of an object onto a surface or plane. This involves the application of a set of principles, procedures and purposes. Broadly speaking, map projection is defined as *the systematic* drawing of a network of parallels and meridians on a plain sheet of paper portraying a part or whole of the *earth's surface*. Naturally, it is scale-dependent and is done in accordance with a set of geometric and mathematical principles to satisfy certain objectives of the user.

Hence, map projection is a device by which the curved surface of the earth is represented on a flat plane. The operational process essentially involves dimensional transformation, i.e., a 2-dimensional representation of the 3-dimensional figure of the earth. This produces deformations which are inevitable because the surface of the generating globe and the surface or plane of projection are not *geometrically applicable*.

Mathematically, the general equations describing such transformation in map projection are:

$$u = f_1 (\lambda, \phi) \quad \text{... (i)}$$
$$v = f_2 (\lambda, \phi) \quad \text{... (ii)}$$

where, λ, ϕ define the coordinates of positions on the original 3-dimensional surface, u, v describe the corresponding coordinates on the transformed 2-dimensional plane and f_1 and f_2 are *real, single-valued, continuous and differentiable* functions of λ and ϕ in certain domains so that the Jacobian determinant does not vanish:

$$J = \begin{vmatrix} \frac{\partial u}{\partial \lambda}, \frac{\partial u}{\partial \phi} \\ \frac{\partial v}{\partial \lambda}, \frac{\partial v}{\partial \phi} \end{vmatrix} \neq 0$$

On the transformed plane, the rectangular coordinates (x, y) and polar coordinates (ρ, A) of the same point with geographical coordinates (λ, ϕ) or spherical coordinates (θ, r) are given by:

$$x = f_1 (\lambda, \phi) \quad \text{or} \quad \rho = f_1 (\theta, r)$$
$$y = f_2 (\lambda, \phi) \qquad\quad A = f_2 (\theta, r)$$

Therefore, x and y or r and A are specific functions of latitude and longitude and one single point (λ, ϕ) or (θ, r) on the earth is represented by one and the only point (x, y) or (ρ, A) on the map. Thus, map projection is *reversible* and *unique*.

SCALE FACTOR

Fundamentally, map projection is a 2-step process in which the earth is first reduced to a generating globe of a desired size and then the generating globe is projected onto a plane. The transformation of a globe onto a plane is identical to the problem of trying to make the skin of an orange exactly coplaner and coincident with a table top without *contortions*, *stretching* and even *tearing*. Hence *deformation* and *distortions* are inevitable in map projection.

The scale in which the generating globe (a 3-dimensional figure) is conceptualised is called the *principal scale*. On maps it is correctly maintained only at selected points or lines (i.e., the point of tangency or the lines of contact of the projection plane or developable surface with the generating globe). Elsewhere on the map where distortions occur, the principal scale becomes significantly different from that in which the map is actually generated. The scale of the resultant map is termed as the *real scale*. It is the differential stretching and contortions of the generating globe that make the real scale unequal at each and every point on the map. Hence, on the resultant map, a one-to-one correspondence for all points is a practical impossibility. The ratio between the principal scale and the real scale at any point on the map is called the *scale factor* at that point. Mathematically speaking,

Scale Factor,

$$SF = \frac{\text{Denominator of the Principal Scale}}{\text{Denominator of the Real Scale}}$$

Radial Scale Factor and Tangential Scale Factor

Tissot's (1850) *law of deformation* states that '... *at each point* of the spherical surface there exists at least two perpendicular directions which reappear at right angles to each other on the projection, although all other angles at that point may be altered from their *original disposition*.' On a map these two directions are as follows—one along a parallel and the other along a meridian. The scale factor measured along a parallel is called the *parallel scale factor* or *tangential scale factor* while that measured along a meridian is called the *meridional scale factor* or *radial scale factor*. The equations for derivation are:

Tangential Scale Factor (TSF)

$$= \frac{\left(\begin{array}{c}\text{Denominator of the Principal Scale}\\ \text{along a parallel } (\phi)\end{array}\right)}{\left(\begin{array}{c}\text{Denominator of the Real Scale}\\ \text{along the same parallel } (\phi)\end{array}\right)}$$

$$= \frac{\text{Length of a parallel on globe } (L_{\phi g})}{\text{Length of the same parallel on map } (L_{\phi m})}$$

Hence, along a parallel ϕ, tangential scale factor,

$$TSF = \frac{L_{\phi g}}{L_{\phi m}} \qquad \text{... (i)}$$

and tangential scale is expressed by

$$1 : TSF \qquad \text{... (ii)}$$

Radial Scale Factor (RSF)

$$= \frac{\left(\begin{array}{c}\text{Denominator of the Principal Scale}\\ \text{along a meridian } (\lambda)\end{array}\right)}{\left(\begin{array}{c}\text{Denominator of the Real Scale}\\ \text{along the same meridian } (\lambda)\end{array}\right)}$$

$$= \frac{\text{Length of a meridian on globe } (L_{\lambda g})}{\text{Length of the same meridian on map } (L_{\lambda m})}$$

Hence along a meridian λ, radial scale factor,

$$RSF = \frac{L_{\lambda g}}{L_{\lambda m}} \qquad \text{... (iii)}$$

and radial scale is expressed by

$$1: RSF \qquad \text{... (iv)}$$

DEFORMATION

Along the two principal directions, it is the balance of the scale factors that determines the nature and magnitude of deformations on a projection. There are four principal types of deformations. These are deformations in *area, shape, distance* and *direction,* which are mutually exclusive in nature. On a projection transformation, scale factors are simple vectors, their products and resultants determine the specific property of a projection. On the basis of this, projections are classified into five types:

1. Equal-Area Projections

In these, the area of a segment on the generating globe is truly preserved on the corresponding segment of the graticules. At any point of such projections, the product of the two scale factors is unity, or, in other words,

$$RSF \times TSF = 1$$

These are also called *authalic, homolographic* or *equivalent projections.*

2. Orthomorphic Projections

In these, the shape of a segment on the generating globe is truly preserved on the corresponding segment of the graticules. At any point of such projections, the two scale factors are exactly equal in magnitude. The necessary condition of orthomorphism is, therefore, the equality of scales along the two principal directions, i.e.,

$$RSF = TSF$$

These are also known as *true-shape* or *conformal projections.*

3. Equidistant Projections

In these, the distance between any two points on the generating globe is truly preserved between the corresponding points on the graticules.

4. Azimuthal Projections

In these, the azimuth (true direction) of any point from the centre of projection is truly preserved on projection.

5. Aphylactic Projections

In these, neither of the above four properties is truly and fully preserved. Such projections are genetically neither azimuthal nor equidistant, nor equivalent and nor conformal.

CLASSIFICATION OF MAP PROJECTIONS

Map projections are fundamentally classified based on their *extrinsic* and *intrinsic* properties. The extrinsic properties include the exogenic parameters of transformation, i.e., the *nature of datum surface, plane or surface of projection* and the *transformation process* involved. The intrinsic properties include the *specific property* preserved, the *geometric combination of parallels* and *meridians* and the final *shape of the graticules* (Table 2.1). *Direct projections*

Table 2.1 Classification of Map Projections

Criteria	Parameter	Classes/Sub-classes		
1. EXTRINSIC	A. Datum Surface	I. Direct or Spheroidal Projection		
		II. Double or Spherical Projection		
		III. Triple Projection		
	B. Plane or Surface of Projection	1st Order (plane)	2nd Order (aspect)	3rd Order (case)
		I. Planar	a. Tangent	i. Normal
		II. Conical	b. Secant	ii. Transverse
		III. Cylindrical	c. Polysuperficial	iii. Oblique
	C. Method of Projection	I. Perspective		a. Gnomonic
		II. Semiperspective		b. Stereographic
		III. Non-perspective		c. Orthographic
		IV. Conventional		d. Equidistant
				e. Equal Area
2. INTRINSIC	A. Properties (preserved)	I. Azimuthal		
		II. Equidistant		
		III. Equivalent or Authalic or Homolographic		
		IV. Orthomorphic or Conformal		
		V. Aphylactic		
	B. Appearance of Parallels and Meridians	I. Both parallels and meridians are straight lines		
		II. Parallels are straight lines and meridians are regular curves		
		III. Parallels are regular curves and meridians are straight lines		
		IV. Both parallels and meridians are regular curves		
		V. Parallels are concentric circles and meridians are regular curves		
		VI. Parallels are concentric circles and meridians are radiating straight lines		
		VII. Parallels are irregular curves and meridians are radiating straight lines		
		VIII. Both parallels and meridians are irregular curves		
	C. Geometric shape	I. Rectangular		
		II. Circular		
		III. Elliptical		
		IV. Parabolic		
		V. Butterfly		
		VI. Others		

are when the projection is directly made from an oblate spheroid. *Double projections* are when the projection is made from a sphere originally transformed from an oblate spheroid.

TERMINOLOGIES IN MAP PROJECTION

Geographical Coordinates

The location of a point on the earth's surface is usually determined with respect to an imaginary reference framework constituted by a fixed pair of mutually intersecting perpendicular lines, i.e., the *equator* and the *prime meridian*. The ordinate values corresponding to the y-axis are called *latitudes* (ϕ) while the abscissa values corresponding to the x-axis are called *longitudes* (λ) (Fig. 2.1). These define the geographical coordinates of a place on the earth's surface as, $P \equiv (\lambda, \phi)$.

Such coordinates can easily be transformed into rectangular (x, y) and polar (r, z) coordinates with the help of mathematical principles:

where, $x = f(\lambda)$, $r = f(\lambda, \phi)$,

$y = f(\phi)$, $z = f(\lambda, \phi)$.

From Fig. 2.1,

$$x = r \sin z, \qquad y = r \cos z,$$

$$r = \sqrt{(x^2 + y^2)} \qquad z = \tan^{-1} (x/y)$$

The points at which the rotational axis intersects the earth's surface are called *poles*. The imaginary circle circumscribing the earth being equidistant from the poles and drawn through the centre with a plane perpendicular to the axis is called the *equator*. The *latitude* of a place is the angle produced by the line joining the place and the earth's centre on the equatorial plane at the centre of it. It is measured either to the north or to the south of equator and is accordingly specified as °N or

Fig. 2.1 Location of a Point by Geographical and Cartesian Coordinates

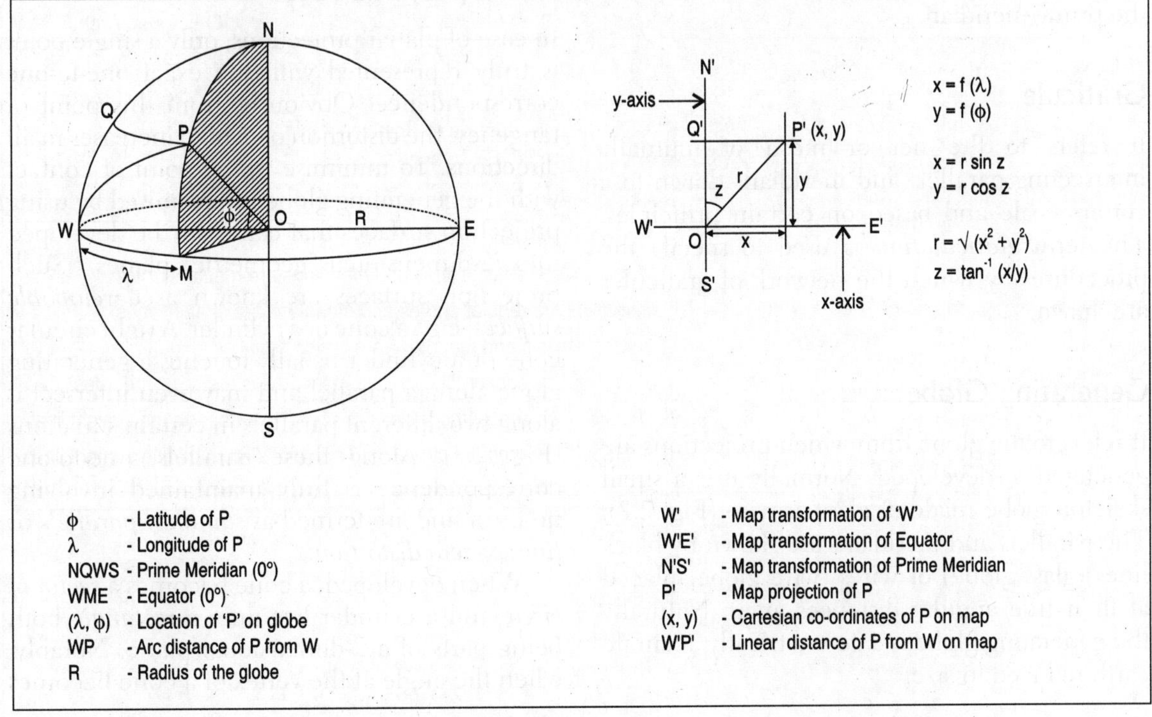

°S. The *latitude* of the equator is 0° and that of the poles are 90°N and 90°S. Through each latitude, circles parallel to the equator and centred on the polar axis may be imagined. These are called *parallels of latitudes* or simply *parallels*. Of the parallels only the equator is a great circle. The radius and the length of the parallels gradually decrease from its maxima at the equator to its minima at the poles.

The semicircular lines joining the two poles and intersecting the parallels at right angles are called *meridians* or *lines of longitudes*. All meridians are equal in size. Of these, the one that passes through Greenwich is taken as the reference line and is called the *prime meridian*. The *longitude* of a place is described as the angle subtended by the meridional plane passing through a place on the plane of the prime meridian, i.e., 0° at the centre of the earth. It is measured either to the east or to the west of the prime meridian and is accordingly specified as °E and °W. As the earth is round, 180°E and 180°W represent the same meridian antipodal to the prime meridian.

Graticule

It refers to the net or mesh of mutually intersecting parallels and meridians drawn to a certain scale and based on certain principles. The term *graticulation* is used to specify the procedures by which the network of graticules are drawn.

Generating Globe

It refers to the globe from which projections are generated or developed. Normally it is a small skeleton globe made of glass or wire (Fig. 2.2). The parallels and meridians are shown by black lines (glass globe) or wires (wire globe) placed at their true angular distances apart. Naturally the generating globe is a geometrically accurate earth reduced in size.

Projection Plane

It is a 2-dimensional geometric plane upon which the parallels and meridians are usually projected. In case of a perspective planar projection, the projection plane touches the generating globe at a single point (Fig. 2.2).

Fig. 2.2 Projection Plane and Generating Globe

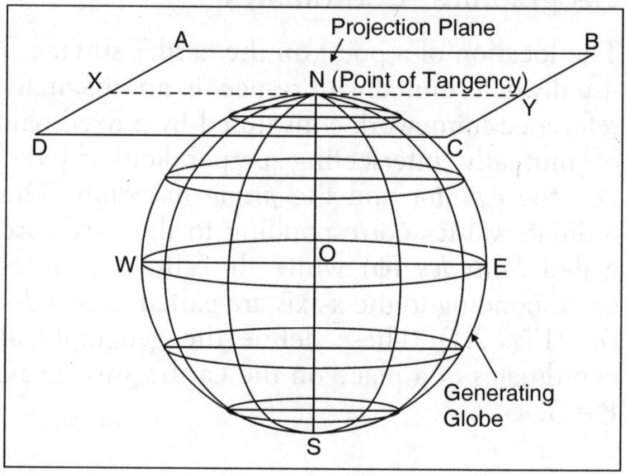

Developable Surface

In case of planar projections, only a single point is truly represented with the exact one-to-one correspondence. Obviously, from this point of tangency, the distortion on a map increases in all directions. To minimise it, the point of contact with the generating globe is maximised by using projection surfaces that can easily be developed into 2-dimensional geometric planes. Such projection surfaces are known as *developable surfaces*, e.g., a cone or a cylinder. A right circular cone or a cylinder usually touches a generating globe along a parallel and may even intersect it along two different parallels in certain situations (Fig. 2.3). Along these parallels, one-to-one correspondence is truly maintained involving no error and are termed as *standard parallels* or *lines of zero distortion*.

When developed, a cone becomes a *sector of circle* and a cylinder becomes a *rectangle* both being parts of a 2-dimensional plane. Notably, when the angle at the vertex of a cone becomes

Fig. 2.3 Developable Surfaces: Cone and Cylinder

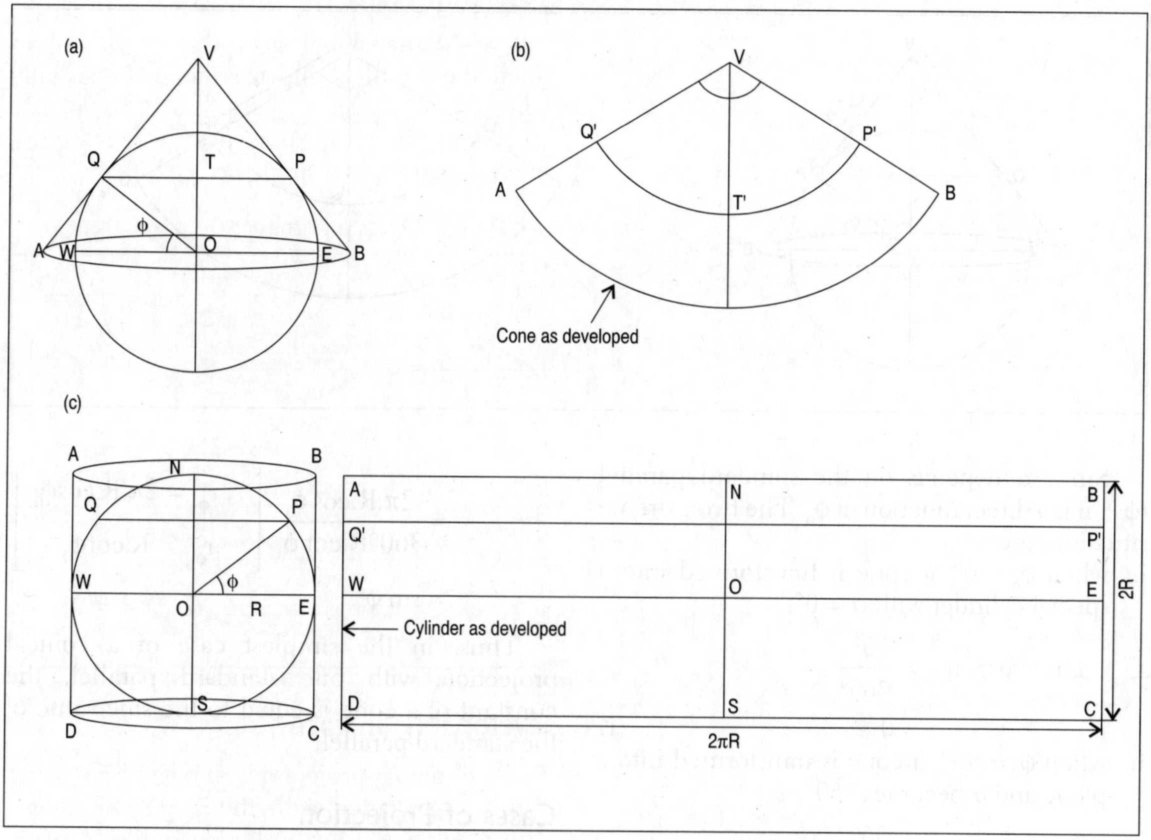

180°, the cone is developed into a projection plane touching the generating globe at a single point only. Again, when the apex of a cone lies at infinity, the cone is developed into a cylinder touching the generating globe along the equator.

Central Meridian

For a given longitudinal extension, it refers to that meridian, *which lies exactly at the median* or *middle-most* position of that extension. It has only constructional importance and is normally drawn as a straight line. The mesh of graticules on one side of the central meridian (CM) is in fact the mirror image of the other side.

Standard Parallel

The parallel(s), *along which a projection plane or a developable surface touch*(es) or *intersect*(s) *the generating globe*, are called *standard parallel*(s). Along the standard parallels, the tangential scale is essentially 1:1. Hence, these are always the *lines of zero distortion.*

Constant of a Cone

It is defined as the *ratio between the angle at the vertex* or apex of a cone when developed (α) and the angle at the pole of the generating globe (360°) (Fig. 2.4).

Therefore, the constant of a *cone*, $n = \dfrac{\alpha}{360°}$

Fig. 2.4 Constant of a Cone

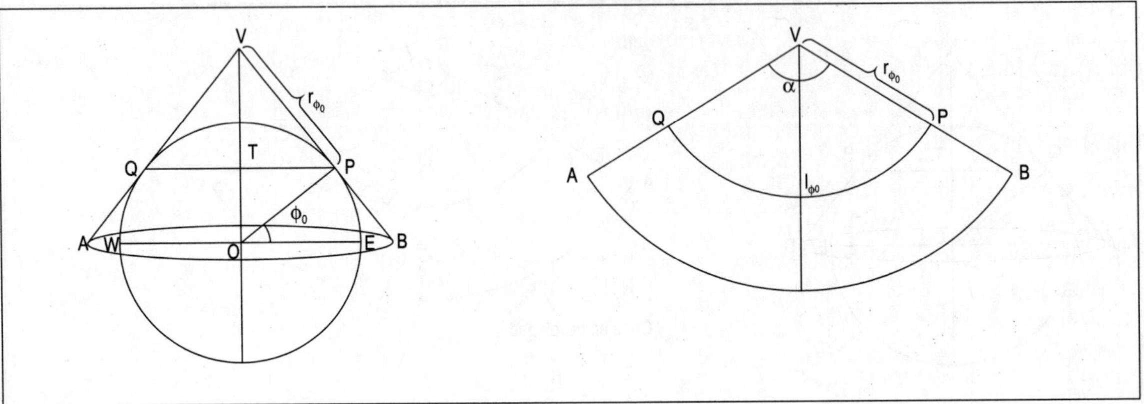

Since α depends on the standard parallel (ϕ_0), n is a direct function of ϕ_0. The two extreme situations are:

i. when $\phi_0 = 0°$, a cone is transformed into a special cylinder with $\alpha = 0°$.

$$\text{Therefore, n} = \frac{0°}{360°} = 0 \qquad \ldots \text{(i)}$$

ii. when $\phi_0 = 90°$, a cone is transformed into a plane and α becomes 360°.

$$\text{Therefore, n} = \frac{360°}{360°} = 1 \qquad \ldots \text{(ii)}$$

Hence, $0 \le n \le 1$ is the boundary condition of the constant of a cone.

From Fig. 2.4,

$$\alpha.\, r_{\phi_0} = l_{\phi_0}$$

$$\text{or, } \alpha = \frac{l_{\phi_0}}{r_{\phi_0}}$$

Therefore, n

$$= \frac{\alpha}{360°}$$

$$= \frac{l_{\phi_0}}{360°.r_{\phi_0}}$$

$$= \frac{2\pi.R.\cos\phi_0}{360°R.\cot\phi_0} \quad \left[\begin{array}{l} \therefore l_{\phi_0} = 2\pi.R.\cos\phi_0 \\ r_{\phi_0} = R.\cot\phi_0 \end{array}\right]$$

$$= \sin\phi_0.$$

Thus, in the simplest case of a conical projection with one standard parallel, the constant of a cone is equal to the *sine* value of the standard parallel.

Cases of Projection

This is defined as the geometric relation expressed as the angle between the axis of symmetry of the plane of projection and the polar axis of the globe, (β). There can be three cases of projection—*normal, transverse* and *oblique*. In the normal case, $\beta = 0°$, in the transverse case $\beta = 90°$ and in the oblique case $90° > \beta > 0°$.

Aspects of Projection

This refers to the *attitude* of the plane or the surface of projection. The plane of projection may be tangent at one point only or intersect along a parallel circle. A polyhedric surface, (i.e., one which has more than one plane) may be tangent at a number of points. Similarly, a cone or a cylinder may be tangent along a parallel or may interest along two parallels. A polyconic or polycylindrical surface may even be chosen for a projection. The main objective is to maximise the points of contact in order to minimise the cumulative deformation.

Perspective Projections

In these, graticules are drawn from a transparent generating globe made of glass with the help of a light source. Rays emerging from the sources cast shadows of parallels and meridians on the projection plane, e.g., *Gnomonic projection, Stereographic projection, Orthographic projection* and the *Simple Conic projection with I standard parallel.*

Semi-perspective Projections

In these, one set of intersecting lines is geometrically projected and the other set drawn purely to suit a desired property.

Non-perspective Projections

In these, projection is done in accordance to a consistent mathematical principle to satisfy certain objectives.

Conventional Projections

These are non-perspective projections constructed following a set of conventions purely based on mathematical operations postulated by a cartographer to portray the whole globe with certain objectives.

The Great Circle

If a plane intersects a sphere, the resulting section of the curved surface which is traced on the plane, is a circle. If the intersecting plane passes through the center of a sphere, the resulting section is a circle, whose radius is the largest which can occur and is equal to the

Fig. 2.5 Cases and Types of Map Projection

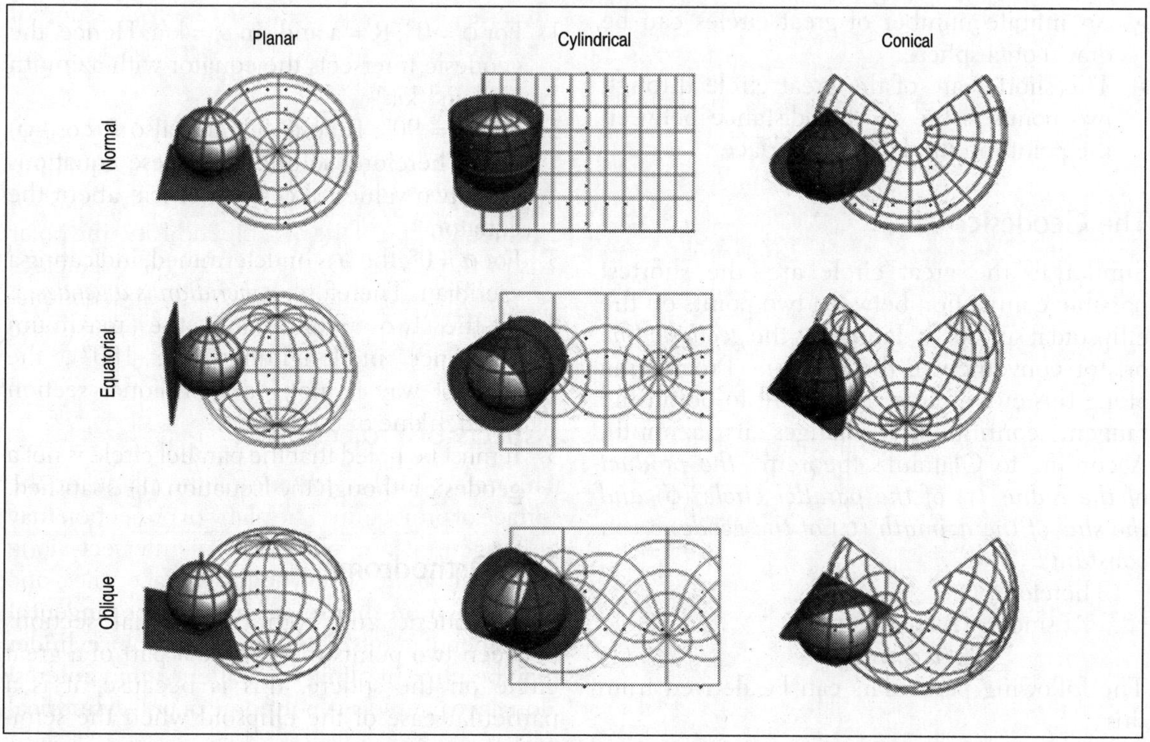

radius of the sphere itself. This is defined as a *great circle*. Thus a meridian is a part of a great circle. The equator is the only parallel which is a great circle and all other parallels are *small circles*. If the plane does not pass through the centre of the sphere, the radius of the resulting circle is less than that of the sphere. This is called a small circle. The special features of great circle are:

- The axis of two or more great circles cannot coincide.
- Intersecting great circles bisect each other.
- The plane of a great circle divides a sphere into two equal halves.
- The section of all great circles passes through the centre of the sphere; therefore the centre of the sphere is the common centre of all the great circles.
- Only one great circle can be drawn through any two points on the spherical surface which are not diametrically opposite to one another.
- An infinite number of great circles can be drawn through a single point.
- An infinite number of great circles can be drawn on a sphere.
- The shorter arc of the great circle through two points is the shortest distance between the points on the spherical surface.

The Geodesic

Similar to the great circle arc, the shortest possible connection between two points on the ellipsoidal surface is defined as the *geodetic line* or, for convenience, the *geodesic*. Progressing along this curved line from point to point, the tangent continuously changes its azimuth. According to Clairaut's theorem, *'the product of the radius* (r) *of the parallel circle* (ϕ) *and the sine of the azimuth* (α) *of the geodesic is a constant'*.

Therefore,

$$r \sin\alpha = (R \cos\phi).\sin\alpha = k \text{ (a constant)} \quad \ldots (1)$$

The following particulars can be derived from this,

Fig. 2.6 The Geodesic

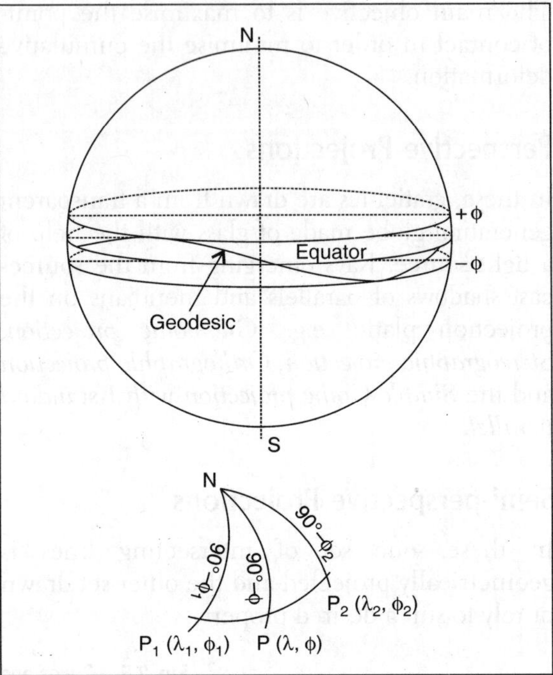

1. For $\phi = 0°$, $R = a$ and $\sin\alpha = k/a$. Hence, the geodesic intersects the equator with azimuth $\alpha = \sin^{-1}(k/a)$
2. For $\alpha = 90°$, $R\cos\phi = k$ and also $R\cos(-\phi) = k$. Therefore, solutions of these equations yield two values of ϕ symmetrical about the equator.
3. For $\alpha = 0°$, the ϕ is undetermined, indicating a meridian. Therefore, *a meridian is a geodesic.*
4. If the two points have the maximum difference in longitude (i.e., 180°), the shortest way is along a meridional section through one of the poles.
5. It must be noted that the parallel circle is not a geodesic although the Equation (1) is satisfied.

The Orthodrome

The geodesic and both the normal sections between two points merge into a part of a great circle on the sphere. this is because, it is a particular case of the ellipsoid when the semi-major and the semi-minor axes become equal. Hence the shortest connection between two points on the sphere is part of a great circle. This curve, however, is not portrayed as a straight line in any but a *Gnomonic* projection. This projected curve is called the *orthodrome* and is of great significance in *great circle navigation*, and *finding direction by wireless*. The orthodrome can be plotted on a map in two ways:

1. *By computing* the latitude (ϕ) and longitude (λ) of various points of the line progressing from the starting point to the terminal point by spherical trigonometry, plotting them on a map, and connecting them with a continuous curve.
2. *By drawing* a linear connection between the two points in a map of Gnomonic projection. The latitude and longitude of the intersection points with the meridians and parallels can then be transferred to a map of any other non-*Gnomonic* projection generating the orthodrome curve.

The length of the orthodrome between two points P_1 (λ_1, ϕ_1) and P_2 (λ_2, ϕ_2) is given by,
$ds = R.\beta / \rho$
where R = radius of sphere, β = the central angle subtended by the arc of the orthodrome in degrees, $\rho = 57.29578°$ or $57°17'44.8''$
β is computed from the equation,
$\cos\beta = \sin\phi_1.\sin\phi_2 + \cos\phi_1.\cos\phi_2.\cos(\lambda_2 - \lambda_1)$

The Rhumbline or Loxodrome

The rhumbline between two points P_1 (λ_1, ϕ_1) and P_2 (λ_2, ϕ_2) is a curve intersecting the meridians at a constant azimuth angle, α. The equation of the curve is generated from the differential equation,

$$d\lambda = \tan\alpha.(1/\cos\phi).d\phi \quad \text{... (2)}$$

Integrating this from P_1 to P_2,

$$\tan\alpha = \frac{\lambda_2 - \lambda_1}{\ln\tan\left(\frac{90° + \phi_2}{2}\right) - \ln\tan\left(\frac{90° + \phi_1}{2}\right)}$$

Therefore, any point P (λ, ϕ) of the curve is

Fig. 2.7 The Rhumbline

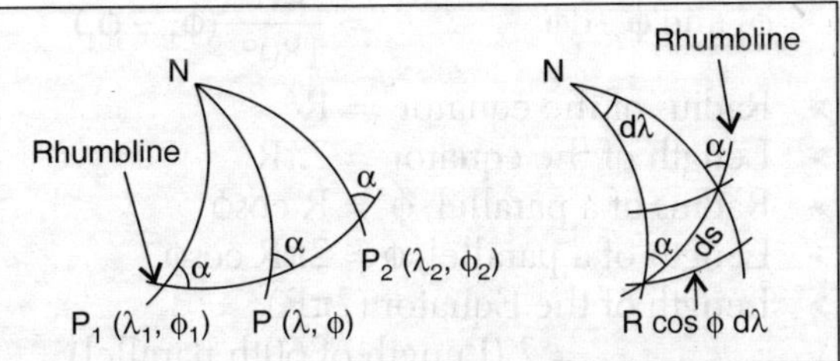

now computed by integrating with boundaries λ_1 and λ. Thus,

$$\lambda - \lambda_1 = \tan\alpha\left[\ln\tan\left(\frac{90° + \phi}{2}\right) - \ln\tan\left(\frac{90° + \phi_1}{2}\right)\right]$$

From the above, the following particulars about the rhumbline or loxodrome can be deduced:

1. If $\phi_1 = \phi_2$, $\tan\alpha$ = infinity, or $\alpha = 90°$, rendering a parallel circle
2. If $\lambda_1 = \lambda_2$, $\alpha = 0°$, it yields a meridian
3. If $\phi = 90°$, $\lambda - \lambda_1$ becomes infinity. Hence, the curve spirals towards the pole as an asymptotic point

The distance between the points P_1 and P_2 along the loxodrome is given by,

$$P_1P_2 = \int_{P_1}^{P_2} ds = R.\frac{1}{\cos\alpha}.d\phi$$
$$= R(\phi_2 - \phi_1)\sec\alpha$$

The rhumbline is important in navigation particularly because it is projected as a straight line between points in the *conformal cylindrical projection of Mercator*. In *Gnomonic projection*, the *orthodrome* is a *straight line* and the *loxodrome* is a *curve* but in Mercator projection, the *loxodrome* is a *straight line* and the *orthodrome* is a *curve*.

MEASUREMENTS ON A GLOBE (radius = R)

- Radius of a meridian = R
- Length of a meridian = πR
- Arc length of meridian between parallels

ϕ_1 and ϕ_2, $d\phi = \dfrac{\pi R}{180°}(\phi_2 \sim \phi_1)$

- Radius of the equator $= R$
- Length of the equator $= 2\pi R$
- Radius of a parallel, $\phi = R\cos\phi$
- Length of a parallel, $\phi = 2\pi R\cos\phi$
- Length of the Equator $(2\pi R)$
- $= 2$ (Length of 60th parallel)
- Arc length of parallel ϕ, between meridians λ_1 and λ_2, $d\lambda = \dfrac{2\pi R.\cos\phi}{360°}(\lambda_1 \sim \lambda_2)$
- Arc distance of a parallel (ϕ) from the equator $= R.\phi^c$
- Arc distance of a parallel ϕ from the poles $= R\,(90° - \phi)^c$
- Surface area of the globe $= 4\pi R^2$
- Area of a hemisphere $= 2\pi R^2$
- Height of a parallel (ϕ) from the equator $= R\sin\phi$
- Surface area enclosed by the equator and a parallel (ϕ) $= 2\pi R^2\sin\phi$
- Surface area enclosed by a pole and a parallel (ϕ) $= 2\pi R^2(1 - \sin\phi)$
- Surface area enclosed by two parallels ϕ_1 and $\phi_2 = 2\pi R^2(\sin\phi_2 - \sin\phi_1)$
- Equatorial radius = 6,378.1 km (3963.1 miles)
  Polar radius = 6,356.8 km (3,950.1 miles)
  Flattening = 0.0033528 (1:298.25722)
  Circumference = 40075.02 km (Equatorial)
  Circumference = 40007.86 km (Polar)
  Surface Area = 5.10072×10^8 km^2
  Volume = 1.08321×10^{12} km^3

For all practical class room purposes, the mean radius of the earth is taken as 6,400 km (4,000 miles) or 640×10^6 cm (250×10^6 inch).

POLAR ZENITHAL GNOMONIC PROJECTION

Principle

In this projection, a 2-dimensional plane of projection touches the generating globe at either of the poles. It is a perspective projection. The source of light lies at the centre of the generating globe (Fig. 2.8). Parallels are projected as concentric circles of varying radius while the meridians are projected as straight lines converging at the pole.

Fig. 2.8 Principles of Polar Zenithal Gnomonic Projection

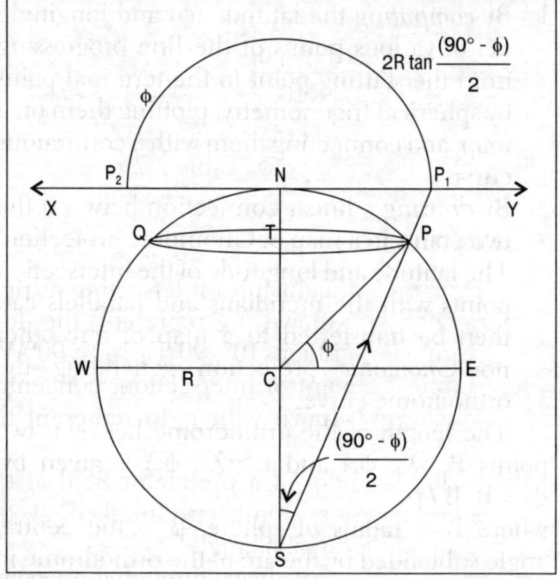

In Fig. 2.8, let XY be the projection plane and C the source of light. PQ is a parallel projected on XY as a circle of radius NP_1 or NP_2.

Radius CE = CW = CP = CN = CS = R

XY is tangent at N.

$\therefore \angle XNC = \angle YNC = 90°$

and ΔNP_1C is a rt $\angle\Delta$ and XY || WE

$\therefore \angle NP_1C = P_1CE = \phi$

From the rt $\angle\Delta NP_1C$, $\dfrac{NP_1}{CN} = \cot\phi$

or, $NP_1 = CN.\cot\phi$

$= R.\cot\phi$

$\therefore$ Radius of any parallel (ϕ) $= R.\cot\phi$

Theory

i. Radius of the generating globe, R = Actual radius of the earth ÷ Denominator of R.F.

ii. Radius of any parallel (ϕ), r_ϕ = R.cot ϕ

Example

Draw graticules at 10° interval on scale 1:234 × 10^6 for the extension 90°S – 40°S around the pole.

Calculation

i. $R = \dfrac{640 \times 10^6 \text{ cm}}{234 \times 10^6}$

$= 2.7350$ cm

ii. $r_\phi = 2.7350 \cot\phi$ (Table 2.2)

Construction

i. A pair of straight lines intersecting at right angles are drawn to represent the four cardinal meridians (0°, 90°E, 180°, 90°W).

ii. From the point of intersection, concentric circles are drawn with r_ϕ to represent the parallels.

iii. With the help of a protractor held at the pole, division points are marked at the required interval of angle.

iv. Straight lines are drawn through these points joining the poles to represent the meridians.

v. The graticules are then properly labelled (Fig. 2.9).

Properties

i. Parallells are concentric circles of varying radius.

ii. Interparallel distance increases rapidly towards the equator.

iii. The equator cannot be represented in this projection.

Fig. 2.9 Polar Zenithal Gnomonic Projection

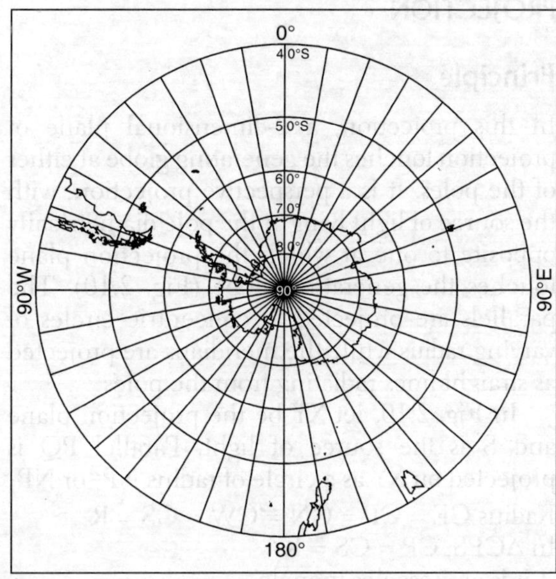

iv. Meridians are straight lines radiating from the poles at true angular distances apart.

v. It is an azimuthal projection as the azimuth of a point at the poles is truly preserved.

vi. All great circles appear as straight lines as their planes pass through the centre of the globe, where the source of light is kept.

vii. The shortest distance between two points is represented by a straight line while a rhumbline is represented by a *nautilus* (i.e., a curve analogous to that of a snail's back).

viii. Deformation increases rapidly towards the margin of the map.

ix. It is useful to the navigators and is suited to small areas around the pole.

Table 2.2 Computation of r_ϕ

ϕ	40° S	50° S	60° S	70° S	80° S	90° S
r_ϕ = 2.7350 cot ϕ (cm)	3.2594	2.2949	1.5790	0.9954	0.4823	0

POLAR ZENITHAL STEREOGRAPHIC PROJECTION

Principle

In this projection, a 2-dimensional plane of projection touches the generating globe at either of the poles. It is a perspective projection, with the source of light lying at the pole diametrically opposite to one at which the projection plane touches the generating globe (Fig. 2.10). The parallels are projected as concentric circles of varying radius while the meridians are projected as straight lines radiating from the poles.

In Fig. 2.10, let XY be the projection plane and S is the source of light. Parallel PQ is projected on XY as a circle of radius NP_1 or NP_2

Radius CE = CP = CN = CW = CS = R

In ΔCPS, CP = CS = R

∴ it is an isosceles triangle

and $\angle CSP = \angle CPS$

$\angle NCP = \angle NCE - \angle PCE = (90° - \phi)$

$\angle NCP$ being an external angle at C of ΔCPS,

$\angle CPS + \angle CSP = \angle NCP$

or, $2\angle CSP = (90° - \phi)$

or, $\angle CPS = \left(\dfrac{90° - \phi}{2}\right)$

XY is tangent at N.

∴ ΔNP_1S is a right angled triangle

and $\dfrac{NP_1}{NS} = \tan \angle CSP$

or, $NP_1 = NS.\tan \angle CSP$

$= 2R \tan\left(\dfrac{90° - \phi}{2}\right)$

∴ Radius of any parallel $\phi = 2R.\tan\left(\dfrac{90° - \phi}{2}\right)$

Theory

i. Radius of the generating globe, R = Actual radius of the earth ÷ Denominator of R.F.

Fig. 2.10 Principles of Polar Zenithal Stereographic Projection

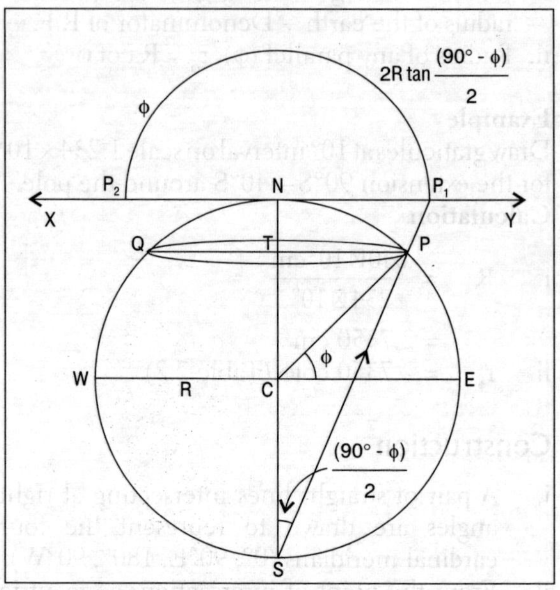

ii. Radius of any parallel (ϕ),

$$r_\phi = 2R.\tan\left(\frac{90° - \phi}{2}\right)$$

Example

Draw graticules at 10° interval on scale 1:184 × 10⁶ for the extension 90°S—40°S around the pole.

Calculation

i. $R = \dfrac{640 \times 10^6 \text{ cm}}{184 \times 10^6}$

$= 3.4783$ cm

ii. $r_\phi = 2 \times 3.4783 \tan\left(\dfrac{90° - \phi}{2}\right)$

$= 6.9566 \tan\left(\dfrac{90° - \phi}{2}\right)$ (Table 2.3)

Construction

Similar to Polar Zenithal Gnomonic Projection (Fig. 2.11).

Fig. 2.11 Polar Zenithal Stereographic Projection

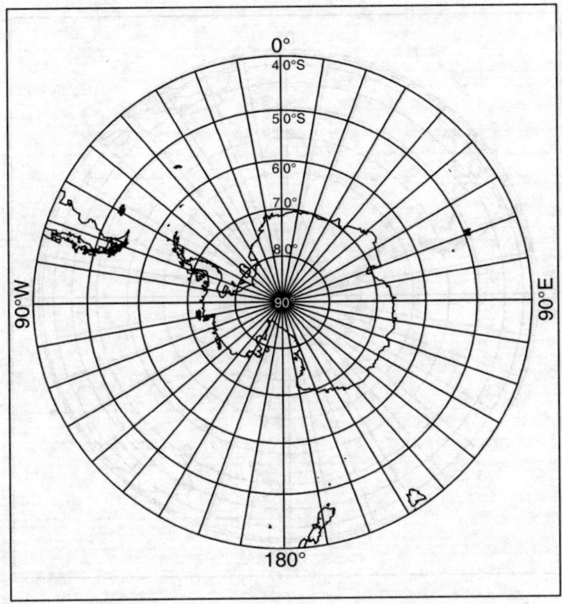

Properties

i. Parallels are represented by concentric circles of varying radius.
ii. Inter-parallel distance gradually increases toward the equator.
iii. Meridians are straight lines radiating from the poles at true azimuth apart.
iv. The direction between any two points is preserved.
v. At any point, the radial scale is equal to the tangential scale.
vi. It is an orthomorphic projection, i.e., the shape of a map is truly preserved.
vii. It is commonly used for the map of the *world* in hemispheres.

POLAR ZENITHAL ORTHOGRAPHIC PROJECTION

Principle

In this projection, a 2-dimensional plane of projection touches the generating globe at either of the poles. It is a perspective projection and the source of light lies at infinity (Fig. 2.12). Rays of light passing through the parallels become incident on the projection plane at right angles. The parallels are projected as concentric circles of varying radius while the meridians are projected as straight lines at true azimuth apart at the poles.

Fig. 2.12 Principles of Polar Zenithal Orthographic Projection

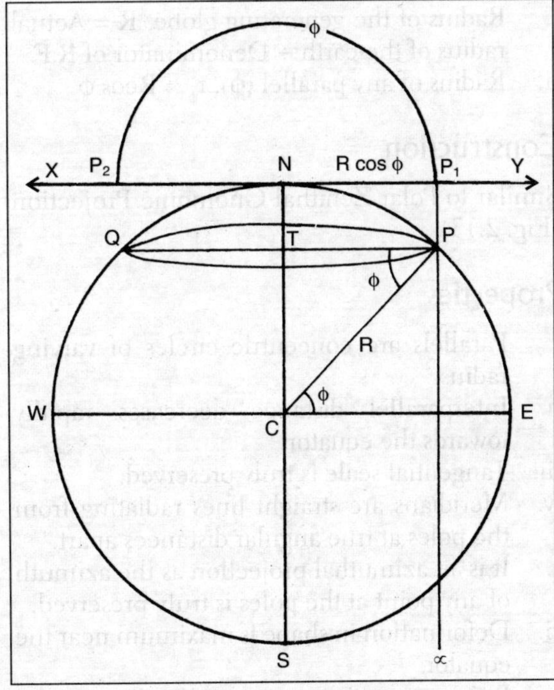

Table 2.3 Computation of r_ϕ

φ	40° S	50° S	60° S	70° S	80° S	90° S
$r_\phi = 6.9566 \tan\left(\dfrac{90° - \phi}{2}\right)$ cm	3.2439	2.5320	1.8640	1.2266	0.6086	0

In Fig. 2.12, let XY be the projection plane. The source of light is at infinity (μ). Parallel PQ is projected on XY as a circle of radius NP_1.

Radius, CE = CW = CP = CN = CS = R

XY is tangent at N

∴ XY || PQ || WE,

NP_1 = PT = radius of the projected parallel ϕ and

∠TPC = ∠PCE = ϕ

From rt ∠ΔTPC, $\frac{PT}{PC} = \cos\phi$

or, $PT = PC.\cos\phi$

$= R\cos\phi$

∴ radius of any parallel ϕ = $R\cos\phi$

Theory

i. Radius of the generating globe, R = Actual radius of the earth ÷ Denominator of R.F.
ii. Radius of any parallel (ϕ), $r_\phi = R\cos\phi$

Construction

Similar to Polar Zenithal Gnomonic Projection (Fig. 2.13).

Properties

i. Parallels are concentric circles of varying radius.
ii. Interparallel distance decreases rapidly towards the equator.
iii. Tangential scale is truly preserved.
iv. Meridians are straight lines radiating from the poles at true angular distances apart.
v. It is an azimuthal projection as the azimuth of any point at the poles is truly preserved.
vi. Deformation in shape is maximum near the equator.
vii. It is a very important projection to the astronomers who use it for plotting celestial observations.

Fig. 2.13 Polar Zenithal Orthographic Projection

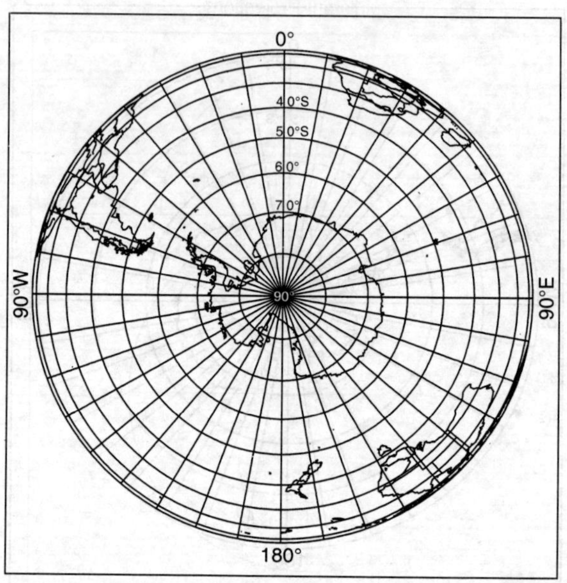

Example

Draw graticules at 10° interval on Scale 1:197 × 10^6 for the extension 90°S – 0° around the pole.

Calculation

i. $R = \frac{640 \times 10^6 \text{ cm}}{197 \times 10^6}$

$= 3.2487$ cm

ii. $r_\phi = 3.2487 \cos\phi$ (Table 2.4)

POLAR ZENITHAL EQUIDISTANT PROJECTION

Principle

In this projection, a 2-dimensional plane of projection touches the generating globe at either of the poles. The radial scale along a meridian is truly maintained so that parallels are equidistant on it (Fig. 2.14). True azimuth at the poles is

Table 2.4 Computation of r_ϕ

ϕ	30° S	40° S	50° S	60° S	70° S	80° S	90° S
$r_\phi = 3.2487 \cos\phi$ (cm)	2.8135	2.4886	2.0882	1.6244	1.1111	0.5641	0

Fig. 2.14 Principle of Polar Zenithal Equidistant Projection

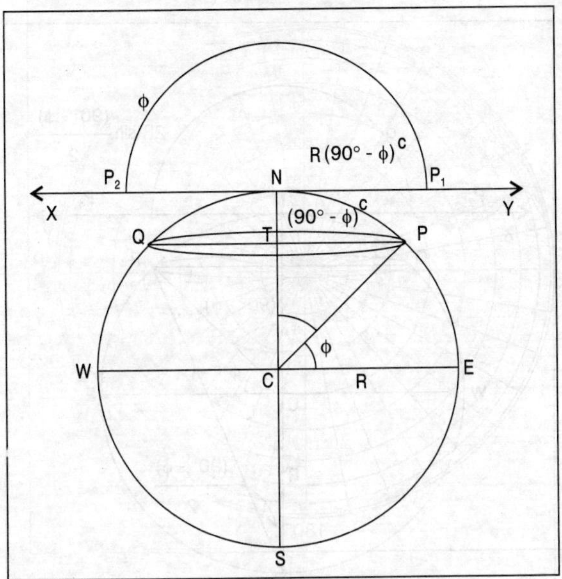

Fig. 2.15 Polar Zenithal Equidistant Projection

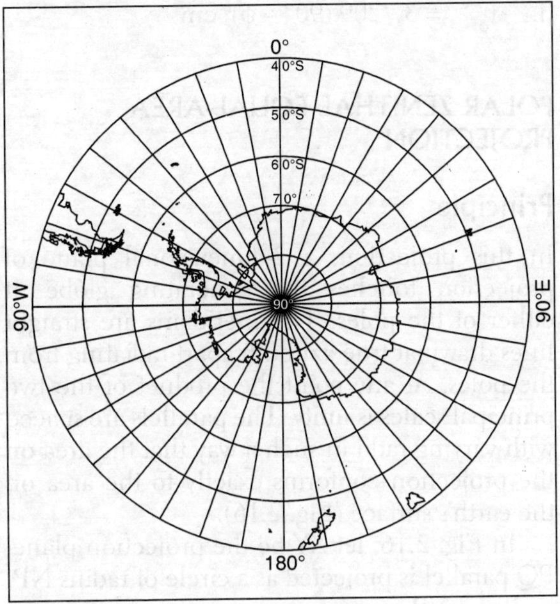

preserved in spacing the meridians. The parallels are concentric circles while the meridians are straight lines radiating from the poles.

In Fig. 2.14, let XY be the projection plane. PQ parallel is projected as circle of radius NP_1 or NP_2.

Radius CE = CP = CN = CW = CS = R

NP_1 = Arc distance NP

$\angle NCP \quad = \angle NCE - \angle PCE$

$\quad = (90° - \phi)$.

From the segment ∠NCP,

arc NP $= CP.(\angle NCP)^c$

$\quad = R\,(90° - \phi)^c$.

∴ Radius of any parallel $\phi = R(90° - \phi)^c$.

Theory

i. Radius of the generating globe, R = Actual radius of the earth ÷ Denominator of R.F.

ii. Radius of any parallel (φ), $r_\phi = R.(90° - \phi)^c$

Construction

Similar to Polar Zenithal Gnomonic Projection (Fig. 2.15)

Properties

i. It is a non-perspective projection.

ii. Distance between any two points is truly maintained.

iii. Parallels are equispaced on a meridian.

iv. Parallels are concentric circles of varying radius.

v. Meridians are straight lines radiating from the pole at true angular distances apart.

vi. It is an azimuthal projection as the azimuth (true direction) of any point at the pole is truly maintained.

vii. Tangential scale increases rapidly toward the equator.

vii. It is used for small areas around the pole within the 60th parallel.

Example

Draw graticules at 10° interval on scale 1:176 × 10^6 for the extension 90°S—30°S around the pole.

Calculation

i. $R \quad = \dfrac{640 \times 10^6 \text{ cm}}{172 \times 10^6}$

$= 3.7209$ cm.

ii. $r_\phi = 3.7209\ (90° - \phi)^c$ cm

POLAR ZENITHAL EQUAL-AREA PROJECTION

Principle

In this projection, a 2-dimensional plane of projection touches the generating globe at either of the poles. The meridians are straight lines drawn at true azimuth apart radiating from the poles. At any point the product of the two principal scales is unity. The parallels are spaced with varying radii in such a way that the area on the projection conforms exactly to the area on the earth's surface (Fig. 2.16).

In Fig. 2.16, let XY be the projection plane. PQ parallel is projected as a circle of radius NP_1 or chord NP.

Radius, CE = CP = CN = CW = CS = R

The area contained within parallel ϕ and the equator on the globe,

$= WQPE$

$= WE.TC$

$= 2\pi R\ R.\sin\phi$

$= 2\pi R^2 \sin\phi$

$$\left[\begin{array}{ll} \therefore WE & = 2\pi R \\ \text{and } TC & = CP.\sin\phi \\ & = R.\sin\phi \end{array}\right]$$

The area between parallel, ϕ and the poles on the globe,

$= WQNPE - WQTPE$

$=$ Area of the hemisphere $- 2\pi R^2 \sin\phi$

$= 2\pi R^2 - 2\pi R^2 \sin\phi$

$= 2\pi R^2 (1 - \sin\phi)$

$= 2\pi R^2 [1 - \cos(90° - \phi)]$

$$= 2\pi R^2 .2\sin^2\left(\frac{90° - \phi}{2}\right)$$

$$= 4\pi R^2 \sin^2\left(\frac{90° - \phi}{2}\right)$$

Fig. 2.16 Principles of Polar Zenithal Equal-Area Projection

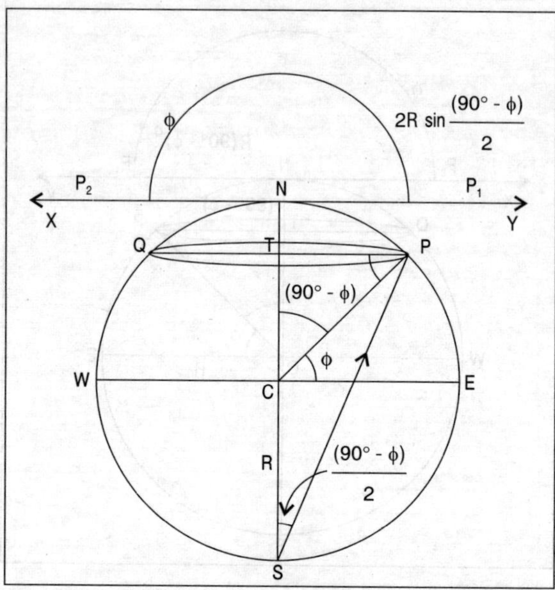

This is represented on the projection by a circle with radius NP_1 and centre at N.

$$\therefore \pi(NP_1)^2 = 4\pi R^2 \sin^2\left(\frac{90° - \phi}{2}\right)$$

$$\text{or, } NP_1 = 2R\sin\left(\frac{90° - \phi}{2}\right)$$

$\therefore$ Radius of any parallel, ϕ

$$= 2R\sin\left(\frac{90° - \phi}{2}\right)$$

Theory

i. Radius of the generating globe, R = Actual radius of the earth ÷ Denominator of the R.F.

ii. Radius of any parallel (ϕ),

Table 2.5 Computation of r_ϕ

ϕ	30° S	40° S	50° S	60° S	70° S	80° S	90° S
$r_\phi = 3.7209\ (90° - \phi)^c$ cm	3.8965	3.2471	2.5977	1.9483	1.2988	0.6494	0

$$r_\phi = 2R\sin\left(\frac{90° - \phi}{2}\right)$$

Construction

Similar to the Polar Zenithal Gnomonic Projection (Fig. 2.17).

Fig. 2.17 Polar Zenithal Equal-Area Projection

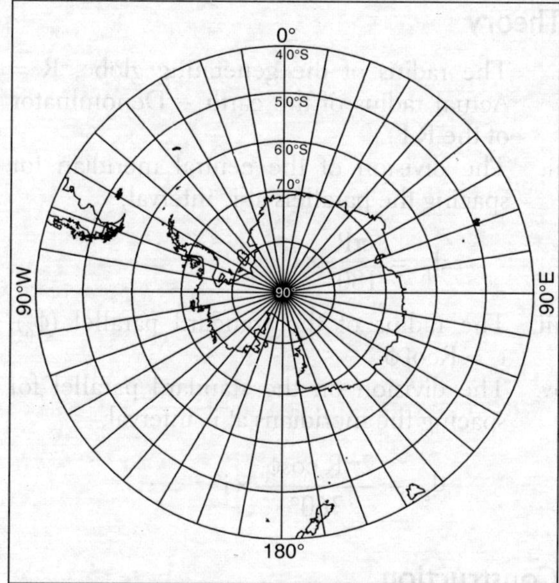

Properties

i. It is a non-perspective projection.
ii. Parallels are concentric circles of varying radius.
iii. Interparallel distance decreases gradually toward the equator.
iv. Meridians are straight lines radiating from the poles at true angular distances apart.
v. It is an azimuthal projection as the azimuth of any point at the poles is truly maintained.
vi. It is an equal-area projection as at any point the product of tangential and radial scale is unity.
vii. It is most commonly used to show polar areas in a world atlas. (Fig. 2.17).

Example

Draw graticules at 10° interval on scale 1:167 × 10^6 for the extension 40°S—90°S around the pole.

Calculation

i. $R = \dfrac{640 \times 10^6 \text{ cm}}{167 \times 10^6}$

$= 3.8323$ cm.

ii. $r_\phi = 2 \times 3.8323.\sin\left(\dfrac{90° - \phi}{2}\right)$

$= 7.6646 \sin\left(\dfrac{90° - \phi}{2}\right)$

SIMPLE CONICAL PROJECTION WITH I STANDARD PARALLEL

Principle

In this projection, a simple right circular cone touches the generating globe along a parallel. This is the parallel along which distortions of any kind is nil and is known as the *standard parallel*. It is a perspective projection in which the parallels and meridians are projected directly on the inner surface of the cone with respect to a light source at the centre of the generating globe (Fig. 2.18).

Table 2.6 Computation of r_ϕ

φ	30° S	40° S	50° S	60° S	70° S	80° S	90° S
$r_\phi = 7.6646 \sin\left(\dfrac{90° - \phi}{2}\right)$ cm	3.8323	3.2392	2.6214	1.9837	1.3309	0.6680	0

Fig. 2.18 Principles of Simple Conical Projection with I Standard Parallel

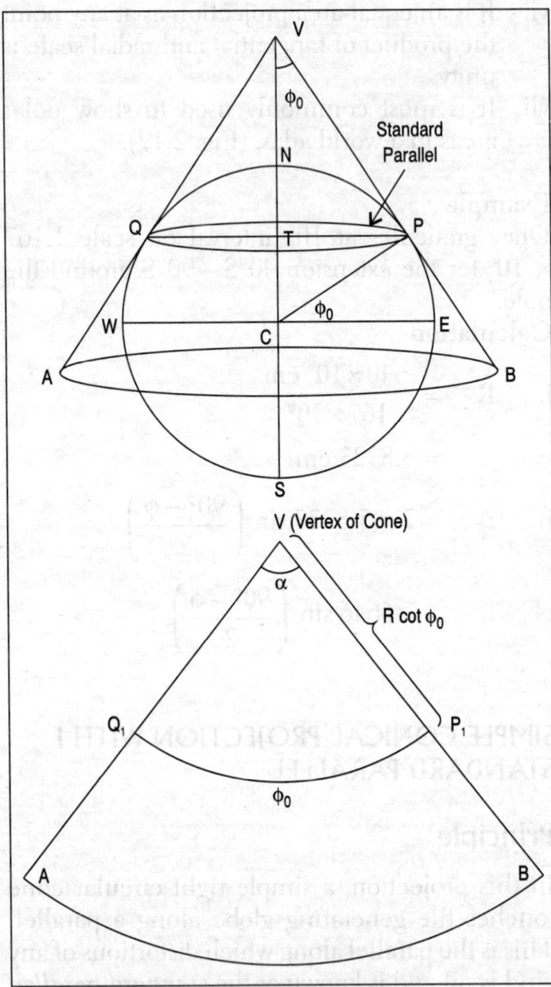

In Fig. 2.18, let cone VAB be the projection plane which touches the generating globe along PQ (ϕ_0). Therefore standard parallel PQ is projected as an arc of circle with radius VP or VP_1.

Radius, CE = CP = CN = CQ = CW = CS = R

VB is tangent at P and

$\therefore$ $\angle VPC = \angle CPB = 90°$

and ΔVPC is a right angled triangle

$$\angle PVC = [180° - (\angle VPC + \angle VCP)]$$
$$= [180° - (90° + \angle VCP)]$$
$$= (90° - \angle VCP)$$
$$= \angle VCE - \angle VCP \; [\because \angle VCE = 90°]$$
$$= \angle PCE$$
$$= \phi_0$$

From the rt $\angle\Delta PVC$, $\frac{VP}{PC} = \cot\phi_0$

or, $VP = PC.\cot\phi_0$

$= R\cot\phi_0$

$\therefore$ Radius of the standard parallel, $r_0 = R\cot\phi_0$

Theory

i. The radius of the generating globe, R = Actual radius of the earth ÷ Denominator of the R.F.

ii. The division of the central meridian for spacing the parallels at i° interval,

$$d_1 = \frac{\pi R}{180°} \times i°$$

iii. The radius of the standard parallel (ϕ_0), $r_0 = R\cot\phi_0$

iv. The division on the standard parallel for spacing the meridians at i° interval,

$$d_2 = \frac{2\pi R\cos\phi_0}{360°} \times i°$$

Construction

i. A straight line is drawn vertically through the centre of the paper to represent the central meridian.

ii. It is then divided by d_1 for spacing the parallels.

iii. An arc of circle is then drawn through the standard parallel mark with radius and centre on the central meridian (produced if necessary).

iv. Concentric arcs of circle are then drawn through each division on the central meridian to represent other parallels.

v. The standard parallel is divided by d_2 on both sides on the central meridian for spacing the meridians.

vi. Straight lines are drawn through each of

these division points joining the centre of the arcs to represent the meridians.

vii. The graticules are then properly labelled (Fig. 2.19).

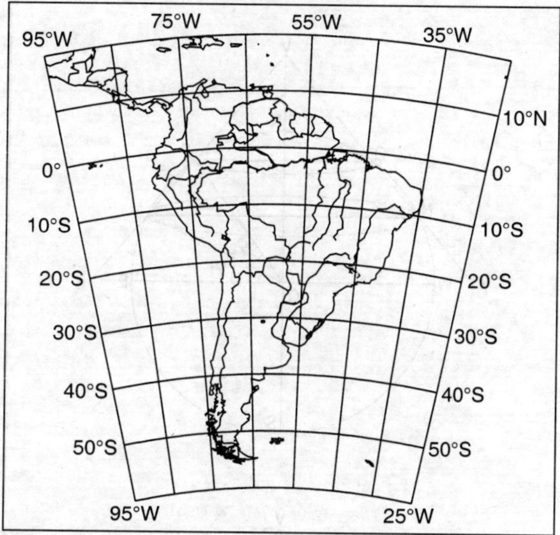

Fig. 2.19 Simple Conical Projection with I Standard Parallel

Properties

i. Genetically it is a perspective projection. The parallels are concentric arcs of circles truly spaced on the central meridian.

ii. Poles are also represented by arcs in this projection.

iii. Radial scale is true along all the meridians.

iv. Meridians are straight lines truly spaced on the standard parallel and converging at the vertex of the cone.

v. Tangential scale is true along the standard parallel only.

vi. Deformation is positive towards the equator- and negative deformation towards the pole.

vii. It is an aphylactic projection, i.e., one that maintains neither area nor shape.

viii. It is suitable for smaller countries of mid-latitude or temperate regions.

Example

Draw graticules at 10° interval on scale 1:150 × 10^6 for the extensions, 20°N—60°S and 25°W—95°W.

Calculation

i. $R = \dfrac{640\times10^6 \text{ cm}}{150\times10^6}$

$= 4.2666$ cm

ii. Parallels to be drawn: 20°N, 10°N, 0°, 10°S, 20°S, 30°S, 40°S, 50°S, 60°S. The standard parallel (ϕ_0) chosen is 20°S.

iii. $d = \dfrac{\pi\times4.2666}{180°}\times10°$

$= 0.7466$ cm

iv. $r_o = 4.2666\cot 20°$

$= 11.7224$ cm

v. $d_1 = \dfrac{2\pi.4.2666\cos 20°}{360°}\times10°$

$= 0.6998$ cm

SIMPLE CONICAL PROJECTION WITH II STANDARD PARALLELS

Principle

In this projection, a simple right circular cone is taken as the projection plane. Two circles of the cone correspond to two different parallels on the generating globe and form an ordinary cone independent of the globe (Fig. 2.20). These are the standard parallels which are so selected as to cover two-thirds of the latitudinal extent of the area to be mapped. The parallels appear as concentric arcs of a circle while the meridians appear as straight lines converging at the vertex of the cone.

In Fig. 2.20, standard parallels MN(ϕ_1) and PQ(ϕ_2) are projected as arcs of circle M_1N_1 and P_1Q_1 with radii r_1 and r_2 respectively

$VP_1.\alpha$ = arc P_1Q_1

= true length of ϕ_2 parallel on globe

or, $r_2\alpha = 2\pi R\cos\phi_2$

$$\therefore\ r_2 = \frac{2\pi R\cos\phi_2}{\alpha} \quad \ldots \text{(i)}$$

$VM_1.\alpha$ = arc M_1N_1

Fig. 2.20 Principles of Simple Conical Projection with II Standard Parallels

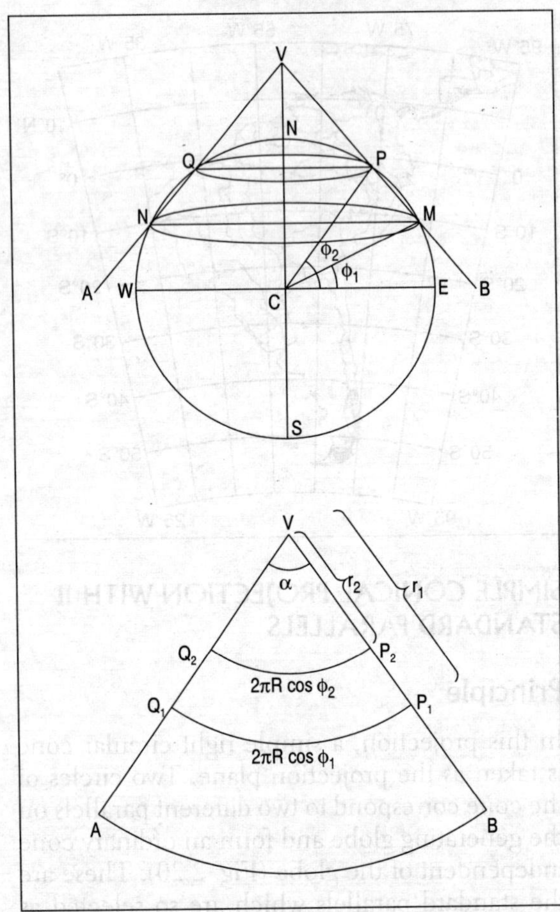

$= $ true length of ϕ_1 parallel on globe

or, $r_1 . \alpha = 2\pi R \cos\phi_1$

$$\text{or, } r_1 = \frac{2\pi R \cos\phi_1}{\alpha} \quad \ldots \text{(ii)}$$

Now,

$r_1 - r_2 = VM_1 - VP_1$

$= M_1P_1$

$=$ True distance on the central meridian between ϕ_1 and ϕ_2 parallels

$= R(\phi_2 - \phi_1)^c \quad \ldots \text{(iii)}$

From equations (i) and (ii),

$$\frac{r_1}{r_2} = \frac{2\pi R \cos \phi_1 / \alpha}{2\pi R \cos \phi_2 / \alpha}$$

By componendo and dividendo,

$$\frac{r_1}{r_1 - r_2} = \frac{\cos \phi_1}{\cos \phi_1 - \cos \phi_2}$$

$$\text{or, } r_1 = (r_1 - r_2) \frac{\cos \phi_1}{\cos \phi_1 - \cos \phi_2}$$

$$= R(\phi_2 - \phi_1)^c \frac{\cos \phi_1}{\cos \phi_1 - \cos \phi_2}$$

Similarly from equation (i) and (ii)

$$\frac{r_1}{r_2} = \frac{\cos \phi_1}{\cos \phi_2}$$

$$r_2 = (r_1 - r_2) \frac{\cos \phi_2}{\cos \phi_1 - \cos \phi_2}$$

$$= R(\phi_2 - \phi_1)^c \frac{\cos \phi_2}{\cos \phi_1 - \cos \phi_2}$$

Theory

i. The radius of the generating globe, R = Actual radius of the earth ÷ Denominator of R.F.

ii. The division along the central meridian for spacing the parallels at an interval i°,

$$d = \frac{\pi R}{180°} \times i°$$

iii. The radius of the standard parallel, ϕ_1,

$$r_1 = R(\phi_2 - \phi_1)^c \frac{\cos \phi_1}{\cos \phi_1 - \cos \phi_2}$$

iv. The radius of the standard parallel, ϕ_2,

$$r_2 = R(\phi_2 - \phi_1)^c \frac{\cos \phi_2}{\cos \phi_1 - \cos \phi_2}$$

v. The division on the standard parallel, ϕ_1 for spacing the meridians at i° interval,

$$d_1 = \frac{2\pi R \cos\phi_1}{360°} \times i°$$

vi. The division on the standard parallel, ϕ_2 for spacing the meridians at i° interval,

$$d_2 = \frac{2\pi R \cos\phi_2}{360°} \times i°$$

Construction

i. A straight line is drawn vertically through the centre of the paper to represent the central meridian.
ii. It is then divided by d for spacing the parallels.
iii An arc of circle of r_1 radius centred on the central meridian is drawn passing through the ϕ_1 division-mark on the central meridian to represent the first standard parallel.
iv. Similarly, with r_2 radius the standard parallel ϕ_2 is drawn.
v. From the centre of the standard parallels, concentric arcs of circles are drawn through the remaining division-marks to represent the remaining parallels.
vi. By d_1, the standard parallel, (ϕ_1), is divided for spacing the meridians.
vii. Similarly by d_2, the standard parallel, (ϕ_2), is divided for spacing the meridians.
viii Straight lines are drawn through the corresponding division points on the standard parallels to represent the meridians.
ix The graticules are then properly labelled (Fig. 2.21).

Fig. 2.21 Simple Conical Projection with II Standard Parallels

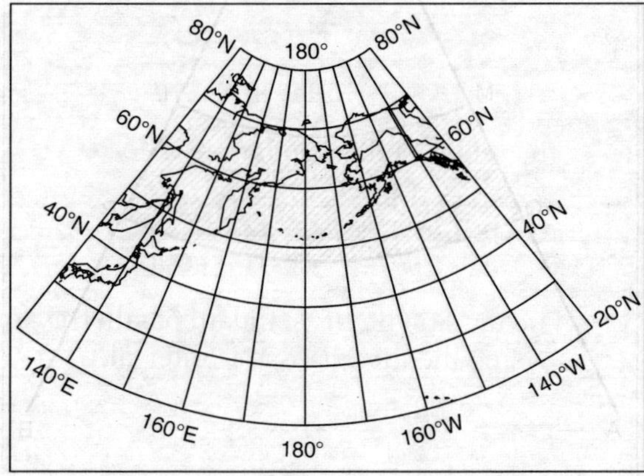

Properties

i. It is a non-perspective projection.
ii. Parallels are concentric arcs of circles truly spaced on the central meridian.
iii. The pole is represented by an arc of circle.
iv. The radial scale is true along the central meridian.
v. Parallels are equidistant from one another.
vi. Meridians are straight lines truly spaced on the standard parallels.
vii. Meridians converge at the vertex of the cone.
viii. The tangential scale is true along the standard parallels.
ix. Deformation is negative between the standard parallels and positive beyond the standard parallels.
x. It is the most suitable projection for mid-latitude countries with latitudinal extent relatively smaller than the longitudinal extent.

Example

Draw graticules at 10° interval on scale 1:160 × 10^6 for the extension, 20°N–80°N and 130°E–130°W.

Calculation

i. $R = \frac{640 \times 10^6 \text{ cm}}{160 \times 10^6}$

$= 4$ cm

ii. $d = \frac{\pi.4}{180°} \times 10°$

$= 0.6981$ cm

iii. For 20°N to 80°N parallels at 10° interval standard parallels chosen are:

$\phi_1 = 20°N + \frac{(80° - 20°)N}{3}$

$= 40°N$

and $\phi_2 = 80°N + \frac{(80° - 20°)N}{3}$

$= 60°N$

iv. Parallels to be drawn are 20°N, 30°N, 40°N,

50°N, 60°N, 70°N and 80°N
Meridians to be drawn are 130°E, 140°E, 150°E, 160°E, 170°E, 180°, 170°W, 160°W, 150°W, 140°W and 130°W

$$r_{40} = 4 \times (60^o - 40^o)^c \frac{\cos 60^o}{\cos 40^o - \cos 60^o}$$

$$= 4.0204 \text{ cm}$$

$$r_{60} = 4 \times (60^o - 40^o)^c \frac{\cos 60^o}{\cos 40^o - \cos 60^o}$$

$$= 2.6241 \text{ cm}$$

v. $d_1 = \frac{2\pi \times 4\cos 40^\circ}{360^\circ} \times 10^\circ$

$$= 0.5348 \text{ cm}$$

vi $d_2 = \frac{2\pi \times 4\cos 60^\circ}{360^\circ} \times 10^\circ$

$$= 0.3491 \text{ cm}$$

CONICAL EQUAL-AREA PROJECTION WITH I STANDARD PARALLEL

Principle

It is a non-perspective projection of a conical group developed by Lambert (1772). In this a simple right circular cone is supposed to touch the generating globe along a selected parallel, called the standard parallel. Parallels are projected as concentric arcs of circles with varying radii. Meridians are projected as straight lines radiating from the vertex of the cone and truly spaced on the standard parallel. The principle is that the area of a segment on projection is made exactly identical to the corresponding area on the generating globe to retain the property of equal area.

Theory

i. Radius of the generating globe, R = Actual radius of the earth ÷ Denominator of the R.F.
ii. Radius of the standard parallel (ϕ_0), r_0 = $R\cot\phi_0$

Fig. 2.22 Principle of Conical Equal Area Projection with I Standard Parallel

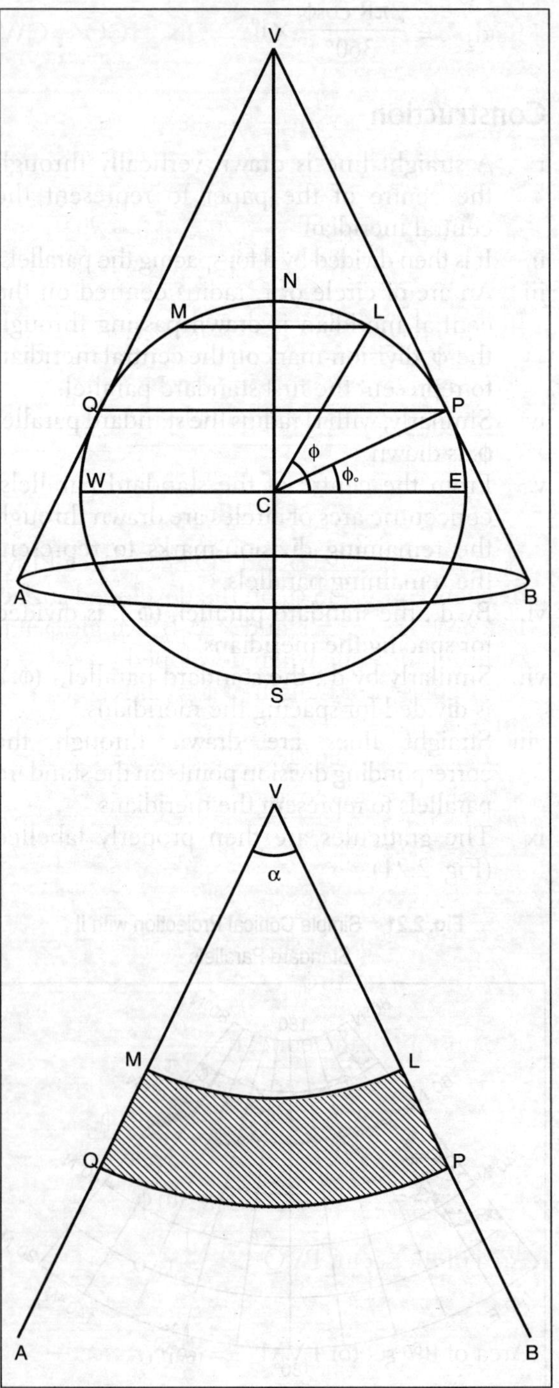

In Fig. 2.22, let the cone VAB touch the generating globe along the standard parallel ϕ_0 ($\angle$ECP).

Radius, CE = CP = CL = CN = CQ = CW = CS = R

VB is tangent at P

$\therefore \angle VPC = \angle CPB = 90°$

ΔVPC is a right $\angle\Delta$

$$\therefore \angle PVC = [180° - (\angle VCP + \angle VPC)]$$
$$= [180° - \{(90° - \phi_0) + 90°\}]$$
$$= \phi_0$$

In the rt $\angle\Delta VPC$, $\dfrac{VP}{PC} = \cot\phi_0$

or, $VP = PC\cot\phi_0 \Rightarrow R\cot\phi_0$

iii. Radius of any parallel (ϕ),

$$r_\phi = R\sqrt{\cot^2\phi_0 - \frac{2\sin\phi}{\sin\phi_0} + 2}$$

Let LM be parallel ϕ and PQ be ϕ_0. Area PLMQ lies between ϕ and ϕ_0 on the developed surface. Let VM = r, VQ = r_0 and $\angle$AVB = angle ϕ at the vertex of the cone when developed.

$r_0 = R\cot\phi_0$ and Length of $\phi_0 = 2\pi R\cos\phi_0$

From Fig 2.20,

$$\alpha . r_0 = 2\pi R\cos\phi_0$$

$$\text{or} \quad \alpha = \frac{2\pi R\cos\phi_0}{r_0}$$

$$= \frac{2\pi R\cos\phi_0}{R\cot\phi_0}$$

$$= 2\pi\sin\phi_0$$

$$\therefore \text{Constant of the cone, } n = \frac{\alpha}{2\pi}$$

$$= \frac{2\pi\sin\phi_0}{2\pi}$$

$$\Rightarrow \sin\phi_0$$

Area of the sector PVQ $= \dfrac{1}{2}r_0^2\alpha$

Area of the sector LVM $= \dfrac{1}{2}r^2\alpha$

Area PLMQ on projection $= \dfrac{1}{2}r_0^2\alpha - \dfrac{1}{2}r^2\alpha$

$$= \frac{\alpha}{2}(r_0^2 - r^2) \quad \ldots \text{(i)}$$

On the globe, the same area

$$= 2\pi R^2(\sin\phi - \sin\phi_0) \quad \ldots \text{(ii)}$$

Hence to maintain the equal area property,

$$\frac{\alpha}{2}(r_0^2 - r^2) = 2\pi R^2(\sin\phi - \sin\phi_0)$$

$$\text{or,} \quad r_0^1 - r^1 = \frac{2\pi R^2(\sin\phi - \sin\phi_0)}{\alpha}$$

$$\text{or,} \quad r^2 = r_0^2 - \frac{2\pi R^2(\sin\phi - \sin\phi_0)}{2\pi\sin\phi_0}$$

$$= R^2\cot^2\phi_0 - \frac{2R^2\sin\phi}{\sin\phi_0} + \frac{2R^2\sin\phi_0}{\sin\phi_0}$$

$$= R^2\left[\cot^2\phi_0 - \frac{2\sin\phi}{\sin\phi_0} + 2\right]$$

$$\therefore \quad r = R\sqrt{\cot^2\phi_0 - \frac{2\sin\phi}{\sin\phi_0} + 2}$$

iv. Division length on standard parallel (ϕ_0), for spacing the meridians at i° interval,

$$d = \frac{2\pi R\cos\phi_0}{360°} \times i°$$

Example

Draw graticules at 5° interval on scale 1: 68 × 10⁶ for the extensions 55°N-85°N and 80°W-10°W.

Calculation

$$\text{i.} \quad R = \frac{640 \times 10^6 \text{ cm}}{68 \times 10^6}$$

$$= 9.4117 \text{ cm}$$

$$\text{ii.} \quad r_0 = R\cot\phi_0$$
$$= 9.4117\cot 70°$$
$$= 3.4256 \text{ cm}$$

$$\text{iii.} \quad d = \frac{2\pi . 9.4117\cos 70°}{360°} \times 5°$$

$$\Rightarrow 0.2809 \text{ cm}$$

Construction

i. A straight line is drawn vertically through the centre of the paper to represent the central meridian.
ii. An arc of circle is drawn with radius r_0 and the centre on the central meridian to represent the standard parallel (ϕ_0).
iii. Concentric arcs of circles are drawn with radii r_ϕ to represent the remaining parallels (Table 2.7).
iv. The standard parallel is then divided with d on both sides of the central meridian.
v. Straight lines are drawn through these division points joining the centre to represent the meridians.
vi. The graticules are then accurately and properly labelled (Fig. 2.23).

Properties

i. It is a non-perspective projection.
ii. The pole is represented by an arc of circle.
iii. The tangential scale is true only along the standard parallel.
iv. The tangential scale increases gradually away from the standard parallel toward the pole.
v. The radial scale decreases gradually away from the standard parallel toward the pole.
vi. At any point the product of the radial and tangential scale is unity.
vii. Interparallel distance gradually decreases toward pole.
viii. It is an equal area projection.
ix. It is suitable for showing the distribution of any element in the mid-latitude countries or regions.

Fig. 2.23 Conical Equal Area Projection with I Standard Parallel

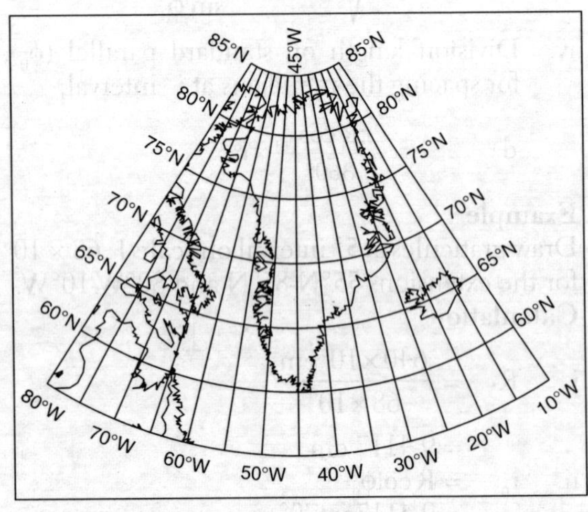

Table 2.7 Computation of r_ϕ

ϕ	55°N	60°N	65°N	70°N	75°N	80°N	85°N
r_ϕ (cm)	5.8703	5.0619	4.2460	3.4256	2.6055	1.7970	1.0403

CONICAL EQUAL-AREA PROJECTION WITH II STANDARD PARALLELS

Principle

It is a non-perspective projection developed by Albers. In this, a right circular cone is supposed to intersect the generating globe along two selected standard parallels. Parallels are projected as concentric arcs of circles and meridians as straight lines radiating from the vertex of the cone and truely spaced on the standard parallels. The principle is that the area of a segment on projection is made exactly identical to the corresponding area on the generating globe to retain equal area property.

Theory

i. Radius of the generating globe, R = Actual radius of the earth ÷ Denominator of the R.F.
ii. Radius of the standard parallels (ϕ_1) and (ϕ_2)

$$r_1 = \frac{2R\cos\phi_1}{\sin\phi_1 + \sin\phi_2}$$

$$r_2 = \frac{2R\cos\phi_2}{\sin\phi_1 + \sin\phi_2}$$

In Fig. 2.24, cone VAB intersects the generating globe along standard parallels, ϕ_1 and ϕ_2. Let PQ be parallel ϕ_1 and LM parallel ϕ_2.

$\therefore \angle PCE = \phi_1$ and $\angle LCE = \phi_2$.

Let radius CS = CE = CP = CL = CN = CW = CS = R.

On the developed cone, let r, r_1 and r_2 be the radii of parallels ϕ, ϕ_1 and ϕ_2 respectively.

$$\therefore \quad r_1.\alpha = 2\pi R\cos\phi_1$$

$$\text{or} \quad r_1 = \frac{2\pi R\cos\phi_1}{\alpha}$$

$$r_2.\alpha = 2\pi R\cos\phi_2$$

Fig. 2.24 Principle Conical Equal Area Projection with II Standard Parallel

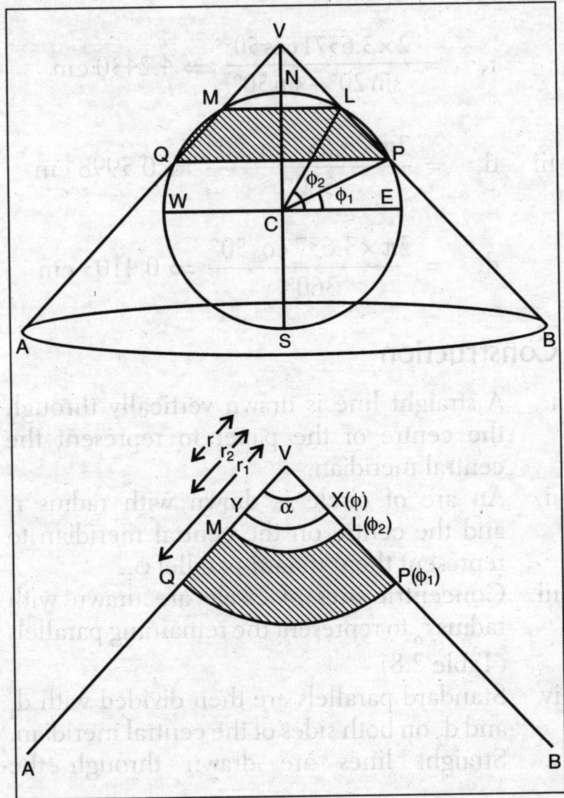

$$\text{or} \quad r_2 = \frac{2\pi R\cos\phi_2}{\alpha}$$

On the developed cone,

$$\text{area PLMQ} = \frac{\alpha}{2}(r_1^2 - r_2^2)$$

On the generating globe the same area

$$= 2\pi R^2(\sin\phi_2 - \sin\phi_1)$$

On principle,

$$\therefore \quad \frac{\alpha}{2}(r_1^2 - r_2^2) = 2\pi R^2(\sin\phi_2 - \sin\phi_1)$$

$$\text{or,} \quad r_1^2 - r_2^2 = \frac{4\pi R^2(\sin\phi_2 - \sin\phi_1)}{\alpha}$$

$$\text{or,} \quad \frac{4\pi^2R^2\cos^2\phi_1}{\alpha^2} - \frac{4\pi^2R^2\cos^2\phi_2}{\alpha^2} = \frac{4\pi R^2(\sin\phi_2 - \sin\phi_1)}{\alpha}$$

$$\text{or,} \quad \frac{4\pi^2R^2}{\alpha^2}(\cos^2\phi_1 - \cos^2\phi_2) = \frac{4\pi R^2}{\alpha}(\sin\phi_2 - \sin\phi_1)$$

$$\text{or,} \quad \alpha = \frac{\pi(\cos^2\phi_1 - \cos^2\phi_2)}{\sin\phi_2 - \sin\phi_1}$$

$$= \frac{\pi(\sin^2\phi_2 - \sin^2\phi_1)}{\sin\phi_2 - \sin\phi_1}$$

$$= \pi(\sin\phi_1 + \sin\phi_2)$$

$$\text{Constant of the cone, } n = \frac{\alpha}{2\pi}$$

$$= \frac{\pi(\sin\phi_1 + \sin\phi_2)}{2\pi}$$

$$= \frac{1}{2}(\sin\phi_1 + \sin\phi_2)$$

$$\text{Hence, } r_1 = \frac{2\pi R\cos\phi_1}{\alpha}$$

$$= \frac{2R\cos\phi_1}{\sin\phi_1 + \sin\phi_2}$$

and $$r_2 = \frac{2\pi R\cos\phi_2}{\alpha}$$

$$= \frac{2R\cos\phi_2}{\sin\phi_1 + \sin\phi_2}$$

iii. Radius of any parallel (ϕ)

$$r_f = \frac{2R}{(\sin\phi_1 + \sin\phi_2)} \times \sqrt{\cos^2\phi_1 - (\sin\phi - \sin\phi_1)(\sin\phi_1 + \sin\phi_2)}$$

On the developed cone, area PXYQ $= \frac{\alpha}{2}(r_1^2 - r^2)$

On the generating globe the same area $= 2\pi R^2(\sin\phi - \sin\phi_1)$

∴ On principle,

$$\frac{\alpha}{2}(r_1^2 - r^2) = 2\pi R^2(\sin\phi - \sin\phi_1)$$

or, $$(r_1^2 - r^2) = \frac{4\pi R^2}{\alpha}(\sin\phi - \sin\phi_1)$$

$$\therefore r^2 = r_1^2 - \frac{4\pi R^2}{\alpha}(\sin\phi - \sin\phi_1)$$

$$= \frac{4R^2\cos^2\phi_1}{(\sin\phi_1 + \sin\phi_2)^2} - \frac{4\pi R^2(\sin\phi - \sin\phi_1}{\pi(\sin\phi_1 + \sin\phi_2)}$$

$$= \frac{4R^2}{(\sin\phi_1 + \sin\phi_2)^2} \times [\cos^2\phi_1 - (\sin\phi - \sin\phi_1)(\sin\phi_1 + \sin\phi_2)]$$

$$\therefore r = \frac{2R}{(\sin\phi_1 + \sin\phi_2)} \times \sqrt{\cos^2\phi_1 - (\sin\phi - \sin\phi_1)(\sin\phi_1 + \sin\phi_2)}$$

iv. Division length on standard parallels (ϕ_1) and (ϕ_2), for spacing the meridians at i° interval,

$$d_1 = \frac{2\pi R\cos\phi_1}{360°} \times i°$$

$$d_2 = \frac{2\pi R\cos\phi_2}{360°} \times i°$$

Example

Draw graticules at 10° interval on scale 1:175 × 10^6 for the extensions 0°—80°S and 70°W—20°E.

Calculation

i. $R = \frac{640 \times 10^6 \text{ cm}}{175 \times 10^6} \Rightarrow 3.6571$ cm

ii. ϕ_1 = 20°S and ϕ_2 = 50°S

$r_1 = \frac{2 \times 3.6571 \cos 20°}{\sin 20° + \sin 50°} \Rightarrow 6.2028$ cm

$r_2 = \frac{2 \times 3.6571 \cos 50°}{\sin 20° + \sin 50°} \Rightarrow 4.2430$ cm

iii. $d_1 = \frac{2\pi \times 3.657 \cos 20°}{360°} \Rightarrow 0.5998$ cm

$d_2 = \frac{2\pi \times 3.657 \cos 50°}{360°} \Rightarrow 0.4103$ cm

Construction

i. A straight line is drawn vertically through the centre of the paper to represent the central meridian.

ii. An arc of circle is drawn with radius r_1 and the centre on the central meridian to represent the standard parallel ϕ_1.

iii. Concentric arcs of circle are drawn with radius r_ϕ to represent the remaining parallels (Table 2.8).

iv. Standard parallels are then divided with d_1 and d_2 on both sides of the central meridian.

v. Straight lines are drawn through the

Table 2.8 Computation of r_ϕ

ϕ	0°	10°S	20°S	30°S	40°S	50°S	60°S	70°S
r_ϕ (cm)	7.4154	6.8267	6.2028	5.5540	4.8942	4.2430	3.6298	3.1014

corresponding division points on the standard parallels joining the centre to represent the meridians.

vi. The graticules are then properly labelled (Fig. 2.25).

Properties

i. It is a non-perspective projection.
ii. The pole is represented by an arc of circle.
iii. The tangential scale is true along the standard parallels.
iv. The radial scale gradually decreases beyond the standard parallels and increases in between.
v. The tangential scale is smaller in between the standard parallels and gradually increases away from these.
vi. At any point on the projection, the product of the radial and tangential scale is unity.
vii. Interparallel distance gradually decreases toward the poles.
viii. It is most suited for showing a distribution in mid-latitude or temperate countries with relatively greater latitudinal extension.

Fig. 2.25 Conical Equal Area Projection with II Standard Parallels

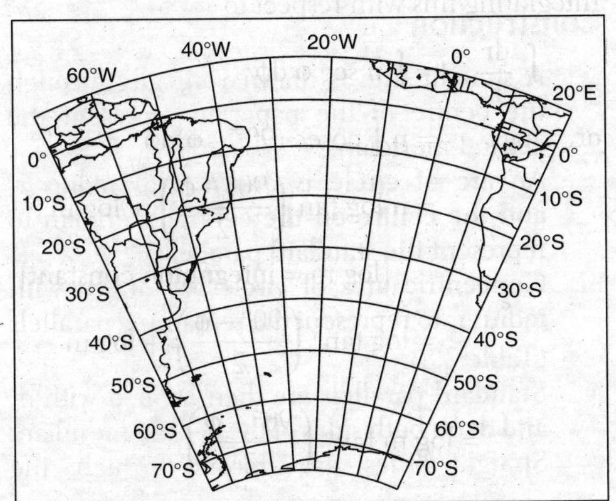

CONICAL ORTHOMORPHIC PROJECTION WITH I STANDARD PARALLEL

Principle

It is a non-perspective projection developed by Lambert (1772). In this, a right circular cone is supposed to touch the generating globe along a parallel called standard parallel. Parallels are projected as concentric arcs of circles with varying radii. Meridians are projected as straight lines radiating from the vertex of the cone and truely spaced on the standard parallel. The principal is that at any point on the map the tangential scale is exactly equal to the radial scale so that the property of orthomorphism (i.e., preservation of shape) is truly maintained.

Theory

i. Radius of the generating globe, R = Actual radius of the earth ÷ Denominator of the R.F.

ii. Radius of the standard parallel (ϕ_0), $r_0 = R\cot\phi_0$

In Fig. 2.26, cone VAB touches the generating globe along PQ (ϕ_0).

Radius, CE = CP = CL = CN = CW = CS = R

VB is tangent at P.

$\therefore\ \angle VPC = 90°$ and

ΔVPC is a rt $\angle\Delta$.

$\therefore\ \angle PVC = [180° - (\angle VCP + \angle VPC)]$

$= [180° - \{(90° - \phi_0) + 90°\}]$

$= \phi_0$

In rt $\angle\Delta VPC$,

$$\frac{VP}{PC} = \cot\phi_0$$

Fig. 2.26 Principle of Conical Orthomorphic Projection with I Standard Parallel

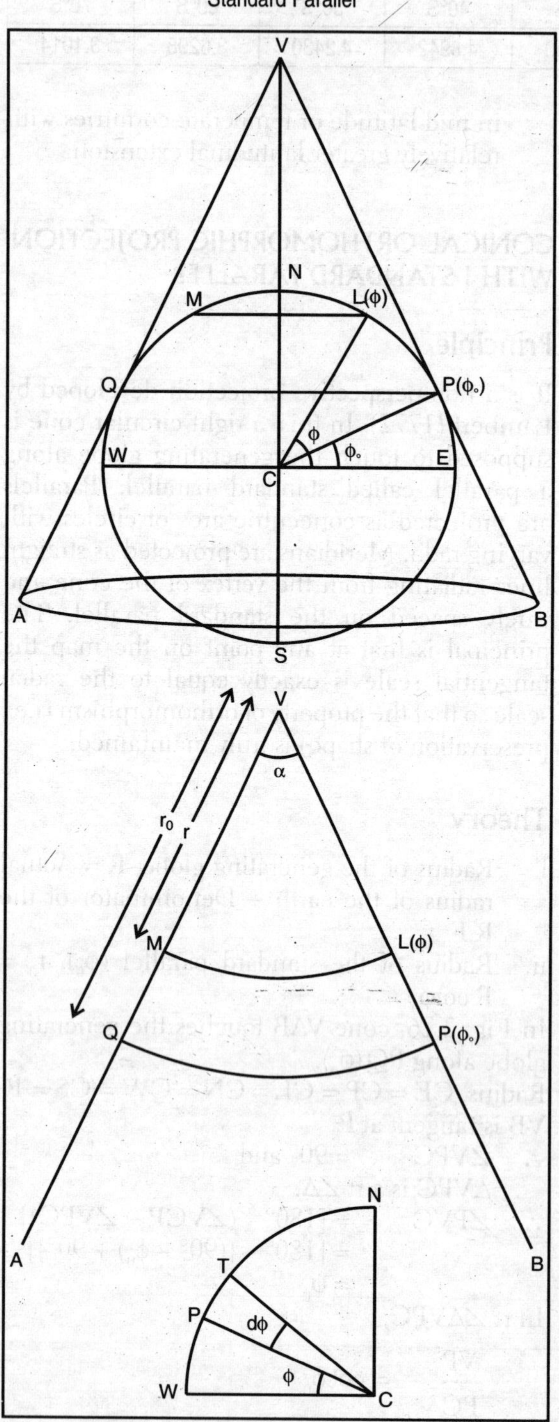

or VP $= PC\cot\phi_0$

∴ $r_0 = R\cot\phi_0$

iii. Radius of any parallel (φ),

$$r_\phi = m\tan^n\left(\frac{90^\circ-\phi}{2}\right)$$

where, $n = \sin\phi_0$

and $$m = \frac{R\cot\phi}{\tan^n\left(\frac{90^\circ-\phi}{2}\right)}$$

Length of parallel (φ) on the globe = 2πRcosφ.

On the developed cone, length of parallel (φ) = r.α

∴ $$\text{Tangential scale along } \phi = \frac{2\pi R\cos\phi}{r\alpha}$$

Let T be parallel at elemental latitude (dφ) away from P(φ).

∴ The true arc distance PT on the generating globe = R.dφ

Let dr = map length corresponding to PT.

∴ $$\text{Radial scale} = \frac{Rd\phi}{dr}$$

On principle, Radial scale = Tangential scale

∴ $$\frac{Rd\phi}{dr} = \frac{2\pi R\cos\phi}{r\alpha}$$

or, $$\frac{dr}{r} = \frac{\alpha}{2\pi}\sec\phi\, d\phi$$

$$= n\sec\phi\, d\phi$$

Integrating this with respect to

$$\int\frac{dr}{r} = \int n\sec\phi\, d\phi$$

or, $$\log r = n\int \operatorname{cosec}(90^\circ-\phi)d\phi$$

$$= n\log\tan\left(\frac{90^\circ-\phi}{2}\right) + \log m$$

[log m = integration constant]

$$= \log\tan^n\left(\frac{90^\circ-\phi}{2}\right) + \log m$$

$$= \log m\tan^n\left(\frac{90^\circ-\phi}{2}\right)$$

$$\therefore \quad r = m \tan^n \left(\frac{90° - \phi}{2}\right)$$

Now, $\alpha . r_0 = 2\pi R \cos\phi_0$

$$\text{or,} \quad \frac{\alpha}{2\pi} = \frac{R \cos \phi_0}{r_0}$$

$$\text{or,} \quad n = \frac{R \cos \phi_0}{R \cot \phi_0} = \sin \phi_0$$

$$\therefore \quad \text{for } \phi_0,\ r_0 = m \tan^n \left(\frac{90° - \phi_0}{2}\right)$$

$$\text{or,} \quad m = \frac{r_0}{\tan^n \left(\frac{90° - \phi_0}{2}\right)} = \frac{R \cot \phi_0}{\tan^n \left(\frac{90° - \phi_0}{2}\right)}$$

iv. Division length for spacing the meridians on the standard parallel (ϕ_0) at i° interval,

$$d = \frac{2\pi R \cos \phi_0}{360} \times i°$$

Example
Draw graticules at 5° interval on scale 1:80 × 10^6 for the extension 10°S—45°S and 110°E—155°E.

Calculation

i. $R = \frac{640 \times 10^6 \text{ cm}}{80 \times 10^6} \Rightarrow 8 \text{ cm}$

ii. $r_0 = 8 \cot 25° \Rightarrow 17.1560 \text{ cm}$

iii. $n = \sin 25° \Rightarrow 0.42262$

$$m = \frac{8 \cot 25°}{\tan^{0.42262}\left(\frac{90° - 25°}{2}\right)} = 20.75734 \text{ cm}$$

$$r_\phi = 20.75734 \times \tan^{0.42262}\left(\frac{90° - \phi}{2}\right)$$

v. $d = \frac{2\pi \times 8 \cos 25°}{360°} \Rightarrow 0.6327 \text{ cm}$

Construction

Same as in Conical Equal Area Projection with I Standard Parallel (Fig. 2.27).

Fig. 2.27 Conical Orthomorphic Projection with I Standard Parallel

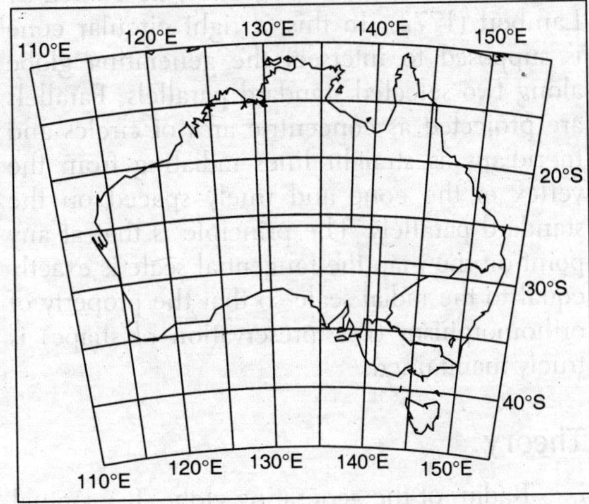

Properties

i. It is a non-perspective projection.
ii. The tangential scale is true only along the standard parallel.

Table 2.9 Computation of r_ϕ

ϕ	10°S	15°S	20°S	25°S	30°S	35°S	40°S	45°S
r_ϕ (cm)	19.2740	18.5593	17.8550	17.1560	16.4570	15.7524	15.0364	14.3022

iii. The scale decreases gradually away from the standard parallel (Table 2.9).
iv. Interparallel distance gradually decreases toward the equator.
v. At any point on the map, tangential scale is exactly equal to the radial scale.
vi. The meridional scale is true only at the point of intersection with the standard parallel.
vii. It is most suitable for showing the exact shape of a country in the mid-latitude or temperate regions.

CONICAL ORTHOMORPHIC PROJECTION WITH II STANDARD PARALLELS

Principle

It is a non-perspective projection developed by Lambert (1772). In this, a right circular cone is supposed to intersect the generating globe along two selected standard parallels. Parallels are projected as concentric arcs of circles and meridians as straight lines radiating from the vertex of the cone and truely spaced on the standard parallels. The principle is that at any point on the map the tangential scale is exactly equal to the radial scale so that the property of orthomorphism (i.e., preservation of shape) is truely maintained.

Theory

i. Radius of the generating globe, R = Actual radius of the earth ÷ Denominator of the R.F.
ii. Radius of any parallel (ϕ),

$$r_\phi = m \tan^n\left(\frac{90° - \phi}{2}\right)$$

where,

$$n = \frac{\log\cos\phi_1 - \log\cos\phi_2}{\log\tan\left(\frac{90-\phi_1}{2}\right) - \log\tan\left(\frac{90-\phi_2}{2}\right)}$$

$$m = \frac{R\cot\phi_1}{\tan^n\left(\frac{90° - \phi_1}{2}\right)}$$

$$= \frac{R\cot\phi_2}{\tan^n\left(\frac{90° - \phi_2}{2}\right)}$$

ϕ_1 and ϕ_2 = standard parallels ($\phi_2 > \phi_1$)

iii. Division length on the standard parallels for spacing the meridians at i° interval,

$$d_1 = \frac{2\pi R\cos\phi_1}{360°} \times i°$$

$$d_2 = \frac{2\pi R\cos\phi_2}{360°} \times i°$$

Fig. 2.28 Principle of Conical Orthomorphic Projection with II Standard Parallels

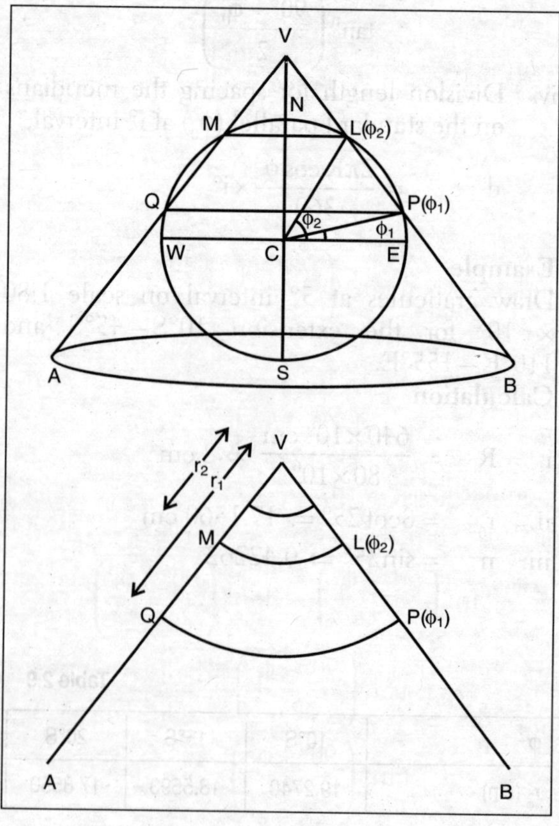

In Fig. 2.28, PQ and LM are respectively parallel to ϕ_1 and ϕ_2

Radius CE = CP = CL = CN = CW = CS = R

On globe, length of $\phi_1 = 2\pi R\cos\phi_1$

length of $\phi_2 = 2\pi R\cos\phi_2$

On map, length of $\phi_1 = r_1.\alpha$

length of $\phi_2 = r_2.\alpha$

$\therefore$ $r_1.\alpha = 2\pi R\cos\phi_1$ and

$r_2.\alpha = 2\pi R\cos\phi_2$

$$\text{or, } r_1 = \frac{2\pi R\cos\phi_1}{\alpha} = \frac{R\cos\phi_1}{\frac{\alpha}{2\pi}} \Rightarrow \frac{R\cos\phi_1}{n}$$

$$\text{or, } r_2 = \frac{2\pi R\cos\phi_2}{\alpha} = \frac{R\cos\phi_2}{\frac{\alpha}{2\pi}} \Rightarrow \frac{R\cos\phi_2}{n}$$

$$\text{Given, } r = m\tan^n\left(\frac{90° - \phi}{2}\right)$$

$$\therefore \quad r_1 = m\tan^n\left(\frac{90° - \phi_1}{2}\right)$$

$$m = \frac{r_1}{\tan^n\left(\frac{90° - \phi_1}{2}\right)} \quad \ldots \text{(i)}$$

$$\text{and } r_2 = m\tan^n\left(\frac{90° - \phi_2}{2}\right)$$

$$m = \frac{r_2}{\tan^n\left(\frac{90° - \phi_2}{2}\right)} \quad \ldots \text{(ii)}$$

Comparing (i) & (ii)

$$\frac{r_1}{\tan^n\left(\frac{90° - \phi_1}{2}\right)} = \frac{r_2}{\tan^n\left(\frac{90° - \phi_2}{2}\right)}$$

$$\text{or, } \frac{r_1}{r_2} = \frac{\tan^n\left(\frac{90° - \phi_1}{2}\right)}{\tan^n\left(\frac{90° - \phi_2}{2}\right)}$$

$$\frac{\frac{R\cos\phi_1}{n}}{\frac{R\cos\phi_2}{n}} = \frac{\tan^n\left(\frac{90° - \phi_1}{2}\right)}{\tan^n\left(\frac{90° - \phi_2}{2}\right)}$$

$$\frac{\cos\phi_1}{\cos\phi_2} = \frac{\tan^n\left(\frac{90° - \phi_1}{2}\right)}{\tan^n\left(\frac{90° - \phi_2}{2}\right)}$$

taking logarithm on both sides.

$$\therefore \log\left(\frac{\cos\phi_1}{\cos\phi_2}\right) = \log\left[\frac{\tan^n\left(\frac{90° - \phi_1}{2}\right)}{\tan^n\left(\frac{90° - \phi_2}{2}\right)}\right]$$

$$\text{or, } \log\cos\phi_1 - \log\cos\phi_2$$

$$= \log\tan^n\left(\frac{90° - \phi_1}{2}\right) - \log\tan^n\left(\frac{90° - \phi_2}{2}\right)$$

$$= n\left[\log\tan\left(\frac{90° - \phi_1}{2}\right) - \log\tan\left(\frac{90° - \phi_2}{2}\right)\right]$$

$$\therefore \quad n = \frac{\log\cos\phi_1 - \log\cos\phi_2}{\log\tan\left(\frac{90° - \phi_1}{2}\right) - \log\tan\left(\frac{90° - \phi_2}{2}\right)}$$

Example

Draw graticules on Conical Orthomorphic Projection with II Standard Parallels at 10° interval on R.F 1:105 × 10^6 for the extension 20°N – 80°N and 40°W – 40°E.

Calculation

$$\text{i.} \quad R = \frac{640 \times 10^6 \text{ cm}}{105 \times 10^6}$$

$$= 6.0952 \text{ cm}$$

$$\text{ii.} \quad n = \frac{\log\cos 40° - \log\cos 60°}{\log\tan\left(\frac{90° - 40°}{2}\right) - \log\tan\left(\frac{90° - 60°}{2}\right)}$$

$$= 0.77003$$

iii. $m = \dfrac{6.0952 \cos 40^\circ}{\tan^{0.77003}\left(\dfrac{90^\circ - 40^\circ}{2}\right)} = \dfrac{6.0952 \cos 60}{\tan^{0.77003}\left(\dfrac{90^\circ - 60^\circ}{2}\right)}$

$= 8.40183$ cm

iv. $d_1 = \dfrac{2\pi \times 6.0952 \cos 40^\circ}{360^\circ} \times 10^\circ \Rightarrow 0.8149$ cm

v. $d_2 = \dfrac{2\pi \times 6.0952 \cos 60^\circ}{360^\circ} \times 10^\circ \Rightarrow 0.5319$ cm

Construction

Same as Conical Equal Area Projection with II Standard Parallels (Fig. 2.29).

Fig. 2.29 Conical Orthomorphic Projection with II Standard Parallels

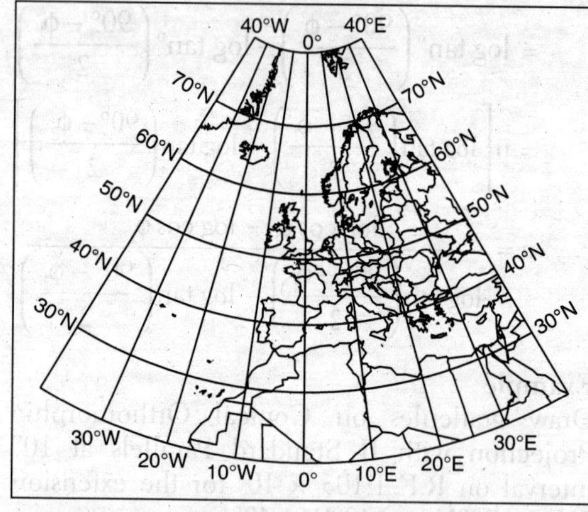

Properties

i. It is a non-perspective projection.
ii. The tangential scale and the radial scale are exactly identical at all points on the map.
iii. The tangential scale is true only along the standard parallels (Table 2.10).
iv. The radial scale is true at the points of intersection of the meridians with the standard parallels.
v. It is an orthomorphic projection that maintains shape.
vi. It is most suitable for showing the correct shape of a country in the mid-latitude or temperate regions.

BONNE'S PROJECTION

Principle

In this projection, a simple right circular cone is supposed to touch the generating globe along the standard parallel. The radial scale is truly preserved along the central meridian and the tangential scale is preserved along all the parallels. Consequently, the parallels are represented as concentric arcs of circles but the meridians appear as smooth curves (Fig. 2.30). This is the modification designed by R Bonne (a French cartographer) to the original Simple Conical Projection with I Standard Parallel.

Theory

i. Radius of the generating globe, R = Actual radius of the earth ÷ Denominator of R.F.
ii. The division on the central meridian for spacing the parallels at i° interval,

$$d = \frac{\pi R}{180^\circ} \times i^\circ$$

iii. Radius of the standard parallel (ϕ_0),

$$r_0 = R \cot\phi_0$$

Table 2.10 Computation of r_ϕ

ϕ	20°N	30°N	40°N	50°N	60°N	70°N	80°N
r_ϕ (cm)	6.38548	5.50396	4.66919	3.85817	3.04759	2.20809	1.28719

Fig. 2.30 Principles of Bonne's Projection

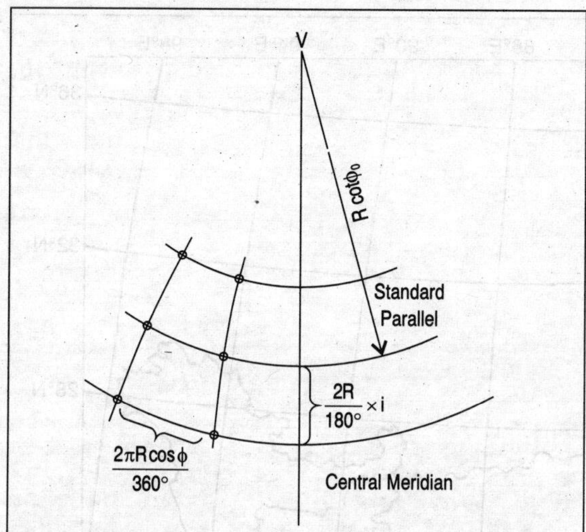

iv. The division on the parallels for spacing the meridians at i° interval,

$$d_\phi = \frac{2\pi R\cos\phi}{360°} \times i°$$

Construction

i. A straight line is drawn vertically through the centre of the paper to represent the central meridian.
ii. It is then divided by d for spacing the parallels.
iii. An arc of circle is then drawn through the standard parallel mark with radius r_0 and centre on the central meridian (produced if necessary).
iv. Concentric arcs of circles are then drawn through each division on the central meridian to represent the remaining parallels.
v. Parallels are then divided with their corresponding division lengths (d_ϕ) on both sides of the central meridian.
vi. Smooth free-hand curves are then drawn through the corresponding division points on the parallels to represent the meridians.
vii. Lastly the graticules are properly labelled.

Properties

i. Parallels are concentric arcs of circles, truly spaced on the central meridian.
ii. Radial scale is true only along the central meridian.
iii. The tangential scale is true along all the meridians.
iv. Excepting the central meridian, all are regular curves concave towards the centre.
v. At any point the product of the two principal scales is unity.
vi. It is an equal-area projection.
vii. Since both parallels and meridians are curves, their intersection appears to be acute near the poles and obtuse near the lower latitudes.
viii. For small and compact countries with nearly equal latitudinal and longitudinal extensions, angular deformation is relatively less, i.e., distortion in shape is less.
ix. It is used for countries like France, Netherlands, Switzerland, Belgium, India, etc. (Fig. 2.31).

Example

Draw graticules at 4° interval on scale, $1:25\times10^6$ for the extension, 66°E—98°E and 8°N—40°N.

Calculation

i. $r = \frac{640\times10^6 \text{ cm}}{25\times10^6} \Rightarrow 25.6 \text{ cm}$

ii. $d = \frac{\pi\times25.6}{180°} \times 4° \Rightarrow 1.7872 \text{ cm}$

iii. Parallels to be drawn are: 8°N, 12°N, 16°N, 20°N, 24°N, 28°N, 32°N, 36°N, 40°N.
Standard parallel (ϕ_0) chosen = 24°N

$r_0 = 25.6.\cot 24° \Rightarrow 57.4985$ cm

iv. $d_\phi = \frac{2\pi\times25.6\cos\phi}{360°} \times 4°$

$= 1.7872 \cos\phi$ (Table 2.11)

Fig. 2.31 Bonne's Projection

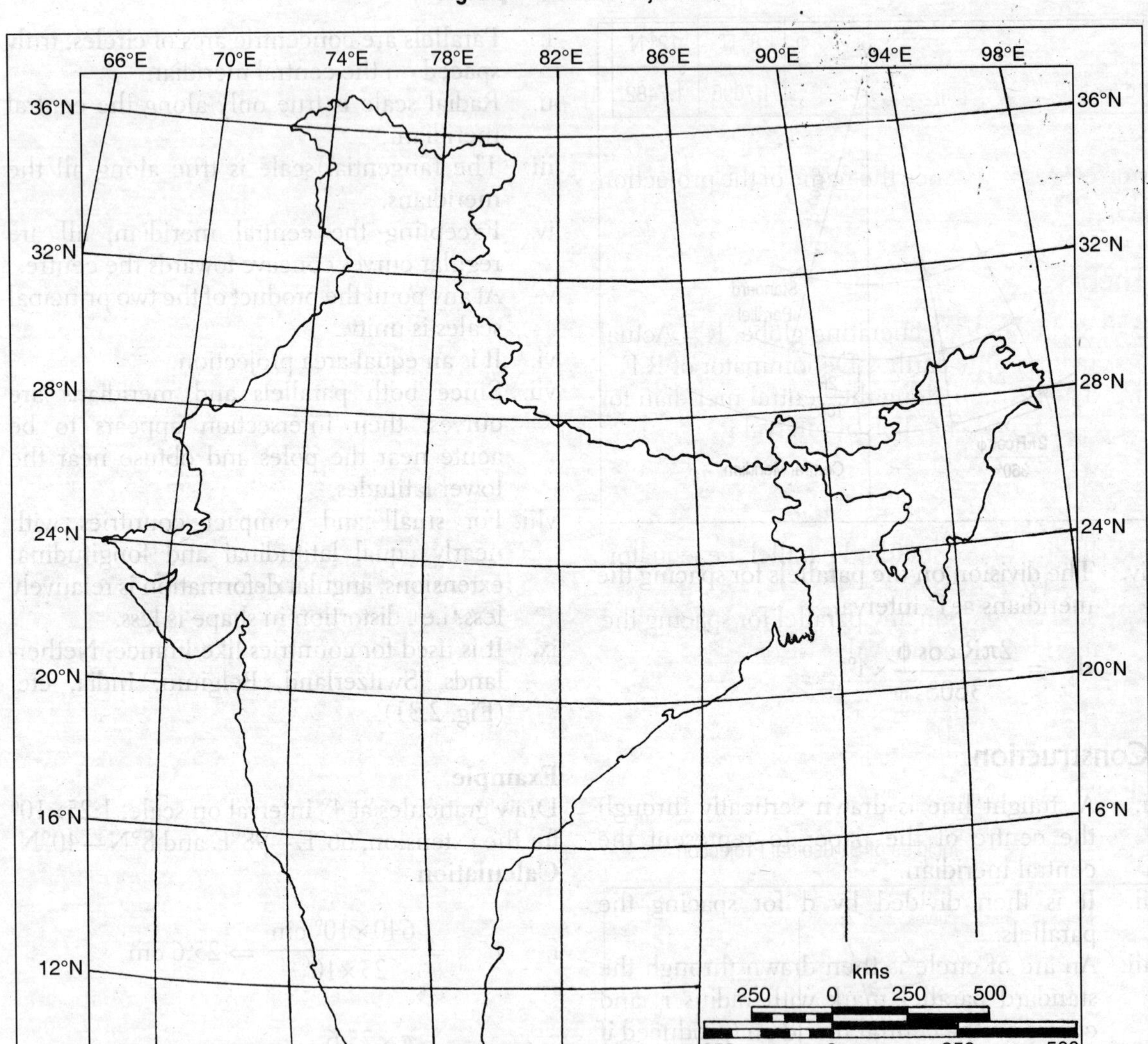

SINUSOIDAL PROJECTION

Principle

It is a special case of Bonne's projection. Sanson, a French cartographer, and Flamsteed, a British astronomer Royal, have together designed this projection in which a right circular cone touches the generating globe along the equator (0°). Hence, the equator is the standard parallel with infinite radius. This, in effect, is a straight line perpendicular to the central meridian. Obviously all the parallels become straight lines parallel to the equator. As the tangential scale along all the parallels is truly preserved, meridians become

Table 2.11 Computation of d_ϕ

ϕ	8° N	12° N	16° N	20° N	24° N	28° N	32° N	36° N	40°N
d_ϕ = 1.7872 cos ϕ (cm)	1.7698	1.7482	1.7180	1.6794	1.6327	1.5780	1.5156	1.4459	1.3691

sine curves and hence the name of the projection (Fig. 2.32).

Theory

i. Radius of the generating globe, R = Actual radius of the earth ÷ Denominator of R.F.

ii. The division along the central meridian for spacing the parallels at interval i°,

$$d = \frac{\pi R}{180^\circ} \times i^\circ$$

iii. Radius of the standard parallel, i.e. equator, $r_0 = R \cot 0^\circ \Rightarrow \infty$

iv. The division on any parallel for spacing the meridian at i° interval,

$$d_\phi = \frac{2\pi R \cos\phi}{360^\circ} \times i^\circ$$

Fig. 2.32 Principles of Sinusoidal Projection

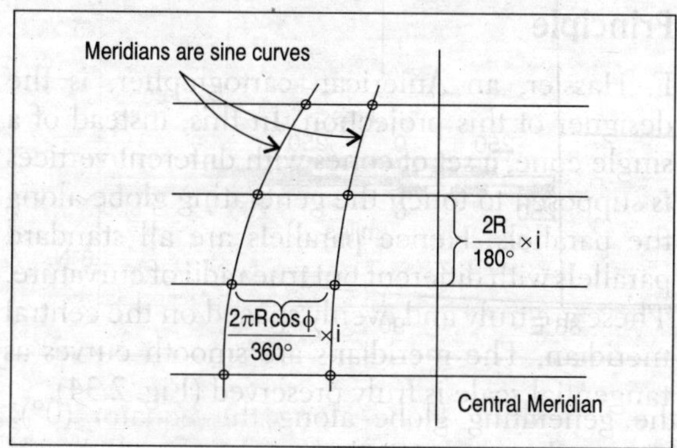

Construction

i. A straight line is drawn vertically through the centre of the paper to represent the central meridian.

ii. It is then divided by d for spacing the parallels along it.

iii. Through each of these division points, straight lines are drawn perpendicular to the central meridian to represent the parallels.

iv. Each parallel is then divided by corresponding division lengths, d_ϕ, on both sides of the central meridian (Table 2.12).

v. Free-hand smooth curves are drawn through the corresponding division points on the parallels to represent the meridians.

vi. The graticule is then accurately and properly labelled (Fig. 2.33).

Properties

i. Parallels are straight lines, parallel to each other and evenly spaced on the central meridian.

ii. The radial scale is true only along the central meridian.

iii. Except for the central meridian, all are *sine curves*.

iv. The tangential scale is true at all points.

v. At any point, the product of the two principal scales is unity.

vi. It is an equal-area projection.

vii. The shape is highly distorted near the poles and towards the margins of the map.

Table 2.12 Computation of d_ϕ

ϕ	0°	15°N/S	30°N/S	45°N/S	60°N/S	75°N/S	90°N/S
d_ϕ = 0.5585 cos ϕ (cm)	0.5585	0.5395	0.4837	0.3949	0.2792	0.1445	0

Fig. 2.33 Sinusoidal Projection

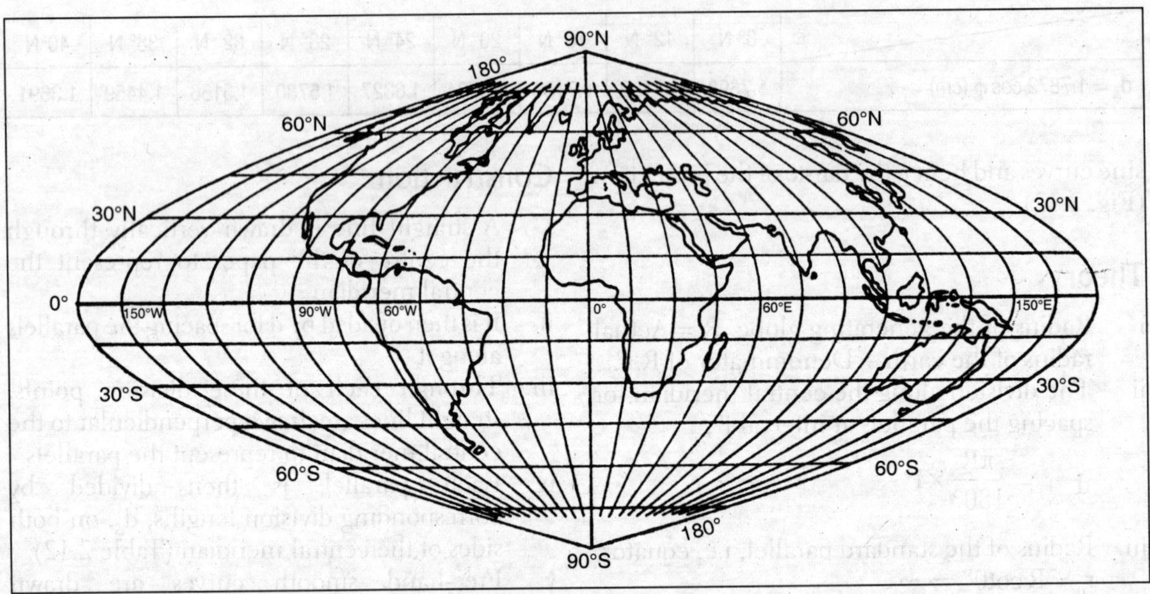

viii. The central meridian is the shortest with true length.

ix. Meridians intersect the equator at right angles but intersect other parallels at acute angles; hence shape is less disturbed near low latitudes.

x. Meridians are all bisected by the equator.

xi. It is suitable for countries or continents near the equator, e.g., South America, Africa, India, Sri Lanka, Bangladesh, Australia. It is also used for world maps.

Example

Draw graticules at 15° interval on Scale, 1:300 × 10^6 for the whole globe.

Calculation

i. $R = \frac{640 \times 10^6 \text{ cm}}{300 \times 10^6} \Rightarrow 2.1333$ cm

ii. $d = \frac{2.1333\pi}{180°} \times 15° \Rightarrow 0.5585$ cm

iii. $r_0 = R \cot 0° = \infty.$

iv. $d_\phi = \frac{2\pi \times 2.1333 \cos \phi}{360°} \times 15° = 0.5585 \cos\phi$ (Table 2.12).

POLYCONIC PROJECTION

Principle

F. Hassler, an American cartographer, is the designer of this projection. In this, instead of a single cone, a set of cones with different vertices is supposed to touch the generating globe along the parallels. Hence parallels are all standard parallels with different but true radii of curvature. These are truly and evenly spaced on the central meridian. The meridians are smooth curves as tangential scale is truly preserved (Fig. 2.34).

Theory

i. The radius of the generating globe, R = Actual radius of the earth ÷ Denominator of R.F.

ii. The division on the central meridian for spacing the parallels at i° interval,

Fig. 2.34 Principles of Polyconic Projection

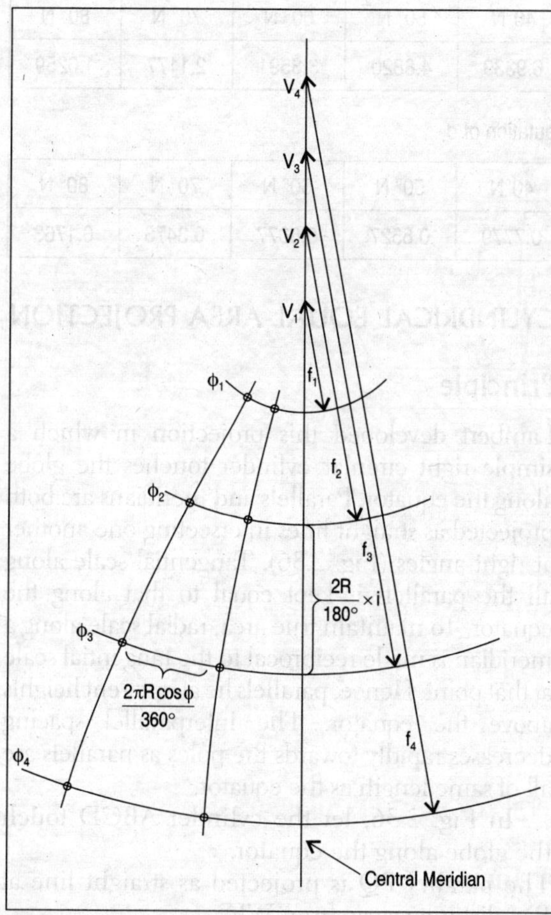

$$d = \frac{\pi R}{180°} \times i°$$

iii. Radius of a parallel (ϕ), r_ϕ = Rcotϕ.

iv. Divisions on all parallel (ϕ) for spacing the meridians at i° interval,

$$d_\phi = \frac{2\pi R.\cos\phi}{360°} \times i°$$

Construction

i. A straight line is drawn vertically through the centre of the paper to represent the central meridian.

ii. It is then divided by d for spacing the parallels.

iii. Non-concentric arcs of circles are drawn through each of these division points with corresponding radius (r_ϕ) and centre on the central meridian to represent the parallels.

iv. Each parallel is then divided by the corresponding division lengths, d_ϕ.

v. Free-hand smooth curves are drawn through the corresponding division points on the parallels to represent the meridians.

vi. The graticules are then accurately and properly labelled (Fig. 2.35).

Properties

i. All are standard parallels with true dimensions.

ii. Parallels are non-concentric arcs of circles with the radius of curvature decreasing toward the poles.

iii. Parallels are truly and evenly spaced only along the central meridian.

iv. The radial scale increases away from the central meridian.

Fig. 2.35 Polyconic Projection

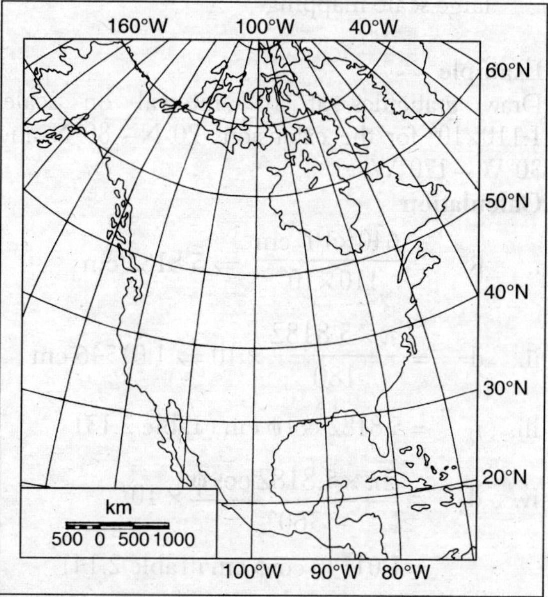

Table 2.13 Computation of r_ϕ

ϕ	20° N	30° N	40°N	50° N	60° N	70° N	80° N
$r_\phi = 5.8182 \cot\phi$ (cm)	15.9854	10.0774	6.9339	4.8820	3.3591	2.1177	1.0259

Table 2.14 Computation of d_ϕ

ϕ	20° N	30° N	40°N	50° N	60° N	70° N	80° N
$d_\phi = 1.001546 \cos\phi$ (cm)	0.9542	0.8794	0.7779	0.6527	0.5077	0.3473	0.1763

v. Meridians are smooth curves truly spaced on the parallels.
vi. The tangential scale is true along the parallels.
vii. The east-west distance between any two points is correct along the parallels but the north-south distance between any two points is correct only along the central meridian.
vii. Away from the central meridian, deformation increases towards the margins of the map.
ix. It is an aphylactic projection.
x. This projection can be drawn with more than one central meridian in an interrupted fashion. Hence, it is largely used for topographical survey sheets for accurate large scale mapping.

Example

Draw graticules at 10° interval on scale, 1:110×10⁶ for the extensions 20°N–80°N and 30°W–170°W.

Calculation

i. $R = \dfrac{640\times10^6 \text{ cm}}{110\times10^6} \Rightarrow 5.8182$ cm

ii. $d = \dfrac{\pi\times5.8182}{180}\times 10 \Rightarrow 1.01546$ cm

iii. $r_\phi = 5.8182 \cot\phi$ cm (Table 2.13)

iv. $d_\phi = \dfrac{2\pi\times5.8182\cos\phi}{360°}\times 10°$

$= 1.01546 \cos\phi$ cm (Table 2.14)

CYLINDRICAL EQUAL-AREA PROJECTION

Principle

Lambert developed this projection in which a simple right circular cylinder touches the globe along the equator. Parallels and meridians are both projected as straight lines intersecting one another at right angles (Fig. 2.36). Tangential scale along all the parallels is kept equal to that along the equator. To maintain true area, radial scale along a meridian is made reciprocal to the tangential scale at that point. Hence, parallels lie at different heights above the equator. The interparallel spacing decreases rapidly towards the poles as parallels are all of same length as the equator.

In Fig. 2.36, let the cylinder ABCD touch the globe along the equator.

The parallel PQ is projected as straight line at PM distance away from WE.

$P_1Q_1 \parallel WE$ and $\angle POM = \phi$

Length of parallel (ϕ) on globe = $2\pi R\cos\phi$.

Length of parallel (ϕ) on projection = $2\pi R$

$$\therefore \text{tangential scale} = \frac{2\pi R}{2\pi R\cos\phi} = \sec\phi \quad \text{... (i)}$$

Let S be another parallel at $d\phi$ angle away from P(ϕ)

$\therefore$ the true angular distance of $d\phi$ on globe = $R.d\phi$

Let dy be the corresponding linear distance of $d\phi$ from the equator on projection.

$$\therefore \text{radial scale} = \frac{dy}{R.d\phi} \quad \text{... (ii)}$$

Fig. 2.36 Principles of Cylindrical Equal-Area Projection

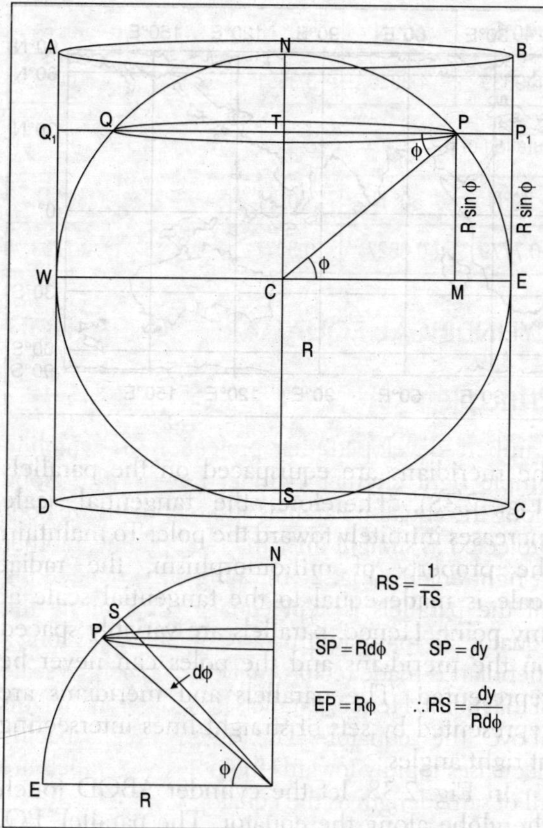

Since it is an equal-area projection,

radial scale × tangential scale = 1

$$\text{or,} \quad \frac{dy}{R.d\phi}.\sec\phi = 1$$

$$\text{or,} \quad dy = R\cos\phi.d\phi$$

By integration,

$$\int dy = R\int \cos\phi.d\phi$$

$$\therefore \quad y = R\sin\phi$$

Theory

i. Radius of the generating globe, R = Actual radius of the earth ÷ Denominator of R.F.

ii. Division along the equator for spacing the meridians at i° interval,

$$d = \frac{2\pi R}{360^\circ} \times i^\circ$$

iii. Height of any parallel above equator,

$$y_\phi = R\sin\phi$$

Construction

i. A straight line is drawn horizontally through the centre of the paper to represent the equator.

ii. It is then divided by d for spacing the meridians.

iii. Through each of these division points, straight lines are drawn perpendicular to the equator to represent the meridians.

iv. On the central meridian, heights of different parallels (y_ϕ) from the equator are marked.

v. Through each of these points, straight lines are drawn perpendicular to the central meridian to represent the parallels.

vi. The graticules are then properly labelled (Fig. 2.37).

Properties

i. Parallels are represented by a set of parallel straight lines.

ii. Parallels are of same the length as the equator ($2\pi R$).

iii. Parallels are variably spaced on the meridians.

iv. Interparallel spacing decreases rapidly toward the pole.

v. The tangential scale rapidly increases poleward and is infinity at the poles.

vi. Meridians are parallel straight lines truly spaced on the equator.

vii. Meridians are of same length equal to the diameter of the globe (2R).

viii. The intermeridian spacing is uniform on all the parallels.

ix. The pole is represented by a straight line of length $2\pi R$.

x. At any point, the product of the two principal scales is unity.

Fig. 2.37 Cylindrical Equal-Area Projection

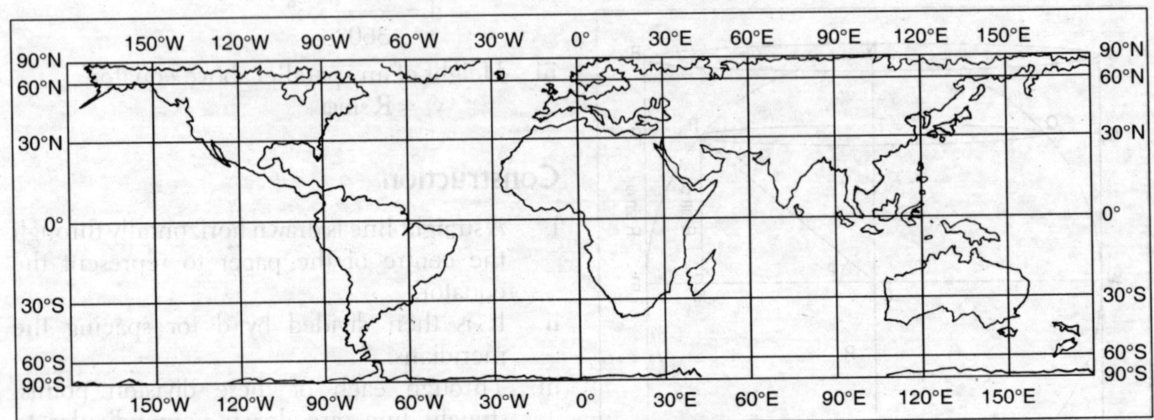

Table 2.15 Computation of y_ϕ

ϕ	30° N/S	60°N/S	90°N/S
$y_\phi = 2.1695 \sin\phi$ (cm)	1.0848	1.8788	2.1695

xi. It is an equal-area projection.

xii. The shape is largely distorted near the poles.

Example

Draw graticules at 30° interval on scale, 1:295 × 10^6 for the whole globe.

Calculation

i. $R = \dfrac{640\times10^6 \text{ cm}}{295\times10^6} \Rightarrow 2.1695$ cm

ii. $d = \dfrac{2\pi\times2.1695}{360°}\times30° \Rightarrow 1.1359$ cm

iii. $y_\phi = 2.1695 \sin\phi$ cm (Table 2.15).

MERCATOR'S PROJECTION

Principle

This is a cylindrical orthomorphic projection designed by Flemish, Mercator and Wright. In this, a simple right circular cylinder touches the globe along the equator. All the parallels are of the same length equal to that of the equator and the meridians are equispaced on the parallels (Fig. 2.38). Therefore, the tangential scale increases infinitely toward the pole. To maintain the property of orthomorphism, the radial scale is made equal to the tangential scale at any point. Hence, parallels are variably spaced on the meridians and the poles can never be represented. The parallels and meridians are represented by sets of straight lines intersecting at right angles.

In Fig. 2.38, let the cylinder ABCD touch the globe along the equator. The parallel, PQ, is projected as straight line at PM distance away from WE. $P_1Q_1 \| WE$ and $\angle POM = \phi$

Let S be another parallel at $d\phi$ angle away from $P(\phi)$

$\therefore$ true angular distance of $d\phi$ on globe = $R.d\phi$

Let dy be the corresponding linear distance of $d\phi$ from the equator on projection.

$$\therefore \quad \text{radial scale} = \frac{dy}{R.d\phi} \quad \text{... (i)}$$

Length of parallel (ϕ) on globe = $2\pi R \cos\phi$

Length of parallel (ϕ) on projection = $2\pi R$

$$\therefore \quad \text{tangential scale} = \frac{2\pi R}{2\pi R\cos\phi} \quad \text{... (ii)}$$

$$= \sec\phi \quad \text{... (iii)}$$

Since it is an orthomorphic projection,

radial scale = tangential scale

Fig. 2.38 Principles of Mercator's Projection

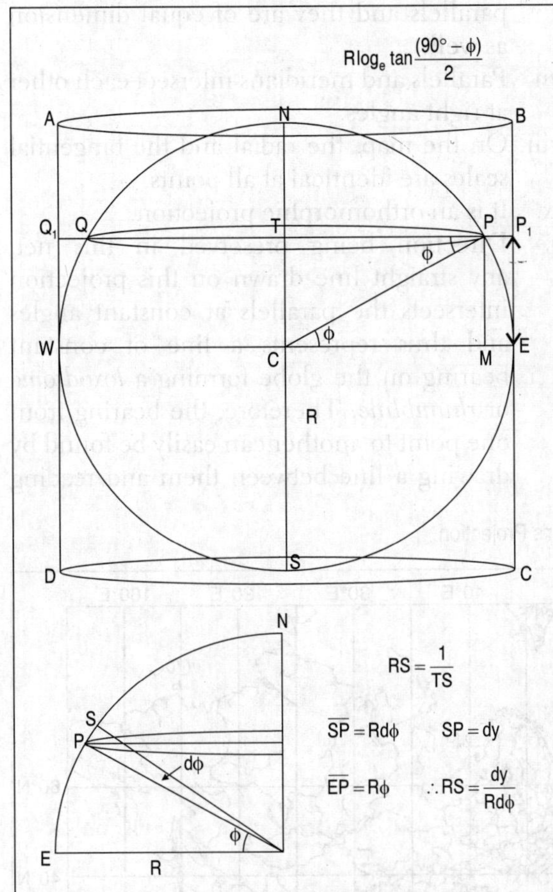

or, $\dfrac{dy}{R.d\phi} = \sec\phi$

or $dy = R\sec\phi.d\phi$

By integration and simplification

$\int dy = R\int \sec\phi.d\phi$

or, $y = R\log_e \tan\left(\dfrac{90^\circ - \phi}{2}\right)$

$= 2.3026\ R \log\tan\left(\dfrac{90^\circ - \phi}{2}\right)$

Theory

i. Radius of the generating globe, R = Actual radius of the earth ÷ Denominator of R.F.

ii. The height of any parallel (ϕ) above the equator,

$$y_\phi = 2.3026\ R\log\tan\left(\frac{90^\circ - \phi}{2}\right)$$

iii. Divisions on the equator for spacing the meridians at i° interval, $d = \dfrac{2\pi R}{360^\circ} \times i^\circ$

Example

Draw graticules at 20° interval on scale, 1:295 × 10^6 for the whole globe.

Calculation

i. $R = \dfrac{640 \times 10^6 \text{ cm}}{295 \times 10^6} \Rightarrow 2.1695$ cm

ii. $d = \dfrac{2\pi \times 2.1695}{360^\circ} \Rightarrow 0.7573$ cm

iii. $y_\phi = 2.3026 \times 2.1695 \log\tan\left(\dfrac{90^\circ - \phi}{2}\right)$ (Table 2.16)

Table 2.16 Computation of y_ϕ

ϕ	20°N/S	40°N/S	60°N/S	80°N/S
$\left(\dfrac{90^\circ - \phi}{2}\right)$	55°	65°	75°	85°
y_ϕ (cm)	0.7732	1.6551	2.8571	5.2854

Construction

i. A straight line is drawn horizontally through the centre of the paper to represent the equator.

ii. It is then divided by d for spacing the meridians.

iii. Through each of these division points, straight lines are drawn perpendicular to the equator to represent the meridians.

iv. On the central meridian, heights of different parallels (y_ϕ) from the equator are marked.

v. Through each of these points, straight lines are drawn perpendicular to the central meridian to represent the parallels.

vi. The graticules are then properly labelled (Fig. 2.39).

Properties

i. Parallels are represented by a set of parallel straight lines.
ii. All parallels are of the same length as the equator.
iii. Parallels are variably spaced on a meridian; interparallel distance increases away from the equator.
iv. The poles cannot be represented in this projection.
v. Meridians are represented by a set of parallel straight lines truly spaced on the equator only.
vi. Meridians are equispaced on all the parallels and they are of equal dimension as well.
vii Parallels and meridians intersect each other at right angles.
viii. On the map, the radial and the tangential scales are identical at all points.
ix. It is an orthomorphic projection.
x. Direction being preserved in this net, any straight line drawn on this projection intersects the parallels at constant angles and thus represents a line of constant bearing on the globe forming a *loxodrome* or *rhumbline*. Therefore, the bearing from one point to another can easily be found by drawing a line between them and reading

Fig. 2.39 Mercator's Projection

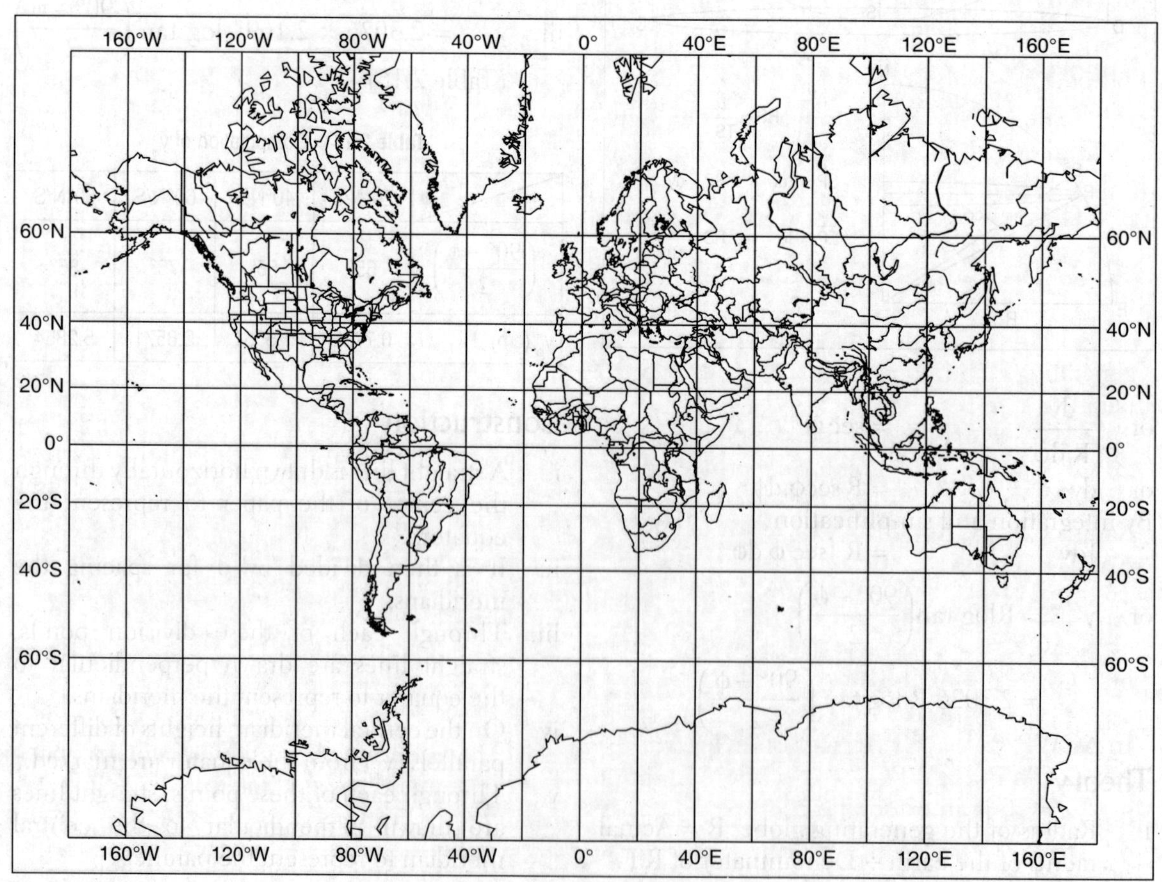

off the angle it makes with the meridians. Hence, it is very useful to sailors.

xi. It is suitable for world maps showing wind circulation patterns, ocean circulation patterns, routes, drainage patterns, etc.

GALL'S STEREOGRAPHIC PROJECTION

Principle

It is a cylindrical projection of secant type in which a right circular cylinder intersects the generating globe at 45°N and 45°S parallels. Parallels are stereographically projected as straight lines at varying distances away from the equator (i.e., from a point on the sphere opposite to the surface being mapped). Meridians are projected as evenly and truly spaced straight lines on the 45°N and 45°S parallels (Fig. 2.40).

Fig. 2.40 Principles of Gall's Stereographic Projection

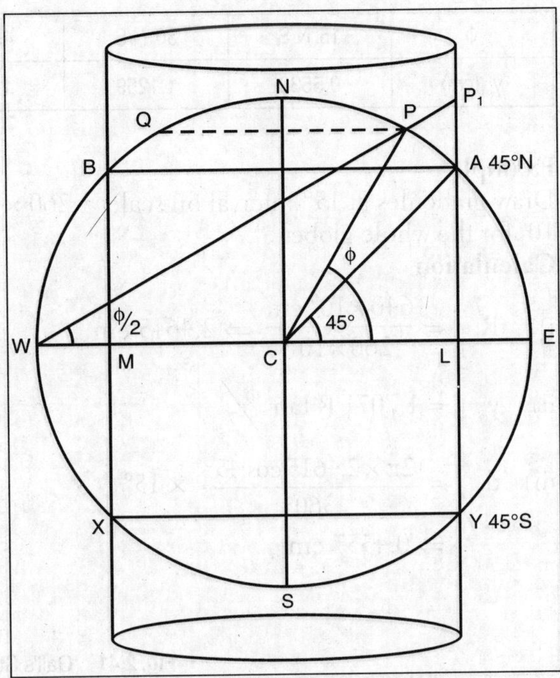

Theory

i. Radius of the generating globe, R = Actual radius of the earth ÷ Denominator of the R.F.

ii. Height of any parallel (ϕ) above the equator,

$$y_\phi = 1.7071\, R \tan \frac{\phi}{2}$$

In Fig. 2.40, the cylinder intersects the generating globe along 45°N (AB) and 45°S (XY) parallel.

∴ ∠ACE = 45° PQ is the parallel ϕ so that ∠PCE = ϕ. It is stereographically projected from W at P′.

Radius CE = CA = CP = CN = CW = CS = R

ΔACL is a rt∠Δ.

$$\therefore \quad \frac{CL}{AC} = \cos 45^\circ$$

$$\text{or,} \quad CL = AC.\cos 45^\circ = R\cos 45^\circ$$

In ΔCWP, CW = CP. It is an isosceles Δ.

∴ ∠CWP = ∠WPC

∠PCL is an external ∠.

∴ ∠CPW + ∠WPC = ∠PCL

or, 2∠CWP = ϕ

$$\text{or,} \quad \angle CWP = \frac{\phi}{2}$$

ΔP_1LW is a right∠Δ.

$$\therefore \quad \frac{P_1L}{WL} = \tan\angle P_1WL$$

$$\text{or,} \quad P_1L = WL\tan\angle P_1WL$$

$$= (WC + CL)\tan\angle CWP$$

$$= (R + R\cos 45^\circ)\tan\frac{\phi}{2}$$

$$= (1 + \cos 45^\circ)\, R \tan\frac{\phi}{2}$$

$$\therefore \quad y_\phi = 1.7071\, R \tan\frac{\phi}{2}$$

iii. Division length on the equator for spacing the meridians at i° inteval,

$$d = \frac{\text{Length of the equator}}{360^\circ} \times i^\circ \Rightarrow \frac{2\pi R \cos 45^\circ}{360^\circ} \times i^\circ$$

Table 2.17 Computation of y_θ

ϕ	15°N/S	30°N/S	45°N/S	60°N/S	75°N/S	90°N/S
y_θ (cm)	0.5532	1.1259	1.7405	2.4260	3.2243	4.2020

Example

Draw graticules at 15° interval on scale, 1:260 × 10^6 for the whole globe.

Calculation

i) $R = \frac{640\times10^6 \text{ cm}}{260\times10^6} \Rightarrow 2.4615 \text{ cm}$

ii) $y_\phi = 1.7071 \text{ R} \tan \phi/2$

iii) $d = \frac{2\pi\times2.4615\cos 45°}{360°} \times 15°$

$\Rightarrow 0.4557 \text{ cm}$

Construction

Same as Mercator Projection (Fig. 2.41).

Properties

i. It is a perspective projection of the cylindrical group.
ii. The tangential scale is true along the 45°N and 45°S parallels, which are essentially standard parallels.
iii. Parallels are straight lines of length 1.85πR.
iv. Meridians are straight lines of length 1.7071R.
v. Parallels and meridians intersect at 90°.

Fig. 2.41 Gall's Stereographic Projection

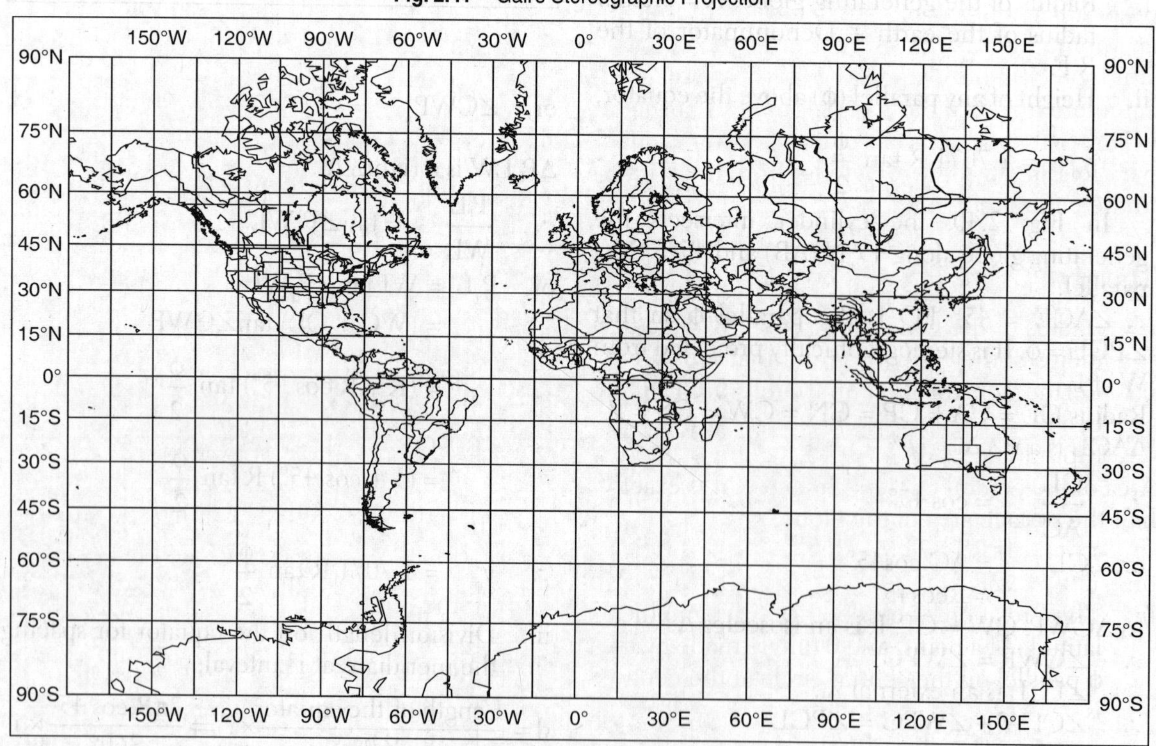

vi. On the equatorial sides of the forty-fives, the tangential scale is too small where as on the polar sides it is too large.
vii. The radial scale is everywhere too large except between the standard parallels, where it is too small.
viii. It is neither equal-area nor orthomorphic, but serves to make a fair map of the world.

MOLLWEIDE'S PROJECTION

Principle

This is an elliptical projection with homolographic property, devised by K Mollweide (1805). The central meridian is a straight line exactly half the length of the equator. The 90th meridian forms a complete circle while the remaining meridians form ellipses. Meridians are equispaced on the parallels. Parallels are projected as horizontal straight lines perpendicular to the central meridian and lie at varying distances away from the equator (Fig. 2.42). The distance of any parallel ϕ is calculated on the basis of the angle (θ) made by the parallel f on the *central circle*. Hence parallels are doubly projected—from the ellipse to the central circle.

Theory

i. Radius of the generating globe, R = Actual radius of the earth ÷ Denominator of R.F.
ii. Let the radius of the central circle (NCFSDA) be r

From Fig. 2.42,

Area of the central circle = πr^2 since, it is exactly half the surface area of the globe,

$\therefore \quad \pi r^2 = 4\pi R^2 \div 2$

or, $\quad r = \sqrt{2}R$ … (i)

iii. The relation between ϕ (geographical latitude of a point) and θ (angle made by the ϕ parallel on the central circle at the centre) is

$$\pi \sin\phi = 2\theta + \sin 2\theta$$

Fig. 2.42 Principles of Mollweide's Projection

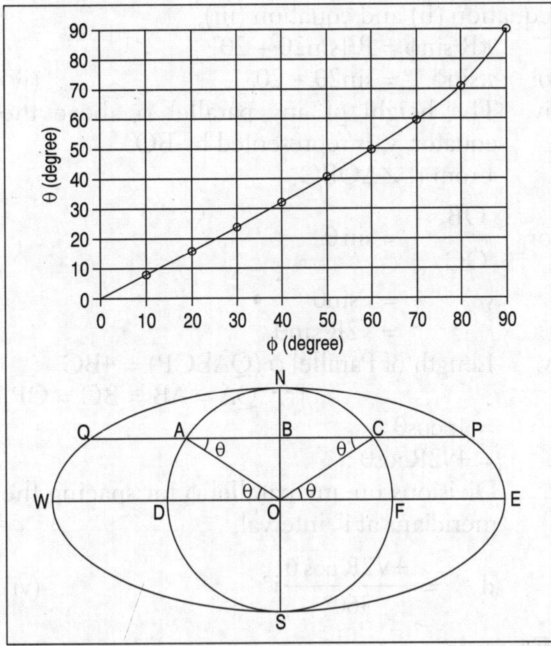

Area within the PQ(ϕ) parallel and the equator on globe = $2\pi R^2 \sin\phi$

The same area within the central circle,

$$= \frac{1}{2}.2\pi R^2 \sin\phi$$

$$= \pi R^2 \sin\phi \qquad \ldots \text{(ii)}$$

Now, on projection, area between the central circle (between ϕ parallel and the equator)

= area ABOD + area BCFO

= 2 area BCFO [since the two are equal]

$= 2\,[\Delta BCO + \angle COF]$

$= 2\,[\tfrac{1}{2}BC.OB + \tfrac{1}{2}OF^2.\theta]$

$= BC.OB + r^2.\theta\ [\because\ OF = r]$

$$\text{In}\left[\begin{array}{r}\text{rt}\angle OBC, OB = OC\sin\theta \Rightarrow r\sin\theta \\ \text{and } BC = OC\sin\theta \Rightarrow r\cos\theta\end{array}\right]$$

$= r\sin\theta.r\cos\theta + r^2\theta$

$= r^2\sin\theta.\cos\theta + r^2\theta$

$= 2R^2\sin\theta.\cos\theta + 2R^2\theta$ [from eqn. (i)]

$= R^2[2\sin\theta.\cos\theta + 2\theta]$

$= R^2[\sin 2\theta + 2\theta]$ … (iii)

Now, to preserve the equal-area property, from equation (ii) and equation (iii),

$\pi R^2 \sin\phi = R^2[\sin 2\theta + 2\theta]$

or, $\pi \sin\phi = \sin 2\theta + 2\theta$ ... (iv)

iv. The height of any parallel ϕ above the equator, y, is represented by BO

From rt $\angle\Delta$OBC,

or, $\dfrac{OB}{OC} = \sin\theta$

$\therefore$ $y = r\sin\theta$

$= \sqrt{2}R\sin\theta$

v. Length of Parallel ϕ (QABCP) = 4BC

[$\because$ QA = AB = BC = CP]

$= 4r\cos\theta$

$= 4\sqrt{2}R\cos\theta$

$\therefore$ Divisions on any parallel ϕ for spacing the meridians at i° interval,

$d = \dfrac{4\sqrt{2}R\cos\theta}{360°} i°$ ... (vi)

Example

Draw graticules at 30° interval on scale, 1:280 × 10^6 for the whole globe.

Calculation

i. $R = \dfrac{640 \times 10^6 \text{ cm}}{280 \times 10^6} \times i°$

$= 2.2857$ cm

ii. $r = \sqrt{2}R$

$= 3.2325$ cm

iii. Calculation of θ corresponding to ϕ. This can be done from a *graph* showing the $\theta - \phi$ curve drawn from the relation, $\pi\sin\phi = 2\theta + \sin 2\theta$ (Fig. 2.42). This can also be calculated by trial and error method. (Table 2.18).

iv. $y = 2.2857\sqrt{2}\sin\theta$ (Table 2.19)

v. $d = \dfrac{4\sqrt{2} \times 2.2857\cos\theta}{360°} \times 30°$ (Table 2.20)

Construction

i. A straight line of length ($4\sqrt{2}R$) is drawn horizontally through the centre of the paper to represent the equator.

ii. Through its mid-point, another straight line of length $\sqrt{2}R$ is drawn at right angles to represent the central meridian.

iii. A circle of radius $\sqrt{2}R$ is drawn with its centre at the intersection of the equator and central meridian to represent the 90th meridian.

iv. Divisions for spacing the parallels are marked on the central meridian with y from the equator.

v. Through each of these, straight lines are drawn parallel to the equator to represent the parallels of latitude.

Table 2.18 Computation of θ corresponding to ϕ

ϕ	10°N/S	20°N/S	30°N/S	40°N/S	50°N/S	60°N/S	70°N/S	80°N/S	90°N/S
$\pi\sin\phi$	0.54553	1.07448	1.57079	2.01937	2.40660	2.72070	2.95213	3.09386	3.14159
θ	7°51'48"	15°47'03"	23°49'36"	32°04'16"	40°37'45"	49°40'30"	59°31'34"	70°58'38"	90°
$2\theta+\sin 2\theta$	0.54553	1.07448	1.57079	2.01937	2.40660	2.72070	2.95213	3.09386	3.14159

Table 2.19 Computation of y

ϕ	30°N/S	60°N/S	90°N/S
θ	23°49'36"	49°40'30"	90°
y (cm)	1.3058	2.4644	3.2324

Table 2.20 Computation of d

ϕ	0°N/S	30°N/S	60°N/S
θ	0	23°49'36"	49°40'30"
d (cm)	1.0775	0.9857	0.6973

Fig. 2.42 Principles of Mollweide's Projection

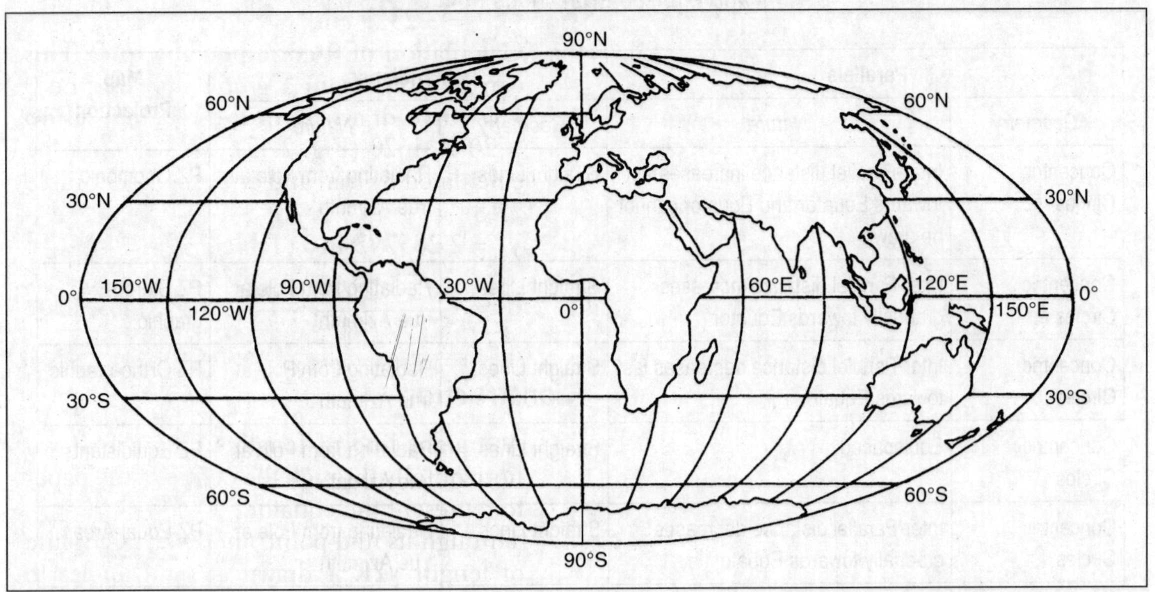

vi Parallels are divided with the corresponding d for spacing the meridians.

vii. Smooth curves are then drawn through the corresponding division points on the parallels to represent the remaining meridians.

viii. The graticules are then properly labelled (Fig. 2.42).

Properties

i. It is an elliptical and double projection.

ii. It is an equal-area projection.

iii. Parallels are straight lines variably spaced on the central meridian.

iv. Interparallel spacing gradually decreases poleward.

v. The central meridian is a straight line half in length ($2\sqrt{2}R$) as that of the equator ($4\sqrt{2}R$).

vi. The 90th meridian forms a circle with radius $\sqrt{2}R$ and area equivalent to one hemisphere.

vii. The remaining meridians are ellipses of varying eccentricity.

ix. The distortion of shape near the poles is relatively less.

x. At any point the product of the two principal scales is unity.

ix. Parallels are a little smaller near the equator but relatively large near the poles corresponding to their true lengths.

xii. It is extensively used for world maps.

Identification of Map Projections from Graticules

Parallels		Meridians		Map Projection
Geometry	*Nature*	*Geometry*	*Nature*	
Concentric Circles	Inter-Parallel distance increases fast towards Equator and Equator cannot be drawn	Straight Lines	Radiating from Pole at true Azimuth	PZ Gnomonic
Concentric Circles	Inter-Parallel distance increases gradually towards Equator	Straight Lines	Radiating from Pole at true Azimuth	PZ Stereo-graphic
Concentric Circles	Inter-Parallel distance decreases fast towards Equator	Straight Lines	Radiating from Pole at true Azimuth	PZ Ortho-graphic
Concentric Circles	Equispaced	Straight Lines	Radiating from Pole at true Azimuth	PZ Equidistant
Concentric Circles	Inter-Parallel distance decreases gradually towards Equator	Straight Lines	Radiating from Pole at true Azimuth	PZ Equal-Area
Concentric Arcs	Equispaced	Straight Lines	Radiating from the Vertex of Cone	Simple Conic I
Concentric Arcs	Equispaced	Only CM = Straight Line	Smooth Curves	Bonne's
Non-Concentric Arcs	Equispaced on the CM only	Only CM = Straight Line	Smooth Curves	Polyconic
Straight Lines	Equispaced	Only CM = Straight Line	Sine Curves	Sinusoidal
Straight Lines	Inter-Parallel distance decreases fast towards Pole	Parallel Straight Lines	Equispaced	Cylindrical Equal Area
Straight Lines	Inter-Parallel distance increases fast towards Pole and Pole can't be drawn	Parallel Straight Lines	Equispaced	Mercator
Straight Lines	Inter-Parallel distance slowly decreases towards Poles	Only CM = Straight Line	90th Meridians = Circle Others = Ellipses	Moll-weide's

EXERCISE

1. What is meant by the term projection?
2. Define map projection.
3. Explain the terms: *poles, Equator, great circle, latitude, longitude, parallels, meridians, Prime Meridian, graticule, generating globe, projection plane, developable surface, standard parallel, central meridian, standard parallel, constant of a cone.*
4. Explain why deformation is inevitable in map projection.
5. Explain why map projection is reversible and unique.
6. Explain the terms: *principal scale, real scale, scale factor, TSF, RSF, deformations, authalic/homolographic/equivalent projections, conformal/orthomorphic projections, azimuthal projections, equidistant projections, aphylactic projections, double projections, perspective projections, non-perspective projections, conventional projections, loxodrome, rhumbline, orthodrome, great circle distance, Tissot's indicatrix.*
7. What are the extrinsic properties of map projection?
8. Classify map projection based on the datum surface.
9. Classify map projection based on the projection plane.
10. Classify map projection based on the methods of projection.
11. Classify map projection based on the properties of projection.
12. Classify map projection based on the appearance of parallels and meridians.
13. Classify map projection based on the geometric shapes of graticules.
14. Classify map projection based on the cases of map projection.
15. State the principles, theories, construction and properties of P. Z. Gnomonic Projection.
16. State the principles, theories, construction and properties of P. Z. Orthographic Projection.
17. State the principles, theories, construction and properties of P. Z. Stereographic Projection.
18. State the principles, theories, construction and properties of P. Z. Equidistant Projection.
19. State the principles, theories, construction and properties of P. Z. Equal-Area Projection.
20. State the principles, theories, construction and properties of S. Conical Projection with I Standard Parallel.
21. State the principles, theories, construction and properties of S. Conical Projection with II Standard Parallels.
22. State the principles, theories, construction and properties of Conical-Equal Area Projection with I Standard Parallel.
23. State the principles, theories, construction and properties of Conical-Equal Area Projection with II Standard Parallels.
24. State the principles, theories, construction and properties of Conical Orthomorphic Projection with I Standard Parallel.
25. State the Principles, theories, construction and properties of Conical Orthomorphic Projection with II Standard Parallels.
26. State the principles, theories, construction and properties of Bonne's Projection.
27. State the principles, theories, construction and properties of Sinusoidal Projection.
28. State the principles, theories, construction and properties of Polyconic Projection.
29. State the principles, theories, construction and properties of Cylindrical Equal-Area Projection.
30. State the principles, theories, construction and properties of Cylindrical Orthomorphic Projection.
31. State the principles, theories, construction and properties of Gall's Stereographic Projection.
32. State the principles, theories, construction and properties of Mollweide's Projection.
33. Compare the polar group of map projections.
34. Compare the conical group of map projections.

35. Compare the cylindrical group of map projections.
36. Compare the conventional group of map projections.
37. Explain why the gnomonic chart is important.
38. Explain how Bonne's Projection is a modification of S. Conical Projection with I Standard Parallel.
39. Explain how Sinusoidal Projection is a modification of Bonne's Projection.
40. Explain how Polyconic Projection is a modification of S. Conical Projection with I Standard Parallel.
41. Explain the principles of the Stereographic Projections and its use.
42. Explain the popularity of the Mercator's Projection among navigators.
43. Draw a net of graticules on P. Z. Gnomonic Projection for the Northern Hemisphere up to 50°N at 5° interval on scale, 1:100,000,000. Find the distance and azimuth of B (65°E, 85°N) with respect to A (0, 50°N) from the graticule.
44. Draw a net of graticules on P. Z. Stereographic Projection for the Southern Hemisphere at 10° interval on scale, 1:150,000,000. Find the distance and azimuth of Z (160°E, 30°S) with respect to A (30°E, 70°S) from the graticule.
45. Draw a net of graticules on P. Z. Orthographic Projection for the Northern Hemisphere at 15° interval on scale, 1:100,000,000. Find the distance and azimuth of P (180°, 20°N) with respect to Q (30°W, 30°N) from the graticule.
46. Draw a net of graticules on P. Z. Equidistance Projection for the Southern Hemisphere at 10° interval on scale, 1:150,000,000. Find the distance and azimuth of L (130°W, 20°N) with respect to M (90°E, 30°N) from the graticule.
47. Draw a net of graticules on P. Z. Equal-Area Projection for the Northern Hemisphere at 10° interval on scale, 1:130,000,000. Find the distance and azimuth of X (170°W, 80°N) with respect to Y (110°W, 20°N) from the graticule.
46. Draw a net of graticules on a suitable Map Projection for the extensions 25°N-85°N and 30°E-120°E at 10° interval on scale, 1:100,000,000. Find the distance and azimuth of X (30°E, 25°N) with respect to Y (120°E, 85°N) from the graticule.
49. Draw a net of graticules on a suitable Map Projection for the extensions 30°S-80°S and 155°E-135°W at 5° interval on scale, 1:90,000,000. Find the distance and azimuth of P (155°E, 30°S) with respect to Q (135°W, 80°S) from the graticule.
50. Draw a net of graticules on a suitable Map Projection for the extensions 0°-90°N and 10°W-80°E at 10° interval on scale, 1:80,000,000. Find the distance and azimuth of R (10°W, 10°N) with respect to S (70°E, 80°N) from the graticule.
51. Draw a net of graticules on a suitable Map Projection for the extensions 10°S-70°S and 20°W-110° W at 10° interval on scale, 1:85,000,000. Find the distance and azimuth of L (30°W, 20°S) with respect to M (100°W, 60°S) from the graticule.
52. Draw a net of graticules on a suitable Map Projection for the extensions 40°N-80°N and 110°E-20°W at 5° interval on scale, 1:110,000,000. Find the distance and azimuth of A (100°E, 45°N) with respect to B (10°W, 75°N) from the graticule.
53. Draw a net of graticules on a suitable Map Projection for the extensions 35°S-85°S and 20°W-40°E at 5° interval on scale, 1:95,000,000. Find the distance and azimuth of C (10°W, 80°S) with respect to D (40°E, 40°S) from the graticule.
54. Draw a net of graticules on a suitable Map Projection for the extensions 0°-90°N and 30°E-120°E at 10° interval on Scale, 1:100,000,000. Find the distance and azimuth of X (40°E, 10°N) with respect to C (110°E, 80°N) from the graticule.
55. Draw a net of graticules on a suitable Map Projection for the extensions 0°-90°S and 30°W-120°W at 10° interval on scale, 1:120,000,000. Find the distance and azimuth of A (30°W, 0°) with respect to M (110°W, 80°S) from the graticule.

56. Draw a net of graticules on a suitable Map Projection for mapping the World at 10° interval on scale, 1:175,000,000. Find the distance and azimuth of the extreme north-eastern point with respect to X (0°, 0°) from the graticule.
57. Draw a net of graticules on a suitable Map Projection for Eastern Hemisphere at 10° interval on scale, 1:85,000,000. Find the distance and azimuth of the extreme north-western point with respect to X (0°, 0°) from the graticule.
58. Draw a net of graticules on a suitable Map Projection for mapping the World upto 80°N at 10° interval on scale, 1:200,000,000. Find the distance and azimuth of the NW-SE diagonal of the graticule.
59. Draw a net of graticules on a suitable Map Projection for mapping the World at 15° interval on scale, 1:170,000,000. Find the distance and azimuth of the SW-NE diagonal from the graticule.

3

Surveying

HIGHLIGHTS

- Classification of Surveys
- Basic Principles of Surveying
- Measurement of Baseline
- Procedures of Locating a Point
- Surveying Terminology
- Surveying Instruments and Accessories
- Traverse Survey
- Traversing by Prismatic Compass
- Traversing by Theodolite
- Traversing by Plane Table
- Contouring
- Trigonometric Levelling

Surveying may be regarded as the most fundamental base of the *art* and *science* of map making. It is defined as the art of taking such measurements as will determine the relative positions of points on the surface of the earth so that the size and shape of a portion of the earth's surface may be ascertained and delineated on a map or plan. Obviously, it is a process of determining the positions of points on a horizontal plane. Contrary to this, the term *levelling* concerns the determination of the relative positions of points on a vertical plane. In a more comprehensive sense, surveying includes levelling. Since the ancient times, this particular art has become increasingly important in the preparation of sections, plans and large scale maps.

CLASSIFICATION OF SURVEYS

Primarily, surveying may be divided into two general classes—*geodetic* and *plane*. In the former, large distances and areas are normally covered and therefore the curvature of the earth's surface is taken into account. In the latter, the measured dimensions are smaller and therefore the curvature of the earth's surface is neglected. Based upon the methods employed, surveying is of two types—*triangulation* and *traversing*. In the former, the area to be surveyed is first divided into a network of traingles, the dimensions of which are computed from (i) a single line, measured directly, called the *baseline* and (ii) the three angles of each of the triangles measured accurately. In traversing,

Classification of Survey

1. Based on the **Surface Curvature** of Earth—
   a) Plane
   b) Geodetic, e.g., Terrestrial, Photogrammetry, and Remote Sensing
2. Based on the **Methods**—
   a) Traversing
   b) Triangulation
   c) Contouring
3. Based on the **Nature of Field**—
   a) Land
   b) Marine/Hydrographic
   c) Astronomical
4. Based on the **Objectives** of Survey—
   a) Archaeological
   b) Geological
   c) Mine
   d) Meteorological
   e) Hydrological
   f) Topographical
   g) Cadastral
   h) Engineering
   i) Military
   j) Social
   k) Economic
   l) Environment
5. Based on **Instruments** (used)—
   a) Traditional, e.g., Chain, Plane Table, Compass, Level, Clinometer, Sextant, Tacheometer, Theodolite
   b) Modern e.g., GPS, Total Station, Satellite Navigation/ Imaging

the framework consists of a series of connected lines, the lengths and directions of which are progressively measured either in a clockwise or in an anticlockwise direction. In contour survey, leveling is done along with traversing and triangulation and then plotted on a suitable scale to draw the isopleths of heights above sea level at a certain interval.

Based on the objectives of survey, there may be a set of almost twelve different types, e.g., *Archaeological, Geological, Mine, Meteorological, Hydrological, Topographical, Cadastral, Engineering, Military, Social, Economic, and Environment. Surveying may be Chain, Plane Table, Compass, Level, Clinometer, Sextant, Tacheometer, Theodolite, GPS, Total Station,* and *Satellite* depending on the instruments/equipment used.

Basic Principles of Surveying

- **Always work from the whole to the part**
- **Location of a point on the ground is fixed by:**
  - **Triangulation** (angular measurements from a fixed baseline)
  - **Trilateration** (only linear measurements from a fixed baseline)
  - **Polar coordinates** (measurements of distance and bearing)
  - **Rectangular coordinates** (only linear measurements along orthogonal directions)

BASIC PRINCIPLES OF SURVEYING

The fundamental rules of surveying are:

i. *Always work from the whole to the part.* This implies 'precise control surveying' as the first consideration, followed by 'subsidiary detail surveying'.

ii. *The position of a point is then fixed geometrically by ground measurements* (linear, angular, or both). It involves first, the precise measurement of a reference line, called the baseline. Second, a point P on a plane is then fixed relative to the baseline, AB, in one of the following ways: traversing, triangulation,and trilateration to be plotted either by *polar coordinates* or *rectangular coordinates.*

Measurement of Baseline

It refers to both the linear and angular measurements of a baseline. Linear measurement is taken with the help of either a chain or a tape, while angular measurements may be taken either

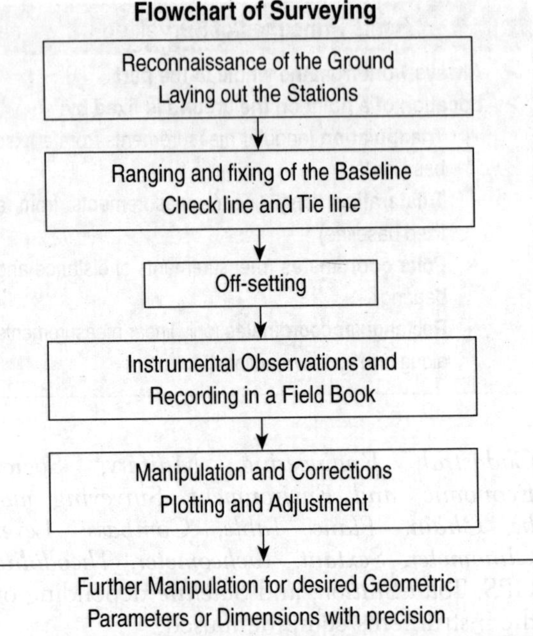

with a prismatic compass, sextant, theodolite or a tacheometer. About any measurement, the following points should be kept in mind:

- There is no such thing as exact measurement. All measurements contain some errors, the magnitude of which depends on the instruments used and the ability of the observer to use them.
- As the true value is never known, the degree of accuracy or its precision can only be quoted as a relative accuracy. For instance, an error of 1 cm in 100 m represents a relative accuracy of 1/10,000.
- Where readings are taken on a graduated scale to the nearest subdivision, the maximum error in estimation will be ± ½ division.
- Repeated measurements increase the accuracy by $\sqrt{\mathbf{n}}$, where n = the number of repetitions (however, this cannot be applied indefinitely). Agreement between repeated measurements does not imply accuracy but only consistency.

While taking linear measurements one must be very careful about the standardisation of the tape, its alignment and deformation, reading and marking the tape, recording working temperature, tension, sag, elevation, slope and grid reduction. For every linear measurement, the following corrections are commonly made. The need for their application, however, depends on the accuracy required:

1. In all cases:
   a. Standardisation
   b. Slope
2. For a relative accuracy of 1/5000+:
   a. Temperature
   b. Tension
   c. Sag
3. For special cases, 1/50000 + accuracy:
   a. Reduction to Mean Sea Level
   b. Reduction to Grid

a. Standardisation

Where the length of the chain or tape does not agree with its nominal value, a correction must be made to the recorded value of a measured quantity. The following rules apply:

- If the tape is too long, the measurements will be too short and corrections will be positive.
- If the tape is too short, the measurements will be too long and corrections will be negative.

If the length of a tape of nominal length, l is $(l+\delta l)$, the error per unit length $= \pm \frac{\delta l}{l}$. If the measured length is l and the true length is l_t, then

$$l_t = l\left[1 - \left(\frac{\delta l}{l}\right)\right]$$

b. Slope

This may be based on either the angle of inclination (a) or the difference in level (d) between the ends of a line. Let the length AC measured = l, horizontal length AB = h and correction to the measured length = c.

Therefore, AB = AC cos α or h = l cos α

Now, c = l – h
= l – l cos α
= l(1 - cos α)

If the difference in level is known,

Fig. 3.1 Correction due to Slope

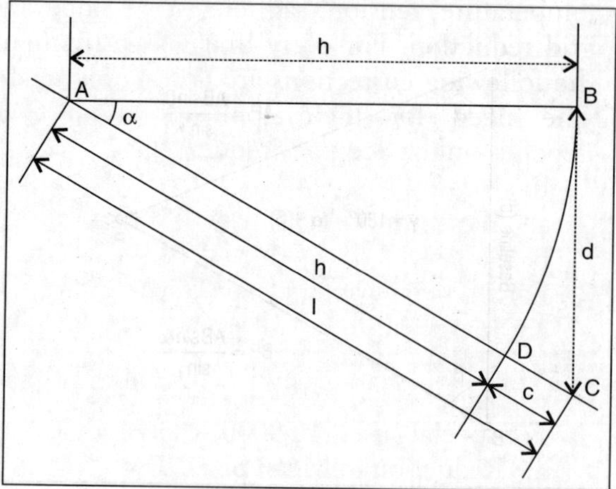

$$c = l - \sqrt{(l^2 - d^2)}$$

$$= l - l\left(1 - \frac{d^2}{l^2}\right)^{1/2}$$

$$= l\left[\left\{1 - \left(1 - \frac{d^2}{2l^2} + \frac{d^4}{8l^4}\cdots\right)\right\}\right]$$

$$= \frac{d^2}{2l} \text{ (neglecting other terms)}$$

c. Temperature

The measuring band is normally standardised at a given temperature (t_s). If during field survey the temperature of the band is recorded as (t_m), then the band will either expand or contract. The correction to the measured length is given as:

$c = l\alpha(t_m - t_s)$, where α = coefficient of linear expansion of the metal of the band.

The co-efficient of linear expansion (α) of a solid is defined as the 'increase in length per unit length of the solid when its temperature changes by 1°'. For a steel tape, α ranges between 5.9×10^{-6}/°F and 6.8×10^{-6}/°F or 10.6×10^{-6}/°C and 12.2×10^{-6}/°C. For an invarband tape, the value ranges between 3×10^{-7}/°F and 4×10^{-7}/°F or 5.4×10^{-7}/°C and 7.2×10^{-7}/°C.

d. Tension

The measuring band is standardised at a given tension (T_s). If during field survey the applied tension is (T_m), then the tape will expand or contract in accordance with Hooke's Law. The required correction factor is given as:

$$c = \frac{l(T_m - T_s)}{A.E}$$ where A = cross sectional area of the tape and E = Young's modulus of elasticity.

For a steel tape, E = 20 to 22×10^5 kgf/sq.cm or 19.3 to 20.7×10^{10} N/sq.m. For an invarband tape, E = 14 to 15.5×10^5 kgf/sq.cm or 13.8 to 15.2×10^{10} N/sq.m.

e. Sag

If the measuring band is standardised on the flat and used in catenary, the general equation for correction is given as:

$$c = -\frac{W^2 l}{24T^2}$$ (where T = tension applied)

Fig. 3.2 Correction due to Sagging

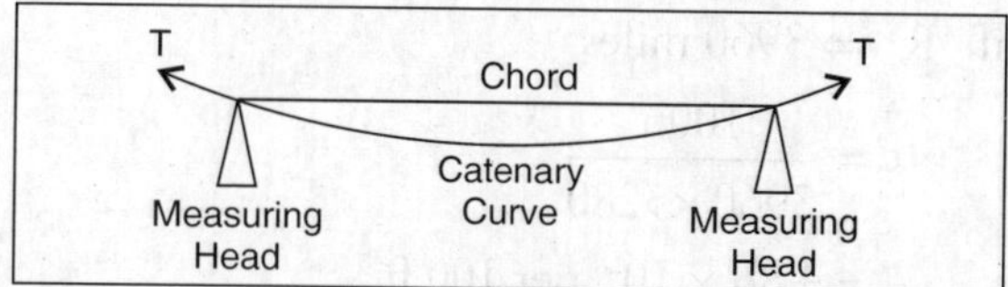

f. Reduction to MSL

If the length at the mean sea level is L and h = height of line above or below mean sea level, then

$$\frac{L}{lm} = \frac{R}{(R \pm h)}$$

or, $$\frac{lm.R}{(R \pm h)}$$

if $L = lm \mp c$,

$$c = lm \mp \frac{lm.h}{(R \pm h)}$$

$$= lm\left[1 \mp \frac{R}{(R \pm h)}\right]$$

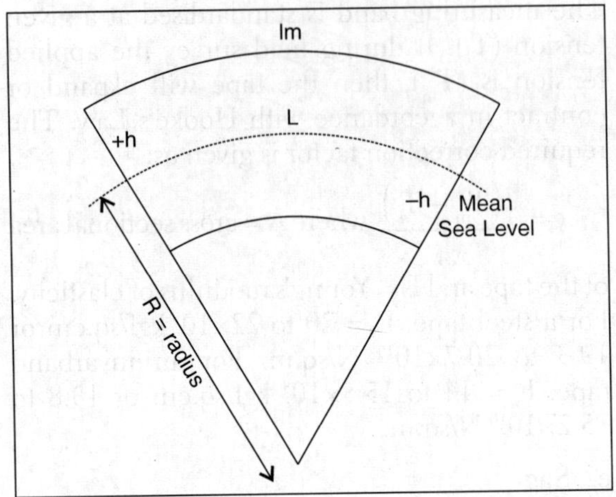

Fig. 3.3 Reduction to MSL

$$= \mp \frac{lm.h}{(R \pm h)}$$

As h is small compared with R,

$$c = \pm \frac{lm.h}{R}$$

if R $\Rightarrow$ 3960 miles

$$c = \frac{100h}{3960 \times 5280}$$

$$= 4.8h \times 10^{-6} \text{ per 100 ft}$$

g. Reduction to Grid

The local scale factor (F) depends upon the properties of projection used. For OS Maps, it is given as:

$F = 0.9996013\ (1+1.23E^2 \times 10^{-8})$, where E = Easting from the true origin in km.

PROCEDURES OF LOCATING A POINT

a. Triangulation

It concerns angular measurements from a fixed baseline. The length AB is known and the angles α and β are measured.

Therefore,

$$\angle ABP = 180^\circ - (\alpha+\beta) \Rightarrow \gamma \text{ (say)}.$$

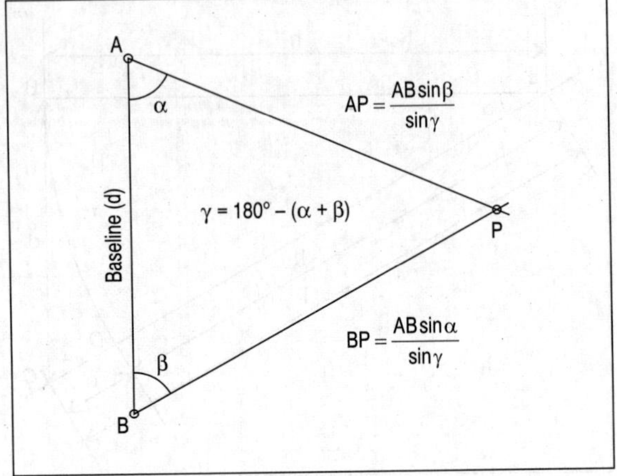

Fig. 3.4 Locating a Point by Angular Measurements

By applying the sine-rule,

$$AP = \frac{AB.\sin\beta}{\sin\gamma} \text{ and } BP = \frac{AB.\sin\alpha}{\sin\gamma}$$

AP and BP being known, the position of P with respect to A and B can be fixed precisely.

b. Trilateration

It concerns linear measurements only. The length of the baseline AB is measured first. Then the lengths AP and BP are measured and plotted. The position of P is always fixed

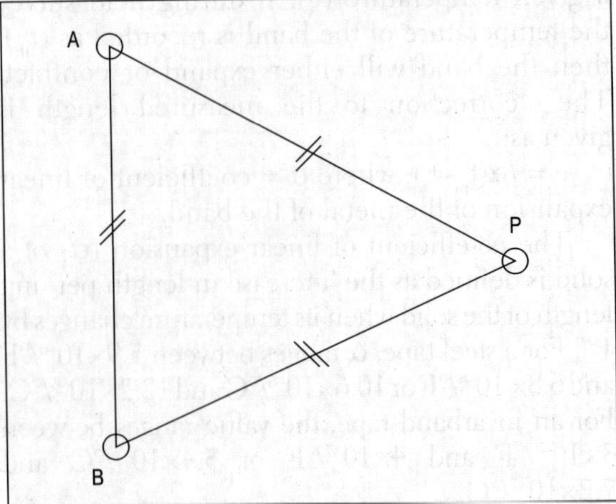

Fig. 3.5 Locating a Point by Linear Measurements

Fig. 3.6 Locating a Point by Polar Measurements

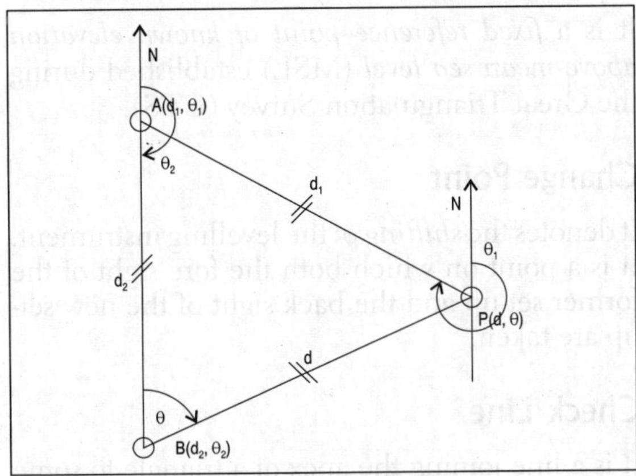

provided AP + BP > AB. This technique replaces triangulation with the use of microwave measuring equipment. In chain surveying also, this is a very useful procedure for locating a point.

c. Polar Coordinates

It involves linear and angular measurements. The distances and magnetic bearings of the lines, BP, PA and AB are precisely measured. Necessary corrections can be applied to check the lengths, magnetic bearings as well as internal angles of the triangle ΔABP. Let the final figures denoting the lengths and bearings of the lines BP, PA and AB be respectively (d, θ), (d_1, θ_1) and (d_2, θ_2). These give the polar coordinates of the points P with respect to AB, of A with respect to PB and of B with respect to AP. This is a very useful technique in traversing, setting out and plotting by protractor.

d. Rectangular Coordinates

It involves linear measurements only at right angles. The pair of orthogonal axes (x and y) meet at the origin. From point P, a perpendicular is dropped on the x-axis. The perpendicular distances of P on x-axis and y-axis are precisely measured as x_1 and y_1. These give the rectangular coordinates of P with respect to the origin of the reference frame of axes. This

Fig. 3.7 Locating a Point by Rectangular Measurements

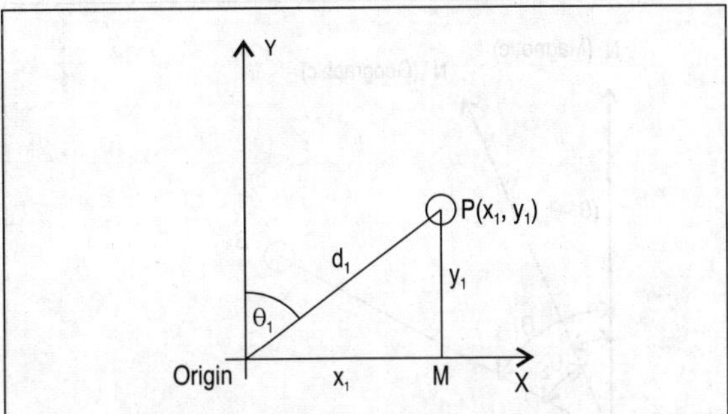

is very often used in taking offsets, setting out and plotting. Mathematically, the polar and rectangular coordinates are reversible and can easily be transformed as:

$$x_1 = d_1 \sin\theta_1$$

$$y_1 = d_1 \cos\theta_1$$

$$d_1 = \sqrt{\left(x_1^2 + y_1^2\right)}$$

$$\theta_1 = \tan^{-1}\frac{x_1}{y_1}$$

SURVEYING TERMINOLOGY

Axis of the Bubble Tube

It is the line tangential to the longitudinal curve of the level tube at its centre.

Axis of the Telescope

It is the line joining the *optical centre* of the object glass to the *centre of the eye piece* of the telescope.

Baseline

It is the *fixed line* or *control line* of a known length in a triangulation survey.

Bearings of a Line

It is the *horizontal angle* which the line makes with a reference direction or meridian, always measured *clockwise* from the line of

Fig. 3.8 Magnetic and True Bearings of a Line

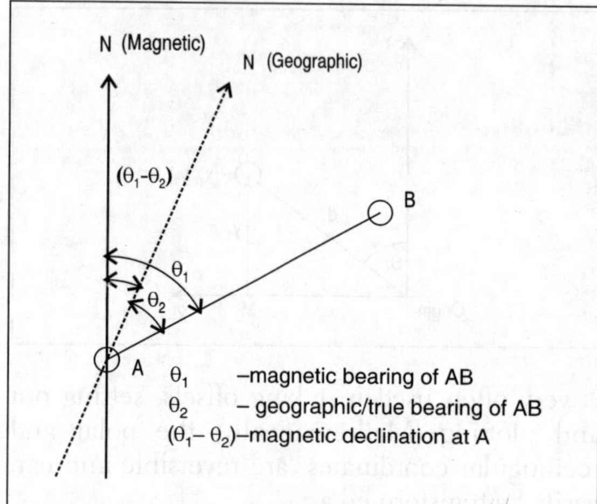

reference. The bearing of a line may be (i) a *true bearing* or azimuth, if the reference line is a geographical meridian (i.e., the line passing through the given point and the geographical north and south poles), (ii) a *magnetic bearing*, if the reference line is the magnetic meridian, passing through the given point and the magnetic north and south poles and (iii) an *arbitrary bearing*, if the reference line is any line fixed and conceived on the ground (Fig. 3.8) during actual survey.

Bench Mark

It is a *fixed reference point of known elevation above mean sea level* (MSL) established during the Great Triangulation Survey (GTS).

Change Point

It denotes the *shifting* of the levelling instrument. It is a point on which both the fore sight of the former set-up and the back sight of the new set-up are taken.

Check Line

It is a line joining the apex of a triangle to some fixed point on the opposite side. It is also known as a *proof line* and is measured to determine the degree of accuracy in a tringulation survey.

Cross-sectioning

It refers to that levelling operation in which the outline of the ground transverse to and on either side of a given line is determined.

Datum

It refers to any arbitrarily assumed *level surface* or line from which vertical distances or elevations are measured (Fig. 3.9).

Fig. 3.9 Level Surface, Plumb Line, Horizontal Line and Vertical Line

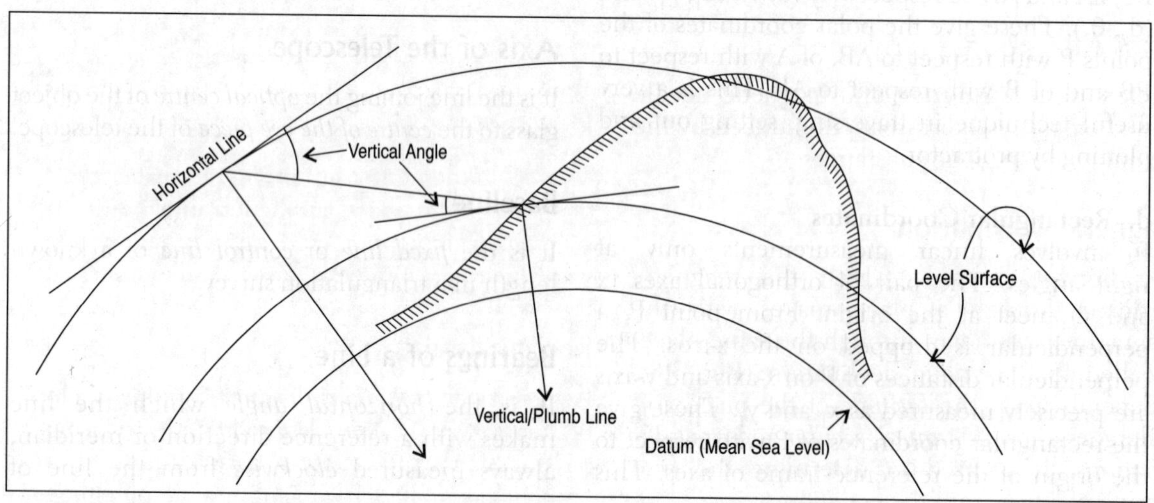

Face Left and Face Right Observations

These are respectively the observations made with the face of the vertical circle on the left or right of the observer.

Horizontal Plane

It refers to the plane tangential to the level surface at a specific point. A *horizontal* line is any line contained in the horizontal plane. It is a straight line tangent to the level line at a particular point (Fig. 3.9).

Level Surface

It refers to any surface parallel to the mean spheroidal surface of the earth. Therefore, it is a curved surface containing points equidistant from the centre of the earth and is normal to the *plumb line* at all points (Fig. 3.9). Any line contained in the level surface is called a level line.

Line of Collimation

It is the line passing through the intersection of the cross hairs to the optical centre of the object glass of the telescope. When the instrument is perfectly levelled, the line of collimation revolves around the vertical axis in a horizontal plane, called the *plane of collimation*.

Local Attraction

It refers to the *deflection* of a magnetic needle caused by external disturbances induced by the proximity of magnetic substances.

Magnetic Declination

The horizontal angle which the magnetic meridian through a place makes with the geographical meridian through the same place, is called the magnetic declination (Fig. 3.8). It depends upon the *latitude* of the place and undergoes *diurnal*, *annual*, *secular* and *irregular* variations. It is used to determine the true bearing of a line from the equation:

True Bearing = Magnetic Bearing ± Magnetic Declination

('+' when the declination is East and '–' when the declination is West.)

Offsets

These are *lateral* measurements (perpendicular or oblique) taken to locate the interior details with respect to the main survey lines.

Orientation

It is the process by which a plane table at each successive station is made parallel to the position which it occupied at the first station. It is done either with the help of a magnetic needle (trough compass) or by back sighting.

Plan

It is the representation of the ground and the objects upon it as projected on a horizontal plane, i.e., a plain sheet of paper. The scale of a plan is essentially very large, e.g., 1:10 to 1:100.

Profile Levelling

It refers to that levelling operation in which the elevations of points of known distances apart along a line are determined and thus an accurate outline of the ground is obtained. It is also known as *longitudinal levelling* or *sectioning*.

Ranging

It refers to the *setting* out of intermediate points on a straight line between the terminal points. It is normally done prior to chaining either directly with the eye or indirectly with the theodolite.

Reconnaissance

It refers to the *preliminary inspection* of the area to be surveyed. It is extremely important in (i) laying out the stations for either a traverse frame or a triangulation frame, (ii) ranging and (iii) offsetting.

Reciprocal Levelling

It refers to that levelling operation in which the difference in elevation between two points is precisely determined by two sets of observations, one each at a point.

Stations

These are the *ground points* defined by the nodes of triangles or junctions of a traverse.

Tie Line

It refers to a line joining some fixed points termed as *tie stations* on the main survey lines. It serves as a *check line* and also enables the surveyor to locate the interior details far away from the main survey lines.

Trunian Axis

It refers to the axis about which the telescope can be rotated in a vertical plane.

Vertical Plane

It is the plane containing vertical lines or plumb lines. The angle between the two intersecting lines in a vertical plane is called a *vertical angle.*

Work of a Surveyor

It consists of three chapters: *field work, office work* and *maintenance of instruments.* Field work includes the measurement of distances and angles, locating the details, recording the data in field books with field notes and setting out. Office work consists of the preparation of plans and sections from field data, computation of areas and volumes and designing the structure. Maintenance of instruments refers to the operating skill of the surveyor in carefully handling the instruments.

Whole Circle and Reduced Bearings

Whole circle bearing (WCB) refers to the bearings expressed in *whole circle system.* In this, the bearing of a line is always measured clockwise from the north line. The resultant amount may take any value between 0° and 360°. In contrast, the reduced bearing (RB) of a line refers to the bearings expressed in a quadrantal system in which it is measured as a horizontal angle from either the north or the south line (whichever is closer) towards either the east or the west line (whichever is closer). The magnitude of reduced bearing may take any value between 0° and 90°. In designating such bearings, quadrants are essentially mentioned. The relations between the two bearings are mentioned in Fig. 3.10.

Fore and Back Bearings

The *fore* or *forward* bearing of a line refers to that which is measured in the direction of the progress of the survey, while the bearing taken in the opposite direction is called *back* or *backward*

Fig. 3.10 Whole Circle and Reduced Bearings of a Line

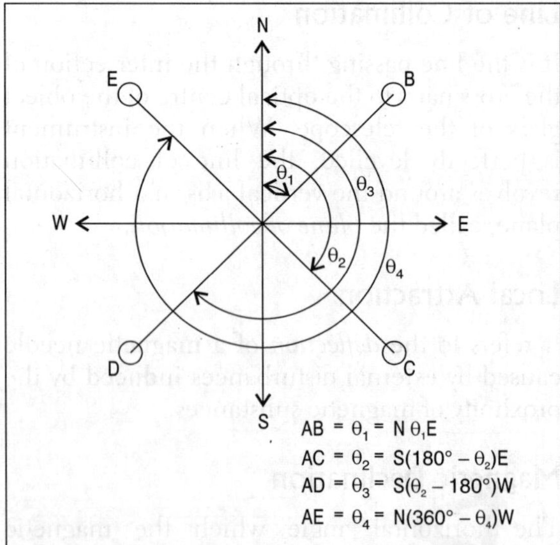

Quadrant	WCB(θ)	RB
I	$0° \le \theta_1 \le 90°$	$N\theta_1E$
II	$90° \le \theta_2 \le 180°$	$S(180° - \theta_2)E$
III	$180° \le \theta_3 \le 270°$	$S(\theta_3 - 180°)W$
IV	$270° \le \theta_4 \le 360°$	$N(360° - \theta_4)W$

Fig. 3.11 Fore and Back Bearings of a Line

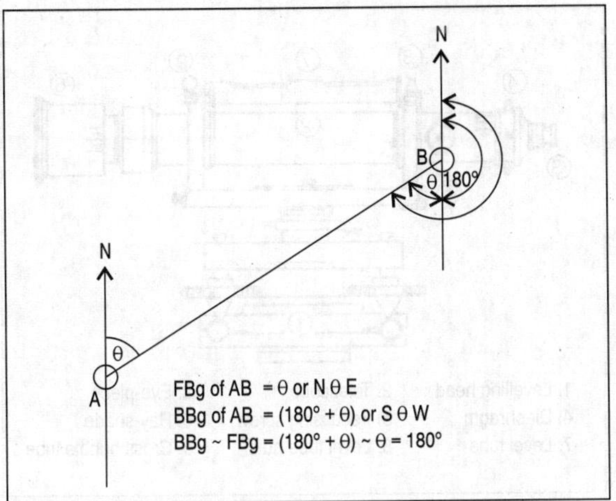

bearing. The two bearings of a line (in a whole circle system) are related by the equations:

Back Bearing = Fore Bearing ±180°
('+' if FB <180° and '–', if FB >180°
or, Back Bearing ~ Fore Bearing = 180°

In a quadrantal system, the difference between the magnitude of the fore and back bearings of a line is 0°. Since they lie in diametrically opposite quadrants, a difference of 180° between the two is noted (Fig. 3.11).

SURVEYING INSTRUMENTS AND ACCESSORIES

Chain

A chain (Fig. 3.12) is composed of 100 or 150 pieces of galvanised mild steel wire (4 mm in diameter) called *links*. The ends of the links are bent into loops and connected by *oval rings* to make the chain more flexible and the joints of the links are usually open. The ends of the chain are provided with *brass handles* with *swivel joints* so that during use they do not get twisted or entangled. The outer side of the handles is the *zero point*. A *link length* is the distance between the centres of two consecutive middle rings. The end links include the handles. *Metallic tags* or *indicators* of distinctive patterns are fixed at regular link intervals to facilitate quick reading. Four types of chains are commonly used:

(i) Gunter's Chain (66 ft long with 100 links),
(ii) Revenue Chain (33 ft long with 16 links),
(iii) Engineer's Chain (100 ft long with 100 links) and
(iv) Metric Chain (20 m long with 100 links and 30 m long with 150 links).

Cross Staff

It is used to find the foot of a perpendicular from an offset to the line and also to set out a right angle from a point on the line. The simplest form is an *open cross staff*. The head is a wooden block with 15 cm side length and 4 cm depth with two fine saw cuts at right angles to each other. The disc is furnished with two pairs of vertical slits at the ends of the saw cuts, thus producing two perpendicularly intersecting lines of sight. The head is fixed on the top of an ironshod wooden staff with 1.2 m to 1.5 m length (Fig. 3.13). While ranging out, one line of sight coincides

Fig. 3.12 Chain

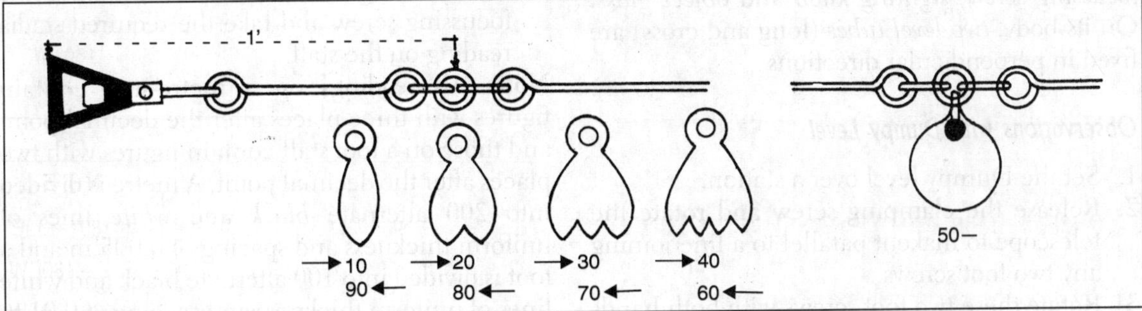

Fig. 3.13 Cross Staff

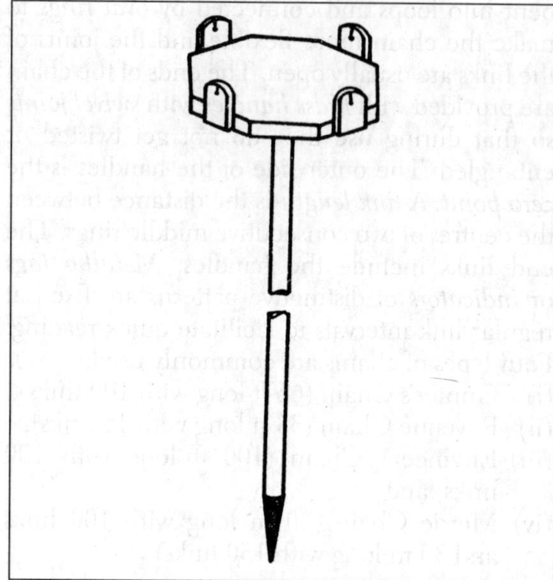

Fig. 3.14 Dumpy Level

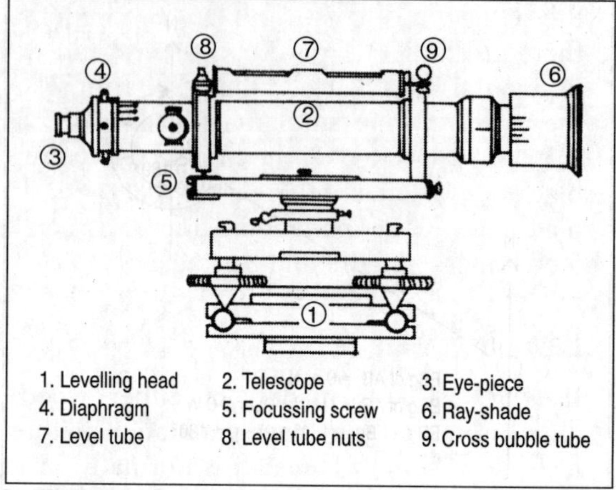

with the chain line at a point and the other fixes the perpendicular line to be set out.

Dumpy Level

Of the levelling instruments, it (Fig. 3.14) is the most simple, compact and stable one. The levelling head consists of a *tribrach* and a *trivet* with three arms each carrying a *foot* or *a levelling screw*. A *telescope* is rigidly fixed to its supports fitted to the spindles which are attached to the central hollow of the tribrach. The rotation of the telescope around the vertical axis is regulated by a *clamp* and a *slow motion tangential* screw. The telescope contains an *eye piece, diaphragm, focussing screw, sighting knob* and *object glass.* On its body, *two level tubes* (long and cross) are fixed in perpendicular directions.

Observations with Dumpy Level

1. Set the Dumpy level over a station.
2. Release the clamping screw and rotate the telescope to make it parallel to a line joining any two foot screws.
3. Rotate these two foot screws with both hands simultaneously either inward or outward until the bubble in the level tube is centred.
4. The telescope is then rotated to make it perpendicular to the line joining the previous two foot screws and passing through the remaining foot screws. Rotate this single foot screw until the bubble in the level tube is centred.
5. The three foot-screws make an equilateral triangle. Step 2 to Step 5 is repeated twice separately for the remaining two sides of the triangle. This series of operations make the Dumpy level perfectly levelled at the given station.
8. Rotate the telescope until the line of collimation passes through the station with the staff held vertically over a station. Clamp the tangential screw and use slow motion tangential screws for precise sighting.
7. Look through the eye piece, adjust the focussing screw and take the required stadia reading on the staff.

Note: All readings on a metre-staff contain figures with three places after the decimal point and those on a foot staff contain figures with two places after the decimal point. A metre is divided into 200 alternate *black* and *white* lines of uniform thickness and spacing of 0.005 m and a foot is divided into 100 alternate black and white lines of uniform thickness and spacing of 0.01 ft.

A reading of 1.235 m = 1 is written in red on the right, = 2 is written in black on the left and there are 7 (0.035 = 0.005×7) black and white horizontal lines between the 1.200 m mark and the middle stadia. Similarly, a reading of 5.64 ft = 5 is written in red on the right, = 6 is written in black on the left and there are 4 (0.04 = 0.01×4) black and white horizontal lines between the 5.60 ft mark and the middle stadia.

Levelling Staff

It (Fig. 3.15) is made of either wood or aluminium and may have graduations either in feet or in metres. The staff is normally 75 mm wide, 18 mm thick and 4 m long. It can be *folded* or may have *sop with telescopic* arrangements. There is a brass cap at each end of the staff. In a foot staff, each foot is divided into 100 equal divisions while in a metre-staff, each metre is divided into 200 equal divisions. On its silver white face, the graduations are numbered in red and black while the smallest divisions are coloured *black* and *white*.

Fig. 3.15 Levelling Staff

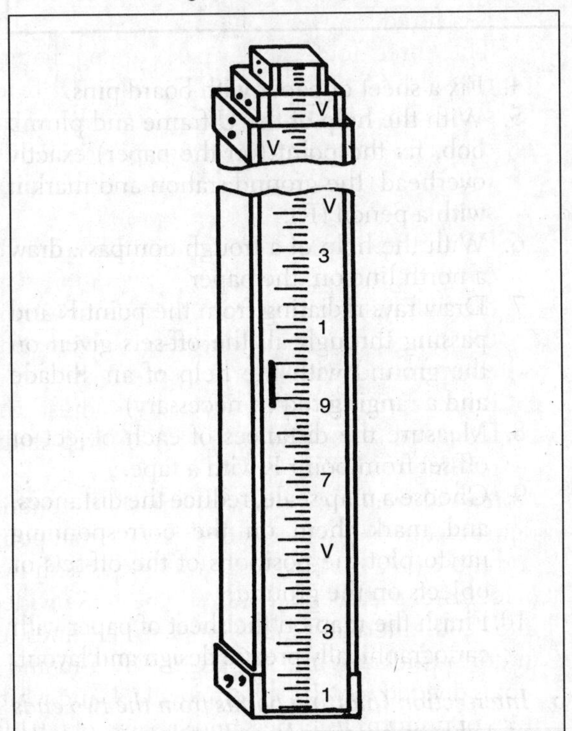

Fig. 3.16 Optical Square

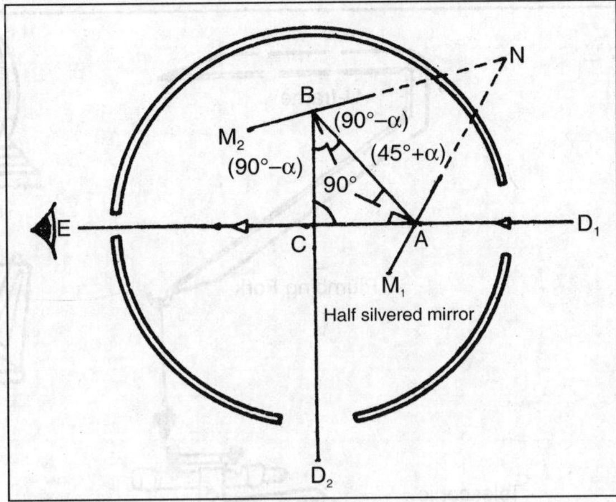

Optical Square

It consists of a circular metal box (5 cm diameter and 1.25 cm deep) in which two mirrors fitted at 45° are fixed in a frame that can easily be adjusted by moving a screw behind it (Fig. 3.16). One of the mirrors is half silvered and is called the *horizon glass* while the other is fully silvered and is called the *index glass*. On the rim of the box, two small holes (eye hole and window hole) lie diametrically opposite to each other. Its working is based on the principle that the *amount of deflection of rays after two successive reflections in the mirror intersecting at 45° is equal to* 90°. In a perpendicular set out, the image of a ranging rod on the silvered portion of the horizon glass exactly coincides with a ranging rod seen through the unsilvered portion of it.

Plane Table

A plane table (Fig. 3.17) consists of (i) a *wooden drawing board* (40 cm × 30 cm to 75 cm × 60 cm) mounted on *a tripod* such that it can be levelled, rotated and clamped in any position and (ii) an *alidade* (50 cm long) with bevelled

Fig. 3.17 Plane Table with Accessories

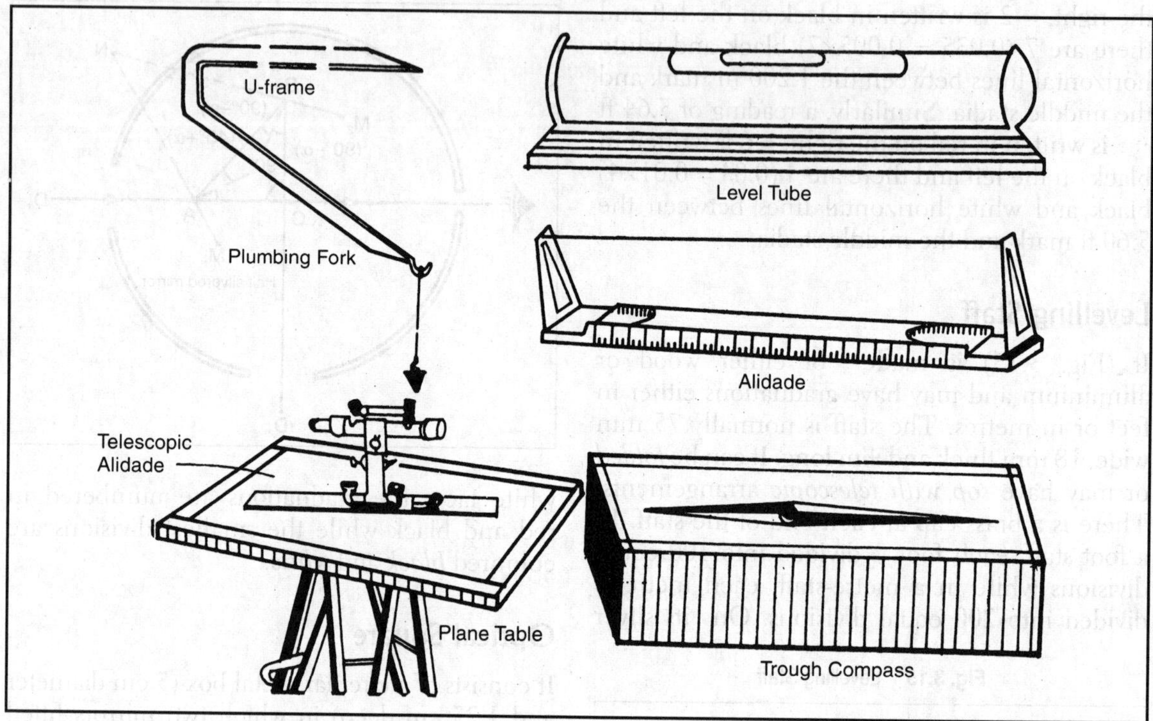

or fiducial edge. The alidade may be fitted with *sight vanes* at both ends or fitted with a *telescope* as well. The *fiducial edge* may be graduated. The telescope is normally provided with a vertical circle and a level tube. The other accessories of a plane table are—a *level tube* for levelling, a *trough compass* for orientation, a *plumbing fork* or *U-frame* with a *plumb bob* for centering.

Observations with Plane Table

a. *Radiation* (taking measurements for polar coordinates from a single point and direct plotting)

Problem: *To plot a number of given objects and off-sets on the ground by radiation*

1. Set the tripod up exactly overhead the station of observation.
2. Mount the table and fix it firmly.
3. With the help of a level tube, level the table.
4. Fix a sheet of paper with board pins.
5. With the help of the U-frame and plumb bob, fix the point (on the paper) exactly overhead the ground station and mark it with a pencil (R).
6. With the help of a trough compass, draw a north line on the paper.
7. Draw rays radiating from the point R and passing through all the off-sets given on the ground with the help of an alidade and a ranging rod (if necessary).
8. Measure the distances of each object or off-set from point R with a tape.
9. Choose a map scale, reduce the distances, and mark them on the corresponding ray to plot the positions of the off-sets or objects on the ground.
10. Finish the map on the sheet of paper with cartographically precise design and layout.

b. *Intersection (plotting points from the two ends of a chosen baseline by radiation)*

Fig. 3.18 Closed Traverse Survey by Plane Table

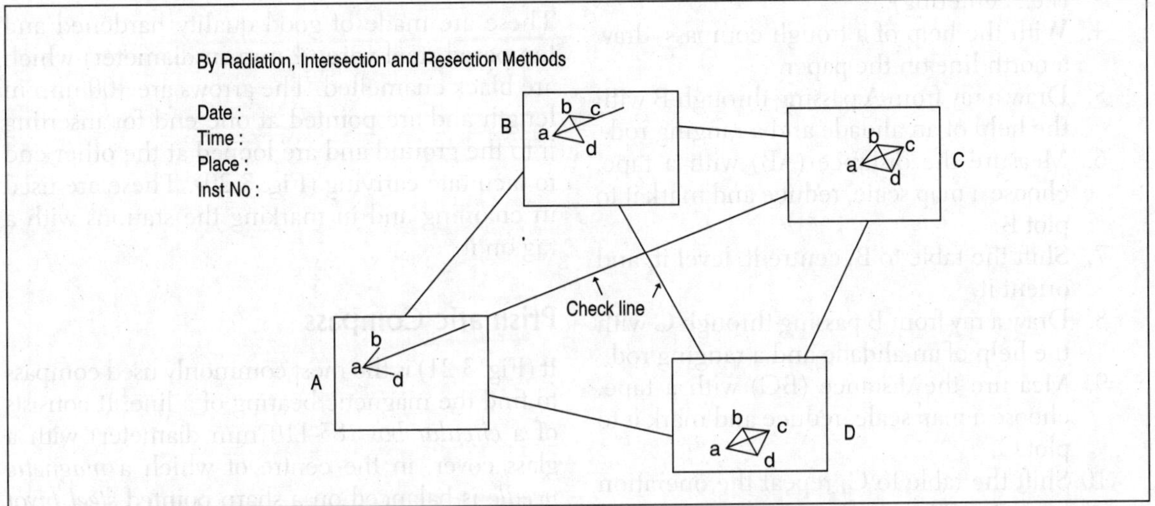

Problem: *To find the distance and direction of a line joining two points or objects or off-sets (P and Q) given on the ground*

1. Select a baseline a – b with convenient length on the ground.
2. Set the tripod exactly overhead a, mount the table and fix it firmly. With the help of level tube, level the table and fix a sheet of paper with board pins.
3. With the help of the U-frame and plumb bob fix the point a on the paper exactly overhead the ground station and mark it (i.e., centring).
4. With the help of trough compass, draw a north line on the paper.
5. Draw a ray from a passing through b with the help of an alidade and a ranging rod.
6. Measure the distance with a tape, choose a map scale, reduce and mark it to plot b.
7. Draw rays from a passing once through P and then through Q with the help of alidade and ranging rod.
8. Shift the table to b, set the tripod up exactly overhead b, mount the table and fix it firmly.
9. With the help of a level tube, level the table.
10. With the help of the U-frame and plumb bob, fix the point a on the paper exactly overhead the ground station by adjusting the table only.
11. Place the trough compass along the north line on the paper and rotate the table to match its previous orientation.
12. Place the alidade along the line a-b and look through to a to verify its exact previous orientation.
13. Draw rays from b passing once through P and then through Q with the help of an alidade and a ranging rod.
14. The corresponding rays from a and b intersect to plot the points P and Q.
15. Measure the required distance and direction of the line P-Q.
16. Finish the map on the sheet of paper with cartographically precise design and layout.

c. *Traversing (consecutively and successively plotting points of a given traverse)*

Problem: *To plot the traverse ADCD given on the ground*

1. Set the tripod exactly overhead A, mount the table and fix it firmly.
2. With the help of a level tube, level the table and fix a sheet of paper with board pins.
3. With the help of the U-frame and Plumb bob, fix the point A on the paper exactly

overhead the ground station a and mark it (i.e., centering).

4. With the help of a trough compass, draw a north line on the paper.
5. Draw a ray from A passing through B with the help of an alidade and a ranging rod.
6. Measure the distance (AB) with a tape, choose a map scale, reduce and mark it to plot B.
7. Shift the table to B, centre it, level it, and orient it.
8. Draw a ray from B passing through C with the help of an alidade and a ranging rod.
9. Measure the distance (BC) with a tape, choose a map scale, reduce and mark it to plot C.
10. Shift the table to C, repeat the operation from step 7 to step 9 to plot D.
11. Shift the table to D, repeat the operation from step 7 to step 9 to plot A.
12. The corresponding rays from a and b intersect to plot the points P and Q.
13. Measure the required distance and direction of the line P-Q.
14. Finish the map on the sheet of paper with cartographically precise design and layout.

Plumb Bob

These are top shaped bobs made of brass or steel with a device to attach strings of thread (Fig. 3.19) A plumb bob is used for accurate centering and testing the verticality of a ranging rod and staff.

Fig. 3.19 Plumb Bob

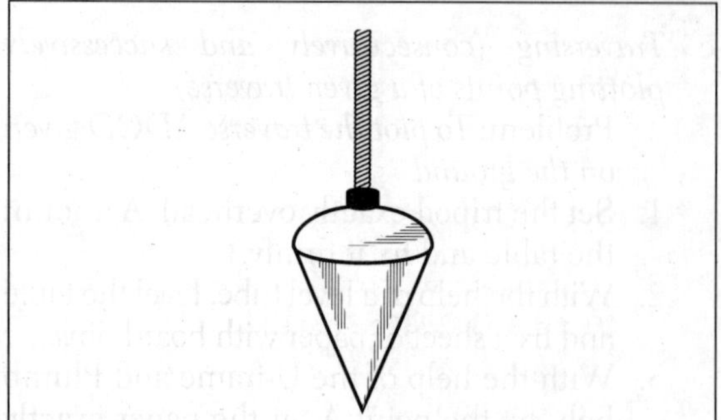

Pin

These are made of good quality hardened and tempered steel wire (4 mm in diameter) which are black enamelled. The arrows are 400 mm in length and are pointed at one end for inserting into the ground and are looped at the other end to facilitate carrying (Fig. 3.20). These are used in chaining and in marking the stations with a tag on it.

Prismatic Compass

It (Fig. 3.21) is the most commonly used compass to find the magnetic bearing of a line. It consists of a *circular box* (85-110 mm diameter) with a glass cover, in the centre of which a *magnetic needle* is balanced on a sharp pointed *steel pivot* with the help of an *agate cap*. The *needle* carries an *aluminium ring* with graduations in degrees and half degrees, which are written in an inverted style. The *object vane* and the *focussing stud* for the prism are fitted at diametrically opposite points. The former consists of a hinged metal frame in the centre of which a horse hair or a fine wire is

Fig. 3.20 Pin

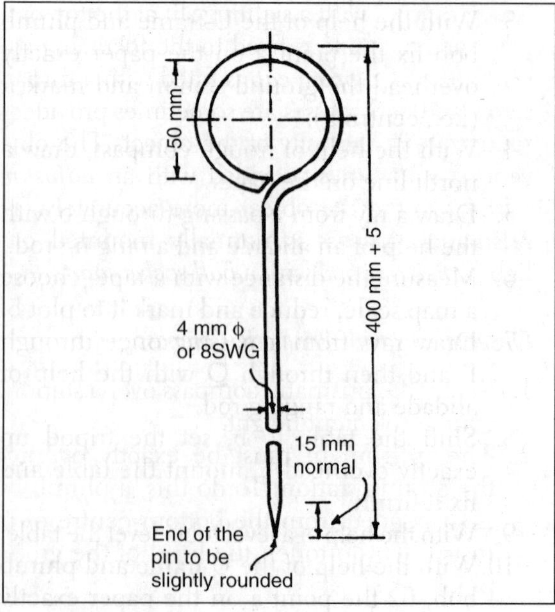

Fig. 3.21 Prismatic Compass

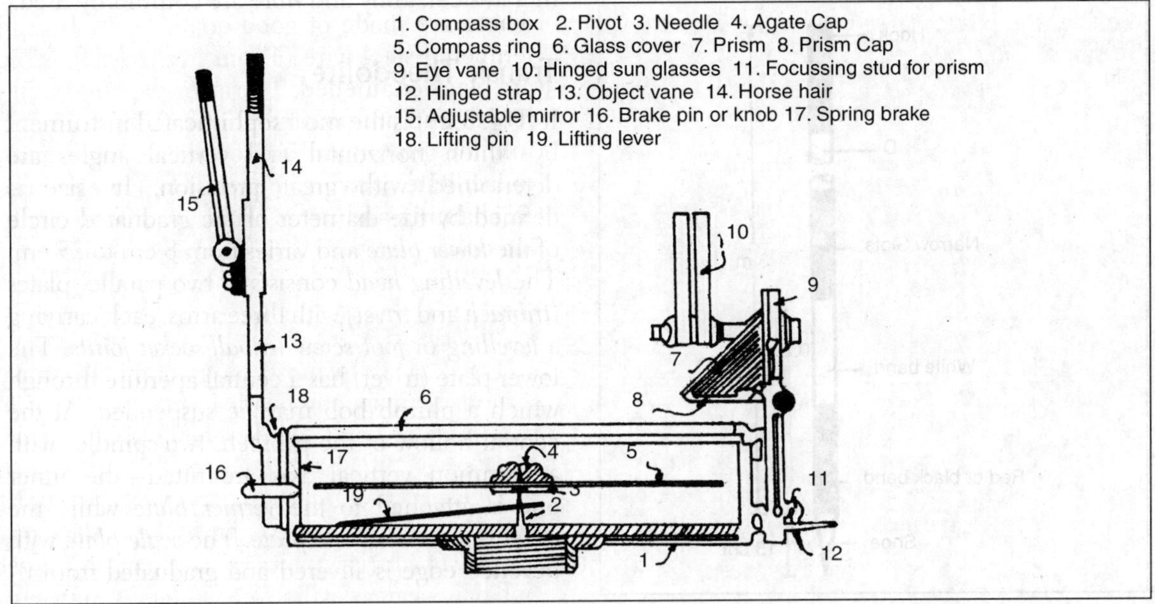

stretched. When it is folded on the lid, it presses the *lifting pin* separating it from the pin. The oscillation of the needle can be quickly checked by the inward pressing of the brake pin fitted at the base of the vane. A *reflecting prism* (right angled isosceles type) with a sighting slit at the top of the perpendicular side is fixed to the focussing stud. On the base, *a prism cap* is fitted with a turnable screw and dark glasses are sometimes provided to reduce the luminosity of the object. The object vane is sometimes provided with an *adjustable mirror* to sight the object more accurately. The prismatic compass is normally mounted on a light *wooden tripod* with a *ball-socket joint.*

Observations with Prismatic Compass

1. Set up the prismatic compass over a station at your convenient height.
2. The instrument must be exactly overhead the ground station. To do this, a plumb bob is suspended from the bottom centre of the instrument to touch the head of the ground pin.
3. Unfold the object vane and also the eye vane.
4. Level the instrument either by hand or by using a round pencil so that the compass ring moves freely.
5. Rotate the instrument so that the line of sight (i.e., the line joining the centre of the eye piece and the horse hair) joins the station with the ranging rod held vertically over another station.
6. Look through the eye piece and take the reading in degrees and minutes at the intersection of the horse hair / ranging rod with the compass ring.

Note: As the least count of the circular scale is 0° 30′, readings are all in full degrees or may contain a term of 30′only. Ranging rods should be held vertically over the ground stations.

Ranging Rod

These are rods of circular or octagonal cross-sections (3 cm in diameter), made of well seasoned wood or light steel. Normally the length of a rod is 2 m with 10 equal divisions which are alternately

Fig. 3.22 Ranging Rod

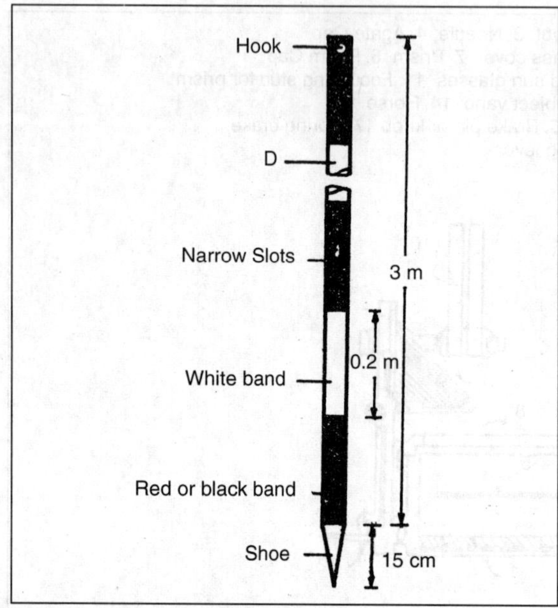

coloured in black and white (Fig. 3.22). These are used in station marking and setting out lines.

Tape

A tape (Fig. 3.23) is the most common and widely used equipment for measuring short lengths. Normally tapes of 10 m, 20 m, 30 m or 50 m dimensions are supplied in a leather or metal case with a corrosion resistant metal finish fitted with a *winding device*. The width of a tape varies from 6 mm to 16 mm. Tapes of cloth, linen, metal, steel, alloy and fibre are commonly used.

Fig. 3.23 Tape

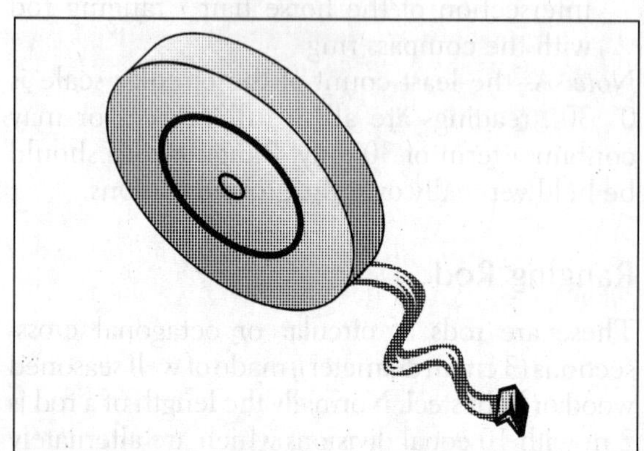

Transit Theodolite

It (Fig. 3.24) is the most sophisticated instrument by which horizontal and vertical angles are determined with great precision. Its size is defined by the diameter of the graduated circle of the *lower plate* and varies from 8 cm to 25 cm. The *levelling head* consists of two parallel plates (*tribrach* and *trivet*) with three arms, each carrying a *levelling* or *foot screw* in *ball-socket joints*. The lower plate (trivet) has a central aperture through which a plumb bob may be suspended. At the central hollow of the tribrach, two spindles with a common vertical axis are fitted—the inner one is attached to the *vernier plate* while the outer one to the *scale plate*. The *scale plate* with bevelled edge is silvered and graduated from 0°

Fig. 3.24 Transit Theodolite

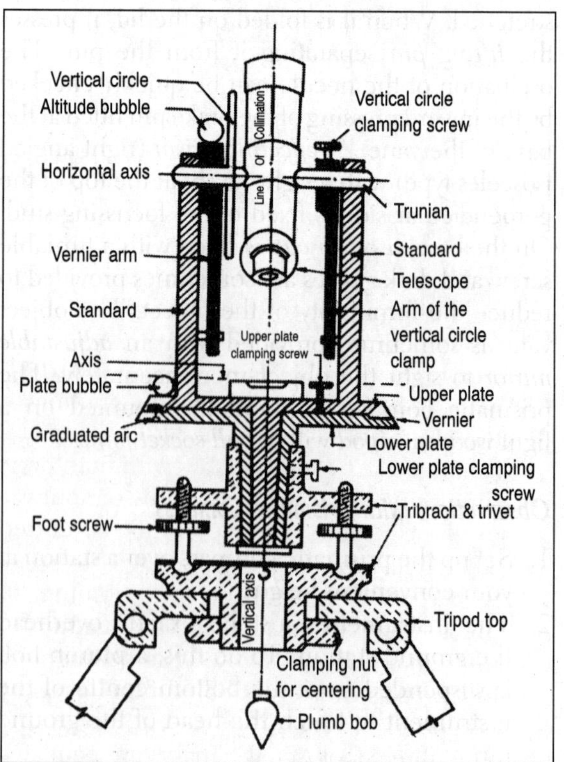

to 360° with a least count of 0°20' in a clockwise direction. It is provided with *a clamp* and *a slow motion tangential screw*. The upper vernier plate carries two *double verniers* (VA and VB) fixed at diametrically opposite sides and are provided with *clamps* and *slow motion tangential screws* and *vernier aids*. A *level tube* is attached to the outer surface of the horizontal plates. Two upright standards or *A-frames* stand upon the vernier plate to support the *trunian axis*, at the centre of and at right angle to which a *telescope* is fitted. It contains an *eye piece, objective, diaphragm, focussing screw* and *sighting knobs*. A *vertical circle* firmly attached to the telescope moves with it. It is bevelled, silvered and graduated from 0° to 180° with a least count of 0°20'. The *index bar* (*T-frame*) which is centred on the horizontal axis carries two double verniers (VC and VD) at the ends of the index arm and are provided with a *clamp* and a *slow motion tangential screw*. The clipping arm is provided with a *fork*, two *screws* and an *altitude bubble tube*, fitted at the top of the frame. The instrument is mounted on a tripod made of solid legs of wood or light steel or aluminium.

Observations with Theodolite

1. Set up the theodolite over a station at your convenient height with the help of a plumb bob.
2. Release all clamping screws.
3. Rotate the telescope on the vertical plane until the reading on VC and VD is 0°–180° and clamp it. Rotate it on the horizontal plane to make it parallel to a line joining any two foot-screws.
4. Rotate these two foot screws with both hands simultaneously either inward or outward until the bubble in the level tube is centred.
5. The telescope is then rotated to make it perpendicular to the line joining the previous two foot screws and passing through the remaining foot screws. Rotate this single foot screw until the bubble in the level tube is centred.
6. The three foot screws make an equilateral triangle. Step 3 to Step 5 is repeated twice separately for the remaining two sides of the triangle. This series of operations make the theodolite perfectly levelled at the given station.
7. Rotate the telescope and look through the eye piece until the line of collimation passes through the station with the staff held vertically over another station. Clamp all the screws of the upper and lower plates and use the slow motion tangential screws for precise sighting.
8. Look through the eye piece, adjust the focussing screw and take the required readings on both the staff and the vernier circles (VC–VD or VA–VD).

Note: For measuring an angle, readings must be taken on VC–VD (vertical) or VA–VB (horizontal) for both the left face and the right face. The value of a single angle is thus an average of the four readings. For a reading:

i. Find the least count (LC) of the main scale, number of vernier divisions (N) and then the vernier constant (VC).
ii. Take the main scale reading (MSR) and find the vernier division that coincides with the main scale division (i).
iii. Hence, the required angle = (MSR + VC × i).

TRAVERSE SURVEY

A traverse consists of a series of related points or stations which, when connected by angular and linear values, form a *framework*. Therefore, a *traverse survey* is one in which the points of the traverse frame are plotted by means of polar coordinates derived from the measurements of the lengths and bearings (directions) of the series of connected lines. The lengths are usually measured with a tape while directions by a prismatic compass. The purpose is to control the subsequent details, i.e., the fixing of the specific points to which details can be related. The accuracy of the control survey must be superior to that of the subsidiary survey in general.

A traverse may be of two types: (a) *closed*, when a complete circuit is made, i.e., the initial and final

Fig. 3.25 Types of Traverse

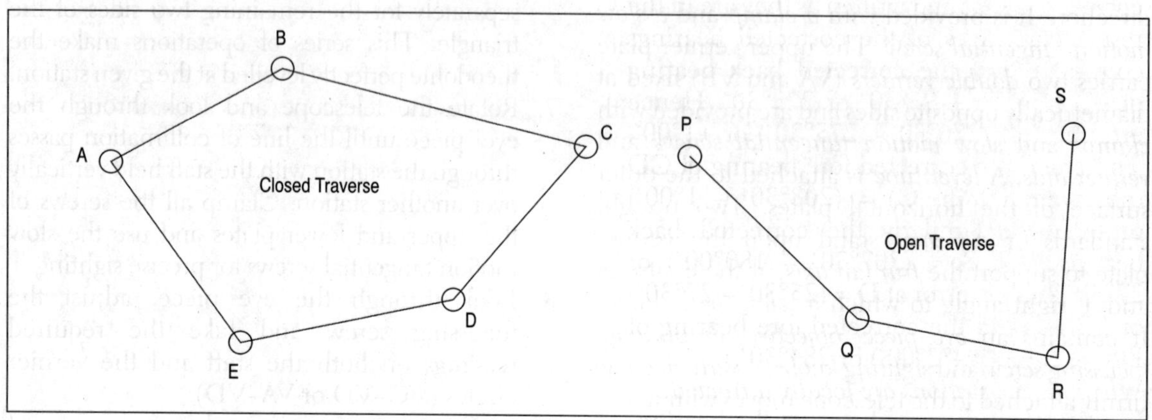

points coincide and (b) *open*, when the traverse framework does not form a closed polygon (Fig. 3.25). Of the several methods of traversing, the most important ones are: (a) prismatic compass traversing, (b) theodolite traversing and (c) plane table traversing.

Traversing by a Prismatic Compass

Loose needle traversing, as it is called, is mainly applied in reconnaissance or exploratory surveys. It is most advantageous because: (i) surveying is quick, (ii) each line is independent—errors tend to compensate and (iii) the bearing of a line can be observed at any point along the line. However, lack of accuracy and vulnerability to local attractions lead to different sources of errors which are the common demerits of such surveys. The instruments and accessories required are: a prismatic compass, a tripod, a plumb bob, a tape, a set of pins and ranging rods.

Procedure

i. Ranging rods are to be fixed at all the stations marked by pins with tags specifying the stations.
ii. A *rough sketch* of the traverse is then drawn on the field book and the date, time, place and instrument number are recorded.
iii. Distances of the lines are then measured with a tape and noted in the field book.
iv. At each station, the prismatic compass is carefully centred and levelled. From each station, the bearings of the two connected lines are then observed and recorded in the appropriate columns as *forward* bearings (measured in the direction of survey) and *backward* bearings (measured in the opposite direction of survey) either in the form of reduced or whole circle bearings.

COMPUTATION

Correction of Bearings

The observed bearings are first corrected by adjusting the local attractions of the stations. The error can easily be detected from the difference of the fore and back bearings of a line. This difference should be exactly 180° and any deviation from this indicates local attraction at both ends of the line. The correction of bearings may be done as follows:

i. *When at least two stations are free from local attraction*:
In the field book (Table 3.1) the bearings of AB is exactly 180° apart. Therefore A and B stations are free of local attractions. Obviously, all the readings taken from these

two stations (circled in the field book) are correct readings and written as they are in the respective places of the corrected bearings column. Now, the corrected back bearing of BC = 180° + 95°30′ or 275°30′. Hence error at C = (276°30′ – 275°30′) or +1°00′. Therefore, the corrected fore bearing of CD (observation from C) = (208°30′ – 1°00′) or 207°30′. Similarly the corrected back bearing of CD = (207°30′ – 180°00′) or 27°30′; hence, error at D = (25°30′ – 27°30′) or – 2°00′ and the corrected fore bearing of DA = (282°30′ + 2°00′) or 284°30′.

ii. *When all the stations are locally attracted*: In this case difference between the bearings of none of the lines is 180° (Table 3.2). Therefore, all the stations are affected by local attraction. Correction is made by equally distributing the error at both ends.

The amount of error, therefore, may vary along the two lines from a single station. This is the demerit of this system in general. The general rules of correction are:

a. If e (or error) is *negative*, deduct $e/2$ from the smaller readings and add $e/2$ to the larger one.
b. If e is *positive*, deduct $e/2$ from the larger readings and add $e/2$ to the smaller one.

Checks on Closed Traverse by Angles

For a closed polygon (Table 3.3), the following checks may be applied (Fig. 3.26):

i. the sum of the internal angles,

$$\sum^{n}\theta_1 = (2n - 4) \times 90°$$

where n = number of angles or sides.

Table 3.1 Field Book
Closed Traverse Survey by Prismatic Compass

Place: Date:
Inst. No.: Time:

Line	Length (ft)	Observed Bearings		Difference	Corrected Bearings		Remarks
		Fore	Back		Fore	Back	
AB	33.1	40°00'	220°00'	180°	40°00'	220°00'	i. Errors at C=+1°00' and D=-2°00'
BC	35.3	95°30'	276°30'	181°	95°30'	275°30'	
CD	36.4	208°30'	25°30'	183°	207°30'	27°30'	ii. A and B free from local attractions
DA	42.2	282°30'	104°30'	178°	284°30'	104°30'	
							iii. Survey was done clockwise

Table 3.2 Field Book
Closed Traverse Survey by Prismatic Compass

Place: Date:
Inst. No.: Time:

Line	Length (ft)	Observed Bearing		Difference (d)	Error (d–180°)	Error/2	Corrected Bearing		Remarks
		Fore	Back				Fore	Back	
AB	33.1	40°00'	219°00'	179°	–1°	–0°30'	39°30'	219°30'	i. all stations locally attracted.
BC	35.3	95°30'	276°30'	181°	+1°	+0°30'	96°00'	276°00'	
CD	36.4	208°30'	25°30'	183°	+3°	+1°30'	207°00'	27°00'	
DA	42.2	282°30'	104°30'	178°	–2°	–1°00'	283°00'	103°30'	ii. survey done clockwise

ii. the sum of the external angles,

$$\sum^{n}\alpha_1 = (2n + 4) \times 90^\circ$$

iii. the sum of the deflection angles,

$$\sum^{n}\beta_i = 360^\circ \text{ and}$$

iv. $\sum^{n}(\beta_i + \theta_i) = n \times 180^\circ$

Included angles at a point formed by two lines is computed from the bearings of the two lines measured from the same point. When the whole circle bearing (i_1 and i_2: Fig. 3.27) are given, the internal or included angle at that point is simply the difference between the two bearings.

From Fig. 3.27,

∠BAC = Fore Bearing of AC–Fore Bearing of AB
= $(i_2 - i_1)$
= Back Bearing of the preceding line – Fore Bearing of the next line (measured in the direction of survey).

However, if the difference of the bearings exceeds 180°, it is generally an external angle and the internal angle is found by subtracting it from 360°, i.e., internal angle = 360° – $(i_2 - i_1)$.

When quadrantal bearings are given, the general rules of finding internal angles are:

i. If the lines lie in the same quadrant, the internal angle is simply the difference between the two reduced bearings.
ii. If the lines lie in the two adjacent quadrants on the same side of the meridian, the internal angle = (180° – sum of the two reduced bearings).
iii. If the lines lie in two adjacent quadrants on the opposite sides of the meridians, the internal angle is simply the sum of the two reduced bearings.
iv. If the lines lie in the two distant quadrants (diagonally opposite), the internal angle = (180° – difference of the two reduced bearings).

As a rule,
external angle (α) = [360° – internal angle (θ)] and
deflection angle (β) = [180° – internal angle (θ)].

Plotting the Traverse

The two most commonly applied methods of plotting are the Parallel Meridian Method and the Included Angle Method.

Parallel Meridian Method

In this, distances and the corrected fore bearings of each line are considered. At first, a map scale is selected and the ground distances are reduced. A small straight line is drawn vertically to represent the magnetic meridian. On it a

Fig. 3.26 Closed Traverse

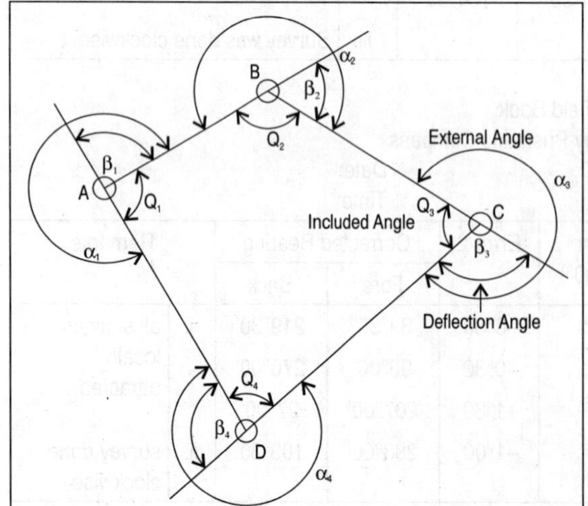

Fig. 3.27 Determination of Included Angles

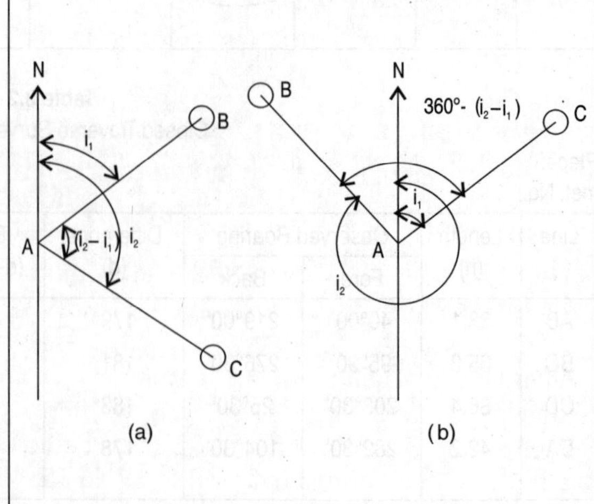

Table 3.3 Computation of Angles and Checking

Line	Corrected Bearings		Angles at	Internal Angle	External Angle	Deflection Angle	(β + θ)	Remarks
	Fore	Back						
[1]	[2]	[3]	[4]	(θ) [5]	(α) [6]	(β) [7]	[8]	
AB	40°00'	220°00'						
			B	124°30'	235°30'	55°30'	180°00'	
BC	95°30'	275°30'						
			C	68°00'	292°00'	112°00'	180°00'	All checks are applied and found satisfactory
CD	207°30'	27°30'						
			D	103°00'	257°00'	77°00'	180°00'	
DA	284°30'	104°30'						
			A	64°30'	295°30'	115°30'	180°00'	
				360°00'	1080°00'	360°00'	720°00'	

$\sum^{4} \theta_i = (2 \times 4 - 4) \times 90^\circ = 360^\circ$ (Cl. 5)

$\sum^{4} \alpha_i = (2 \times 4 + 4) \times 90^\circ = 1080^\circ$ (Cl. 6)

$\sum^{4} (\beta_i + \theta i) = 4 \times 180^\circ = 720^\circ$ (Cl. 8)

point A is first marked and then the fore bearing of the first line (AB) is plotted with the help of a protractor and its length is marked off; thus the position of station B is fixed. Through B another line parallel to the magnetic meridian at A is drawn and the position of station C is similarly plotted with the help of the length and bearing of the line BC. The process is repeated at each station until all the lines are drawn (Fig. 3.28).

Included Angle Method

In this, the magnetic meridian is drawn through the initial point (A). Then the point B is plotted with the reduced length of AB (by a scale) and its correct fore bearing (by a protractor). At B, the included angle (from the table showing angles) is drawn with the help of a protractor and the length of BC is measured off to fix the point C. This process is repeated at all stations, until all the lines are plotted (Fig. 3.29).

Adjustment of Closing Error

After plotting the closed traverse data, there are often errors. The final point does not coincide with the initial point. This error is called *closing error*. To make the traverse consistent, the closing error must be distributed by adjusting the lengths and bearings following either the Bowditch, Wilson or Smirnoff methods. Of these methods, the Bowditch method is more widely used because of its simplicity. It is based on the principle that the magnitude of error is proportional to the square root of distance. The closing error is distributed graphically by shifting each station by an amount proportional to the distance from the initial point parallel to the direction of the closing error (Fig. 3.28). To do this, a straight line is drawn horizontally and the points are marked off with their reduced distances. At the closing point (A´), a perpendicular straight line is drawn and the amount of closing error (AA´) is measured off. The top of this line (A´) is joined to the starting point (A) on the horizontal line and from each point on the horizontal line perpendiculars are drawn. These perpendicular segments (BB´, CC´, DD´, etc.) are the proportionate closing errors to be distributed at the respective points. At each point on the plotted traverse, small straight lines parallel to the direction of closing

Fig. 3.28 Closed Traverse Survey by Prismatic Compass (plotted by Parallel Meridian Method)

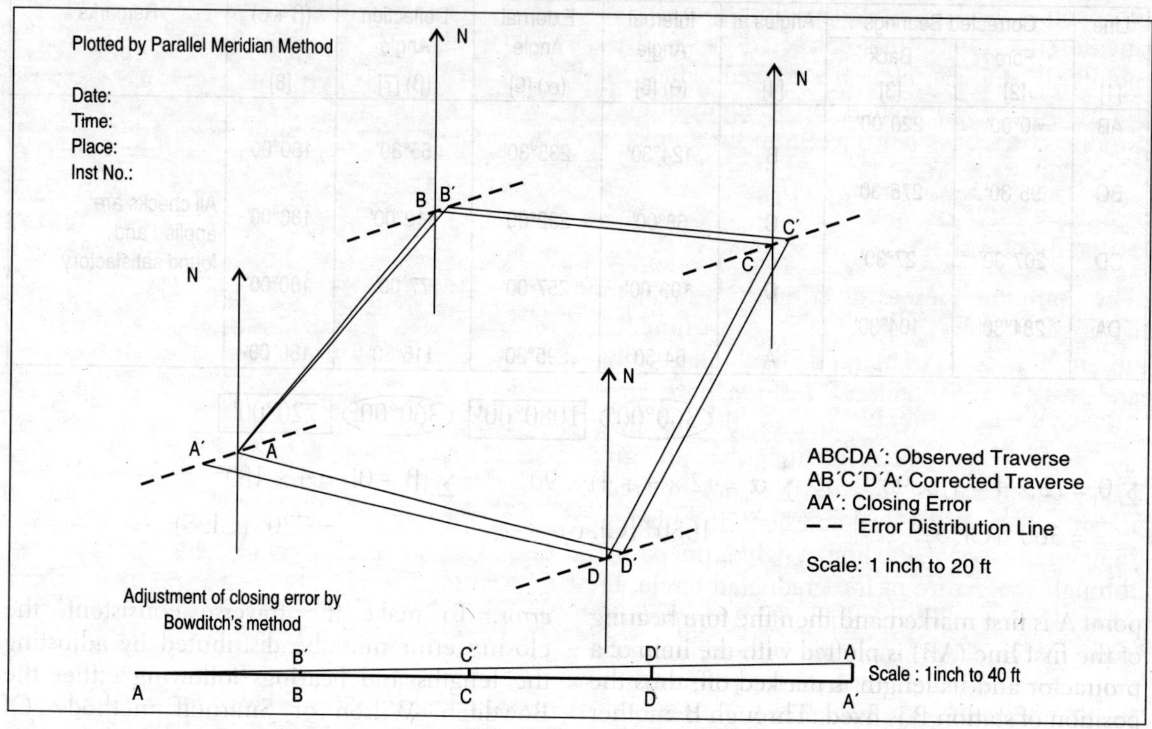

Fig. 3.29 Closed Traverse Survey by Prismatic Compass (plotted by Included Angle Method)

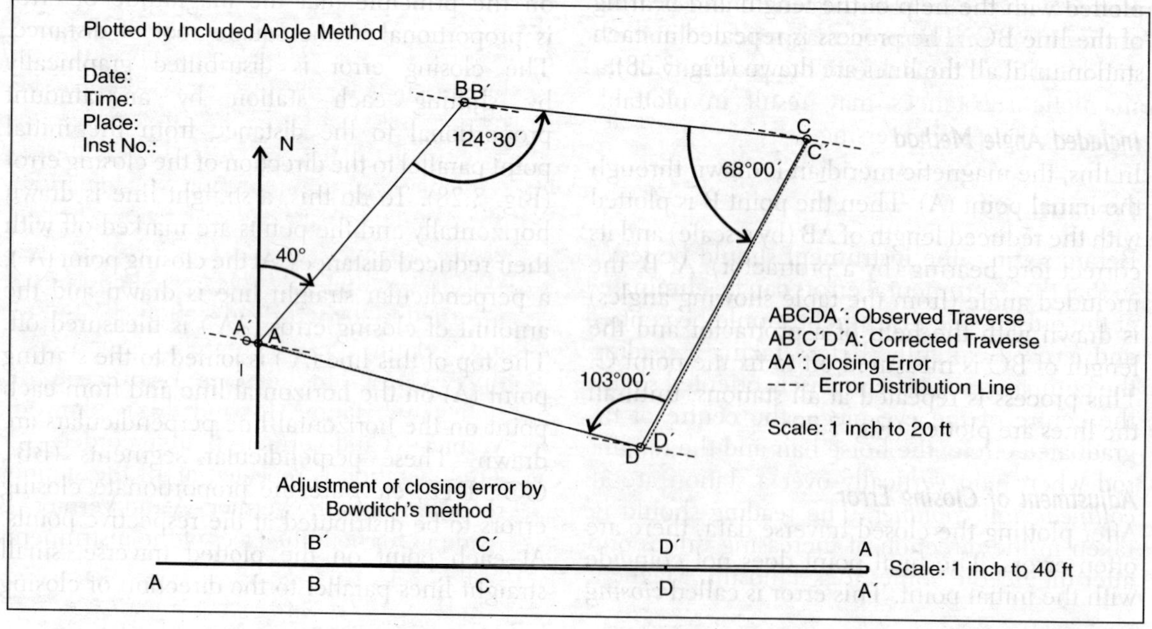

error (error distribution line) are drawn. Then on each of these, respective closing errors are measured off and the corrected or adjusted points (B´, C´, D´, etc.) are marked. These are then successively joined by straight lines to produce the corrected traverse in the form of a closed polygon (AB´C´D´... A).

Sources of Errors

The common sources of errors in a prismatic compass traversing are instrumental, observational due to manipulation and sighting, and external influences. *Instrumental errors* may be of several types, such as, the needle is not perfectly straight or is sluggish with some lost magnetism, the needle may not be freely moving, the pivot is dull and bent, the plane of sight is not vertical, the line of sight is not passing through the centre of the graduated circle, the vertical hair is loose and thick, etc. *Manipulation and sighting errors* may be due to improper centering and levelling, imperfect bisection of the station with eye piece and horse hair, and carelessness in reading the graduated circle and recording the data properly in the field book. *External influences* like magnetic changes in the atmosphere, variations in magnetic declinations and local attractions due to the proximity of the magnetic substances may result in plottable errors in compass traversing.

Precautions

Before setting, the instrument should be tested so that the instrumental errors can be eliminated at the outset. The instrument should be centred and *levelled* carefully. Before taking a reading, the compass should be precisely oriented so that the centre of the eye piece, the centre of the graduated circle, the horse hair and the ranging rod when held vertically over a station are all *collinear* and *coplaner*. The reading should be taken in the direction of increasing values only after the needle comes to a standstill. Magnetic substances of any sort should be avoided about the surveyor and around the traverse.

Traversing by a Theodolite

Where accuracy is required, traversing is done with a theodolite. The linear distances are measured with a tape while the horizontal angles (internal or external or deflection) are measured with a theodolite with great precision. Normally internal angles are measured in case of a closed traverse. The magnetic bearing of only the initial line is measured with the help of a trough compass and theodolite or a prismatic compass. The set of instruments and accessories required are: a transit theodolite, a tripod, a prismatic compass, a tape and a set of pins and ranging rods.

Procedure

i. Stations are first selected and pins with tags are fixed at each station.
ii. The distance of each line is measured with a tape.
iii. A theodolite is set up at the initial station (A) and it is then precisely centred and perfectly levelled (Table 3.4, Fig. 3.27).
iv. With 0°- setting in Vernier - A, the telescope is turned to the magnetic north as indicated by the needle of the trough compass by using the lower clamp and tangential screw. The upper clamp is then released and the telescope is turned to station B. The reading on Vernier - A gives the magnetic bearing of AB.
v. With 0°- setting in Vernier - A, the telescope is turned to the left hand station (B) and the lower clamp is fixed. Now after releasing the upper clamp, the telescope is turned to the right hand station (D) and the readings on Vernier - A and Vernier - B are recorded.
vi. The operation is repeated at all the stations recording both the Vernier - A and Vernier - B readings for both the faces of the instruments (Table 3.4).

Table 3.4 Field Book
Closed Traverse Survey by Transit Theodolite

Place:
Inst. No.:

Date:
Time:

Instrument at	Object sighted	Face of instrument	Horizontal Circle Reading VA	VB	Internal Angle	Mean Angle	Corrected Angle	Remarks
A	Magnetic North Direction	L	0°	180°	40°28'	40°30'		Magnetic Bearing of AB.
		R	0°	180°	40°28'			
	B	L	40°28'	220°28'	40°32'			
		R	40°32'	220°32'	40°32'			
A	B	L	0°	180°	64°34'	64°32'	64°33'	AB = 33.1 ft
		R	0°	180°	64°34'			BC = 35.3 ft
	D	L	64°34'	244°34'	64°30'			CD = 36.4 ft
		R	64°30'	244°30'	64°30'			DA = 42.2 ft
B	C	L	0°	180°	124°10'	124°12'	124°13'	Magnetic Bearings
		R	0°	180°	124°10'			AB = 40°30'
	A	L	124°10'	304°10'	124°14'			BC = 96°17'
		R	124°14'	304°14'	124°14'			CD = 207°28' DA = 285°03'
C	D	L	0°	180°	68°47'	68°48'	68°49'	Reduced Bearings
		R	0°	180°	68°47'			AB = N40°30'E
	B	L	68°47'	248°47'	68°49'			BC = S83°43'E CD = S27°28'W
		R	68°49'	248°49	68°49'			DA = S74°57'W
D	C	L	0°	180°	102°23'	102°24'	102°25'	Survey
		R	0°	180°	102°23'			was done
	A	L	102°23'	282°23'	102°25'			clockwise
		R	102°25'	282°25'	102°25'			
						Σ359°56'	Σ360°00'	

Note: A deficit of 4' is equally distributed at the four stations.

COMPUTATION

Correction of Internal Angles

The sum of all the internal angles should be (2n – 4) right angles; any deficit or surplus from this should be divided and equally distributed among the angles (Table 3.4).

To Find the Magnetic Bearings of Each Line

The magnetic bearings of each line can be calculated from the internal angles and the bearing of the first line (rules discussed in Prismatic Compass Traversing).

Fig. 3.30 Computing the Rectangular Coordinates of a Point

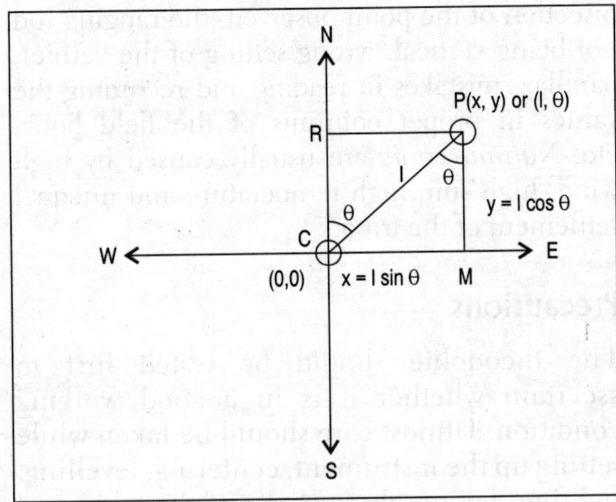

To Find the Coordinates of Points

The magnetic bearings of each line is first transformed into quadrantal form. From the lengths and reduced bearings, *latitudes* and *departures* of each line are evaluated (Fig. 3.30). The latitude of a line (CP) is its projection on the N-S line (CR or PM) while departure is its projection on the E-W line (CM or PR). Latitude (l cos θ) may be positive (+) or negative (–) depending respectively on whether it is *northing* (on the north line) or *southing* (on the south line). Similarly, departures (l sin θ) may be positive (+) or negative (–) based respectively on whether it is *easting* (on the east line) or *westing* (on the west line). This Cartesian transformation is very important because for a closed traverse,

$$\Sigma \text{ latitudes} = 0 \text{ or } \Sigma(l\cos\theta) = 0 \text{ and}$$
$$\Sigma \text{ departures} = 0 \text{ or } \Sigma(l\sin\theta) = 0$$

On the basis of this check, latitudes and departures are corrected. This is done in proportion with the total latitudes and total departures or lengths of the concerned sides of the traverse (Table 3.5).

After correction, the consecutive coordinates of each point are calculated successively by algebraically summing up the latitudes (y-coordinates) and departures (x-coordinates) assuming that the initial point (A) lies at the origin, C (0,0). However, independent coordinates are more advantageous in plotting the traverse. It is calculated in the same way as that of the consecutive coordinates but the origin is shifted in such a way that the entire traverse is plotted in the positive quadrant (NE).

Plotting the Traverse

With the help of the lengths and magnetic bearings of lines, a traverse may be plotted by the *parallel meridian method* (Fig. 3.9). It may also be plotted by the included angle method considering the lengths, magnetic bearings of the first line and included angles. These plottings

Table 3.5 Computation of Coordinates

Line	Length (ft)	Reduced Bearing	Observed		Corrected		Points	Consecutive Coordinates		Independent Coordinates	
			Latitude	Departure	Latitude	Departure					
	[l]	[θ]	l cosθ	l sinθ	l cosθ	l sinθ		x	y	x	y
AB	33.1	N40°30'E	+25.169	+21.497	+25.176	+21.728	A	0	0	+20.000	20.000
BC	35.3	S83°43'E	–3.863	+35.088	–3.856	+35.335	B	+21.728	+25.176	+41.728	+45.176
CD	36.4	S27°28'W	–32.296	–16.789	–32.287	–16.568	C	+57.063	+21.320	+77.063	+41.320
DA	42.2	N74°57'W	+10.958	–40.752	+10.967	–40.495	D	+40.495	–10.967	+60.495	+9.033
			Σ–0.032	–0.956	0	0	A	0	0	+20.000	+20.000

Note: Latitudes and departures are corrected as per Bowditch's Rule on the basis of lengths of lines.

need closing error adjustments. However, the traverse may be most accurately plotted by Cartesian coordinates (either consecutive or independent) with reference to the origin defined by a pair of orthogonally intersecting straight lines (Fig. 3.31).

Sources of Errors

The common sources of errors in a theodolite traversing are instrumental, observational and natural. *Instrumental errors* may be due to imperfect adjustment of the plate levels, the collimation line not being perpendicular to the horizontal axis, the horizontal axis not being perpendicular to the vertical axis, non-parallelism of the axis of the telescope and the collimation line, eccentricity of centres (inner and outer axis) and of verniers and errors of graduation, etc. *Observational errors* include inaccurate centering, imperfect levelling, wrong manipulation of the tangent screw, inaccurate bisection of the point observed, the ranging rod not being vertical, wrong setting of the vernier, parallax, mistakes in reading and recording the values in proper columns of the field book, etc. *Natural errors* are usually caused by high wind, high sun, high temperature and unequal settlement of the tripod.

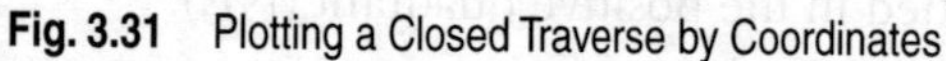

Fig. 3.31 Plotting a Closed Traverse by Coordinates

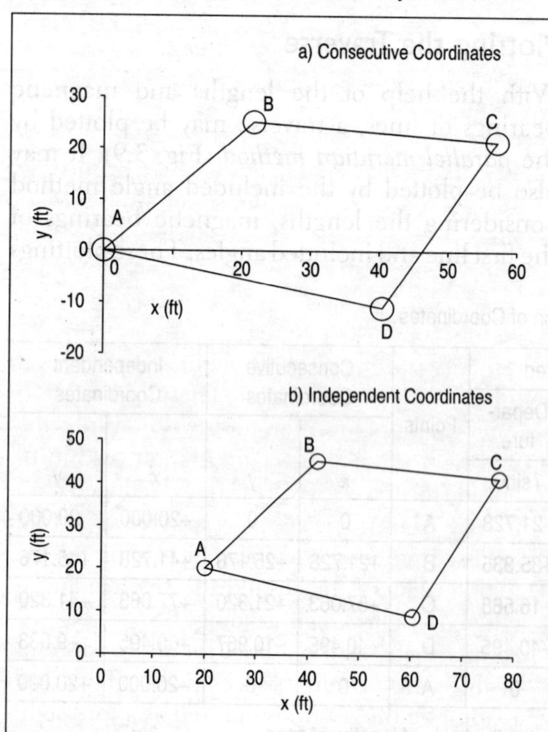

Precautions

The theodolite should be tested first to ascertain whether it is in a good working condition. Utmost care should be taken while setting up the instrument, centering, levelling, sighting, taking observations of the same and recording in the field book. Surveying during high sun, high wind and rain or snow should be avoided.

Traversing by a Plane Table

The set of instruments and accessories required for this are a plane table, a tripod, a U-frame, a plumb bob, a trough compass, a level tube, a tape, ranging rods and pins. There are three different processes of locating a point by plane table—*radiation*, *intersection* and *resection*. All these are involved in traversing procedures (which have already been discussed).

Sources of Errors

The common sources of errors in a plane table traversing are *inaccurate centering*, *imperfect levelling* and *incorrect orientation*. Besides, errors may appear if:

i. Rays do not radiate from or converge at the station point due to wrong fixing of the alidade.
ii. Inaccurate intersection or resection gives wrong points while plotting due to wrong sighting of the object.
iii. Non-verticality of the ranging rods effect the whole survey process.

Precautions

A set of precautions should be taken so that the errors may be eliminated entirely. These are:

i. The surface of the board should be a perfect plane with paper well stretched and perpendicular to the vertical axis.
ii. The fiducial edge must be a smooth straight line and the sight vanes should be perpendicular to the base of the alidade.
iii. Care must be taken while centering, levelling, orientating, sighting and drawing the rays so that each of these processes becomes free from any undesired and forced errors.
iv. Ranging rods must be held vertical on the ground stations.

CONTOURING

A *contour* is an imaginary line joining places with equal elevation above the sea level. Technically, it is defined as the line of intersection of a level surface with the surface of the ground. The vertical distance between two consecutive contours is called *contour interval* while the horizontal distance between the two is known as *horizontal equivalent*. Normally, the nature of the ground, the purpose and extent of the survey, the map scale and the amount of time and financial investment involved, together determine the contour interval. The smaller the map scale, the larger the contour interval and the smaller the interval, the larger the amount of field and office works. Contour maps are useful for engineering, hydrological and geomorphological studies.

Methods

The preparation of a contour plan or map requires first, the surveying and plotting of a traverse plan, second, levelling operations to find the reduced levels of all the points on the traverse and finally, the interpolation of the contours on the traverse plan.

a) *Traverse Survey*
If the area to be contoured is not very extensive, a traverse may be so laid that it consists of a *set of radial lines diverging from the apparently highest or lowest point of the area*. The lengths and directions of each line may be fixed by the traverse surveying with either a prismatic compass or a theodolite. The points at a regular distance apart on each line are then marked on the ground by pegs.

In some cases, the area is divided into rectangular or square grids, each corner of which is marked on the ground by pegs. The position of these points is then fixed by traversing.

b) *Levelling Operations*
The equipment for levelling consists of a level (commonly a Dumpy level), a tripod, a levelling staff, a tape and a well laid out and neatly drawn field book for recording the staff readings, distances and field notes.

Procedure

i. The instrument is first set up at a convenient height and at a place from where a maximum number of stations can be sighted.
ii. With the help of the bubble tube(s), the instrument is then perfectly levelled by turning the foot screws right in and left out in positions when the telescope is: a) parallel to a line through two foot screws and b) perpendicular to it. This operaion is successively repeated taking other foot screws until the bubble remains stationary at the centre of the level tube for any position of the telescope.
iii. The telescope is then directed towards the staff held vertically over the station (within or outside the traverse) with a known reduced level (Bench Mark). The eye piece and the object are focussed properly and the staff reading for the middle stadia is taken and the first reading is entered as a back sight reading (BS).
iv. Similarly the staff readings for all the stations visible are taken successively. The last reading of the set up is entered as a fore sight

reading (FS) while all others as intermediate sight readings (IS).

v. The instrument is then shifted to some other convenient position(s) from where the readings of the remaining stations (not covered in the first set up) can be taken. In this set up, the first reading is taken on the staff held at the last station of the former set up. It is certainly a back sight reading. This station is called a change point (CP).

vi. Following the same procedure, staff readings are then taken on the remaining stations.

Computation

The reduced levels of the stations can be calculated in either of the two common methods—the collimation method and the rise and fall method.

The Collimation Method

In this, the reduced levels are evaluated by subtracting the staff readings from the corresponding height of collimation (Table 3.6). The common rules are:

a. HC = BS + RL of BM
b. RL of a Station = HC – IS or FS
c. For a new set up, i.e., at change points, new heights of collimation are evaluated as HC of a CP = BS + RL of the CP

Arithmetical checks are done to avoid any ambiguity and confusion. The common checks are:

$$\Sigma BS \sim \Sigma FS = \text{Last RL} \sim \text{First RL} \quad \text{... (i)}$$

$$\Sigma(HC \times \text{No. of Applications}) = \Sigma IS + \Sigma FS + \Sigma RLs - \text{First RL} \quad \text{... (ii)}$$

The Rise and Fall Method

In this, the differences in levels between consecutive points are determined by comparing each point with that immediately preceding it. It is entered as a *rise* or as a *fall* accordingly as the staff reading at the point is smaller or greater than that at the preceding point (Table 3.7). The reduced levels are then found by adding the rise to and subtracting the fall from the reduced level of the preceding point. A complete arithmetical check can be applied on the accuracy of calculation as:

$$\Sigma BS \sim \Sigma FS = \Sigma \text{Rise} \sim \Sigma \text{Fall} = \text{Last RL} \sim \text{First RL}$$

Note: HC = Height of Collimation; BS = Back Sight Reading; IS = Intermediate Sight Reading; FS = Fore Sight Reading; CP = Change Point; BM = Bench Mark.

Interpolation of Contours

On the plotted traverse, the reduced levels of each point is written and the desired contours at a chosen interval are then interpolated either by *eye estimation* or by *arithmetical calculation*. In the second method, the ground is assumed to be uniformly sloping between two points (with known distances and reduced levels). Hence, the location of a desired contour depends on the general slope vector, i.e., the amount of relief between the points with respect to the distance and the direction of rise or fall.

In Fig. 3.32, let A and B be the two points with reduced levels, R_a and R_b respectively. The ground slopes from A to B such that $R_a > R_b$. Thus PM = $(R_c - R_b)$ and AC = $(R_a - R_b)$

Let the map distance between A and B be *d* and the desired contour be R_c at P which is *x* distance away from B. Therefore, the right angled ΔPBM and ΔABC are similar. Hence,

$$\frac{PB}{PM} = \frac{AB}{AC}$$

$$\text{or,} \quad PB = \frac{AB}{AC} \times PM$$

$$\therefore x = \frac{d(R_c - R_b)}{(R_a - R_b)}$$

By applying this simple method, the location of a desired contour between any two known

Table 3.6 Field Book

Determination of Reduced Levels by Dumpy Level

Date : Time : Place : Inst. No. : *Collimation Method*

Stations	Staff Reading (ft)			Height of Collimation (ft)	Reduced Levels (ft)	Remark
	BS	IS	FS			
A	09.87			34.87	025.00	BM
B		04.56			030.31	
C		02.32			032.55	
D	08.76		1.23	42.40	033.64	CP
E		06.18			036.22	
F		03.57			038.83	
G	11.65		1.94	52.11	040.46	CP
H		05.83			046.28	
I			1.57		050.54	
Σ	30.28	22.46	4.74		333.83	

Arithmetical Check:

(i) $\Sigma BS \sim \Sigma FS = 30.28 - 4.74 = 25.54$; $RL_I \sim RL_A = 50.54 - 25.00 = \mathit{25.54}$

(ii) $\Sigma HC \times$ No. of Applications $= 34.87 \times 3 + 42.40 \times 3 + 52.11 \times 2 = \mathit{336.03}$; $\Sigma IS + \Sigma FS + \Sigma RL - RL_A = \mathit{336.03}$

Table 3.7 Field Book

Determination of Reduced Levels by Dumpy Level

Date : Time : Place : Inst. No. : *Rise and Fall Method*

Stations	Staff Readings (ft)			Rise (ft)	Fall (ft)	Reduced Levels (ft)	Remarks
	BS	IS	FS				
A	9.87					25.00	BM
B		4.56		5.31		30.31	
C		2.32		2.24		32.55	
D	8.76		1.23	1.09		33.64	CP
E		6.18		2.58		36.22	
F		3.57		2.61		38.83	
G	11.65		1.94	1.63		40.46	CP
H		5.83		5.82		46.28	
I			1.57	4.26		50.54	
Σ	30.28		4.74	25.54	—		

Arithmetical Check:

$\Sigma BS \sim \Sigma FS = 30.28 - 4.74 = \mathit{25.54}$; $\Sigma Rise \sim \Sigma Fall = 25.54 - 0 = \mathit{25.54}$; $\Sigma RL_I \sim \Sigma RL_A = 50.54 - 25.00 = \mathit{25.54}$

elevation levels can be fixed. Through these interpolated points of identical elevation, a freehand smooth line is drawn to represent the desired contour (Fig. 3.32 and 3.33).

Sources of Errors

The common sources of errors in contouring are *instrumental, observational* and *natural*. Sources of errors in compass traversing have already been discussed. However, in levelling operations, the instrumental errors consist of a defective foot screw, a defective bubble tube, imperfect adjustment of the telescope and bubble, a faulty focussing tube, erroneous divisions on the staff and so on. Improper and careless levelling, parallax in sighting, non-verticality of the staff and mistakes in reading and recording the values in the field

Fig. 3.32 Interpolation of Contours

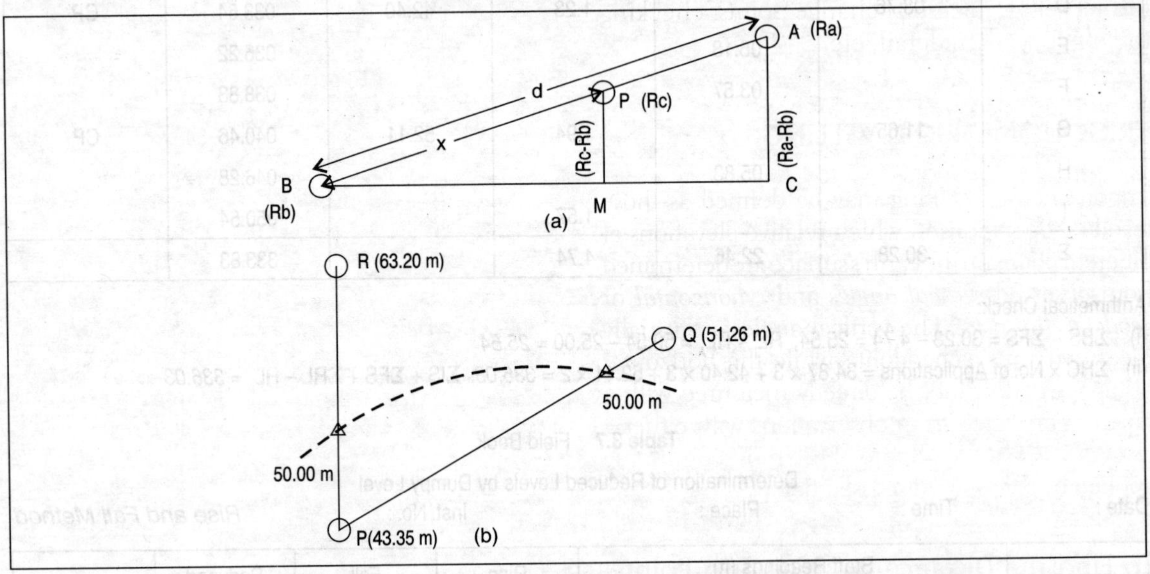

Fig. 3.33 Preparation of a Contour Plan by Traversing and Levelling

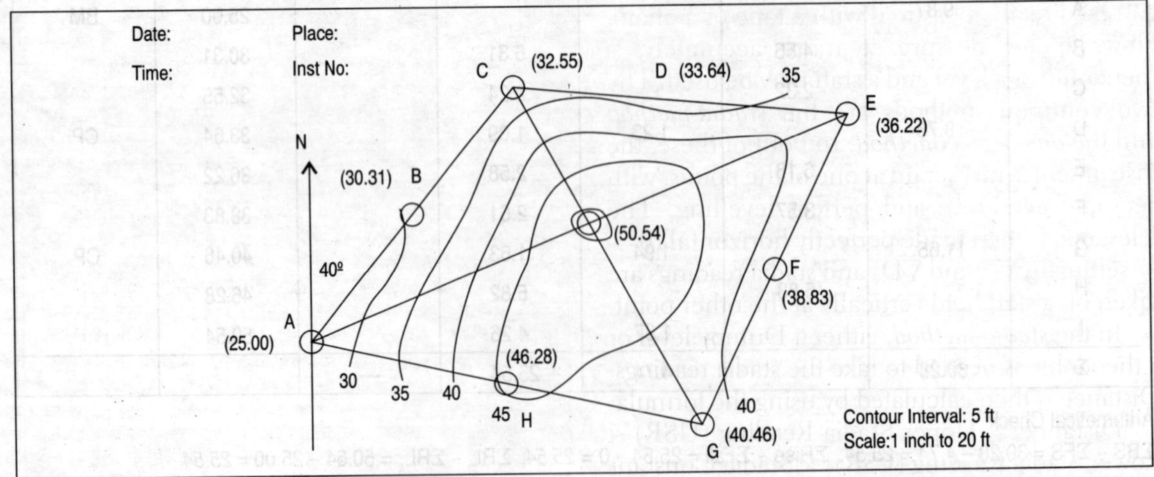

book are the common observational errors. The natural sources of errors are high wind, high sun, high temperature, curvature and refraction.

Precautions

At every stage, surveying should be done very carefully. Special precautions should be taken while levelling the instrument, reading the staff, holding the staff vertically over the stations, taking back sight reading at the last station of the former set up in case of change points, checking the office work in detail, etc.

TRIGONOMETRIC LEVELLING

Trigonometric levelling may be defined as those levelling operations in which relative elevations of different points on the earth's surface are determined from observed *vertical angles* and *horizontal* or *geodetic distances*. The vertical angles are usually measured with a theodolite while the horizontal distances may either be directly measured with a tape or computed from observations with either a level or a theodolite. Some of the most commonly applied methods are discussed below.

To Find the Distance between Two Points

On the ground, a distance between two points can be directly measured with a tape or a chain. However to measure it more accurately, a theodolite or a level and a staff may be used. The two common methods are the *stadia-method* and the *one-degree method*. In both of these, the instrument is first set up at one of the points, with accurate centering and perfect levelling. The telescope is then made perfectly horizontal (0° – 0° setting in VC and VD) and stadia readings are taken on a staff held vertically at the other point.

In the *stadia-method*, either a Dumpy level or a theodolite is needed to take the stadia readings. Distance is then calculated by using the formula:

Distance = [Upper Stadia Reading (USR) ~ Lower Stadia Reading (LSR)] × Stadia Constant

As per the certification of the instuments (Dumpy level and Theodolite), the value of stadia constant is 100). Therefore,

Distance (AB) = (USR ~ LSR) ×100.

In the *one degree method*, a theodolite is normally used. In this, middle stadia readings (MSR) on a staff held vertically over a point from a station of observation are taken for two particular situations: i) when the telescope is absolutely horizontal and ii) when it is inclined upward by exactly 1°. This is done by rotating the telescope with the help of the vertical slow motion screws and simultaneously looking through the vernier aid in either VC or VD (Fig. 3.34b).

Let,

the MSR (AB) $= h_1$, when the telescope is horizontal and

the MSR (AC) $= h_2$, when the telescope is 1° inclined upward.

Therefore,

$BC = h_2 - h_1 \Rightarrow h$ (say)

Since ∠ABC is a rt △Δ,

$$\frac{AB}{BC} = \cot 1^\circ$$

Fig. 3.34 Determination of Distance (a) Stadia Method (b) One Degree Method

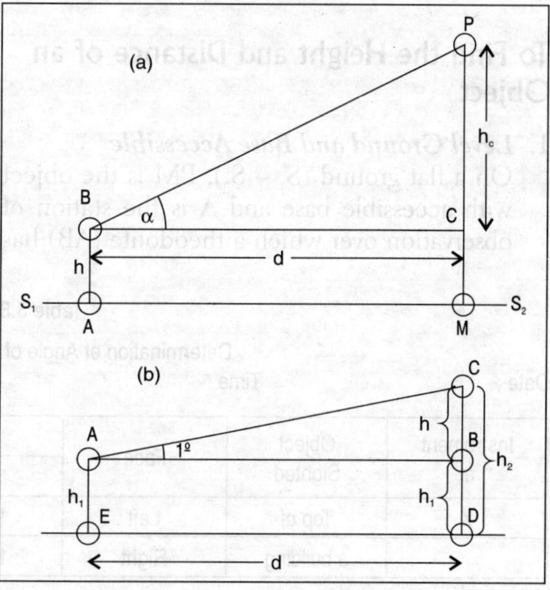

or, AB = BC cot1° ⇒ h cot1°

Hence,

the distance (ED) = cot 1° × h

To Find the Vertical Angle with a Theodolite

A vertical angle is the angle which the inclined line of sight on the vertical plane makes with the horizontal. It may be either an angle of elevation or an angle of depression. The procedures of measurement are:

i. The instrument is set up at the given station of observation with precise centering and perfect levelling.
ii. The telescope is then rotated horizontally until the line of sight bisects the object. Both the vertical and horizontal plates are then clamped. With the help of the vertical slow motion screws, the telescope is moved on the vertical plane to bisect the desired point with great precision.
iii. The vertical circle readings on both VC and VD are taken for both *face left* and *face right* positions and noted in the field book. The average of these give the required vertical angle (Table 3.8).

To Find the Height and Distance of an Object

1. ***Level Ground and Base Accessible***

On a flat ground ($S_1 - S_2$), PM is the object with accessible base and A is the station of observation over which a theodolite (AB) has been set at a height (h) above the ground (Fig. 3.32). Therefore, BC is the line of collimation.

Procedure

i. After proper centering and levelling of the theodolite at A, its height (h) is measured with a tape or a graduated staff.
ii. The horizontal distance, AM (d), is then measured either directly with a tape or by computations from observations of the stadia readings on a staff held vertically at the base of the object.
iii. The angle of elevation (α) of the top of the object (P) is then found from the observations of the vertical circle readings by the theodolite.

Computation

From Fig. 3.34,

AM || BC and AB || PM.

Therefore,

CM = AB = h and BC = AM = d.

d = (USR ~ LSR) × Stadia Constant

From the rt∠ ΔPBC,

$PC\ (h_c) = BC \tan \alpha$

$= d \tan \alpha.$

Hence,

i. The horizontal distance of the object from the station of observation is d,
ii. The height of the object above collimation is h_c,
iii. The height of the object above ground is $(h_c + h)$ and
iv. The slanting distance (AP) of the object is $\sqrt{(h_c + h)^2 + d^2}$

Table 3.8 Field Book

Determination of Angle of Elevation by Transit Theodolite

Date : Time : Place : Inst. No. :

Instrument at	Object Sighted	Face	Vertical Circle Reading		Mean Angle	Grand Mean Angle (α)
			VC	VD		
X	Top of a building	Left	10°24'	10°28'	10°26'	10°27'
		Right	10°26'	10°30'	10°28'	

Fig. 3.35 Determination of Height and Distance of an Object

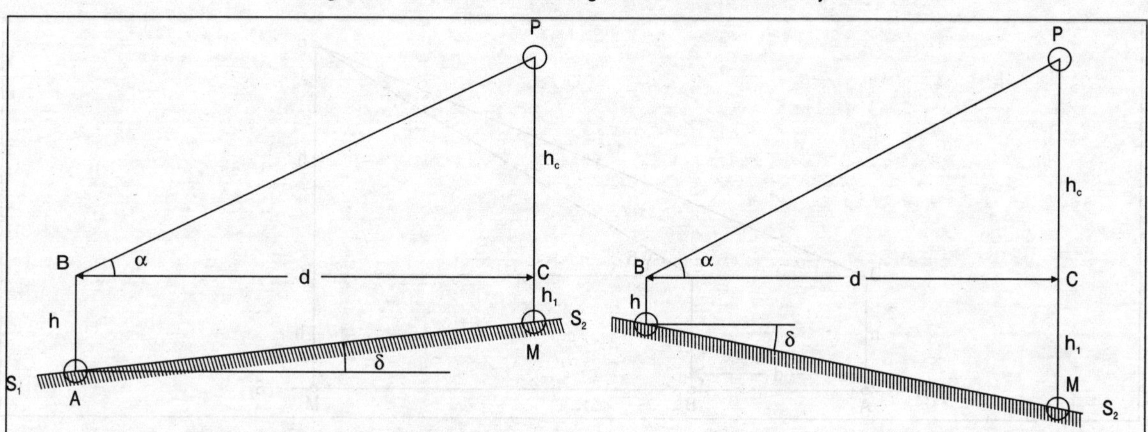

2. ***Sloping Ground and Base Accessible***
   On a sloping ground (S_1 - S_2), PM is the object with an accessible base and A is the station of observation over which a theodolite AB is set up at height, h. Therefore, BC is the line of collimation (Fig. 3.35).

Procedure

i. After proper centering and levelling of the theodolite at A, its height is measured with a tape or a graduated staff.
ii. The horizontal distance, BC (d), is then measured by observations of the stadia readings on a staff held vertically at the base of the object.
iii. The angle of elevation (α) of the top of the object (P) is then found from the observations of the vertical circle readings by a theodolite.

Computation

Horizontal distance,
d = (USR ~ LSR) × Stadia Constant
From Fig. 3.35, ΔPBC is rt△Δ. Therefore,

$$PC = BC.\tan\alpha$$
$$= d.\tan\alpha$$

$$\text{Again, } \tan\delta = \frac{(h_1 - h)}{d}$$

$$\delta = \tan^{-1}\left[\frac{(h_1 - h)}{d}\right] \text{ where, } \delta = \text{slope}$$

Hence,
i. The horizontal distance of the object from A = d.
ii. The height of the object above collimation, h_c = PC.
iii. The height of the object above its base = ($h_c + h_1$).
iv. The height of the object above A = (h_c + h).
v. The slope of the ground = δ°.

3. ***Level Ground and Base Inaccessible; Object and Stations—collinear and coplaner; Instrument heights same at both Stations.***
   On a flat ground (S_1 – S_2), PM is the object with inaccessible base. A and B are the two stations at d distance apart such that A, B and M are collinear on the ground. Therefore, the instruments at A and B and the object lie on the same vertical plane. At both the stations, instruments are set up at the same height (h) (Fig. 3.36). Therefore, DEC represents the common and fixed line of collimation.

Procedure

i. After proper centering and levelling of the theodolite at A, its height (h) is measured with a tape or a graduated staff.

Fig. 3.36 Determination of Height and Distance of an Object

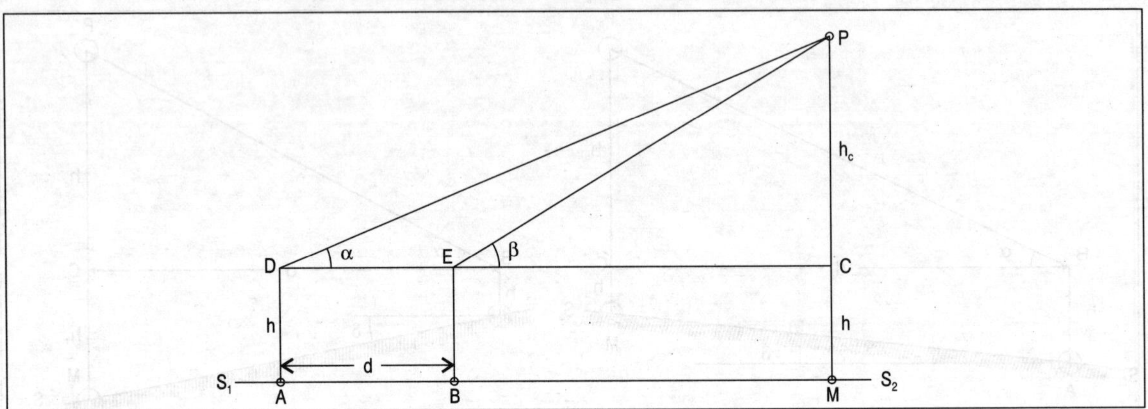

ii. The angle of elevation (α) of the top of the object (P) at A is measured from the observations of the vertical circle readings of a theodolite.
iii. With the horizontal plate fixed, another station of observation (B) is selected on the ground by looking through the telescope such that d or AB ≥ 5 m.
iv. The distance, AB (d), is measured with a tape.
v. The instrument is then shifted to B, set up at height (h), carefully centred and levelled and
vi. The angle of elevation (β) of the top of the object (P) at B is then found from the vertical circle readings by a theodolite.

Computation

From Fig. 3.36,
AM || DC and AD || BE || MC.
Therefore, AD = BE = MC = h and AB = DE = d.
From the rt △ΔPDC, DC = PC.cotα ... (i)
from the rt △ΔPEC, EC = PC.cotβ ... (ii)
As D, E and C are collinear and coplaner

DC – EC = DE

or, PC.cotα – PC.cotβ = d

$$\therefore \quad PC\ (h_c) = \frac{d}{\cot\alpha - \cot\beta}$$

Once PC is found, EC can be determined from the equation (ii)

Hence,
i. The horizontal distance of the object from B = EC and from A = (d + EC).
ii. The height of the object above collimation, h_c = PC.
iii. The height of the object above ground, H = (h_c + h).

4. ***Level Ground and Base Inaccessible; Object and Stations—collinear and coplaner; Instrument heights different at both Stations.*** On a flat ground (S_1 – S_2), PM is the object with inaccessible base. A and B are in the two stations d distance apart such that A, B and M are collinear on the ground. Instruments are set up at different heights (h_1 at A and h_2 at B) so that the survey operation becomes easier (Fig. 3.37).

Procedure

i. After proper centering and levelling of the instrument at A, its height (h_1) is measured with a tape or a graduated staff.
ii. The angle of elevation (α) of the top of the object (P) at A is measured from the observations of the vertical circle readings.
iii. With the horizontal plate fixed, station B is selected on the ground by looking through the telescope such that AB (d) ≥ 5m.

Fig. 3.37 Determination of Height and Distance of an Object

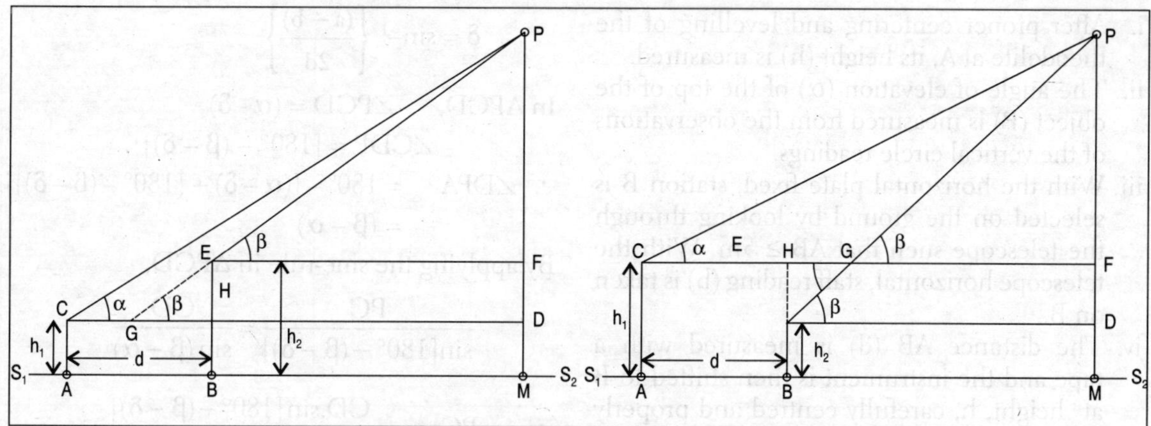

iv. The distance, d is measured with a tape; the instrument is then set up at B, at height, h_2, carefully centred and properly levelled.

v. The angle of elevation (β) of the top of the object (P) at B is then measured from the vertical circle readings by a theodolite.

Computation

From Fig. 3.37,

AC ∥ BE ∥ PM and AM ∥ CD ∥ EF.

Therefore, AC = BH = MD = h_1,

BE = MF = h_2, AB = CH = d, ∠EGH = ∠PEF = α and EH = EB − HB = $h_2 - h_1$.

From the rt ∠ΔPCD, CD = PD.cot α ...(i)

from the rt ∠ΔPGD, GD = PD.cot β ...(ii)

and from the rt. ∠ΔEGH,

GH = EH.cotβ ⇒ $(h_2 - h_1)$.cotβ ...(iii)

As G lies on the collimation line CD,

CD − GD = CG

or, PD.cotα − PD cotβ = CH ± GH

[- if $h_2 > h_1$ and + if $h_1 > h_2$.]

$$\text{or, } PD = \frac{CH \mp GH}{\cot\alpha - \cot\beta}$$

$$= \frac{d \mp (h_2 - h_1)\cot\beta}{\cot\alpha - \cot\beta}$$

Hence,

i. The horizontal distance of the object from A = CD and from B = (CD − d)

ii. The height of the object above ground, H = $(PD + h_1)$.

5. ***Sloping Ground and Base Inaccessible; Object and Stations—Collinear and Coplaner; Instrument Heights Same at Both Stations.***

On a sloping ground $(S_1 - S_2)$, PM is the object with inaccessible base. A and B are the two stations at d distance apart such that A, B and M are collinear on the ground. Instruments are set up at the same height, h, on both the stations (Fig. 3.38).

Fig. 3.38 Determination of Height and Distance of an Object

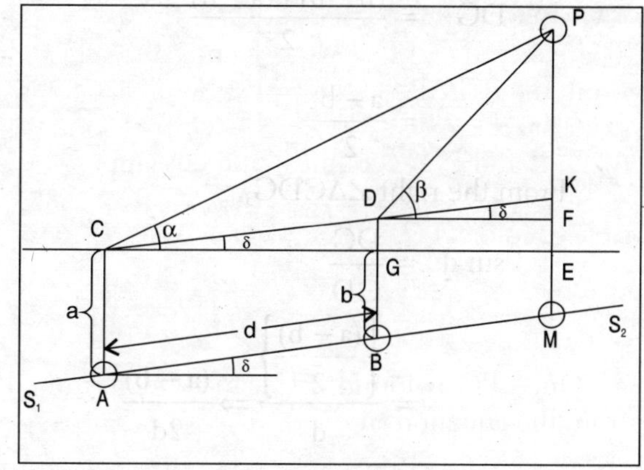

Procedure

i. After proper centering and levelling of the theodolite at A, its height (h) is measured.
ii. The angle of elevation (α) of the top of the object (P) is measured from the observations of the vertical circle readings.
iii. With the horizontal plate fixed, station B is selected on the ground by looking through the telescope such that AB ≥ 5m. With the telescope horizontal, staff reading (b) is taken on B.
iv. The distance AB (d) is measured with a tape and the instrument is then shifted to B at height, h, carefully centred and properly levelled.
v. With the telescope horizontal, staff reading is taken over A (a) and then the angle of elevation (β) of the top of the object (P) is measured from the vertical circle readings by a theodolite.

Computation

Instrument A:

Staff Reading on A = h
Staff Reading on B = b

Level difference = (h – b)

Instrument at B:

Staff Reading on A = a
Staff Reading on B = h

Level difference = (a – h)

∴ true difference in level between A and B,

$$DG = \frac{(h-b)+(a-h)}{2}$$

$$= \frac{a-b}{2}$$

From the right ∠ΔCDG,

$$\sin d = \frac{DG}{CD}$$

$$= \frac{\left\{\frac{(a-b)}{2}\right\}}{d} \Rightarrow \frac{(a-b)}{2d}$$

Hence the slope of the ground,

$$\delta = \sin^{-1}\left\{\frac{(a-b)}{2d}\right\}$$

In ΔPCD, $\angle PCD = (\alpha - \delta)$

$$\angle CDP = [180° - (\beta - \delta)]$$

$$\therefore \angle DPA = 180° - [(\alpha-\delta) + \{180° - (\beta-\delta)\}]$$
$$= (\beta - \alpha)$$

By applying the sine-rule in ΔPCD,

$$\frac{PC}{\sin[180° - (\beta-\delta)]} = \frac{CD}{\sin(\beta-\alpha)}$$

$$\therefore PC = \frac{CD.\sin[180° - (\beta-\delta)]}{\sin(\beta-\alpha)}$$

$$= \frac{d.\sin(\beta-\delta)}{\sin(\beta-\alpha)} \quad [\because CD = AB = d]$$

From rt ∠ΔPCE,

$$PE = PC.\sin\alpha$$

$$= \frac{d.\sin\alpha.\sin(\beta-\delta)}{\sin(\beta-\alpha)}$$

$$\text{and } CE = PC.\cos\alpha$$

$$= \frac{d.\cos\alpha.\sin(\beta-\delta)}{\sin(\beta-\alpha)}$$

from the rt∠ ΔCKE, CK = CE.cosδ

Hence,

i. The height of the object above A = (PE + h) and the same above B = (PE + b).
ii. The distance of the object from A, AM = CK and the same from B, BM = (AM – d).

6. *Level ground and Base Inaccessible; Object and Stations Form a Ground Triangle; Instrument Heights Same at Both Stations.* On a flat ground ($S_1 - S_2$), PM is the object with inaccessible base. A and B are the two stations d distance apart such that ABM forms a ground triangle. At both the stations, instruments are set up at the same heights (h) (Fig. 3.39). Thus CDN forms a triangle on the collimation plane.

Fig. 3.39 Determination of Height and Distance of an Object

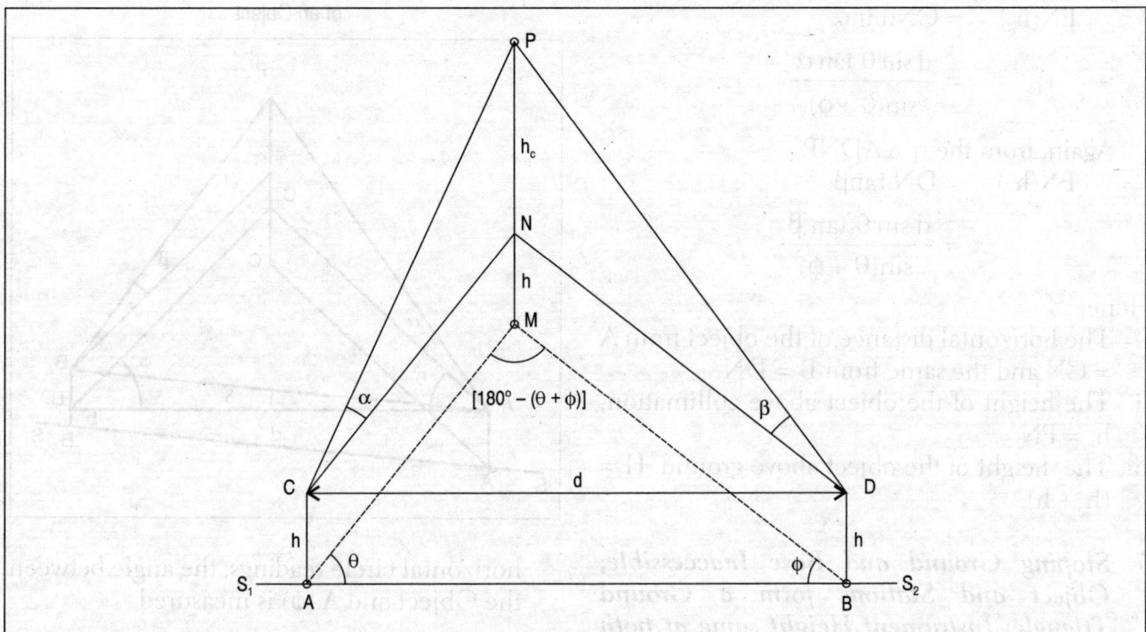

Procedure

i. After proper centering and levelling of the theodolite at A, its height (h) is measured with a tape or a graduated staff.
ii. The angle of elevation (α) of the top of the object (P) at A is found from the observations of the vertical circle readings.
iii. The horizontal angle (θ) of the ground triangle at A is found from the observations of the horizontal circle readings with the telescope focused respectively on the object and the staff held over B.
iv. The instrument is then shifted to B and the steps (i–iii) are followed to find the horizontal angle (ϕ) of the ground triangle at B and the angle of elevation (b) of the top of the object (P) at B.
v. The horizontal distance AB (d) is measured with a tape (d ≮ 5m).

Computation

From Fig. 3.39,
AC || DB || PM, AB || CD, AM || CN and BM || DN. Therefore, AC = BD = MN = h and AB = CD = d. As the planes containing the Δs ABM and CDN are horizontal and parallel, therefore ∠MAB = ∠NCD = θ and ∠ABM = ∠CDN = ϕ.

Therefore, ∠AMB = ∠CND = [180° – (θ + ϕ)]

By applying the sine-rule in ΔCND,

$$\frac{CN}{\sin\phi} = \frac{CD}{\sin[180° - (\theta + \phi)]} \text{ and}$$

$$\frac{DN}{\sin\phi} = \frac{CD}{\sin[180° - (\theta + \phi)]} \text{ and}$$

$$\therefore \quad CN = \frac{d \sin\phi}{\sin(\theta + \phi)} \text{ and}$$

$$DN = \frac{d \sin\theta}{\sin(\theta + \phi)}$$

From the right CNP,

$$PN(h_c) = CN.\tan\alpha$$

$$= \frac{d \sin\theta \tan\alpha}{\sin(\theta + \phi)}$$

Again, from the rt $\triangle\Delta DNP$

$$PN(h_c) = DN.\tan\beta$$

$$= \frac{d \sin\theta \tan\beta}{\sin(\theta + \phi)}$$

Hence,

i. The horizontal distance of the object from A = CN and the same from B = DN.
ii. The height of the object above collimation, h_c = PN.
iii. The height of the object above ground, H = $(h_c + h)$.

7. ***Sloping Ground and Base Inaccessible; Object and Stations form a Ground Triangle; Instrument Height same at both Stations.***

On a sloping ground $(S_1 - S_2)$, PC is the object with inaccessible base. A and B are two stations at d distance apart (d ≥ 5m), such that these together with the object form a triangle. At both stations, the instrument is set at the same height h (AA_1, or BB_2). Ground slopes from B to A at angle δ.

Procedure

i. After proper centering and levelling of the theodolite at A, its height (h) is measured with a tape. AB is measured with a tape (d).
ii. With the telescope perfectly horizontal at 0° – 0° setting, middle Stadia reading (MSR) is taken on a staff held vertically at B.
iii. The angle of elevation of P (α) is measured from the observation of vertical circle readings. From horizontal circle readings, the angle between object and B, is θ measured.
iv. The instrument is then shifted to B, set up at height, h, and perfectly levelled.
v. The angle of elevation of P (β) is measured from the vertical circle readings. From horizontal circle readings, the angle between the Object and A (ϕ) is measured.

Fig. 3.40 Determination of Height and Distance of an Object

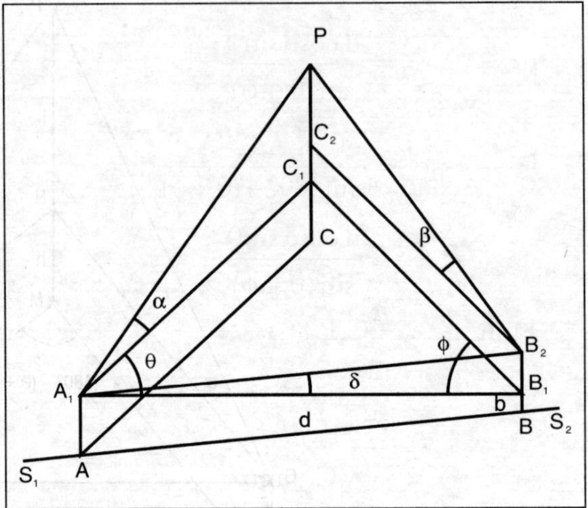

Computation

From Fig. 3.39,

$AA_1 = B_1B_2 = h$

$AB = A_1B_2 = d$

MSR at B $= BB_1 = b$

$B_1B_2 = BB_2 - BB_1 = (h–b)$

From rt $\triangle\Delta A_1B_1B_2$,

$$\sin\delta = \frac{B_1B_2}{A_1B_2} \Rightarrow \frac{(h-b)}{d}$$

$$\therefore \delta = \sin^{-1}\left[\frac{(h-b)}{d}\right]$$

$$\text{Again,} \quad \frac{A_1B_1}{A_1B_2} = \cos\delta$$

$$\text{or,} \quad A_1B_1 = A_1B_2 \cos\delta = d\cos\delta$$

In $\Delta A_1B_1C_1$, $\angle C_1A_1B_1 = \theta$ and $\angle A_1B_1C_1 = \phi$

$\therefore \angle A_1C_1B_1 = 180° - (\theta + \phi)$.

By applying the sine-rule,

$$\frac{A_1C_1}{\sin\phi} = \frac{A_1B_1}{\sin[180° - (\theta + \phi)]}$$

or, $$A_1C_1 = \frac{d\cos\delta\sin\phi}{\sin(\theta + \phi)}$$

and $$\frac{B_1C_1}{\sin\theta} = \frac{A_1B_1}{\sin[180° - (\theta + \phi)]}$$

or, $$B_1C_1 = \frac{d\cos\delta\sin\theta}{\sin(\theta + \phi)}$$

From rt∠ $\Delta A_1C_1P_1$,

$$\frac{PC_1}{A_1C_1} = \tan\alpha$$

or, $$PC1 = A_1C_1\tan\alpha = \frac{d\cos\delta\sin\phi\tan\alpha}{\sin(\theta + \phi)}$$

and from rt $\angle\Delta B_2C_2P$,

$$\frac{PC_2}{B_2C_2} = \tan\beta$$

or, $$PC_2 = B_2C_2\tan\beta = B_1C_1\tan\beta = \frac{d\cos\delta\sin\theta\tan\beta}{\sin(\theta + \phi)}$$

Hence,

i. Distance of the Object from A

$$= A_1C_1 \Rightarrow \frac{d\cos\delta\sin\theta}{\sin(\theta + \phi)}$$

ii. Distance of the Object from B

$$= B_2C_1 = B_1C_1 \Rightarrow \frac{d\cos\delta\sin\theta}{\sin(\theta + \phi)}$$

iii. Height of the Object above A (PC)

$$\delta = PC_1 + CC_1 = \frac{d\cos\delta\sin\theta\tan\beta}{\sin(\theta + \phi)} + h$$

iv. Height of the object above B

$$= PC_2 + BB_2 = \frac{d\cos\delta\sin\theta\tan\beta}{\sin(\theta + \phi)} + h$$

v. Slope of the ground,

$$\delta = \sin^{-1}\left[\frac{(h - b)}{d}\right]$$

8. ***Flat Ground; Three Stations—collinear and at known distances apart; Object and Stations form ground triangles; Instrument Heights same at all Stations; No measurement of Horizontal Angles.***

On a flat ground ($S_1 - S_2$), PM is the object with inaccessible base. A, B and C are three stations lying on a straight line at known distance apart: AB = x, BC = y and AC = (x+y). The theodolite is set up at a height, h_1, at all the stations so that D, E, F and N are coplaner on the collimation (Fig. 3.41).

Procedure

i. The distances, AB and BC, are measured on the ground by means of a tape.
ii. At all the three stations, A, B and C, the instrument is set up at the same height, h_1, with precise centering and levelling. Angle of elevations (α, β and γ) of the top of the object (P) are measured from the observations of the vertical circle readings respectively from stations A, B and C.

Computation

From Fig. 3.41,

$\angle PDN = \alpha$, $\angle PEN = \beta$, $\angle PFN = \gamma$.

Since, AM || DN, BM || EN, CM || FN, AC || DF and AD || BE || CF || MN, AM = DN, BM = EN, CM = FN, AB = DE, BC = EF, AC = DF and AD = BE = CF = MN = h_1.

Let, PN = h. Therefore from the right ∠Δs PDN, PEN and PFN, ND = h.cot a, NE = h.cotβ and FN = h.cotγ.

By applying cosine-rule in Δs DNE and DNF,

$$\cos\angle NDE = \frac{h^2\cot^2\alpha + x^2 - h^2\cot\beta}{2h.x.\cot\alpha}$$

Fig. 3.41 Determination of Height and Distance of an Object

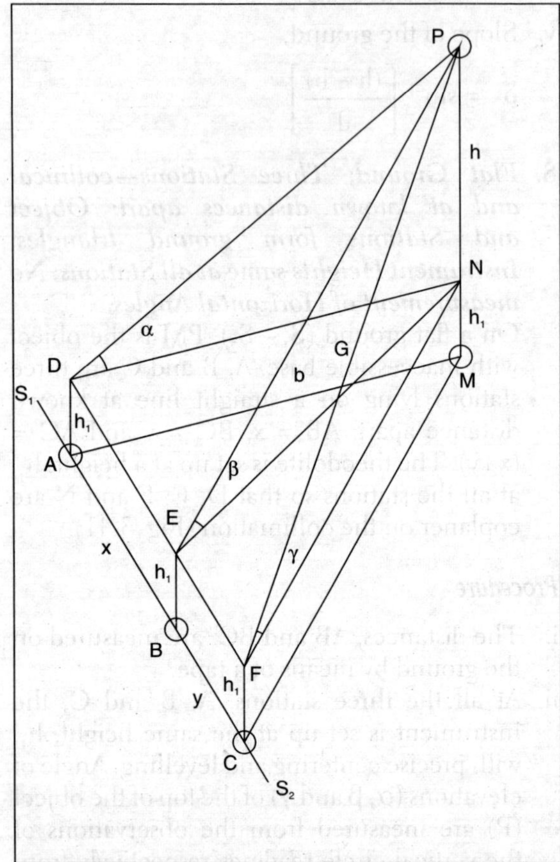

$$\cos \angle NDE = \frac{h^2 \cot^2\alpha + (x+y)^2 - h^2 \cot^2\gamma}{2h.(x+y)\cot\alpha}$$

$$\therefore (x + y)\,[h^2 (\cot^2\alpha - \cot^2\beta) + x^2] = x\,[h^2 (\cot^2\alpha - \cot^2\gamma) + (x + y)^2]$$

$$\text{or, } h^2 [(x + y)(\cot^2\alpha - \cot^2\beta) - x(\cot^2\alpha - \cot^2\gamma)] = x(x + y)^2 - x^2 (x+y)$$

$$h^2 = \frac{(x+y)[x(x+y) - x^2]}{(x+y)(\cot^2\alpha - \cot^2\beta) - x(\cot^2\alpha - \cot^2\gamma)}$$

$$= \frac{xy\,(x+y)}{(x+y)(\cot^2\alpha - \cot^2\beta) - x(\cot^2\alpha - \cot^2\gamma)}$$

$$\therefore h = \left[\frac{xy\,(x+y)}{x(\cot^2\gamma - \cot^2\beta) + y(\cot^2\alpha - \cot^2\beta)}\right]^{1/2}$$

If x = y or B lies at the midpoint of AC the equation becomes,

$$h = \frac{\sqrt{2x}}{\left[\cot^2\gamma - 2.\cot^2\beta + \cot^2\alpha\right]^{1/2}}$$

Hence,

i. The distances of the object from A = h.cotα, from B = h.cotβ and from C = h.cotγ.
ii. The height of the object from collimation = h.
iii. The height of the object above the ground = $(h + h_1)$.

EXERCISE

1. What is surveying? What is levelling? Distinguish between the two.
2. What are the objectives of surveying?
3. What is geodetic surveying?
4. What is plane surveying?
5. Classify surveying with examples.
6. What are the uses of a survey?
7. State the principles of surveying?
8. Explain the corrections applied to the baseline measurements.
9. Explain the methods of locating a point.
10. What is a plan? Distinguish it from a map.
11. Discuss the job of a surveyor.

12. Explain the terms: baseline, ranging, station, checkline, tie line, off-set, reconnaissance, north line, azimuth, magnetic bearing, quadrantal bearing, fore bearing, back bearing, local attraction, magnetic declination, plumb line, level line, level surface, horizontal line, horizontal plane, vertical line, vertical plane, datum, bench mark, line of collimation, plane of collimation, axis of telescope, line of sight, axis of bubble tube, face left and face right observations, trunian axis, orientation, change point, profile levelling, cross-sectioning, reciprocal levelling, link, loop, levelling head, slow motion screw, A-frame, T-Frame, fiducial edge, traverse, checks applied to a closed traverse, rise, fall, reduced level, checks applied to the computations of reduced levels, contour, contour plan, interpolation and extrapolation of contours, slope and gradient.
13. Define the bearing of a line. What are its different types? Give examples.
14. Classify bearing based on the system of expression with examples.
15. Explain how WCB and RB can be converted into one another.
16. What is meant by fore and back bearings of a line? Explain with a diagram how they are related.
17. What is meant by local attraction? Explain how it can be minimised.
18. What is magnetic declination? What are its controls? How is it related to the bearing of a line?
19. Elaborate how linear measurements are taken with a tape.
20. Elaborate how linear measurements are taken with a chain.
21. Explain how a ranging rod should be held in the field.
22. Explain the use of a cross staff.
23. Explain the use of an optical square.
24. Explain the function of a prismatic compass with a neat sketch. Explain also how it is used to find the bearing of a line. State the sources of errors in doing so and explain how it can be minimised.
25. Explain the function of a Dumpy level with a neat sketch. Explain also how it is used to find the stadia readings. State the sources of errors in doing so and explain how it can be minimised.
26. Explain the function of a theodolite with a neat sketch. Explain also how it is used to find the horizontal and vertical angles. State the sources of errors in doing so and explain how it can be minimised.
27. Explain the function of a plane table with a neat sketch. Explain also how it is used to map a portion of the ground. State the sources of errors in doing so and explain how it can be minimised.
28. Explain how traversing is done with a chain, a prismatic compass, a plane table and a theodolite.
29. Explain how triangulation is done with a plane table and a theodolite.
30. Explain how trilateration is done with a plane table and a theodolite.
31. Compare and contrast surveying with a chain, a prismatic compass, a plane table and a theodolite.
32. Compare and contrast levelling with a Dumpy level and a theodolite.
33. Explain how the ground distance of an object is measured with a Dumpy level and a theodolite.
34. Explain how the distance and height of an object with accessible base is measured with a theodolite.
35. Explain how the distance and height of an object with inaccessible base is measured with a theodolite.
36. The distance between two points A and B measured along a slope is 200 m. Find the horizontal distance of AB, when i) the angle of the slope is 8°, ii) the slope is 1 in 5, and iii) the difference of level between A and B is 110 m.
37. A baseline is found to be 10,560 ft long when measured in catenary using a tape 300 ft long which is standard without tension at 60°F. The tape measures in cross-section 1/8 inch × 1/20 inch. If one half of the line is measured at 80°F with an applied tension of 50 lbf, and the bays are approximately equal, find the total correction to be applied to the measured length, given i) coefficient of linear expansion = 6.5×10^{-6}/°F, ii) weight of 1 cubic inch of steel = 0.28 lbf, and iii) Young's Modulus = 29×10^{6} bf/inch^2.

38. A steel tape was exactly 30 m long at 18°C when supported throughout its length under a pull of 10 kg. A line was measured with a tape under a pull of 14 kg and found to be 1001 m. The mean temperature during the measurement was 26°C. Assuming the tape top be supported at every 30 m, compute the true length of the line, given that the cross-sectional area of the tape = 0.04 cm^2, weight of 1 cubic centimetre = 0.0077 kg, the coefficient of linear expansion = 0.0000031/°C and the modulus of elasticity = 2.1×10^6 kg/cm^2.
39. P and Q are two points 150 m apart on the near bank of the river, which flows from east to west. The bearings of the tree on the far bank as observed from P and Q are N50°E and 40°W. Determine the width of the river.
40. Find the reduced bearings: 23°45'56", 112°34'52", 243°15'18", 333°33'33", 289°50'30".
41. Find the WCB: N13°15'32"E, S43°51'E, S63°45'16"W, N72°30'10"W, S14°12'E.
42. Find the fore bearings: 23°51', 123°12', 223°51', 323°00'32", 180°, 359°59'59".
43. Find the back bearings: N81°31'E, S14°34'E, S80°40'10"W, N27°15'W, S43°18'E.
44. The fore and back bearings of AB, BC, CD, and DA were observed to be respectively: 44°30' and 226°30', 124°30' and 303°15', 181°0' and 1°0', and 289°30 and 108°45'. Correct for local attractions and check for the internal angles of the traverse, ABCD.
45. The fore and back bearings of AB, BC, CD and DA were observed to be respectively: S45°30'E and N45°30'W, S60°0'E and N60° 40' W, S5°30' E and N 3°20' W, and N 83°30' W and S85°0'E. Correct for local attractions and check for the internal angles of the traverse, ABCD.
46. The lengths and bearings (fore and back) of the lines of a traverse ABCD, as measured by a tape and a prismatic compass are respectively: AB = 454.9m, 0°30', 180°00', BC = 527.3 m, 270°00', 89°30', CD = 681.0 m, 205°30', 25°00', DA = 831.2 m, 78°30', 259°00. Make necessary corrections for local attractions. Plot the traverse on a suitable scale and make necessary adjustment after Bowditch.
47. Find the azimuth of a line PQ, if its magnetic bearing is 143°12' (given, magnetic declination = 5°51'E).
48. The following consecutive readings were taken with a dumpy level at 30 m interval along a line XY: 3.865, 3.762, 3.453, 2.987, 1.789, 0.854, 3.790, 3.561, 3.112, 2.569, 1.762, 1.111, 1.023, 0.986, 0.855. The level was shifted after 5th and 11th readings. RL of the first station = 32.000 m. Enter the readings on a neatly drawn field book, find the RLs of all the stations and check. Find the gradients of the sections between the CPs. Draw the long profile of XY on a suitable scale and interpret.
49. The following consecutive readings were taken with a dumpy level at 20 m interval along a line AB (N 48°30' E): 4.386, 3.962, 3.853, 2.987, 1.709, 1.012, 3.495, 3.656, 3.512, 2.509, 2.076, 1.811, 1.623, 0.986, 0.822. The level was shifted after 4th and 9th Stations. RL of the first CP = 10.000 m. Enter the reading on a neatly drawn field book, find the RLs of all the stations and check. Find the gradients of the sections. Draw the long profile of AB on a suitable scale and interpret.
50. The following consecutive readings were taken with a dumpy level at 10 m interval with a 4 m staff along a line PQ (248°45'): 0.987, 1.962, 2.853, 2.987, 3.709, 3.902, 0.345, 0.985, 1.656, 1.987, 2.550, 3.076, 3.783, 1.162, 1.986, 2.822, 3.222, 3.789, 3.992. RL of the second CP = 5.000 m. Enter the readings on a neatly drawn field book, find the RLs of the Stations and check. Draw the long profile of PQ on a suitable scale and interpret.
51. The following are the internal angles of a closed traverse ABCDE: ∠A = 78°12'40", ∠B = 168°33'50", ∠C = 84°22'30", ∠D = 115°43'25", ∠E = 90°07'35". Find the magnetic bearings of the lines, given the magnetic bearing of AB = 30°32'.
52. The following are the internal angles measured with a Theodolite for the closed traverse ABCDE: ∠A = 96°50'15", ∠B = 75°03'50", ∠C = 159°05'30", ∠D = 128°33'25", ∠E = 80°17'30". Lengths

of the sides of the traverse are: AB = 70 m, BC = 81 m, CD = 43 m, DE = 48 m, and EA = 114 m. The magnetic bearing of AB, measured with a compass = 30°30'. Find i) the corrected angles and check, ii) magnetic bearings of all the lines, iii) corrected coordinates of the points, and iv) area of the traverse. Plot the traverse on a suitable scale using: i) distances and bearings, ii) latitudes and departures, iii) consecutive coordinates and iv) independent coordinates.

53. A tower has an elevation of 60° from a point due north of it and 45° from a point due south. If the two points are 200 m apart, find the height of the tower and distance from each point of observation.
54. A 4 m staff casts a 1.75 m shadow on a flat ground. What is the elevation of the sun?
55. At a point A, a student observes the elevation of the top of a tower B to be 42°15'. He walks 10 yards up a uniform slope of elevation 12° directly towards the tower, and then finds the elevation of B has increased by 23°09'. Calculate the height of B above A.
56. At two points, 500 m apart on a horizontal plane, observations of the bearing and elevation of an aeroplane are taken simultaneously. At one point the bearing is 41° and the elevation 24°, and at another point the respective readings are 32° and 16°. Find the height of the aeroplane above the plane.
57. Three survey stations X, Y and Z lie in one straight line on the same plane. A series of angles of elevation is taken to the top of a hill that lies to one side of the line XYZ. The angles measured at X, Y and Z are respectively 14°02', 26°34' and 18°26'. The lengths XY = 400 ft and YZ = 250 ft. Find the height of the hill top above X.
58. The altitude of a mountain, observed at the end A of a baseline AB f 2992.5 m, was 19°42' and the horizontal angles at A and B were 127°54' and 33°09'. Find the height of the mountain.
59. An observer on the top of a cliff 20 m a.s.l. observes the angles of depression of two ships at anchor to be 60° and 30° respectively. Find the distance between the ships if the lines joining them point to the cliff base.
60. Two boys are on the opposite sides of a tall tower. The measures of angles of elevation of the top of the tower are 30° and 45°. If the height of the tower is 20 m, find the distance between the boys.
61. A man on the deck of a ship is 10 m above water level. He observes that the angle of the top of a cliff is 38°54' and the angle of the depression of the cliff base is 18°09'. Calculate the distance of the cliff from the ship and the height of the cliff.
62. A tower is 250 ft high. Its shadow is x m shorter when the sun's altitude is 52°54' than when it is 32°12'. Find x.
63. P and Q are two points 500 m apart on a river bank. A temple is situated on the other bank at R, which is directly opposite to P. From P the angle of elevation of the top of the temple is 45°54' and from Q it is 32°3'. Find the height of the temple.
64. Two stations P and Q are situated on the opposite sides of a base line XY (length = 1234 m and direction = 27°54') such that PX = 889.7 m, ∠PXY = 66°30', QX = 1120 m and ∠QXY = 47°18'. Find the rectangular coordinates of P and Q w.r.t. X and also the distance and bearing of PQ.

Unit II

Statistical Analysis

4

Statistical Methods

HIGHLIGHTS

- Some Basic Concepts
- Frequency Distribution
- Cumulative Frequency Distribution
- Graphical Representation of Frequency Distribution
- Description of Data
- Central Tendency
- Dispersion
- Shape of Frequency Curve
- Relationships
- Regression Equation
- Standard Error of Estimate
- Coefficient of Determination
- Correlation Coefficient
- Test of Significance
- Time Series Analysis

In common usage, the term *statistics* usually applies to facts and figures. Technically, it refers to a *branch of applied mathematics* concerned with the science, or perhaps the art, of the collection, presentation, description, analysis, inference, significance, testing and prediction of numerical information. Although the variety of such techniques is almost infinite, they may be broadly divided into seven categories—*univariate*, *bivariate*, *multivariate*, *time-series*, *directional*, *network* and *spatial* (Fig. 4.1).

The *univariate* analysis concerns a single

Fig. 4.1 Techniques of Geographical Data Analysis

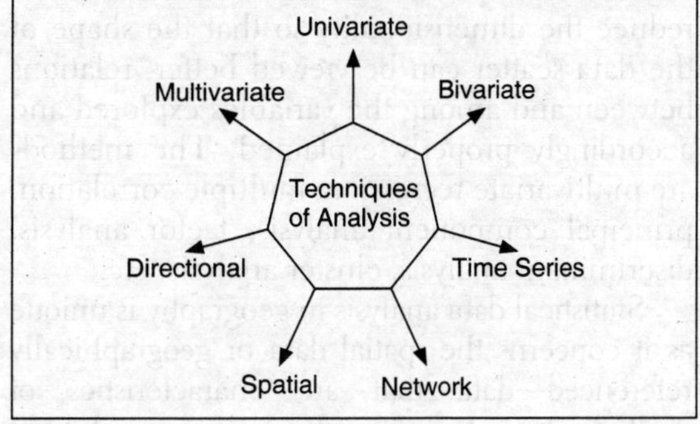

variable and allows the distribution of points along the line to be described and statistically tested (data description: frequency distribution, range, average, spread and shape). In *bivariate* analysis, two variables are analysed together to describe and analyse the shape of the scatter for the purpose of investigating the relationship between the data points and/or the relationship between the variables (bivariate correlation coefficient, regression: linear, polynomial, power, logarithmic and exponential). In *time-series* analysis, the sequence of data in time is explored for orders.

Directional analysis concerns data expressed in terms of angles or azimuth or bearings from north that can be ordinated on a circle. Such data are of two types—directional and oriented, data that can be analysed with measures of central tendency and randomness. *Network* analysis involves the evaluation of parameters like, structure connectivity, optimisation and shortest path analysis etc.

In *spatial* analysis more than two variables are analysed together, two (or three) of which are spatial coordinates: grid references or latitude/longitudes, with or without altitude or depth. The remaining variable is a measurement of geographical interest and is regarded as varying continuously over the space. The data may be imagined as points in three dimensions and are analysed with the objective of constructing a smooth surface to describe the spatial variation.

The *multivariate* analysis includes the general methods applicable to any number of variables analysed simultaneously and is usually applied to more than three variables. The objective is to reduce the dimensionality so that the shape of the data scatter can be viewed better, relations between and among the variables explored and accordingly properly explained. The methods are multivariate regression, multiple correlation, principal component analysis, factor analysis, discriminant analysis, cluster analysis, etc.

Statistical data analysis in geography is unique as it concerns the spatial data or geographically referenced data that are characteristics of multivariate situations. It has two distinct components—locations and attributes at locations. In the geographer's data matrix (GDM), a column represents the variations of attributes of the natural or socio-economic characteristics across some geographical spaces (Fig. 4.2). A row, on the other hand, denotes a specific location in geographic space. Therefore, each cell formed by the rows and columns of the GDM contains a specific item of geographic fact that can be found at a particular location. In a GDM, the columns can be compared to study the nature of spatial variation of the geographic characteristics. By comparing the rows of a GDM, we study the differences among different places, an aspect known in geography as areal differentiation.

Fig. 4.2 Structure of a GDM

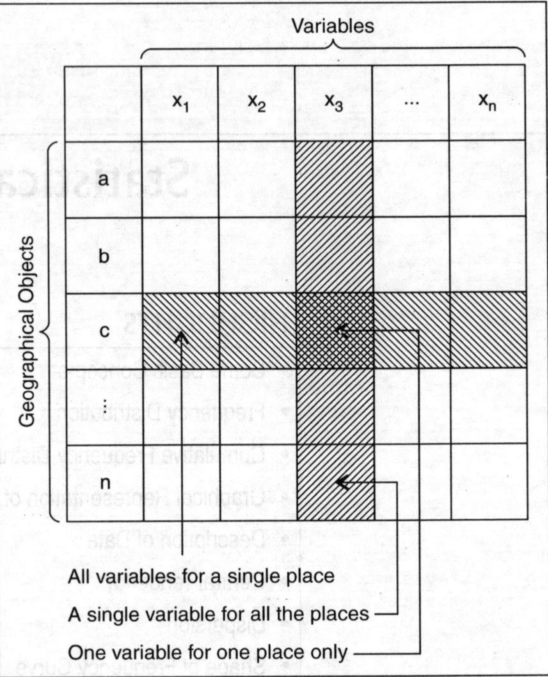

SOME BASIC CONCEPTS

Data

The term *data* means a body of information in numerical form. A set of data arranged in a tabular form is normally referred to as a data matrix.

Raw Data

It refers to the *unclassified* data from which a frequency distribution may be prepared for further analysis (Table 4.1). The arrangement of raw data is unsystematic and irrespective of any order.

Table 4.1 Raw Dataset: Population Density of 125 Districts of a Country, 2001 (person/sq km)

840	671	524	671	687
450	355	320	374	526
422	878	486	436	354
671	392	498	610	496
486	492	317	374	486
436	671	486	452	608
826	311	325	767	354
448	614	538	362	782
110	381	328	494	463
662	752	576	392	448
138	333	330	455	392
536	307	780	751	494
162	381	428	288	486
934	576	612	582	642
195	303	292	436	448
458	470	464	881	597
740	264	655	273	461
472	764	256	346	496
486	260	576	587	486
754	456	242	231	880
436	202	765	448	448
562	588	221	581	346
594	217	486	470	594
416	337	470	229	337
982	555	412	464	732

The Array

It is defined by an arrangement of data in which all items are placed *sequentially* in order of magnitude (Table 4.2). The array helps one to see at a glance the range of values, the nature of their concentration and the degree of continuity of the data. It also gives a rough idea of the distribution.

Table 4.2 Array: Population Density of 125 Districts of a Country, 2001 (person/sq km)

110	337	448	494	642
138	337	448	494	655
162	346	448	496	662
195	346	450	496	671
202	354	452	498	671
217	354	455	524	671
221	355	456	526	671
229	362	458	536	687
231	374	461	538	732
242	374	463	555	740
256	381	464	562	751
260	381	464	576	752
264	392	470	576	754
273	392	470	576	764
288	392	470	581	765
292	412	472	582	767
303	416	486	587	780
307	422	486	588	782
311	428	486	594	826
317	436	486	594	840
320	436	486	597	878
325	436	486	608	880
328	436	486	610	881
330	448	486	612	934
333	448	492	614	982

Variables

The items about which information have been collected are usually referred to as *individuals*. In the data matrix, each column gives the value of a specific property or characteristic that varies from one individual to the other. These are called *variables*. The variables that can take fractional values are called *continuous variables*, e.g., height, length, etc. On the other hand, the variables that can take only whole numbers and not fractions are called *discrete* variables. Commonly these concern count data, e.g., population, frequency, etc.

Class

In organising and summarising the statistical data, a frequency distribution is normally prepared. In this, the whole range of data is divided into some groups defined by intervals and the number of observations falling in each group is stated. The groups are called *class intervals* or simply *classes* and the observations are called *frequencies*. The classes are defined by limits or boundaries on two ends. When one end of the class is not specified, it is called an *open-end class*. Normally in a distribution, all classes are of the same size but in some cases, classes of *unequal* sizes may also be taken, particularly for a *highly dispersed* data.

Class Limits

In a grouped frequency distribution, the classes are defined by pairs of numbers such that the upper end of one class does not coincide with the lower end of the following class. These numbers are called class limits which signify the limits of a class for the purpose of tallying with the original distribution. The smaller value of the pair is called *lower class limit* (LCL), while the larger value is called *upper class limit* (UCL).

Class Boundaries

These are defined as the most extreme values that would ever be included in a class. In fact, the class boundaries are the real class limits. The lower extreme value is called the *lower class boundary* (LCB), while the upper extreme value is called the *upper class boundary* (UCB). For all practical purposes of graphical representation of data, class boundaries are used. The working formula is:

$$\text{Lower Class Boundary} = \left[\text{LCL} - \frac{d}{2}\right] \text{ and}$$

$$\text{Upper Class Boundary} = \left[\text{UCL} + \frac{d}{2}\right]$$

where,

d = Difference between the two successive class limits

Class Width

Width or size of a class is the difference between the lower and upper class boundaries. It is expressed as,

Class Width (w) = (UCB – LCB)

Class Mark

Class mark or mid-value of a class refers to the value that lies exactly at the middle of a class. It is used as the representative value of a class for the purpose of calculation of descriptive measures. The common formula is:

$$\text{Class Mark (x)} = \frac{1}{2}(\text{LCL} + \text{UCL})$$

$$\text{or,} \quad = \frac{1}{2}(\text{LCB} + \text{UCB})$$

Class Frequency

The number of observations contained in a class is known as its *class frequency*. The sum of all class frequencies in a distribution is called the *total frequency*. Classes with zero frequencies are called *empty classes*. Therefore,

$$\text{Total Frequency, } N = \sum_{i=1}^{n} f_i$$

where,

n = Number of class and

f_i = Frequency of the ith class

Relative Frequency

Relative frequency is simply the class frequency (f_i) expressed as a proportion of the total frequency (N) of a distribution. The sum of all relative frequencies in a distribution is equal to unity. In expression form,

$$\text{Relative Frequency, } Rf_i = \left(\frac{f_i}{N}\right) \text{ and } \sum_{i=1}^{n} Rf_i = 1$$

where, n = Number of class.

Frequency Density

Frequency density of a class is defined as its frequency per unit width. It indicates the concentration of frequency in a class. The common expression is:

$$\text{Frequency Density } (f_d) = \frac{\text{Frequency (f)}}{\text{class width (w)}}$$

Description

For ordinary as well as for spatial dataset, *description* means the specific and concise measures of their statistical characteristics. Thus, it implies summarising the nature of a large set of data.

Inference

In most cases, analysis is done with data obtained from samples rather than with all the possible data about a particular situation. Hence, it is assumed that the sample is representative of the whole set of data (or the *population*) from which it has been drawn. The *inferential measures* can help, within certain strictly defined limits, to make statements about the characteristics of a population based only on the sample data.

Significance

An observed difference or relationship between two sets of sample data can be tested for significance: *whether there is a difference within the population from which the samples were drawn or whether the observed differences in the samples appear merely due to chance in the sampling procedure*. Statistical measures can be pursued to find the probability that under certain specified conditions, a *relationship is significant* or that the inferences made on the basis of the samples are valid.

Prediction

This means statements about something in the past or in the future from the analysis of a set of period data. Statistical measures are there to do this within certain limits and with certain probabilities. In situations of deterministic processes, predictions and postdictions can be done with absolute precision. But in multivariate situations, the degree of certainty is less and only some sort of probabilistic prediction is possible.

Measurement Scale

Based on simple scalar systems, Stevens (1959) describes *four* kinds of measurement scales: *nominal*, *ordinal*, *interval* and *ratio*. *Nominal* scaling simply provides a device for labelling or classifying objects rather than measuring the attributes of objects. Nominal data involve information that is purely qualitative as it is presented in the form of names. An *ordinal* scale is one in which a set of events or objects can be ordered from 'most' to 'least' but in which there is no information regarding the actual amount of the measured attributes that separates the objects or events. Although data is expressed in terms of positive integers, these are not valid for any algebraic operations. In *interval* scaling one cannot only rank order objects with respect to a measured attribute but is also able to specify how far apart the magnitudes are from each other. It can be expressed in the real number system but it differs from the ratio scale in that the zero point is not a fundamental termination of the scale. The classic example is temperature measured in °C and °F: the zero points on these scales, though well defined, are not as singular as zero length or zero weight.

The *ratio scale* is the most precise as it uses the real number system that starts from an absolute zero and allows continuous measurement. Units can be converted directly from one system to another and data can be distinguished in various ways. *Absolute* ratio data are the result of direct measurements or additions of units while *relative* ratio data concerns some kind of derived values based on a well defined index. Ratio data may be *continuous*, when the variable can have all values in between with decimals (e.g., elevation, income, etc.) and may be *discrete* when the variable can take only fixed and discrete values (e.g., family size, population density, frequency data, etc.). There may as well be *closed data* (whose attribute values range between two fixed values, e.g., %, etc.) and open data with no such limiting values. Ratio data may be *point data*, concerning attributes at a particular location only, and *period data*, concerning attribute data within a time frame only.

Test Statistic

It refers to a *single value* that provides not only a description of the sample situation but also enables the evaluation of the significance of a relationship between samples or between variables within a particular sample.

Degree of Freedom

It is a number that represents in some way the size of the sample or samples, involved in the test. In some cases, it is simply the sample size while in others it is a value calculated following a definite logic. Hence, it depends upon the particular test concerned.

FREQUENCY DISTRIBUTION

Frequency distribution is a statistical table that shows the values of the variables and the corresponding number of observations or frequencies. Frequency distribution may be of two types—*simple* and *grouped*. In simple frequency distribution, the values of the variables are arranged individually in order of magnitude (Table 4.2). In grouped frequency distribution, the values of the variables are arranged in groups in order of magnitude (Table 4.3). Frequency distribution is generally used to:

i. condense and summarise a huge body of data,
ii. visualise the range, average, spread and shape of the distribution and
iii. approximate the nature of probability distribution in the population.

Construction of Grouped Distribution

This needs a very careful consideration of the nature of the variable and the data. The following steps are carried out in sequential order:

- The total number of classes, into which the frequency distribution is to be divided, is first selected. Actually there is no rigid rule about this. The distribution may appear irregular and meaningless if too many or even too few classes are taken. Basically, it depends upon the nature of data and the objective of the researcher. Normally 6 to 12 classes are taken. In case of a series with smaller numbers of observations, Sturges' formula may be used to solve the problem:

  $$n = 1 + 3.3 \log N$$

  where, n = Number of classes to be selected, and N = The total frequency.
- The *size of the classes* or the *class width* (w) should then be very carefully fixed. It depends upon the nature of data, variables, range and the objective of the researcher. Normally in a distribution, classes of equal size or width are taken. The formula for width determination is,

  $$w = \frac{R}{n}$$

  where, R = Range of data, i.e., the difference between the highest and lowest values of the variable in the series and

  n = Number of classes already fixed.

  From Table 4.2,

  Maximum value = 982

  Minimum value = 110

Fig. 4.3 Cumulative Frequency Distribution

Class Frequency (f_i)		4		12		24		40		16		12		10		5		2		Σf_i=125
Class Boundary	100.5		200.5		300.5		400.5		500.5		600.5		700.5		800.5		900.5		1000.5	
Cumulative Frequency: Less than	0		4		16		40		80		96		108		118		123		125	
Cumulative Frequency: More than	125		121		109		85		45		29		17		7		2		0	

Range = 872

For a 9-class distribution, $w = \frac{872}{9} \Rightarrow 96.8$

It should be noted here that if the value of the range is approximated to its nearest round figure easily divisible by n, class width becomes more comprehensive and easily perceivable. For the above problem, the highest value of 982 may be written as 1,000 and the lowest value of 110 may be written as 100. Therefore, range 1000 –100 = 900 and for a 9-class distribution, $w = \frac{900}{9} \Rightarrow 100$

This kind of approximation may also be done from the value of w, as calculated. A class width of 96.8, can easily be rounded to 100 for practical purposes. Class width is very important because it influences the selection of class limits, class boundaries and class marks which are used not only in charting the distribution but also in calculating the descriptive measures.

- Another important point is that in some special cases there may occur a number of empty classes. This is because the spread of data is much larger and there is marked concentration of values at certain levels. Hence, the distribution is very skewed and the numerical differences between any two values far exceeds the class width chosen. For such a series of data (e.g., size distribution of town or city), usually unequal classes are taken. The width of the larger classes (w_i) is fixed in terms of the size of the initial class (w) chosen, such that, $w_i = x.w$, where x = a positive integer, i.e., 2, 3, 4, etc.
- The next step is to neatly draw a frequency distribution table with (n+1) rows and 6 columns (for equal class width) or 7 columns (for unequal class width). From left to right, the column heads are—*class limits, class boundaries, class marks* (x), *class width* (w), *tally marks* and *frequency* (f_i). In case of unequal classes an additional column with the heading *frequency density* $\left(f_i/w_i\right)$ is drawn.
- From the raw dataset, values are neatly entered in the table. The total frequency of the distribution ($\sum f_i$ or N) is noted at the end of the *frequency column*. A suitable title for the distribution is then added for the table drawn.

CUMULATIVE FREQUENCY DISTRIBUTION

The data of Table 4.3 shows the non-cumulative form of frequency distribution from which the number of districts falling in each class may be ascertained easily. Sometimes, however, it may be useful to know how many or what proportion of districts less than or more than certain specified population density levels exist. For this, frequencies below or above a certain class boundary need to be calculated. These accumulated frequencies above or below a certain class boundary are known as *cumulative frequencies*.

From Fig. 4.3, it is found that 16 districts lie below the density level of 300.5. The *less than cumulative frequency* is 16 and the *more than cumulative frequency* is 109. Thus cumulative frequency distribution may be defined as a statistical table in which the cumulative frequencies below or above the class boundaries are arranged in ascending or descending order (Table 4.4). Sometimes instead of absolute frequencies, percentage frequencies are used in cumulative frequency distribution. These are then called *Cumulative Percentage Distribution* or *Relative Cumulative Frequency Distribution*.

Construction of Cumulative Distribution

From a grouped frequency distribution, cumulative frequency distribution can easily be constructed according to the following steps:

- A table with (n + 2) rows and 7 columns should be neatly drawn, where n = the number of classes. From left to right, the column heads are: *class limit, class boundary, frequency, boundary point, less than cumulative frequency, boundary point* and *more than cumulative frequency* (Table 4.4).
- Non-cumulative frequency distribution is first entered in column 1 to column 3. In column 4 and column 6, all the boundary points of the frequency distribution are entered in ascending order of magnitude.
- Corresponding to each of these boundary points, the less-than and the more-than cumulative frequencies are calculated and accordingly entered in column 5 and column 7 of the table. It should be noted here that the less-than cumulative frequencies corresponding to the highest class boundary and the more-than cumulative frequencies corresponding to the lowest class boundary must be equal to each other and also equal to the total frequency. Again, the sum of the less-than and the more-than cumulative frequencies corresponding to any boundary point is equal to the total frequency of the distribution.

GRAPHICAL REPRESENTATION OF FREQUENCY DISTRIBUTION

The most common diagrams or graphs representing the frequency distributions are:

Table 4.3 Frequency Distribution Table

Class Limit	Class Boundary	Class Width (w)	Class Mark (x)	Tally Marks	Frequency (f_i)
101–200	100.5–200.5	100	150.5	||||	4
201–300	200.5–300.5	100	250.5	卌 卌 ||	12
301–400	300.5–400.5	100	350.5	卌 卌 卌 卌 ||||	24
401–500	400.5–500.5	100	450.5	卌 卌 卌 卌 卌 卌 卌 卌	40
501–600	500.5–600.5	100	550.5	卌 卌 卌 |	16
601–700	600.5–700.5	100	650.5	卌 卌 ||	12
701–800	700.5–800.5	100	750.5	卌 卌	10
801–900	800.5–900.5	100	850.5	卌	5
901–1000	900.5–1000.5	100	950.5	||	2
					$\Sigma f_i = 125$

Display of Frequency Distributions

Univariate Distributions

- All are area diagrams. In a histogram, the frequencies are plotted against class width, and in a polygon or curve, the frequencies are plotted against class marks).
- In a histogram, the area of each rectangle is directly proportional to its class frequency.
- The area enclosed by the frequency polygon is equal to the area enclosed by the histogram.
- The area enclosed by the frequency curve is equal to the area enclosed by the histogram.
- All measures of data description (excepting fractiles) are shown in frequency curves.

Cumulative Frequency Distributions

- The fractiles are plotted in Ogives.
- The range, median and quartiles may also be shown by a Box-and-Leaf diagram or a Dispersion diagram.

Bivariate Distributions

- The relation between variables are shown by scatter plots/graphs.
- The regression equation is plotted to show the best-fit line of estimation.
- The standard error of estimate is plotted to show the dependability of estimates.
- The coefficient of determination is noted to show the goodness of fit.

Distribution	Diagram/Graph
A. Non-cumulative	1. Histogram
	2. Frequency Polygon
	3. Frequency Curve
B. Cumulative	1. Cumulative Frequency Polygons
	2. Cumulative Frequency Curves

Histogram

A *histogram* is the most common form of diagrammatic representation of a grouped frequency distribution. It is an *area diagram* in which class frequencies (f_i) are shown by vertical rectangles, the widths of which are described by the class widths (w_i) defined by the class boundaries. Normally, class boundaries are taken on the abscissa or x-axis and frequencies are taken on the ordinate or y-axis. As the class boundaries are taken to plot the rectangles, these become contiguous to each other (Fig. 4.4).

The basic principle of a histogram is that the area of each rectangle is directly proportional to the class frequency (f_i). Therefore,

$$\textit{Area of a rectangle} \propto f_i$$

or, $$h_i \times w_i = k.f_i$$

where,

Table 4.4 Cumulative Frequency Distribution Table

Class Limit	Class Boundary	Frequency (f_i)	Cumulative Frequency (F)			
			Less than	F	More than	F
101–200	100.5–200.5	4	100.5	0	100.5	125
201–300	200.5–300.5	12	200.5	4	200.5	121
301–400	300.5–400.5	24	300.5	16	300.5	109
401–500	400.5–500.5	40	400.5	40	400.5	85
501–600	500.5–600.5	16	500.5	80	500.5	45
601–700	600.5–700.5	12	600.5	96	600.5	29
701–800	700.5–800.5	10	700.5	108	700.5	17
801–800	800.5–900.5	5	800.5	118	800.5	7
901–1000	900.5–1000.5	2	900.5	123	900.5	2
		Σf_i=125	1000.5	125	1000.5	0

Fig. 4.4 A Histogram

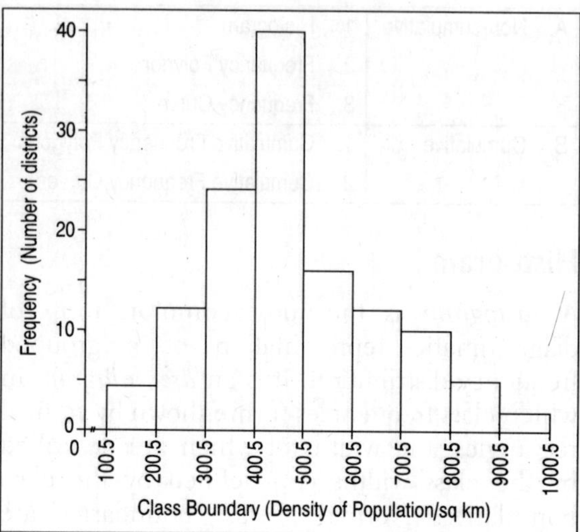

h_i and w_i are respectively the height and the width of the rectangle for the ith class and k = constant of proportionality.

$$\therefore \quad h_1 = \frac{k}{w_i} f_i \qquad \ldots(i)$$

This has two important implications as

a. For a grouped frequency distribution, where all the classes are of the same size (w), the equation (i) becomes,

$$h_i = \left(\frac{k}{w}\right) f_i$$

$$\text{or,} \quad h_i = k_1 f_i \left[\therefore k_1 = \frac{k}{w} = \text{constant}\right]$$

$$\therefore \quad h_i \propto f_i$$

Thus, the height of a rectangle is proportional to the class frequency when all the classes in the distribution are of the same size.

b. For a grouped frequency distribution, where class widths (w_i) vary, the equation (i) becomes,

$$h_i = k\left(\frac{f_i}{w_i}\right)$$

or, $h_i = k.fd_i$, where, fd_i = frequency density of the ith class

$$\therefore \ h_i \propto f.d_i$$

Thus, in case of a frequency distribution with classes of *unequal size*, the *height* of a rectangle becomes *proportional* to the *frequency density* (and not class frequency). And hence, the histogram in such cases must be plotted against the frequency densities (taken on the ordinate or y-axis) and the rectangles appear to be of varying widths.

A histogram gives a visual representation of the relative sizes of the various classes. It also gives an approximate idea of the nature of frequency distribution (i.e., its average, spread and shape). It is normally used for the graphical construction of mode.

Frequency Polygon

A *frequency polygon* is the graphical representation of a grouped frequency distribution. In this, the class frequencies (f_i) are plotted against the class marks (x_i) and the points thus obtained are joined by straight lines. The class marks and frequencies are taken on the abscissa and the ordinate respectively. The basic principles are:

i) The values in a class are evenly distributed throughout the class interval and hence the mid-value of the class (x_i) is taken as representing the class (i) and

ii) The area covered by a histogram is equal to the area included within the polygon. For this, the mid-value of the two empty classes on both ends of the distribution are taken and accordingly joined by straight lines (Fig. 4.5).

Hence, on the x-axis the frequency polygon produces a polygonal traverse whose area is exactly equal to that contained in a histogram. While doing this, it appears that there are items that occur beyond the limits of the observed data. Sometimes it may so happen that the polygon occurs beyond zero on the x-axis—an event which is certainly thematically meaningless but technically correct.

A frequency polygon may be plotted on a histogram as well as separately and individually. For a distribution with unequal classes, frequency density (fd_i) instead of simple frequency (f_i)

Fig. 4.5 A Frequency Polygon

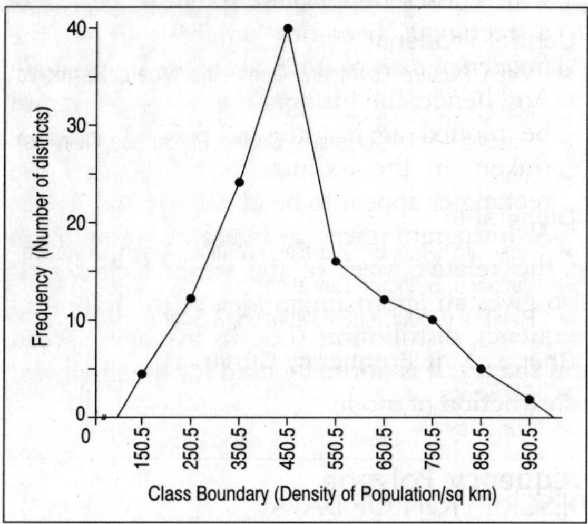

must be plotted against the class marks (x_i). A frequency polygon gives a better idea of the shape of the distribution.

Frequency Curve

A *frequency curve* is simply a modification of a frequency polygon in which the points plotted by class frequencies (f_i) or frequency densities (fd_i) corresponding to the class marks (x_i) are joined by a smooth curve (instead of a series of straight lines). The basic principles are the same as that of a frequency polygon, i.e., the class marks represent the class and the area covered by a histogram is equal to the area included within the frequency curve. However, the smoothening effect produces a better idea of the nature and characteristics of distribution (Fig. 4.6). The measures of central tendency and dispersion are plotted on it. Distributions may be compared more effectively by plotting frequency curves on the same base. It is also useful to determine the normality of a test.

Cumulative Frequency Polygons

A *cumulative frequency polygon* is the graphical representation of the cumulative frequency distribution. In this, the cumulative frequencies (taken on the y-axis) corresponding to the class boundaries (taken on the x-axis) are plotted and the successive points are joined by a straight line. Two separate polygons are normally plotted—the *less-than* type on the basis of the less-than cumulative frequency and the *more-than* type on the basis of the more-than cumulative frequency. The less-than polygon starts from the lowest class boundary on the abscissa and then gradually rises upwards to end at the highest class boundary corresponding to the total frequency (N). Similarly, the more-than polygon starts from the

Fig. 4.6 A Frequency Curve

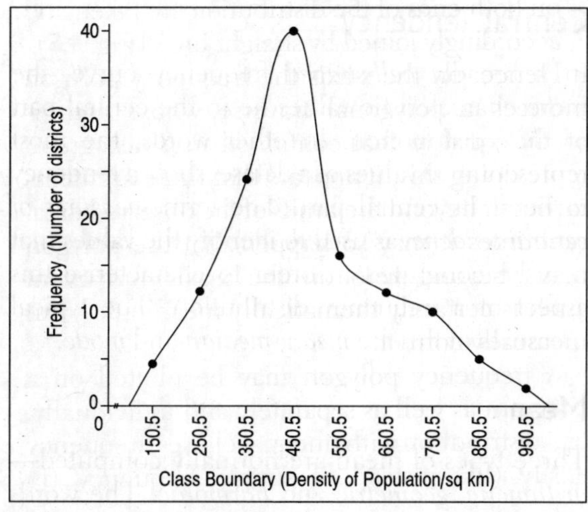

Fig. 4.7 Cumulative Frequency Polygons

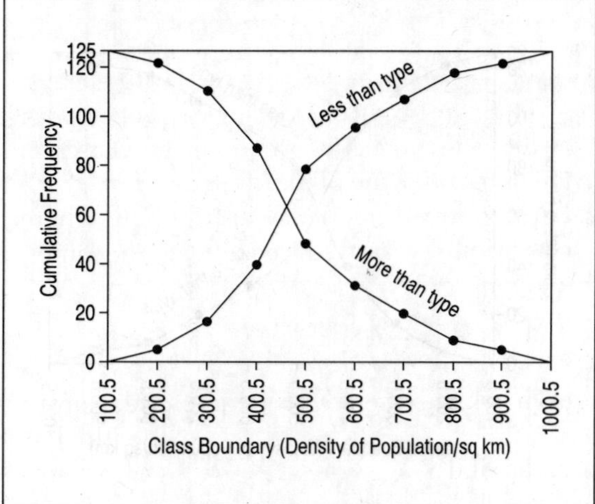

lowest class boundary corresponding to the total frequency (N) and then gradually descends to the highest class boundary on the abscissa. The less-than polygon approximates a broad and elongated S-shape while the more-than polygon is a broad, elongated and inverted S-shape (Fig. 4.7). The *two polygons intersect at the median point* of the distribution. The cumulative frequency polygons are used to find out cumulative frequencies above or below certain specified values. It also helps us to find the number of observations between any two specified values, such as the *fractile values* like median, quartiles, deciles, percentiles, etc., graphically and to compare the sample distribution based on the same values.

Cumulative Frequency Curves

These are similar to the cumulative frequency polygons except that the plotted points are *joined by smooth curves* instead of straight lines.

The less-than cumulative frequency curve and the more-than cumulative frequency curve when plotted together looks like a wine-glass. Thus they are also called *wine-glass curves*.

Both the cumulative frequency graphs (i.e., polygons and curves) are called *ogives* and are equally useful (Fig. 4.8).

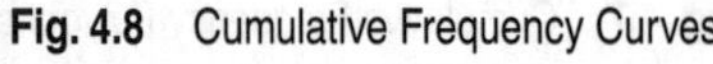

Fig. 4.8 Cumulative Frequency Curves

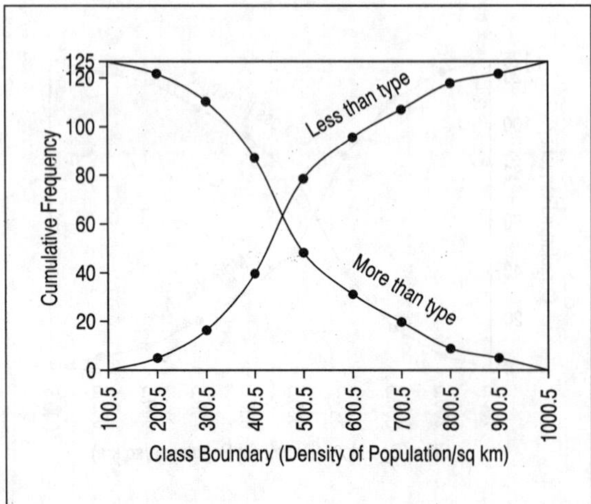

Measures of Data Description

Central Tendency
- Mean (Arithmetic Mean, Geometric Mean, Harmonic Mean)
- Fractiles (Median, Quartiles, Deciles, Percentiles, Quintiles)
- Mode

Dispersion
- Absolute (Range, Quartile Deviation, Mean Deviation, Variance, Standard Deviation)
- Relative (Coefficient of Variation, Z-Score)

Shape of the Frequency Curve
- Skewness
- Kurtosis

DESCRIPTION OF DATA

Average, Spread and Shape

One basic objective of statistics is to provide means for a *concise and consistent description* of a dataset. Concise in the sense that the representative characteristics of a dataset can be described by only a few numbers while consistent in the sense that measures of such concise description is universally applicable. A dataset is primarily described in terms of its three distinctive characteristics—*central tendency* (average), *dispersion* (spread) and *shape* (skewness and kurtosis) of the frequency distribution.

Central Tendency

In a broadly *bell-shaped* frequency curve, the more characteristic values lie in the central part of the distribution. In other words, the most representative values of a dataset show a tendency to lie at its central part. The term *measures of central tendency* is used to identify the values that may be computed in order to characterise this aspect of a frequency distribution. The typical measures of this are *mean, median* and *mode*.

Mean

Three types of mean are normally computed—*arithmetic, geometric* and *harmonic*. The words mean and average refer to the arithmetic mean. The remaining two types are for special use only.

Arithmetic Mean (AM)

For a set of observations, arithmetic mean is defined as their sum divided by the number of observations. Symbolically, arithmetic mean is:

$$\bar{x} = \frac{x_1 + x_2 + ... + x_n}{N}$$

$$= \frac{\sum_{i=1}^{n} x_i}{N}$$

where, N = Total number of observations and x_i = i^{th} value of x – variables.

Thus arithmetic mean ($\bar{x}$) is rigidly defined as it is based on all the observations and can be easily computed. It is readily understood and can be treated algebraically. It is a very stable and reliable average as regards the sampling fluctuations. However, it is highly affected by the presence of even a few unusually large or unusually small observations.

The total annual rainfall (cm) of a place for a period of 10 years is: 97, 100, 95, 85, 115, 112, 102, 106, 87 and 101 cm respectively. The arithmetic mean of this ungrouped series is given by,

$$\bar{x} = \frac{(97 + 100 + 95 + 85 + 115 + 112 + 102 + 106 + 87 + 101)}{10}$$

$$= \frac{1000}{10} \Rightarrow 100 \text{ cm}$$

One *important property* of arithmetic mean is that the algebraic sum of the deviations of the values from the mean is always equal to zero. In other words, $\Sigma(x - \bar{x}) = 0$. This is important because from a chosen or assumed mean value, such deviations can easily be obtained. Mean can now be readily and quickly evaluated saving much labour. Hence,

$$\bar{x} = A_m + \frac{\sum_{i=1}^{n} d_i}{N}$$

where, A_m = Assumed mean, d_i = Deviation of ith value from the assumed mean and N = Total number of observations (Table 4.5).

For a grouped frequency distribution, arithmetic mean is given by,

$$\bar{x} = \frac{\sum_{i=1}^{n} f_i x_i}{\sum_{i=1}^{n} f_i}$$

Table 4.5 Computation of Mean: Assumed Mean Method

X	A_m	$d_i = (x_i - A_m)$	
97		−3	
100		0	
95		−5	
85		−15	
115	100	+15	$\bar{X} = 100 + \frac{0}{100}$
112		+12	
102		+2	= 100
106		+6	
87		−13	
101		+1	
N = 10	$A_m = 100$	$\Sigma d_i = 0$	

Table 4.6 Computation of Arithmetic Mean ($\bar{X}$): Algebraic Method

Class Boundary	Class Marks (x_i)	Frequency (f_i)	$f_i x_i$
100.5–200.5	150.5	4	602
200.5–300.5	250.5	12	3006
300.5–400.5	350.5	24	8412
400.5–500.5	450.5	40	18020
500.5–600.5	550.5	16	8808
600.5–700.5	650.5	12	7806
700.5–800.5	750.5	10	7505
800.5–900.5	850.5	5	4252.5
900.5–1000.5	950.5	2	1901
		$\Sigma f_i = 125$	$\Sigma f_i x_i =$ 60312.5

where, x_i = Class mark of ith class, f_i = Frequency of ith class and n = Number of classes (Table 4.6).

$$\text{Mean, } \bar{x} = \frac{60312.5}{125} \Rightarrow 483 \text{ persons/sq km}$$

Arithmetic mean can also be computed with the help of an *assumed mean* using either the *short* method or the *coding method*. In the short method, the working formula is,

$$\bar{x} = A_m + \left(\sum_{i=1}^{n} f_i d_i \Big/ \sum_{i=1}^{n} f_i \right)$$

where, A_m = Assumed mean, f_i = Frequency of the ith class, d_i = Deviation of ith class mark from the assumed mean or $(x_i - A_m)$ and n = Number of classes (Table 4.7).

$$\bar{x} = 550.5 + \frac{-8500}{125} \Rightarrow 483 \text{ persons/sq km}$$

In the coding method, the deviations are coded in terms of class interval (d'). It is illustrated in Table 4.8. The working formula is,

$$\bar{x} = A_m + \left(\sum_{i=1}^{n} f_i d'_i \Big/ \sum f_i \right) \times w$$

where, A_m = Assumed mean, f_i = Frequency of the ith class, d'_i = Deviation of ith class as coded, w = Class interval and n = Number of classes.

$$\bar{x} = 550.5 + \frac{-85}{125} \times 100 \Rightarrow 483 \text{ persons/sq km}$$

Geometric Mean (GM)

The geometric mean of a set of *n* observations is defined as the nth root of the product of the items. It is defined only when all observations have the same sign and none is zero. For ungrouped series, geometric mean, $GM = \sqrt[n]{x_1 \times x_2 \times \cdots \times x_n}$

$$\text{or, } \log(GM) = \frac{\sum_{i=1}^{n} \log x_i}{N}$$

The geometric mean of the ten annual rainfall data is given by,

$$\text{Geometric Mean} = \sqrt[10]{(97 \times 100 \times 95 \times 85 \times 115 \times 112 \times 106 \times 102 \times 87 \times 101)}$$

$$= \sqrt[10]{9.58466 \times 10^{19}} \Rightarrow 99.57 \text{ cm}$$

In other words, log (GM)

$$= \frac{(\log 97 + \log 100 + \log 95 + \log 85 + \log 115 + \log 112 + \log 102 + \log 106 + \log 87 + \log 101)}{10}$$

Table 4.7 Computation of Arithmetic Mean ($\bar{x}$): Short Method

Class Boundary	Class Marks (x)	Frequency (f_i)	Deviation from A_m(d_i)	$f_i.d_i$
100.5–200.5	150.5	4	– 400	– 1600
200.5–300.5	250.5	12	– 300	– 3600
300.5–400.5	350.5	24	– 200	– 4800
400.5–500.5	450.5	40	– 100	– 4000
500.5–600.5	550.5 A_m	16	0	0
600.5–700.5	650.5	12	+ 10	+ 1200
700.5–800.5	750.5	10	+ 200	+ 2000
800.5–900.5	850.5	5	+ 300	+ 1500
900.5–1000.5	950.5	2	+ 400	+ 800
		Σf_i = 125		– 8500

Table 4.8 Computation of Arithmetic Mean ($\bar{x}$): Coding Method

Class Boundary	Class Marks (x)	Frequency (f_i)	Deviation in terms class interval (d'_i)	$f_i.d'_i$
100.5–200.5	150.5	4	– 4	– 16
200.5–300.5	250.5	12	– 3	– 36
300.5–400.5	350.5	24	– 2	– 48
400.5–500.5	450.5	40	– 1	– 40
500.5–600.5	550.5 A_m	16	0	0
600.5–700.5	650.5	12	+ 1	+ 12
700.5–800.5	750.5	10	+ 2	+ 20
800.5–900.5	850.5	5	+ 3	+ 15
900.5–1000.5	950.5	2	+ 4	+ 8
		$\Sigma f_i = 125$		– 85

$$= \frac{19.98157}{10} \Rightarrow 1.998157$$

$\therefore$ Geometric Mean $= \text{Antilog}\ (1.998157)$
$= 99.57$ cm

In case of *grouped* frequency distribution, the *geometric mean* (GM) $= \sqrt[N]{x_1^{f_1} \times x_2^{f_2} \times \cdots \times x_n^{f_n}}$
or in other words,

$$\log\ (\text{GM}), \quad = \frac{f_1 \log x_1 + f_2 \log x_2 + \ldots + f_n \log x_n}{\Sigma f_i}$$

$$= \sum_{i=1}^{n} f_i \log x_i \Big/ \sum_{i=1}^{n} f_i$$

therefore,

$$\log\ (\text{GM}) \quad = \frac{331.68451}{125} \Rightarrow 2.65347$$

$\therefore$ Geometric Mean= Antilog (2.65347)
= 450 persons/sq km

If a series is symmetrical in the logarithmic sense and the items are evenly distributed within

Table 4.9 Computation of Geometric Mean

Class Boundary	Class Marks (x_i)	Frequency (f_i)	log x_i	f_i.log x_i
100.5–200.5	150.5	4	2.17753	8.71012
200.5–300.5	250.5	12	2.39880	28.78560
300.5–400.5	350.5	24	2.54469	61.07256
400.5–500.5	450.5	40	2.65369	106.14760
500.5–600.5	550.5	10	2.74075	43.85200
600.5–700.5	650.5	12	2.81324	33.75888
700.5–800.5	750.5	10	2.87535	28.75350
800.5–900.5	850.5	5	2.92967	14.64835
900.5–1000.5	950.5	2	2.97795	5.95590
		$\Sigma f_i = 125$		Σ331.68451

the classes geometrically, it is preferable to *use the mid-values of the logarithms of the class limits* rather than that of the class marks. Any series of numbers having the same n and the same Σx have the same *arithmetic mean*. Again any series of numbers having the same n and the same *product* have the same geometric mean. The *product of the ratios* of the values on one side of the geometric mean (GM) to the geometric mean is equal *to the product of the ratios* of the geometric mean to the values on the other side of it. Geometric mean is *useful for averaging* rates, ratios and percentage of change.

Harmonic Mean (HM)

The harmonic mean is defined as the reciprocal of the arithmetic mean of the reciprocals of the values. The algebraic expression is:

$$\text{Harmonic Mean, (HM)} = \frac{1}{(1/x_1 + 1/x_2 + \ldots 1/x_n)/N}$$

$$= \frac{N}{\sum_{i=1}^{n}\left(\frac{1}{x_i}\right)}$$

For the purpose of easy calculation, $\frac{1}{HM}$ is first calculated and then the harmonic mean is found out by taking its reciprocal.

For the above set of rainfall data, HM is

$$\frac{1}{HM} = \left(\tfrac{1}{97} + \tfrac{1}{100} + \tfrac{1}{95} + \tfrac{1}{85} + \tfrac{1}{115} + \tfrac{1}{112} + \tfrac{1}{102} + \tfrac{1}{106} + \tfrac{1}{87} + \tfrac{1}{101}\right) \Big/ 10$$

$$= \frac{0.1008576}{10} \Rightarrow 0.01008576$$

$$\therefore HM = \frac{1}{0.1008576} \Rightarrow 99.15 \text{ cm}$$

In case of *grouped* frequency distribution,

$$\textit{Harmonic Mean} = \sum_{i=1}^{n} f_i \Big/ \sum_{i=1}^{n}\left(\frac{f_i}{x_i}\right)$$

Table 4.10 Computation of Harmonic Mean

Class Boundary	Class Marks (x_i)	Frequency (f_i)	$\frac{f_i}{x_i}$
100.5–200.5	150.5	4	0.026578
200.5–300.5	250.5	12	0.047904
300.5–400.5	350.5	24	0.068474
400.5–500.5	450.5	40	0.088790
500.5–600.5	550.5	16	0.029064
600.5–700.5	650.5	12	0.018447
700.5–800.5	750.5	10	0.013324
800.5–900.5	850.5	5	0.005879
900.5–1000.5	950.5	2	0.002104
		$\Sigma f_i = 125$	$\Sigma = 0.300565$

$$\therefore \text{Harmonic Mean} = \frac{125}{0.300565} \text{ (Table 4.10)}$$

$$= 416 \text{ persons/sq km.}$$

Harmonic mean is *advantageous* when it is necessary to give greater weight to smaller observations and less weight to the larger. Harmonic mean is *useful* when data are given in terms of time, rate, speed and price. For any given set of observations,

$$AM \geq GM \geq HM$$

Median

Median is defined as that value which divides the distribution exactly into two equal halves. This means that an equal number of items lie on either side of it. It is the *real measure* of the central tendency in that it occupies the exact central position of a distribution. It is calculated as follows:

From Ungrouped Series

The given data are arranged in order of magnitude. The middle most value in the array is the *median*. In case of an even number of data the arithmetic mean of the two middle most values are taken as the median.

For the rainfall data, median can be calculated as:
Rainfall (cm):
85, 87, 95, 97, 100, ↓ 101, 102, 106, 112, 115
Median

$$\text{Even number of data: Median} = \frac{100+101}{2} = 100.5 \text{ cm.}$$

From Grouped Data: Graphical Method

Median can also be estimated easily from the cumulative frequency graphs. The *point of intersection* of the less-than type and more-than type cumulative frequency curves or polygons corresponds to the median point (N/2) on the ordinate. From this point, a *perpendicular* is dropped on the abscissa and the value is then interpolated. However, if only one type of cumulative frequency curve or polygon is given, a horizontal line is first drawn from the point N/2 on the ordinate and then the perpendicular on the abscissa is drawn from the intersection of the horizontal line with the cumulative frequency curve or polygon (Fig. 4.9).

Fig. 4.9 Graphical Construction of Median

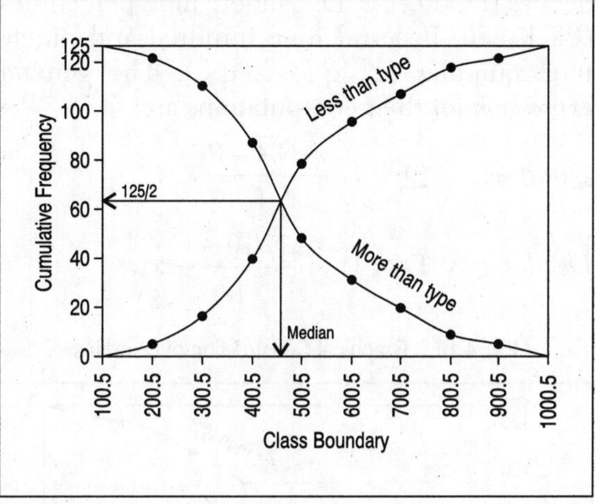

From Grouped Data: Using Formula

In a cumulative frequency distribution table, the median class is first identified by inspecting the position of N/2 in the less-than cumulative frequency column. The working formula is:

$$\text{Median} = L + \frac{\frac{N}{2} - F}{f} \times w$$

where, L = Lower boundary of the median class, f = Frequency of the median class, w = Interval of the median class, N = Total frequency and F = Cumulative frequency upto the median class (Table 4.11).

Median is easily understood and is easy to calculate. It is less reliable because it is readily affected by sampling fluctuations.

Like median, there are several other measures of the frequency distribution, based simply on their position in the series. These are *quartiles*, *deciles*, *percentiles* and *quintiles*. These are collectively known as *partition values* or *fractiles*. Quartiles are values that divide the total number of observations into 4 equal parts (Fig. 4.10). Deciles are values that divide the total number of observations into 10 equal parts (Fig. 4.11) while percentiles are values that divide the distribution into 100 equal parts. Similarly quintiles divide a distribution into 1000 equal parts. Obviously,

Table 4.11 Computation of Median and Fractiles

Class Boundary	Frequency (f_i)	Cumulative Frequency	
		Less than	F
100.5–200.5	4	100.5	0
200.5–300.5	12	200.5	4
300.5–400.5	24	300.5	16 →
400.5–500.5	40	400.5	N/2 → 40 →
500.5–600.5	16	500.5	80 →
600.5–700.5	12	600.5	96 →
700.5–800.5	10	700.5	108 →
800.5–900.5	5	800.5	118
900.5–1000.5	2	900.5	123
	$\Sigma f_i = 125$	1000.5	125

there are three quartiles (Q_1, Q_2 and Q_3), nine deciles (D_1, D_2, ..., D_9), ninety nine percentiles (P_1, P_2, ..., P_{99}) and nine hundred and ninety nine quintiles (q_1, q_2, ..., q_{999}). The general *expression* for their computations are:

Quartiles, $$Q_x = L + \frac{xN/4 - F}{f} \times w$$

Deciles, $$D_x = L + \frac{xN/10 - F}{f} \times w$$

Percentiles, $$P_x = L + \frac{xN/100 - F}{f} \times w$$

Quintiles, $$q_x = L + \frac{xN/1000 - F}{f} \times w$$

The partition values can also be estimated graphically from the cumulative frequency graphs.

$$Q_1 = 300.5 + \frac{31.25 - 16}{24} \times 100 \Rightarrow 364 \text{ pers./sq km.}$$

$$Q_2 = 400.5 + \frac{62.15 - 40}{40} \times 100 \Rightarrow 457 \text{ pers./sq km.}$$

$$Q_3 = 500.5 + \frac{93.75 - 80}{16} \times 100 \Rightarrow 587 \text{ pers./sq km.}$$

$$D_1 = 500.5 + \frac{12.25 - 4}{12} \times 100 \Rightarrow 271 \text{ pers./sq km.}$$

$$D_2 = 300.5 + \frac{25 - 16}{24} \times 100 \Rightarrow 338 \text{ pers./sq km.}$$

$$D_3 = 300.5 + \frac{37 - 16}{24} \times 100 \Rightarrow 390 \text{ pers./sq km.}$$

Fig. 4.10 Graphical Construction of Quartiles

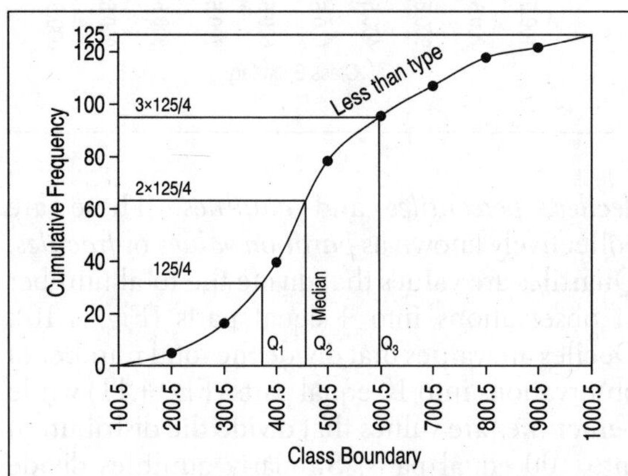

Fig. 4.11 Graphical Construction of Deciles

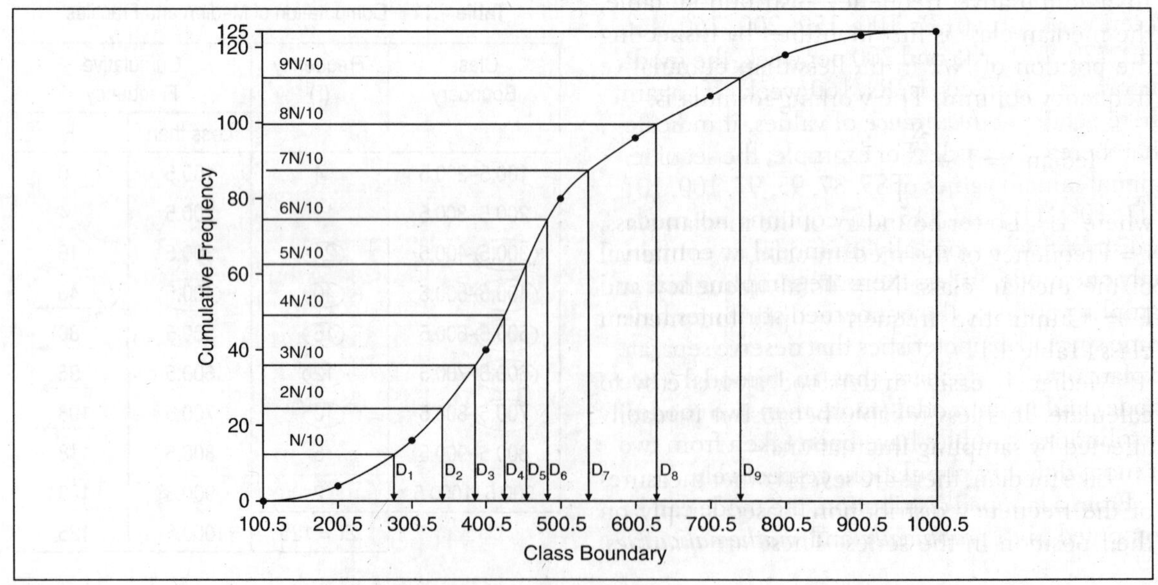

$$D_4 = 500.5 + \frac{50 - 40}{40} \times 100 \quad \Rightarrow 426 \text{ pers./sq km.}$$

$$D_5 = 500.5 + \frac{62.5 - 40}{40} \times 100 \quad \Rightarrow 457 \text{ pers./sq km.}$$

$$D_6 = 500.5 + \frac{75 - 40}{40} \times 100 \quad \Rightarrow 488 \text{ pers./sq km.}$$

$$D_7 = 500.5 + \frac{87 - 80}{16} \times 100 \quad \Rightarrow 547 \text{ pers./sq km.}$$

$$D_8 = 600.5 + \frac{100 - 96}{12} \times 100 \quad \Rightarrow 634 \text{ pers./sq km.}$$

$$D_9 = 700.5 + \frac{112.5 - 108}{10} \times 100 \Rightarrow 746 \text{ pers./sq km.}$$

Mode

The mode of a distribution is the value around which the *items* tend to be most *heavily concentrated*. It is regarded as the *most typical value* of a series and is unaffected by the presence of extreme values.

From an ungrouped dataset, mode can be found by inspection only. It is that value which occurs with *maximum* frequency. For example, if ten workers earn Rs 100, 150, 200, *160*, *160*, 140, 120, *160*, 140, and 200 per week, the modal income of workers is Rs 160/week. If, again, there occurs no *recurrence* of values, it indicates that there is no mode. For example, the set of ten annual rainfall values of 85, 87, 95, 97, 100, 101, 102, 106, 112 and 115 cm contains no mode. A distribution is normally unimodal or contains only one mode, unless there are sampling bias or sampling errors or the concerned attributes have some special characteristics that deserve separate explanation. It appears that in bimodal (two mode) and multi modal (more than two modes) distributions, samples have been taken from two or more different populations respectively.

From a grouped distribution, the mode can be computed both *graphically* and *mathematically*.

Table 4.12 Computation of Mode

Class Boundary	Frequency (f_i)
100.5–200.5	4
200.5–300.5	12
300.5–400.5	24 f_p
L 400.5–500.5	40 f_m
500.5–600.5	16 f_f
600.5–700.5	12
700.5–800.5	10
800.5–900.5	5
900.5–1000.5	2
	$\Sigma f_i = 125$

Graphically, the mode can be plotted and estimated from a histogram. The class containing the longest bar is the modal class. From the two ends of this longest bar, straight lines are drawn on the opposite sides joining the points where the preceding and succeeding bars meet. From the intersection of these two diagonals, a perpendicular is dropped onto the abscissa and the value of the mode is then interpolated (Fig.4.10). Obviously, if the modal class is bound by classes with equal frequency, the mode is given by the class mark of the modal class. But if all the bars are of the same length, there is no mode, whatsoever.

Algebraically, mode is found by the equation,

$$\text{Mode} = L + \frac{(f_m - f_p)}{(f_m - f_p) + (f_m - f_f)} \times w$$

where, f_m = Modal class frequency, f_p = Frequency of the class just preceding the modal class, f_f = Frequency of the class just following the modal class, L = Lower boundary of the modal class and w = Modal class interval.

$$\text{Mode} = 400.5 + \frac{(40 - 24)}{(40 - 24) + (40 - 16)} \times 100$$

$$= 441 \text{ persons/sq km. (Table 4.12)}$$

Fig. 4.12 Graphical Construction of Mode

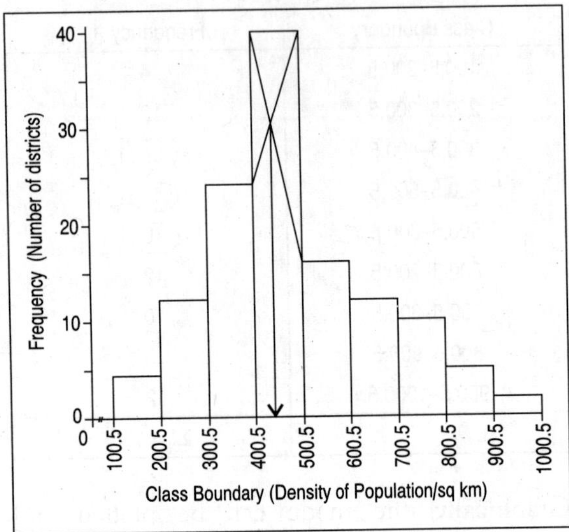

In case of a distribution with unequal class interval, the mode may be found by substituting frequencies by frequency densities.

DISPERSION

Dispersion is concerned with the *spread* of values in a set of data, or, in other words, it measures the *degree of heterogeneity* in the data. An important characteristic about it is that it determines the degree at which the observations vary among themselves. Naturally, when all the values in a distribution are *identical*, dispersion is *zero*. Therefore, with *similar* total frequency, *two* series of data have similar mean but *unequal* dispersion. Hence it is useful in *comparison* as well as in *classification problems*.

Measures of dispersion are those that are designed to compute the exact numerical extent by which individual observations vary on the average. Normally there are two types of such measures—*absolute* and *relative*. Absolute measures are *expressed in the same unit* as that of the observations while relative measures are dimensionless quantity. Relative measures are expressed either as percentages or as proportions and are *better suited to compare* (i) the dispersions of two sets of data, (ii) the relative accuracy of data and (iii) the two sets of data expressed in different units.

Absolute Measures

Range

The range of a set of data is simply the difference between the smallest and the largest values in the data set. It is the *simplest* of all measures of dispersion and does not concern each observation. It is *affected* solely by the extreme values. Symbolically,

Range = Largest value – Smallest value

Example

In the set of 125 density data,
range = (982 – 110) ⇒ 872 persons / sq km.

Quartile Deviation

The quartile deviation is defined as the *semi-interquartile* range. It is dependent on the upper and lower quartile values and does not concern the variability of the largest 25% or the smallest 25% of the observations. However, it takes into account 50% of the observations lying between the first and third quartiles. Symbolically, quartile deviation is given by:

$$QD = \frac{Q_3 - Q_1}{2}$$

Example

In the set of 125 density data,

$$QD = \frac{587 - 364}{2} \Rightarrow 112 \text{ persons/sq km}$$

Mean Deviation

The mean deviation or mean absolute deviation of a set of observations is defined as the *arithmetic mean of the absolute deviations from mean* or any

Table 4.13 Computation of Mean Deviation: Ungrouped Distribution

Rainfall (x_i : cm)	$\bar{x} = \sum x_i / N$	$\lvert x_i - \bar{x} \rvert$
97		3
100		0
95		5
85		15
115	100	15
112		12
102		2
106		6
87		13
101		1
Σ = 1000		Σ = 72

other specified measure of central tendency. It is fairly simple to calculate and easy to understand. It provides a convenient summary of the dispersion of a set of data based on all the values. The working formulae are:

For *ungrouped* distribution,

$$\text{mean deviation, MD} = \frac{\sum_{i=1}^{n} \lvert x_i - \bar{x} \rvert}{N}$$

$$\bar{x} = \frac{1000}{10} \Rightarrow 100 \text{ cm (Table 4.13)}$$

$$\therefore \text{ MD} = \frac{72}{10} \Rightarrow 7.2 \text{ cm}$$

For *grouped* distribution,

$$\text{mean deviation, MD} = \frac{\sum_{i=1}^{n} f_i \lvert x_i - \bar{x} \rvert}{\sum_{i=1}^{n} f_i}$$

$$\bar{x} = \frac{60312.2}{125} \Rightarrow 483 \text{ persons/sq km (Table 4.14)}$$

Mean Deviation,

$$\text{MD} = \frac{17120}{125} \Rightarrow 137 \text{ persons/sq km}$$

Variance

Mean deviation is directly influenced by all the values in the dataset. A value close to the mean contributes little to the mean deviation while a value further away from the mean contributes more. The spread of data becomes relatively more when deviations from the mean are large. This means that even if the mean deviations of the different series of data are identical, their

Table 4.14 Computation of Mean Deviation: Grouped Distribution

Class Boundary	Class Marks (x_i)	Frequency (f_i)	$f_i . x_i$	$\bar{x}$	$x_i - \bar{x}$	$f_i \lvert x_i - \bar{x} \rvert$
100.5–200.5	150.5	4	602		– 332	1328
200.5–300.5	250.5	12	3006		– 232	2784
300.5–400.5	350.5	24	8412		– 132	3168
400.5–500.5	450.5	40	18020		– 32	1280
500.5–600.5	550.5	16	8808	482.5	+ 68	1088
600.5–700.5	650.5	12	7806		+ 168	2016
700.5–800.5	750.5	10	7505		+ 268	2680
800.5–900.5	850.5	5	4252.5		+ 368	1840
900.5–1000.5	950.5	2	1901		+ 468	936
		Σf_i = 125	$\Sigma f_i . x_i$ = 60312.5			Σ = 17120

spread may differ significantly. The measure for this is variance which is defined as *the arithmetic mean of the squared deviation from mean.*

In case of *ungrouped* data,
Variance,

$$\sigma^2 = \frac{\sum_{i=1}^{n}(x_i - \bar{x})^2}{N} \Rightarrow \frac{\sum_{i=1}^{n} x_i^2}{N} - (\bar{x})$$

From Table 4.15,

$$\text{Variance} = \frac{838}{10} \Rightarrow 83.8$$

$$\text{or } \sigma^2 = \frac{100838}{10} \Rightarrow 83.8$$

In case of grouped data,

$$\text{Variance, } \sigma^2 = \frac{\sum_{i=1}^{n} f_i (x_i - \bar{x})^2}{\sum_{i=1}^{n} f_i}$$

or,

$$\sigma^2 = w^2 \left[\left\{ \sum_{i=1}^{n} f_i (d_i')^2 \Big/ \sum_{i=1}^{n} f_i \right\} - \left(\sum_{i=1}^{n} f_i d_i' \Big/ \sum_{i=1}^{n} f_i \right)^2 \right]$$

where, w = Class interval, d_i' = Deviation from mean in the ith class coded in terms of class interval and x_i, f_i, x and n are usual notations.

Table 4.15 Computation of Variance: Ungrouped Data

Rainfall (cm) x_i	$\bar{x}$	$x_i - \bar{x}$	$(x_i - \bar{x})^2$	x_i^2
97		– 3	9	9409
100		0	0	10000
95		– 5	25	9025
85		– 15	225	7225
115	100	+ 15	225	13225
112		+ 12	144	12544
102		+ 2	4	10404
106		+ 6	36	11236
87		– 13	169	7569
101		+ 1	1	10201
Σ = 1000		Σ = 0	Σ = 838	100838

Computation of Variance from *grouped* data:
From Table 4.16 and Table 4.17

$$\text{Variance, } \sigma^2 = \frac{3792000}{125} \Rightarrow 30{,}336$$

$$\text{or, } \sigma^2 = 100^2 \left[\frac{437}{125} - \left(\frac{-85}{125} \right)^2 \right] \Rightarrow 30{,}336$$

Variance is an important measure of dispersion particularly in assessing the variation between two or more samples. *Analysis of variance* (ANOVA) is a very powerful statistical technique that uses the variance to find out whether the number of samples differ significantly from each other or not.

Standard Deviation

Variance is not normally used as a descriptive measure; instead the square root of the variance is taken. It is called standard deviation (σ) which is defined as the *root-mean-squared deviation from mean.* It is superior to variance in that variance may become a large number, even greater than the individual values but in standard deviation it is square-rooted to make it more comprehensive.

In case of *ungrouped* data,
Standard Deviation,

$$\sigma = \sqrt{\frac{\sum_{i=1}^{n}(x_i - \bar{x})^2}{N}} \quad \text{or, } \sigma = \sqrt{\left\{ \frac{\sum_{i=1}^{n} x_i^2}{N} - (\bar{x})^2 \right\}}$$

In case of *grouped* data,

$$\text{Standard Deviation, } \sigma = \sqrt{\frac{\sum_{i=1}^{n} f_i (x_i - \bar{x})^2}{\sum_{i=1}^{n} f_i}}$$

or,

$$\sigma = w \sqrt{\left\{ \sum_{i=1}^{n} f_i (d_i')^2 \Big/ \sum_{i=1}^{n} f_i \right\} - \left(\sum_{i=1}^{n} f_i d_i' \Big/ \sum_{i=1}^{n} f_i \right)^2}$$

From Table 4.16,

$$\bar{x} = \frac{60312.2}{125} \Rightarrow 482.5$$

Table 4.16 Computation of Standard Deviation (σ): Algebraic Method

Class Boundary	Class Marks (x_i)	Frequency (f_i)	$f_i x_i$	$\bar{x}$	$x_i - \bar{x}$	$(x_i - \bar{x})^2$	$f_i(x_i - \bar{x})^2$
100.5–200.5	150.5	4	602		– 332	110224	440896
200.5–300.5	250.5	12	3006		– 232	53824	64588.8
300.5–400.5	350.5	24	8412		– 132	17424	418176
400.5–500.5	450.5	40	18020		– 32	1024	40960
500.5–600.5	550.5	16	8808	482.5	+ 68	4624	73948
600.5–700.5	650.5	12	7806		+ 168	28224	338688
700.5–800.5	750.5	10	7505		+ 268	71824	718240
800.5–900.5	850.5	5	4252.5		+ 368	135424	677120
900.5–1000.5	950.5	2	1901		+ 468	219024	438048
Σ =		*125*	*60312.5*				*3792000*

Table 4.17 Computation of Standard Deviation (σ): Short Method

Class Boundary	Class Marks (x_i)	Frequency (f_i)	d_i'	$f_i \cdot d_i'$	$(d_i')^2$	$f_i(d_i')^2$
100.5–200.5	150.5	4	– 4	–16	16	46
200.5–300.5	250.5	12	– 3	– 36	9	108
300.5–400.5	350.5	24	– 2	– 48	4	96
400.5–500.5	450.5	40	– 1	– 40	1	40
500.5–600.5	550.5	16	0	0	0	0
600.5–700.5	650.5	12	+ 1	+ 12	1	12
700.5–800.5	750.5	10	+ 2	+ 20	4	40
800.5–900.5	850.5	5	+ 3	+ 15	9	45
900.5–1000.5	950.5	2	+ 4	+ 8	16	32
Σ =		125		– 85		437

Standard Deviation,

$$\sigma = \sqrt{3792000/125} \Rightarrow 174 \text{ persons/sq km}$$

From Table 4.17,

$$\bar{x} = 550.5 + \{(-85/125) \times 100\} \Rightarrow 482.5$$

Standard Deviation,

$$\sigma = 100\sqrt{(437/125) - (-85/125)^2}$$

$$= 100\sqrt{3.496 - 0.4624} \Rightarrow 174 \text{ persons/sq km}$$

The value of standard deviation *is always less* than that of variance. For normal distribution, 68.27% of the items lie within a range of $\bar{x} \pm 1\sigma$, 95.45% of the items within $\bar{x} \pm 2\sigma$ and 99.73% within $\bar{x} \pm 3\sigma$.

Standard deviation is thus the most commonly used statistical technique. It also forms the basis of several other parametric statistical methods concerning estimates from samples.

Relative Measures

Coefficient of Variation

In order to compare the spread of two or more distributions, the relative measures of variability are used. It is independent of the units of observations and is defined by,

Nature of Statistical Distribution

Measures of Central Tendency

- (Mean–Mode) = 3(Mean–Median) implies an unimodal distribution of moderate skewness
- Mean > Median > Mode implies a positively skewed distribution
- Mode > Median > Mean implies a negatively skewed distribution
- Mean = Median = Mode implies a normal distribution

Measures of Dispersion

- In a distribution,
  - Standard deviation (σ) ≥ Mean deviation
  - Standard deviation (σ) < Variance
  - Range = 6σ
  - Quartile deviation = $\frac{2\sigma}{3}$
  - Mean deviation = $\frac{4\sigma}{5}$
- In a normal distribution, 68.27% of the frequencies lie within a range ($\bar{x} \pm 1.\sigma$), 95.45% within ($\bar{x} \pm 2.\sigma$) and 99.73% within ($\bar{x} \pm 3.\sigma$)

Coefficient of Variation, CV = $\frac{\sigma}{\bar{x}} \times 100\%$

For example, the reliability of rainfall (cm) for stations P and Q can be compared as follows:

P	100	150	120	60	30	90	70	200	150	80
Q	95	100	105	120	100	110	115	80	115	110

The solved table is:

Stations	Σx_i	$\bar{x}$	σ	CV
P	1050 cm	105 cm	48.01 cm	45.72%
Q	1050 cm	105 cm	11.18 cm	10.65%

Therefore, although the mean rainfall is the me at both the stations, rainfall at Q is more consistent, less variable and more reliable than that at P.

Z–Score

Sample variance and standard deviation are measures of descriptive statistics. They are computed for the purpose of describing a particular set of observations and not for the purpose of making any inference with the parent population. Two distributions with different means and standard deviations are *difficult* to compare unless both the distributions are put into a *standard form*. The process of changing the form of a distribution is called *transformation*. The transformation of raw scores into standard scores by standardisation is called *Z-scores*. It is defined as,

$$\text{Z-Score} = \frac{x_i - \bar{x}}{\sigma}$$

It considers both varying means and varying standard deviations. For example, to compare the growth rates of six towns belonging to six different size-classes during 1991–2001, Z-score can be of great help. The standard scores measure the deviation of the growth rate of a town of the ith class from the ith class mean in terms of the standard deviation (Table 4.18). Therefore, the towns T_2 and T_4 registered significant growth characteristics, T_1 and T_5 registered significant decline while T_3 just sustained its growth during 1991–2001.

SHAPE OF FREQUENCY CURVE

The shape of a frequency curve is the most important of all the characteristics as it brings out the real nature of a distribution. Shape is measured in terms of two geometric properties of a frequency curve—*symmetry* and *peakiness*. Symmetry describes the property of skewness while peakiness the property of kurtosis.

Skewness

The term *skewness* is used to denote the extent of asymmetry in the shape of the frequency

Table 4.18 Computation of Z-Score

City/Town	Class (No)	Growth Rate, 1991–2001 (%)	$\overline{x}$ of the ith Class (%)	σ of the ith Class (%)	Z-Score
T_1	I (10)	4.96	24.31	14.95	– 1.29
T_2	II (25)	67.87	30.62	15.96	+ 2.33
T_3	III (75)	31.63	31.42	19.04	+ 0.01
T_4	IV (125)	114.40	33.20	28.65	+ 2.83
T_5	V (30)	18.32	44.01	25.02	– 1.02
T_6	VI (15)	16.55	29.38	33.37	– 0.38

Fig. 4.13 Nature of Dispersion

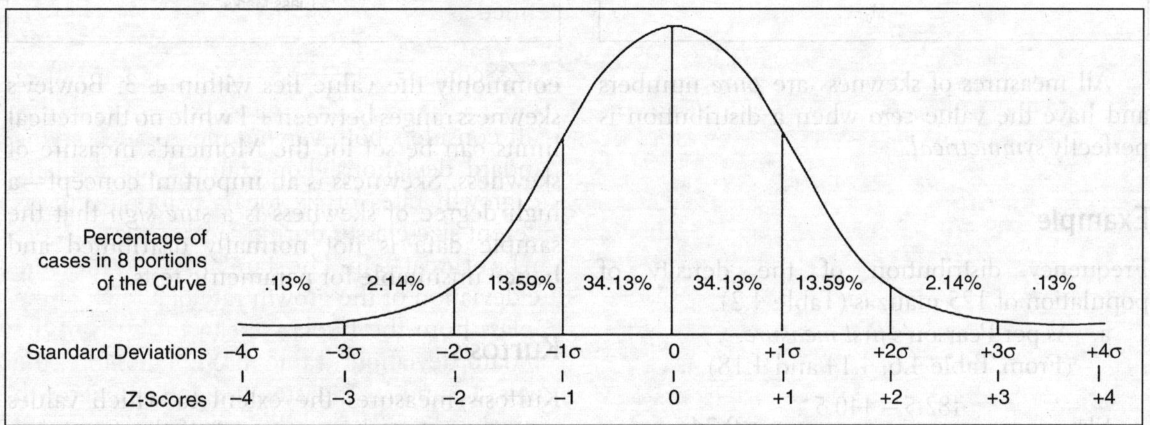

distribution curve. Any *asymmetric* distribution is a *skewed* distribution. Basically, it measures the extent by which the bulk of the values in a distribution are concentrated on one or the other side of the mean. A perfectly symmetrical normal distribution has, therefore, zero skewness.

- If *Mean > Median > Mode*, it is a *positively skewed* distribution with its tail on the right side of the frequency curve.
- If *Mean < Median < Mode*, it is a *negatively skewed* distribution with its tail on the left side of the frequency curve (Fig. 4.14).

It is measured by the following formulas:

a. Pearson's *first measure*:

$$\text{Skewness} = \frac{\text{Mean} - \text{Mode}}{\text{Standard Deviation}}$$

b. Pearson's *second measure*,

$$\text{Skewness} = \frac{3\,(\text{Mean} - \text{Median})}{\text{Standard Deviation}}$$

c. Bowley's measure,

$$\text{Skewness} = \frac{Q_3 + Q_1 - 2Q_2}{Q_3 - Q_1}$$

d. Moment's measure,

$$\text{Skewness} = \frac{m_3}{\sigma^3}$$

$$= \sum_{i=1}^{n}(x_i - \overline{x})^3 \Big/ N\sigma^3 \quad \text{for ungrouped series}$$

$$= \sum_{i=1}^{n} f_i\,(x_i - \overline{x})^3 \Big/ \sum_{i=1}^{n} f_i \,.\, \sigma^3 \quad \text{for grouped series}$$

Fig. 4.14 Shapes of Frequency Distribution

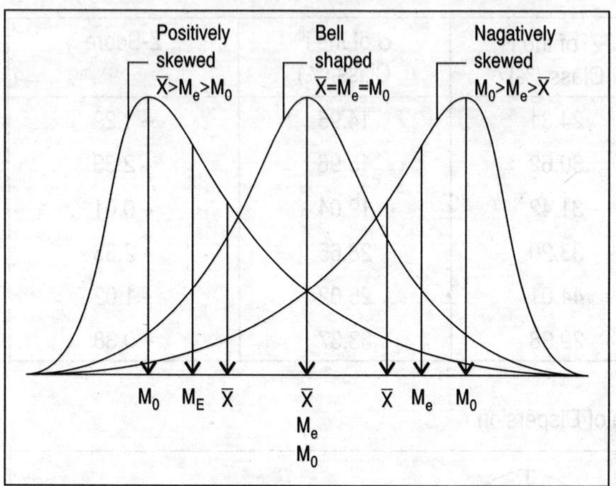

Fig. 4.15 Peakiness of Frequency Distribution

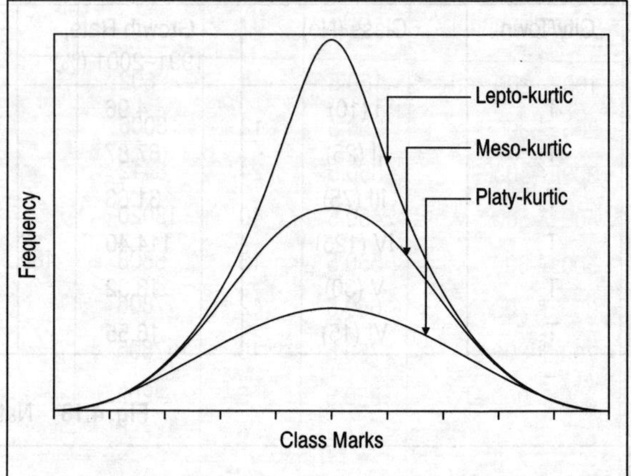

All measures of skewness are *pure* numbers and have the value *zero* when a distribution is perfectly *symmetrical*.

Example

Frequency distribution of the density of population of 125 mauzas (Table 4.2).

a. As per Pearson's *first measure*, (From Table 4.6, 4.14 and 4.18)

$$\text{Skewness} = \frac{482.5 - 440.5}{174.2} \Rightarrow +0.24$$

b. As per Pearson's *second measure*, (From Table 4.6, 4.11 and 4.18)

$$\text{Skewness} = \frac{3(482.5 - 456.75)}{174.2} \Rightarrow +0.44$$

c. As per Bowley's measure, (From Table 4.12)

$$\text{Skewness} = \frac{586.44 + 364.04 - 2(456.75)}{174.2}$$

$$\Rightarrow +0.17$$

d. As per Moment's measure, (From Table 4.19)

$$\text{Skewness} = \frac{3.55871 \times 10^{8})}{125.(174.2)^3} \Rightarrow +0.54$$

According to the measures of Pearson, there are no *theoretical limits* of skewness; most commonly the value lies within ± 3. Bowley's skewness ranges between ± 1 while no theoretical limits can be set for the Moment's measure of skewness. Skewness is an important concept—a high degree of skewness is a *sure sign* that the sample data is not normally distributed and hence unsuitable for parametric tests.

Kurtosis

Kurtosis measures the extent to which values are concentrated in one part of the frequency distribution. Hence it refers to the degree of peakiness of the frequency curve (Fig. 4.15). Two distributions may have the identical average, spread and symmetry but they will certainly differ in degree of peakiness. The Moment's measure of kurtosis is given by:

$$\text{Kurtosis} = \frac{m^4}{\sigma^4}$$

$$\Rightarrow \sum_{i=1}^{n} f_i (x_i - \bar{x})^4 \Big/ N.\sigma^4 \text{(for } \textit{ungrouped} \text{ series)}$$

$$\text{or, } \Rightarrow \sum_{i=1}^{n} f_i.(x_i - \bar{x})^4 \Big/ \left(\sum_{i=1}^{n} fi\right).\sigma^4 \text{ (for grouped series)}$$

$$\text{Kurtosis} = \frac{33.981889 \times 10^{10})}{125(174.2)^4} \text{ (from Table 4.19)}$$

$$\Rightarrow +2.95$$

Table 4.19 Computation of Skewness and Kurtosis

Class Boundary	(x_i)	(f_i)	$f_i x_i$	$\overline{x}$	$x_i - \overline{x}$	$f_i(x_i - \overline{x})^2$	$f_i(x_i - \overline{x})^3$	$f_i(x_i - \overline{x})^4$
100.5–200.5	150.5	4	602		– 332	440896	-1.46377×10^8	4.85973×10^{10}
200.5–300.5	250.5	12	3006		– 232	645888	-1.49846×10^8	3.47642×10^{10}
300.5–400.5	350.5	24	8412		– 132	418176	-0.55199×10^8	0.72862×10^{10}
400.5–500.5	450.5	40	18020		– 32	40960	-0.0131×10^8	0.00419×10^{10}
500.5–600.5	550.5	16	8808	482.5	+ 68	73948	0.0503×10^8	0.03421×10^{10}
600.5–700.5	650.5	12	7806		+ 168	338688	0.56899×10^8	0.95591×10^{10}
700.5–800.5	750.5	10	7505		+ 268	718240	1.92488×10^8	5.15868×10^{10}
800.5–900.5	850.5	5	4252.5		+ 368	677120	2.4918×10^8	9.16983×10^{10}
900.5–1000.5	950.5	2	1901		+ 468	438048	2.05006×10^8	9.59430×10^{10}
Σ =		125	60312.5			3792000	$+3.55871\times10^8$	33.98189×10^{10}

For a *normal distribution* which is neither sharply peaked nor flat-topped, *kurtosis* is 3. This is called a *meso-kurtic* distribution. When kurtosis is less than 3, the frequency curve is flat-topped and the distribution is called *platy-kurtic*. When the frequency curve is sharply peaked and the value of kurtosis is more than 3, the distribution is called *lepto-kurtic*.

RELATIONSHIPS

An useful application of statistics is to describe the relationship between two variables measured from a common sample of individuals. When the relationship is quantitative in nature, the most appropriate statistical method of identifying and measuring the relationship and expressing it in equation form is known as *correlation*.

Normally the relationship between two variables is shown by a *scatter graph*. It represents the scatter of points plotted by the y-values corresponding to the respective x-values. Usually the independent variable is taken on the x-axis and the dependent variable is taken on the y-axis (Fig. 4.16 and Fig. 4.17).

The concept of correlelation involves the measurement of the three basic items, i.e.,

i. An *estimating or regression equation*. This describes the functional relationship between the variables.

Fig. 4.16 Scatter Graph: Negative Relation

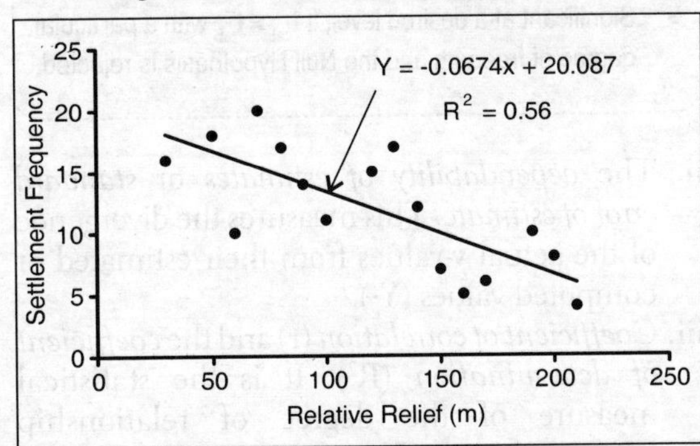

Fig. 4.17 Scatter Graph: Positive Relation

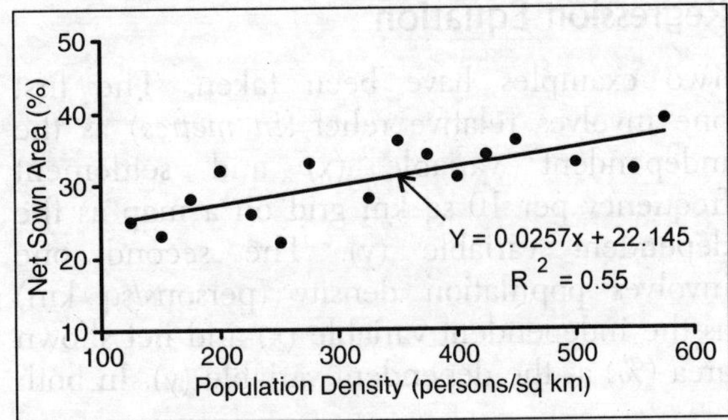

Bivariate Analysis

Linear Regression

- Estimating or Regression Equation:

  $Y_c = a + b.x$

- Dependability or Standard Error of Estimates:

  $S_E = \sqrt{\Sigma(y - Y_c)^2 / N}$

- Coefficient of Determination:

  $R^2 = \left[\Sigma(\bar{Y}_c - \bar{y})^2 / \Sigma(y - y)^2\right]$

  = (Explained Variation) ÷ (Total Variation)

Correlation Coefficient

- Spearman's Rank:

  $r_s = 1 - \left[6\Sigma d^2 / \left(N^3 - N\right)\right]$

- Pearson's Product Moment:

  $r = \left[(\Sigma xy / N) - \sqrt{xy}\right] / \sigma_x . \sigma_y$

Test of Significance

- Student's t-test: $t = \sqrt{\left[r^2 (N-2) / \left(1 - r^2\right)\right]}$
- Significant at a desired level, if $t_{cal} > t_{tab}$ with a particular degree of freedom and the Null Hypothesis is rejected.

ii. The *dependability of estimates* or *standard error of estimate*. This measures the divergence of the actual y-values from their estimated or computed values (Y_c).

iii. *Coefficient of correlation* (r) and the *coefficient of determination* (R^2). It is the statistical measure of the degree of relationship between the variables.

Regression Equation

Two examples have been taken. The first one involves relative relief (*in metres*) as the independent variable (x) and settlement frequency per 10 sq km grid on a map as the dependent variable (y). The second one involves population density (persons/sq km) as the independent variable (x) and net shown area (%) as the dependent variable (y). In both cases, the number of paired data (N) is 16. The scatter graphs explicity express the nature of the relationship—negative in the first case (i.e., a decrease in y-value with an increase in x-value) and positive in the second (i.e., an increase in y-value with an increase in x-value). A straight line can be fitted in each of these, such that

(i) the sum of the vertical deviations of the observed values from the derived straight line is equal to zero [i.e., $\Sigma(y - Y_c) = 0$] and

(ii) the sum of the squares of all such deviations is least from those obtained from any other orientation of the said straight line [i.e., $\Sigma(y - Y_c)^2$ = minimum].

It is known as the *method of least squares*. The general equation of a straight line is given by $Y_c = a + bx$, where a = intercept and b = slope. The normal equations to solve for a and b are:

$$\Sigma y = Na + b\Sigma x \quad \ldots (i)$$

$$\Sigma xy = a\Sigma x + b\Sigma x^2 \quad \ldots (ii)$$

The regression equations solved (Table 4.20 and Table 4.21) are:

$$\begin{array}{llll} & 190 = 16a & + 1950b & \\ & 20000 = 1950a & + 284500b & \\ \text{or} & 11.875 = a & + 121.8750b & \ldots (i) \\ & 10.256 = a & + 145.8974b & \ldots (ii) \\ \hline & 1.619 = & - 24.0224b & (i) - (ii) \end{array}$$

$$\text{or,}\quad b = \frac{1.619}{24.0224} \Rightarrow -0.0674$$

Substituting the value of b in equation (i)

a = 11.875 – 121.8750 (– 0.0674) ⇒ 20.089

Therefore, the regression equation is,

$Y_c = 20.089 - 0.0674x$

$$\begin{array}{llll} & 492 = 16a & + 5350b & \\ & 172325 = 5350a & + 2092500b & \\ \text{or} & 30.75 = a & + 334.3750b & \ldots (i) \\ & 32.21 = a & + 391.125b & \ldots (ii) \\ \hline & -1.46 = & -56.7465b & (i) - (ii) \end{array}$$

$$b = \frac{-1.46}{-56.7465} \Rightarrow +0.0257284$$

Table 4.20 Computation of Regression Parameters

Relative Relief (x: m)	Settlement Frequency /10 km² Grid (y)	xy	x²	y²
30	16	480	900	256
50	18	900	2500	324
60	10	600	3600	100
70	20	1400	4900	400
80	17	1360	6400	289
90	14	1260	8100	196
100	11	1100	10000	121
120	15	1800	14400	225
130	17	2210	16900	289
140	12	1680	19600	144
150	7	1050	22500	49
160	5	800	25600	25
170	6	1020	28900	36
190	10	1900	36100	100
200	8	1600	40000	64
210	4	840	44100	16
1950	*190*	*20000*	*284500*	*2634*

Substituting the value of b in equation (i)

$$a = 30.75 - 334.3750\ (+\ 0.025784) \Rightarrow 22.147$$

Therefore, the regression equation is,

$$Y_c = 22.147 - 0.0257284\ x$$

From these equations, values of y-variable can be regressed for any desired value of x-variable. For instance, the frequency of settlement when relative relief is 125 m would be,

$$Y_c = 20.089 - 0.0674(125) = 11.66 \Rightarrow 12$$

Similarly, the net sown area corresponding to a population density of 300 persons/sq km can be estimated as:

$$Y_c = 22.147 + 0.0257284(300) \Rightarrow 29.87\%$$

Table 4.21 Computation or Regression Parameters

Density (Person/ sq km) (x)	NSA (%) (y)	xy	x²	y²
125	25	3125	15625	625
150	23	3450	22500	529
175	28	4900	30625	784
200	32	6400	40000	1024
225	26	5850	50625	676
250	22	5500	62500	484
275	33	9075	75625	1089
325	28	9100	105625	784
350	36	12600	122500	1296
375	34	12750	140625	1156
400	31	12400	160000	961
425	34	14450	180625	1156
450	36	16200	202500	1296
500	33	16500	250000	1089
550	32	17600	302500	1024
575	39	22425	330625	1521
5350	*492*	*172325*	*2092500*	*15494*

Standard Error of Estimate

In the regression of values, there must be variations because (i) in all samples of 125 m of relative relief, settlement frequency of 11.66 may not be obtained or (ii) everywhere at 300 persons/sq km density level, net sown area cannot always be 29.87%. Rather the regressed values should be thought of as an estimate of the average of the values that may actually be observed. Hence, within a representative sample it would be more useful to find the proportion of places that fall within any desired range of error. The standard deviation of the residuals $(y - Y_c)$ is a good measure of this. It is called the *standard error of estimate*. In equation form,

$$S_E = \sqrt{\Sigma(y - Y_c)^2 / N}$$

In the first case, S_E = 3.212 while in the second case, S_E = 3.201 (Table 4.22 and Table 4.23). The standard error of estimate is a measure of dispersion of all the y-values around the regression equation. It is, therefore, a general or overall measure of dispersion indicating the dependability of estimates. One may always expect to find 68.27% of the total occurrences within ±1 S_E range, 95.45% within ±2 S_E range and 99.73% within ± 3S_E range. Any discrepancy is due to the fact that the sample is small and the scatter is not normally distributed around the regression equation.

From Table 4.22,

$\bar{y}$ = 11.875

$$S_E = \sqrt{\frac{165.105}{16}} \Rightarrow 3.212$$

From Table 4.23,

$\bar{y}$ = – 30.75

$$S_E = \sqrt{\frac{163.959}{16}} \Rightarrow 3.201$$

COEFFICIENT OF DETERMINATION

This is closely related to the regression equation and the standard error of estimate. The sum of the squares of the deviations of y-values from their mean ($\bar{y}$) is defined as total variation $\Sigma(y - \bar{y})^2$. It has two components—one, that can be explained by the regression line and the other, that cannot be explained. The explained variation is the sum of the squares of the deviations of estimated y-values from their mean $\Sigma(Y_c - \bar{y})^2$ while the

Table 4.22 Computation of Total, Explained and Unexplained Variations

Relative Relief (m) (x)	Settlement Frequency/ 10 km² Grid (y)	Y_c = 20.089 –0.0674x	Deviation			Squared Deviation		
			D = $(y - \bar{y})$	D_c = $(Y_c - \bar{y})$	D_s = $(y - Y_c)$	D^2 =	D_c^2 =	D_s^2 =
30	16	18.067	4.125	6.192	– 2.067	17.016	38.341	4.272
50	18	16.719	6.125	4.844	1.281	37.516	23.464	1.641
60	10	16.046	– 1.875	4.171	– 6.046	3.515	17.397	36.554
70	20	15.371	8.125	3.496	4.629	66.016	12.222	21.428
80	17	14.697	5.125	2.822	2.303	26.266	7.964	5.304
90	14	14.023	2.125	2.148	– 0.023	4.515	4.614	0.0005
100	11	13.349	– 0.875	1.474	– 2.349	0.766	2.173	5.518
120	15	12.002	3.125	0.127	2.998	9.766	0.016	8.988
130	17	11.327	5.125	– 0.548	5.673	26.266	0.300	32.183
140	12	10.653	0.125	– 1.222	1.347	0.015	1.493	1.814
150	7	9.979	– 4.875	– 1.896	– 2.979	23.766	3.595	8.874
160	5	9.305	– 6.875	– 2.569	– 4.306	47.266	6.600	18.542
170	6	8.632	– 5.875	– 3.243	– 2.632	34.515	10.517	6.927
190	10	7.284	– 1.875	– 4.591	2.716	3.516	21.077	7.377
200	8	6.609	– 3.875	–5.266	1.391	15.015	27.731	1.935
210	4	5.936	– 7.875	– 5.939	– 1.936	62.016	35.272	3.748
1950	*190*	*190.000*	*0*	*0*	*0*	*377.751*	*212.776*	*165.105*

Table 4.23 Computation of Total, Explained and Unexplained Variations

Density (Persons/ Sq km) (x)	NSA (%)(y)	Y_c = 22.147 + 0.02573 x	Deviation			Squared Deviation		
			D = $(y-\bar{y})$	$D_c = (Y_c-\bar{y})$	$D_s^2 = (y-Y_c)$	D^2	D_c^2	D_s^2
125	25	25.363	– 5.75	– 5.387	– 0.363	33.063	29.020	0.132
150	23	26.006	– 7.75	– 4.744	– 3.006	60.062	22.506	9.036
175	28	26.649	– 2.75	– 4.101	1.351	7.563	16.818	1.825
200	32	27.293	1.25	– 3.457	4.707	1.562	11.951	22.156
225	26	27.936	– 4.75	– 2.814	– 1.936	22.563	7.919	3.748
250	22	28.579	– 8.75	– 2.171	– 6.579	76.562	4.713	43.283
275	33	29.222	2.25	– 1.528	3.778	5.063	2.335	14.273
325	28	30.509	– 2.75	– 0.241	– 2.509	7.562	0.058	6.295
350	36	31.152	5.25	0.402	4.848	27.563	0.162	23.503
375	34	31.795	3.25	1.045	2.205	10.562	1.092	4.862
400	31	32.439	0.25	1.689	– 1.439	0.063	2.853	2.071
425	34	33.082	3.25	2.332	0.918	10.562	5.438	0.843
450	36	33.725	5.25	2.975	2.275	27.563	8.851	5.176
500	33	35.012	2.25	4.262	– 2.012	5.062	18.165	4.048
550	32	36.298	1.25	5.548	– 4.298	1.563	30.780	18.473
575	39	36.942	8.25	6.192	2.058	68.062	38.341	4.235
5350	*492*	*492.002*	*0*	*0.002*	*– 0.002*	*364.961*	*201.002*	*163.959*

unexplained variation is defined by the sum of the squares of the deviations of y-values from their estimated values $\Sigma(y-Y_c)^2$. In equation form,

Unexplained Variation,

$$\Sigma D_s^2 = \Sigma\left(y - Y_c\right)^2 \qquad \ldots \text{(i)}$$

Explained Variation,

$$\Sigma D_c^2 = \Sigma\left(Y_c - \bar{y}\right)^2 \qquad \ldots \text{(ii)}$$

Total Variation,

$$\Sigma D^2 = \Sigma\left(y - \bar{y}\right)^2 \qquad \ldots \text{(iii)}$$

and $$\Sigma D^2 = \Sigma D_s^2 + \Sigma D_c^2 \qquad \ldots \text{(iv)}$$

Coefficient of determination (R^2) is defined as the ratio between the explained variation and the total variation. In other words,

$$R^2 = \frac{\Sigma D_c^2}{\Sigma D^2} \text{ or, } R^2 = 1 - \frac{\Sigma D_s^2}{\Sigma D^2}$$

From Table 4.22,

Variation	Amount	% of Total Variation
Unexplained $\left(\Sigma D_s^2\right)$	165.105	43.71
Explained $\left(\Sigma D_c^2\right)$	212.646	56.29
Total	377.751	100.00

Coefficient of Determination,

$$R^2 = 1 - \left(\Sigma D_s^2 / \Sigma D^2\right)$$

$$= 1 - \frac{165.105)}{377.751} \Rightarrow 0.56$$

Coefficient of Correlation,

$$r = \sqrt{0.56} \Rightarrow \pm 0.75$$

Points to Note: Regression

1. Regression is certainly used to make estimates. To estimate **y**, the regression equation of **y** on **x** in either of the following forms is to be used, i.e.,
   (1) $\mathbf{y - \bar{y} = b_{y.x} (x - \bar{x})}$
   (2) $\mathbf{y_c = a + b.x}$
   where, $b = b_{y.x}$
   $= Cov (x, y) / Var (x)$
   $= r. (\sigma_y / \sigma_x)$
2. Similarly, to estimate **x**, the regression equation of **x** on **y** in either of the following forms is to be used, i.e.,
   (1) $\mathbf{x - \bar{x} = b_{x.y} (y - \bar{y})}$
   (2) $\mathbf{x_c = a' + b'.y}$
   where, $\mathbf{b'} = b_{x.y}$
   $= Cov (x, y) / Var (y)$
   $= \mathbf{r. (\sigma_x / \sigma_y)}$
3. The product of the two regression coefficients is equal to the square of correlation coefficient, i.e., $\mathbf{r^2 = (b_{y.x}) (b_{x.y})}$ **or (b . b′)**
4. Correlation coefficients and the two regression coefficients, all have the same sign, and if r = 0, the regression coefficients are also zero.
5. The regression lines always intersect at the point ($\bar{x}$, $\bar{y}$). The slopes of the regression line of **y** on **x** and the regression line of **x** on **y** are respectively $\mathbf{b_{y.x}}$ (or, b) and $\mathbf{1/ b_{x.y}}$ (or, 1 / b′).
6. The angle θ between the two regression lines is given by:
   $$\tan\theta = \pm \{(1 - r^2)/r\} \{ \sigma_x . \sigma_y / (\sigma_x^2 + \sigma_y^2)\}$$
   Hence, when r = 0, the regression lines are perpendicular to each other, and when **r** = ± 1, the regression lines coincide, the angle between them being zero.
7. When **r = 0**, neither **x** nor **y** can be estimated from the regression equations, and when **r** = ± 1, there is an exact linear relationship between the variables.
8. Regression coefficients are independent of origin but are dependent on scale. In particular, if **u = ax + b**, and **v = cy +d**, then —
   $\mathbf{b_{u.v} = (a / c). b_{x.y}}$ and
   $\mathbf{b_{v.u} = (c / a). b_{y.x}}$
9. The proportion of total variation explained by regression = $\mathbf{R^2}$, which is commonly taken as a measure of goodness of fit.

From Table 4.23,

Variation	Amount	% of Total Variation
Unexplained $\left(\sum D_s^2\right)$	163.959	44.93
Explained $\left(\sum D_c^2\right)$	201.002	55.07
Total	364.961	100.00

Coefficient of Determination,

$$R^2 = 1 - \left(\sum D_s^2 / \sum D^2\right)$$

$$= 1 - \frac{163.959}{364.961} \Rightarrow 0.55$$

Ceofficient of Correlation,

$$r = \sqrt{0.55}$$

$$= \pm 0.74$$

CORRELATION COEFFICIENT

The coefficient of correlation (r) is simply the *square root* of the coefficient of determination (R^2). (Table 4.22 and 4.23). The value of r is always larger than R^2, except in the case where $R^2 = 0$ or $R^2 = 1$. The value of r lies between – 1.0 and + 1.0 (i.e., $+1.0 \geq r \geq -1.0$). A value of + 1.0 indicates a *perfectly direct* or *perfectly positive* correlation and a value of – 1.0 indicates a *perfectly inverse* or *perfectly negative* correlation. A complete absence of correlation is indicated by r = 0. The sign and sense of correlation can easily be obtained by an inspection of the scatter diagram. However, this method is too laborious to use for everyday calculation.

The two most widely used methods of determining r are *Pearson's* method and *Spearman's* method. These two coefficients can be used in the *tests of significance* of a correlation. Thus in random samples, it is possible to estimate the probability that an observed correlation represents a *real* correlation in the population from which the sample data is taken.

Points to Note: Correlation Coefficient

1. The correlation coefficient, **r** is independent of the choice of both origin and scale of observations.
2. The correlation coefficient, **r** is independent of the units of measurements and is simply a pure number.
3. The value of the correlation coefficient, **r** lies between **– 1.0** and **+ 1.0**
4. A value of **r = + 1.0** indicates a perfectly direct or perfectly positive correlation, and a value of **r = – 1.0** indicates a perfectly inverse or perfectly negative correlation. These two extremes indicate equally strong relationship between the variables but with different implications for interpretation.
5. Values near to zero **(r = 0)** indicate a lack of a rectilinear trend.
6. If the data plot as a vertical or horizontal linear scatter, r will also be near zero or unidentified, because in such cases either $\sigma_x = 0$ or $\sigma_y = 0$
7. The correlation coefficient, **r** is simply the square-root of the coefficient of determination **(R^2)** or the proportion of variation explained by regression; hence, r is always larger than **R^2**, except in cases, where **r = 0** or **r = 1**.
8. The correlation coefficient, **r** also helps in estimating the value of the dependent variable, when the value of the independent variable is known. Thus, for a given **x**, the estimated value of **y** can be obtained from the regression equation of **y** on **x**, as follows —

   $$(y - \bar{y}) = r.(\sigma_y/\sigma_x)(x - \bar{x})$$

9. When **r = 0**, neither y nor **x** can be estimated by a linear function of the other variable.
10. Again, when **r = ± 1**, the two regression lines coincide, and the value of one variable can be accurately predicted by the linear regression equation.

Pearson's Product-Moment Correlation Coefficient

It is a parametric measure of the relationship between two variables. The variables must be measured on the interval scale. The method is based on the assumption that both variables have come from the normally distributed populations. In equation form,

$$r = \frac{\frac{\Sigma xy}{N} - \bar{x}.\bar{y}}{\sigma_x \sigma_y}$$

where, r = Product-moment correlation coefficient, N = Number of data-pair, $\bar{x}$ = Mean of x, $\bar{y}$ = mean of y, **Σ**xy = sum of the products of x and y, σ_x = standard deviations of x, σ_y = standard deviations of y.

The numerator $\left(\frac{\Sigma xy}{N} - \bar{x}.\bar{y}\right)$ is simply a measure of the degree to which the two variables vary together. An increase in one is reflected by a proportionate increase or decrease in the other. It is therefore known as the *covariance* of the two variables. It can also be expressed as

$$r = \frac{N\Sigma xy - (\Sigma x)(\Sigma y)}{\sqrt{\{N\Sigma x^2 - (\Sigma x)^2\}\{N\Sigma y^2 - (\Sigma y)^2\}}}$$

a. From Table 4.20,

$\Sigma xy = 20{,}000 \quad N = 16$

$$\bar{x} = \frac{1950}{16} \Rightarrow 121.875$$

$$\bar{y} = \frac{190}{16} \Rightarrow 11.875$$

$$\sigma_x = \sqrt{\frac{284500}{16} - (121.875)^2} \Rightarrow 54.1085$$

$$\sigma_y = \sqrt{\frac{2634}{16} - (11.875)^2} \Rightarrow 4.8589$$

Therefore,

$$r = \frac{(20000/16) - (121.875)(11.875)}{(54.1085)(4.8589)}$$

$$= -0.75$$

b. From Table 4.21,

$\Sigma xy = 172325, N = 16$

$$\bar{x} = \frac{5350}{16} \Rightarrow 334.375$$

$$\bar{y} = \frac{492}{16} \Rightarrow 30.75$$

$$\sigma_x = \sqrt{\frac{2092500}{16} - (334.375)^2} \Rightarrow 137.748$$

$$\sigma_y = \sqrt{\frac{15494}{16} - (30.75)^2} \Rightarrow 4.776$$

Therefore,

$$r = \frac{172325/16 - (334.375)(30.75)}{(137.74836)(4.77624)}$$

$$= +0.74$$

Spearman's Rank Correlation Coefficient

It is a non-parametric measure of the relationship between two sets of ordinal or ranked values. The rank correlation provides a mean of calculating the degree of correspondence between any two sets of rankings. In equation form,

$$r_s = 1 - \frac{6\sum_{i=1}^{n} d^2}{n^3 - n}$$

where, r_s = Rank correlation coefficient, d = Difference of ranks between any two corresponding data and N = Number of data pairs.

The presence of tied ranks may alter the sampling distribution. For a small proportion of tied ranks, the effect is negligible. But in case of large proportions, the following equation should be used:

$$r_s = \frac{(A-B)+(A-C)-\sum_{i=1}^{n} d^2}{2\sqrt{(A-B)(A-C)}}$$

where,

$$A = \frac{N^3 - N}{12},\ B = \frac{\sum_{i=1}^{n}\left(t_x^3 - t_x\right)}{12},\ C = \frac{\sum_{i=1}^{n}\left(t_y^3 - t_y\right)}{12}$$

N = Number of data pairs, t_x = Number of tie in ranks of x, t_y = Number of tie in ranks of y and d = Difference between each x and y ranks

From Table 4.24

$$r_s = 1 - \frac{6(184.5)}{16^3 - 16} \Rightarrow -0.76 \quad \dots \text{(i)}$$

$$A = \frac{16^3 - 16}{12} \Rightarrow 340, \quad B = \frac{0^3 - 0}{12} \Rightarrow 0$$

$$\text{and } C = \frac{10^3 - 10}{12} \Rightarrow 82.5$$

$$\therefore\ r_s = \frac{(340-0)+(340-82.5)-(184.5)}{2\sqrt{(340-0)(340-82.5)}}$$

$$= +0.70 \quad \dots \text{(ii)}$$

Table 4.24 Calculation of Rank Correlation Coefficient

x	Rank (Rx)	y	Rank (Ry)	d = (Rx – Ry)	d^2
125	(1)	25	(3)	– 2	4
150	(2)	23	(2)	– 0	0
175	(3)	28	(5.5)	– 2.5	6.25
200	(4)	32	(8.5)	– 4.5	20.25
225	(5)	26	(4)	+ 1	1
250	(6)	22	(1)	+ 5	25
275	(7)	33	(10.5)	– 3.5	12.25
325	(8)	28	(5.5)	+ 2.5	6.25
350	(9)	36	(14.5)	– 5.5	30.25
375	(10)	34	(12.5)	– 2.5	6.25
400	(11)	31	(7)	+ 4	16
425	(12)	34	(12.5)	– 0.5	0.25
450	(13)	36	(14.5)	– 1.5	2.25
500	(14)	33	(10.5)	+ 3.5	12.25
550	(15)	32	(8.5)	+ 6.5	42.25
575	(16)	39	(16)	0	0
				Σ = 0	Σ=184.5

Table 4.25 Calculation of Rank Correlation Coefficient

x	Rank (Rx)	y	Rank (Ry)	d = (Rx – Ry)	d^2
30	(1)	16	(12)	– 11	121
50	(2)	18	(15)	– 13	169
60	(3)	10	(6.5)	– 3.5	12.25
70	(4)	20	(16)	– 12	144
80	(5)	17	(13.5)	– 8.5	72.25
90	(6)	14	(10)	– 4	16
100	(7)	11	(8)	– 1	1
120	(8)	15	(11)	– 3	9
130	(9)	17	(13.5)	– 4.5	20.25
140	(10)	12	(9)	+ 1	1
150	(11)	7	(4)	+ 7	49
160	(12)	5	(2)	+ 10	100
170	(13)	6	(3)	+ 10	100
190	(14)	10	(6.5)	+ 7.5	56.25
200	(15)	8	(5)	+ 10	100
210	(16)	4	(1)	+ 15	225
				Σ = 0	Σ = 1196

From Table 4.25

$$r_s = 1 - \frac{6(1196)}{16^3 - 16} \Rightarrow -0.76 \quad \text{... (i)}$$

$$A = \frac{16^3 - 16}{12} \Rightarrow 340, \; B = \frac{0^3 - 0}{12} \Rightarrow 0$$

$$\text{and } C = \frac{4^3 - 4}{12} \Rightarrow 5$$

$$\therefore r_s = \frac{(340-0)+(340-5)-(1196)}{2\sqrt{(340-0)(340-5)}}$$

$$= -0.77 \quad \text{... (ii)}$$

TEST OF SIGNIFICANCE

As the value of r satisfies the equation, $+1 \geq r \geq -1$, it can simply be used as a descriptive measure of the degree of correlation between two variables in a sample. Before drawing any valid conclusion, it would be wise to test this for significance, that is, whether this sample correlation coefficient (r) is an accurate measure of the population correlation coefficient (ρ).

The *null hypothesis* of product-moment correlation is that the two sets of data are random samples from two independent and normally distributed populations. Any apparent correlation in the sample data may be due to sampling fluctuations. This means that under null hypothesis,

$$H_0 : \rho = 0$$

The alternative hypothesis states that there is a correlation between the two variables in the population, i.e.,

$H_1 : \rho \neq 0$ (two-tailed test)

$H_1 : \rho > 0$ (one-tailed test)

$H_1 : \rho < 0$ (one-tailed test)

When it merely specifies that there is a correlation without specifying the direction, the test is two-tailed. In a one-tailed test, the direction of correlation (positive or negative) is specified.

In the first example, the sample correlation coefficient between *relative relief* and *settlement frequency* is – 0.75 with (N–2) or 14 degrees of freedom. Therefore, for a one-tailed test ($H_1 : \rho < 0$), the critical value of r at the 0.05 significance level with 14 degrees of freedom is – 0.426 (Appendix). When ρ = 0, the probability of a random sample of 16 individuals producing a coefficient as extreme as $r \leq -0.426$ is 0.05. This is sufficient for the null hypothesis to be rejected at the 0.05 significance level.

In the second example, the sample correlation coefficient between *population density* and *net sown area* is + 0.74 with degrees of freedom (df = N–2 or 14). Therefore, for a one-tailed test ($H_1 : \rho > 0$), the critical value of r at the 0.05 significance level with 14 degrees of freedom is + 0.426 (Appendix). When ρ = 0, the probability of a random sample of 16 individuals producing a coefficient as extreme as $r \geq 0.426$ is 0.05. This is sufficient to enable the null hypothesis to be rejected in favour of a directional alternative hypothesis at the 0.05 significance.

Instead of using the table for Critical Values of Pearson's Product-Moment Correlation Coefficient, significance of **r** can also be tested with the help of a test statistic—Student's t, the critical values of which are given in Appendix 2. The *t* values are computed by the formula,

$$t = \frac{r\sqrt{n-2}}{\sqrt{1-r^2}}$$

In the first example,

$r = -0.75$,

$df = 16 - 2$ or 14

$\therefore t = 4.243$

Therefore, for $H_1 : \rho < 0$, the critical value of t at 0.05 significance level with 14 degrees of freedom is 2.145.

Therefore, at df = 14, $t_{0.05} < 4.243$

As $t_{computed} > t_{tabled}$, the null hypothesis is rejected in favour of alternate hypothesis, i.e., the correlation between the variables, x and y is significant.

To test the significance of r_s, the process is exactly similar. For this, the table for Critical Values of Spearman's Rank Correlation Coefficient ($\mathbf{r}_s$) is to be consulted (Appendix). In this particular analysis, the degrees of freedom is exacty equal to the number of pairs of data, i.e., df = N.

TIME SERIES ANALYSIS

Geographers have a special interest in *time series analysis* because almost all geographical data vary with time. Therefore they need to describe the *nature of the pattern of variation* with time and thereby extract precise information about the *particular trend* so that *estimates can be made with greater confidence*. Pattern of variation is statistically described as movements in time series. This may take four different forms—*secular, periodic, cyclical* and *irregular*.

Fig. 4.18 Time Series Analysis

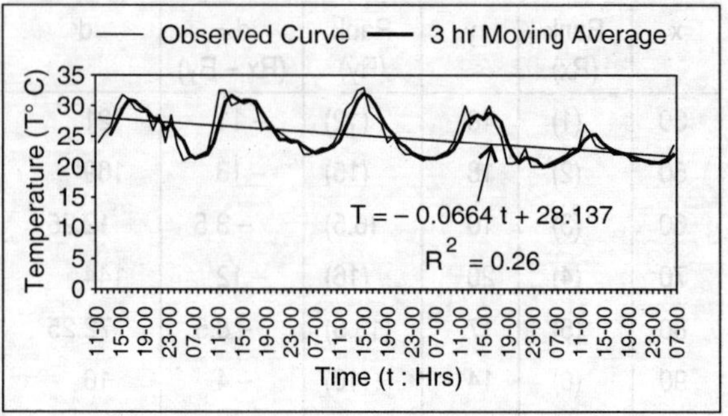

In *a secular* movement, a single trend (either upward or downward) is unusually predominant and virtually no other movement can be discerned. For instance, the growth of population of a country usually produces a distinct upward trend. A *periodic* movement is one which recurs with some degree of regularity, within a definite period. The most frequently studied periodic movement is that which occurs within a year and is known as *seasonal variation* or simply *seasonal*, e.g., production of fruit, milk, etc. *Cyclical* movements are special cases of periodic movements in that they are of *longer duration* (usually more than a year) and also in that they *do not ordinarily exhibit regular periodicity*, e.g., business cycle, climatic cycle, etc. Lastly, the *irregular* variations may be either *episodic* or *accidental*. Episodic variation that takes place as a result of specific events with large scale effects, e.g., earthquake, flood, epidemic, etc. Accidental variations are minor fluctuations not attributable to specific episodes and are too small to merit individual consideration. These are *random* in nature in occurrence.

Time series data, as the name shows, concerns progressive changes in the attribute variables with time. Thus, it is a special case of bivariate analysis, in which the variable 'time' is the independent variable and is plotted on the x- axis, while the other as dependent variable on the y- axis. It may either be plotted as bi-axial line graphs or as bivariate scatter graphs.

Measurement of trend is necessary to

Example: Eye Estimation/Free-hand Method

Year	Production ('000 tons)	Year	Production ('000 tons)	Year	Production ('000 tons)
1998	64	2002	78	2006	88
1999	82	2003	112	2007	100
2000	97	2004	115	2008	146
2001	71	2005	131	2009	150
				2010	120

Source: Compiled by the author

study the behaviour of the time series, to investigate the influence of the seasonal and cyclical components by removing trend from the original series, and for estimating the future position. There are four methods of doing this—*free-hand method, semi-average method, moving average method, and fitting mathematical curves.*

1. In free-hand method the given data are plotted as points on a graph paper against time. The time series data (y_t) are shown along the vertical axis and the time (t) along the horizontal axis. A smooth free-hand curve is then drawn through the scatter of the plotted points, which represents their pattern of movement over time. By eye estimation a straight line is then drawn through these scatter of points to show the trend line as appears to the investigator. The distances of this trend line from the horizontal axis are now read from the graph for all the years giving 'trend values'. The advantages of this method are that a quick estimate of the trend can be easily obtained. However, this method depends too much on individual judgement and is better to avoid.

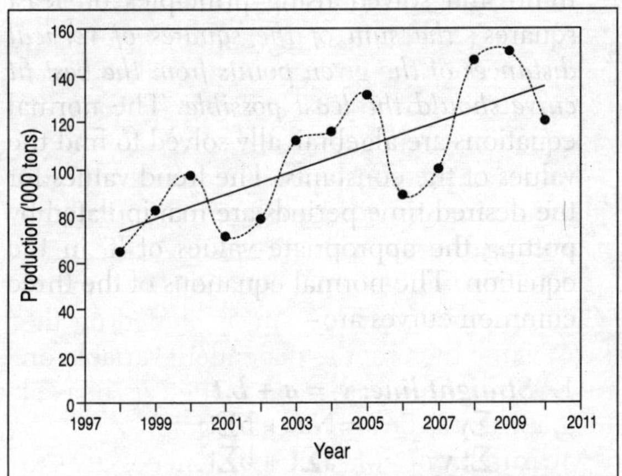

Example: Semi-Average Method

Year	Production ('000 tons)	7-Year Semi Total	Semi-Averages
1998	64		
1999	82		619 / 7 = 88.4
2000	97		
2001	71	619	88.4
2002	78		
2003	112		
2004	115		850 / 7 = 121.4
2005	131		
2006	88		
2007	100	850	121.4
2008	146		
2009	150		
2010	120		

Source: Compiled by the author

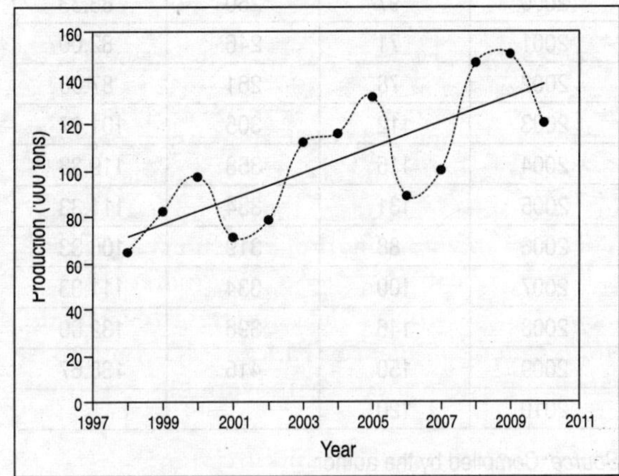

2. In semi-average method, the data is first divided into two parts, and then averages for these two parts are computed. These are then plotted as points on a graph paper against the mid-point of the time interval covered by each part. The straight line joining these two points gives the trend line, and its distances from the horizontal line gives the trend values for different time periods. This method produces quite satisfactory results if the actual trend is a straight line. The semi averages are plotted on a graph paper against the years 2001 and 2007 and a straight line is drawn through these points and extended on both ends (see Example).
3. The moving average method is a very convenient tool for smoothing out fluctuations in time series. In this a series of arithmetic means of successive observations, known as moving averages, are computed and are used as trend values. These are shown against mid-points of time intervals covered by the respective groups. It is a very simple method and needs no complicated mathematical calculations. A disadvantage is that some trend values at the beginning and at the end of the series can not be obtained.

When the period is odd, all moving averages coincide with the given years. But, when the period is even, moving averages fall midway between two successive years. In such cases 2-item moving averages of the moving averages already found are computed for synchronizing them with the given data. It is known as 'centering'. The case of 4 – year moving averages is exemplified below—

Example: 3-Year Moving Average Method

Year	Production ('000 tons)	3-Year Moving Total	3-Year Moving Average
1998	64		
1999	82	243	81.00
2000	97	250	83.33
2001	71	246	82.00
2002	78	261	87.00
2003	112	305	101.67
2004	115	358	119.33
2005	131	334	111.33
2006	88	319	106.33
2007	100	334	111.33
2008	146	396	132.00
2009	150	416	138.67
2010	120		

Source: Compiled by the author

Example: 4-Year Moving Average Method

Year	Value	4-year MT: NC	4-year MA: NC	2-item MT: C	4-year MA:C
2002	105	-	-	-	-
2003	115	410	102.5	-	-
2004	100	385	96.25	198.75	99.375
2005	90	187.50	93.750	365	91.25
2006	80	178.75	89.375	350	87.50
2007	95	171.25	85.625	335	83.75
2008	85	162.50	81.250	315	78.75
2009	75	-	-	-	-
2010	60	-	-	-	-

MT = moving total NC = not centered C = centered

Source: Compiled by the author

4. In curve fitting method, an appropriate type of mathematical equation is selected for trend and solved using principles of 'least squares': *the sum of the squares of vertical distances of the given points from the best fit curve should the least possible*. The normal equations are algebraically solved to find the values of the constants. The trend values for the desired time periods are manipulated by putting the appropriate values of 't' in the equation. The normal equations of the three common curves are—

1. ***Straight line:*** $y_t = a + b.t$

$$\Sigma y = Na + b\Sigma t$$
$$\Sigma t.y = a\Sigma t + b\Sigma t^2$$

4. ***Exponential Curve: log yt = a + b.t***

$$\Sigma \log y = Na + b\Sigma t$$
$$\Sigma(t.\log y) = a\Sigma t + b\Sigma t^2$$

5. ***Power Curve: yt = a.t^b***

$$\Sigma \log y = N.\log a + b\Sigma \log t$$
$$\Sigma(\log t.\log y) = \log a\Sigma \log t + b\Sigma(\log t)^2$$

Example: Fitting a Straight Line or 1st Degree Curve

Year	t	Production (y: '000 tons)	t^2	y.t
1998	–6	64	36	–384
1999	–5	82	25	–410
2000	–4	97	26	–388
2001	–3	71	9	–213
2002	–2	78	4	–156
2003	–1	112	1	–112
2004	0	115	0	0
2005	1	131	1	131
2006	2	88	4	176
2007	3	100	9	300
2008	4	146	16	584
2009	5	150	25	750
2010	6	120	36	720
	Σ= 0	Σ= 1354	Σ= 192	Σ= 998

Source: Compiled by the author

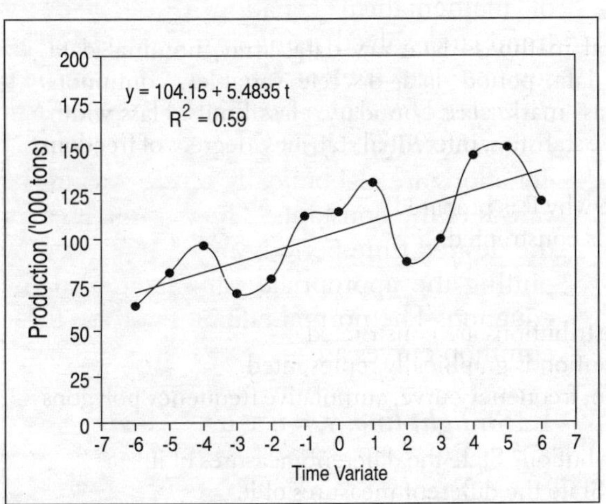

Therefore,

$$1354 = 13a + b.0$$
$$998 = a.0 + 192b$$
$$\text{or, } a = 1354/13$$
$$= 104.15$$
$$\text{or, } b = 998/192$$
$$= 15.197917$$

The trend equation is: $y_t = \mathbf{104.15 + 15.197917t}$

Example: Fitting an Exponential Curve

Year	t	Production (y: '000 tons)	logy	t^2	t.log y
2002	–4	15	1.17609	16	–4.70437
2003	–3	32	1.50515	9	–4.51545
2004	–2	58	1.76343	4	–3.52686
2005	–1	77	1.88649	1	–1.88649
2006	0	98	1.99123	0	0.00000
2007	1	126	2.10037	1	2.10037
2008	2	270	2.43136	4	4.86273
2009	3	372	2.57054	9	7.71163
2010	4	574	2.75891	16	11.03565
	Σ = 0	Σ = 1622	Σ = 18.18358	Σ = 60	Σ = 11.07721

Source: Compiled by the author

Therefore,

$$18.18358 = 9a + b.0$$
$$11.07721 = a.0 + 60b$$

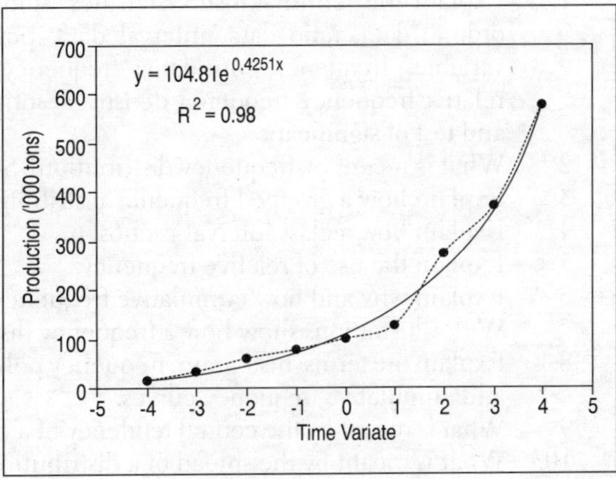

$a = 18.18358 / 9$
$= 2.02039$
$b = 11.07721 / 60$
$= 0.18462$

The trend equation is: $\log y_t = 2.02039 + 0.18462t$

or $y = 104.81e^{0.4251t}$

Example: Fitting a Power Curve

Year	t	Population ('000)	log t	log y	log y.log t	(log t)^2
1996	1	0.8	0.00000	–0.09691	0	0.00000
1997	2	2	0.30103	0.30103	0.090619	0.09062
1998	3	4	0.47712	0.60206	0.287256	0.22764
1999	4	8	0.60206	0.90309	0.543714	0.36248
2000	5	13	0.69897	1.11394	0.778613	0.48856
2001	6	18	0.77815	1.25527	0.976792	0.60552
2002	7	27	0.84510	1.43136	1.209643	0.71419
2003	8	50	0.90309	1.69897	1.534323	0.81557
2004	9	112	0.95424	2.04922	1.955451	0.91058
2005	10	149	1.00000	2.17319	2.173186	1.00000
2006	11	300	1.04139	2.47712	2.579656	1.08450
2007	12	520	1.07918	2.71600	2.93106	1.16463
2008	13	760	1.11394	2.88081	3.209063	1.24087
2009	14	1220	1.14613	3.08636	3.537364	1.31361
2010	15	2500	1.17609	3.39794	3.996288	1.38319
			Σ= 12.11650	Σ= 25.98946	Σ= 25.80303	Σ= 11.40196

Source: Compiled by the author

Therefore,

$25.98946 = 15\log a + 12.11650b$

$25.80803 = 12.11650\log a + 11.40196b$

$\log a = -0.67352$

$a = 0.21207$

$b = 2.97876$

The trend equation is:

$\log P_t = -0.67352 + 2.97876\log t$

or $P = (0.21207)t^{2.97876}$

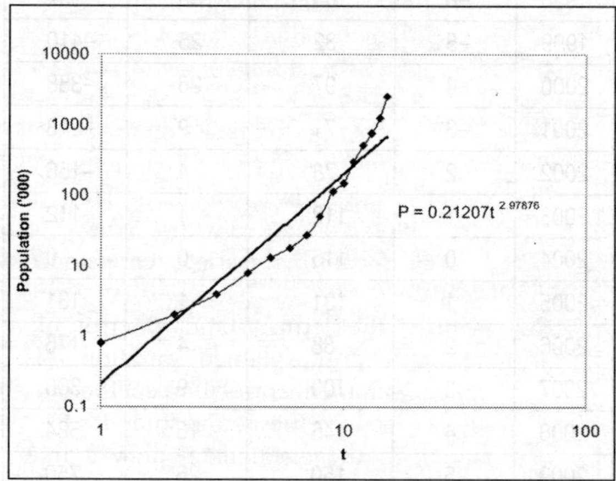

EXERCISE

1. Explain the terms: statistic, statistics, statistical methods, data, raw data, array, nominal data, ordinal data, ratio data, interval data, point data, period data, discrete variables, continuous variables, frequency, cumulative frequency, class mark, class boundary, class limits, class width, relative frequency, frequency density, descriptive statistics, inferential statistics, degrees of freedom and test of significance.
2. What is meant by frequency distribution? State why it is prepared.
3. Explain how a grouped frequency distribution is constructed.
4. Explain how a class interval is chosen.
5. Explain the use of relative frequency.
6. Explain why and how cumulative frequency distributions are constructed.
7. With illustrations show how a frequency distribution is graphically represented.
8. Explain the terms: histogram, frequency polygon, frequency curve, cumulative frequency polygons and cumulative frequency curves.
9. What is meant by the central tendency of a distribution? State the different measures of it.
10. What is meant by the spread of a distribution? State the different measures of it.

11. What does the shape of a distribution mean? State the different measures of it.
12. Explain the terms: average, arithmetic mean, geometric mean, harmonic mean, median and mode.
13. Give the properties of arithmetic mean.
14. Give the properties of median.
15. Give the properties of mode.
16. State the relationship between mean, median and mode.
17. What are fractiles? State its different measures.
18. State how the unimodal, bimodal and multimodal distributions can be explained.
19. Explain the terms: range, quartile deviation, mean deviation, variance, standard deviation, variablity and standard score.
20. State the relations between standard deviation and variance, standard deviation and mean deviation, standard deviation and range, standard deviation and quartile deviation.
21. Explain the frequency ranges in a normal distribution in terms of mean and standard deviation.
22. Explain how reliability of occurrence can be measured and interpreted.
23. Explain the situation in which a data needs to be transformed and standardized.
24. Divide a distribution based on the symmetry of frequency curve and state its measures.
25. Divide a distribution based on the peakiness of the frequency curve and state its measures.
26. What is meant by correlation? How it is shown? Which three basic items is the concept of correlation concerned with?
27. What is meant by bivariate regression? Explain how a bivariate linear equation of regression is solved.
28. Explain the terms: standard error of estimate, coefficient of determination, coefficient of correlation, unexplained variation, explained variation, total variation, residuals, Pearson's product-moment correlation coefficient, Spearman's rank correlation coefficient, null hypothesis, alternative hypothesis, and student's t.
29. Based on the given data, draw a frequency distribution table and a cumulative frequency distribution table.
    1. Relative Relief (m): 40, 60, 80, 100, 120, 140, 160, 180, 200, 220, 240, 260, 280, 300, 320, 340, 360, 380, 400, 420, 440, 460, 480, 500, 520, 540, 45, 50, 55, 58, 65, 68, 72, 90, 110, 125, 135, 155, 165, 210, 230, 250, 275, 285, 310, 330, 355, 345, 360, 365, 370, 372, 275, 378, 425, 450, 465, 475, 485, 510.
    2. Spot Heights (m): 125, 190, 242, 200, 220, 230, 236, 256, 255, 275, 300, 276, 287, 298, 320, 335, 336, 345, 321, 310, 337, 330, 350, 365, 378, 382, 371, 390, 377, 400, 421, 438, 442, 455, 478, 498.
30. Based on the given data, draw a histogram, a frequency polygon, a frequency curve, cumulative frequency polygons and cumulative frequency curves.

Mean fainfall (cm)	0–25	25–50	50–75	75–100	100–125	125–150
No. of Stations	6	14	25	112	35	8

Population density	201–400	401–600	601–800	801–1000	1001–1200	1201–1400
(persons/sq km)	12	34	125	80	25	8

31. Based on the given data, find the values of central tendency and graphically show it (Q 29 and Q 30).
32. Based on the given data, find the values of dispersion and graphically show it (Q 29 and Q 30)
33. Based on the given data, find the shape of the distribution and graphically show it (Q 29 and Q 30).

34. Based on the given data, draw a scatter diagram and interpret it.

X	452	458	465	468	470	472	474	476	480	485	488	492	494	497	498	500
Y	3.1	3.3	3.8	4.6	6.7	5.5	6.5	4.9	5.1	5.4	6.3	6.6	7.0	7.1	7.3	7.9

X	20	40	60	80	100	120	140	160	180	200	220	240	260	280	300	320
Y	252	200	210	225	180	165	125	110	136	148	130	118	110	102	100	85

35. Based on the given data, find the equation for best fit for linear regression (Q 34).
36. Based on the given data, find the values of correlation coefficient and test its significance (Q 34).

5

Presentation of Geographical Data

HIGHLIGHTS

- Graphs
- Biaxial Graphs
- Triaxial Graphs
- Multiaxial Graphs
- Special Graphs
- 1-Dimensional Diagrams
- 2-Dimensional Diagrams
- 3-Dimensional Diagrams
- Map

Geographical data record the locations and characteristics of natural features or human activities that occur on or near the earth's surface. It has two important properties—the reference to geographic space and the representation at geographic scale. Based on nature and use, it falls into three categories—the geodetic control network (GCN), the topographic base (tBase) and the geographic overlays (GrO). The GCN provides a geographical framework by which different sets of geographical data can be spatially cross referenced with one another. The tBase comprise the elevation database of the earth's surface, while the GrO, the thematic database. Thus thematic maps are prepared with attribute database as graphic overlays on a reference map generated by GCN and tBase.

Geographical data is recorded using four different means—*numerical values*,. *alpha-numeric characters* or *texts*, *symbols* or *signs*, and *signals*. These are found in four basic forms as—*document form*, *numeric form*, *graphical form* and *digital form*. These are transformed through *structuring*, *formatting*, *conversion* and *modelling* to build information. *Document* data refers to the qualitative information, written as a text. There is huge geographical information in document form, from which necessary information can be collected and applied at higher levels by way of format conversion. In the *numeric* form, data are systematically organised in tables and semi-tables, from which analysis and presentation are easier. Data in *graphical* form may occur as graphs, diagrams, maps, data products, aerial photographs and satellite imagery. In *digital form*, data is represented by three basic models—*raster*, *vector* and *surface*. From all these, information at a particular

Fig. 5.1 Tabular presentation

Title → Changing Pattern of Cropping Intensity in the Hoogly District, West Bengal

Blocks	Year #			
	1971	1981	1991	2001
Arambagh	135.05	198.26	226.03	220.10
Khanakul – I	126.06	190.63	195.27	312.54
Khanakul – II	101.98	127.29	194.57	201.89
Pursurah	196.91	245.81	290.04	302.88
Goghat – I	105.66	210.56	184.20	163.15
Goghat – II	101.30	150.81	165.87	228.88
Chanditala – I	100.75	133.11	174.68	232.77
Chanditala – II	110.23	138.44	231.41	187.33
Jangipara	104.87	145.80	253.56	207.26
Serampur – Uttarpara	138.67	121.44	255.54	153.36
Polba – Dadpur	121.07	136.79	216.84	179.92
Dhaniakhali	149.11	171.71	203.42	245.85

Stub (Blocks column) · **Box head** (Year # / 1971–2001) · **Body** (figures)

Foot note → @ : District Average; # : All Figures in percentage; ¶ : 1st Column = Spatial Units; ‡ : Remaining Columns = Relating to Attribute Data

Source note → *Source:* Compiled by the author from Statistical Abstract of Hoogly District, 1971, 1981, 1991, and 2001

location or at a particular instant can be easily extracted for further manipulation and analysis.

Geographical data needs to be *scientifically visualised* for appropriate *cartographic communication*. Therefore, *cartographic presentation* of geographical data is important to explore the nature of data, their pattern of variations over time and space, to manipulate and analyse the data set, to communicate the knowledge of spatial information, to identify and classify the real world objects, to perceive relationships, to formulate principles, to introduce personal values and beliefs and thereby intelligence or wisdom, which finally enriches our life and society for a particular decision-making.

Thus, it helps us understand better the world around us by enabling us to develop *spatial intelligence* for *logical decision making* and it is this collective intelligence of mankind, which actually drives the evolution of human civilisation.

Geographers deal with a multitude of statistical information and data. These are represented in four distinct ways—a *textual form*, a *tabular form*, a *semi-tabular form* and a *graphical form*. (Table 5.1). The advantage of a textual representation is that the author can draw attention to and emphasise upon certain figures. A *tabular representation*, on the other hand, is

Table 5.1 Forms of Geographical Data Representation

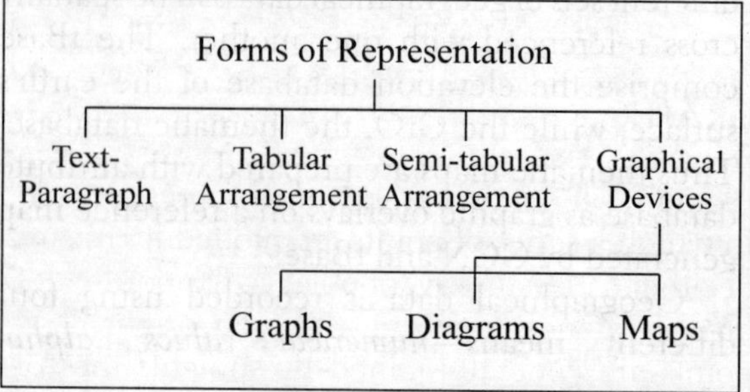

much superior to the textual representation since it is more concise and brief.

A table generally consists of four essential components—*a title* or *heading*, *a stub* (the left hand column and its heading), *a box head* (the other columns and their headings) and *a body*. There may also be a *prefatory note*, a *footnote* and a *source note*. A *semi-tabular* presentation is a combination of the two and has no great advantage in general. A *graphical* representation, on the other hand, is more useful and effective for a quick visualisation of a limited amount of information.

A simple, attractive and well-drawn graph, diagram or map is *easy to understand* and often serves as *a tool* for further geographical analysis. Based on the objectives of analysis, suitable data is collected and arranged. The choice of a particular graphical device depends upon the type of data (*nominal*, *ordinal*, *interval* and *ratio* data), ease of calculation, ease of construction and the amount of space and time available to the cartographer.

All computations are presented in the form of a *well designed table* which includes all the contextual items. These are *title*, *legend* or *index*, *scale* (of maps, diagrams and graphs separately), *north lines* (in case of maps), *border lines*, *guide line* (wherever necessary) and *source notes* (if not original). They should be properly labelled and coloured or shaded for visual effect. The design of the lettering—its *type*, *size*, *direction*, *spacing* and *boldness* must reflect the importance or the degree of emphasis laid on the item to be shown. All these determine the cartographic quality of the graphical composition and are vital to the geographers.

GRAPHS

A graph is the most conventional representation in which a series of points are plotted by means of either cartesian (x, y) or polar (r, q) coordinates on a reference frame of a chosen origin. The points are generally joined by straight lines of various styles. There are several types of graphs (Table 5.2) from which the one most suitable is chosen.

Graphical Presentations

- **Biaxial:** Rectangular reference frame; y- values are plotted against values as points and are successively joined either by smooth curves or by straight lines.
- **Triaxial:** An equilateral triangle is drawn as the reference frame; sets of x, y and z values taken along the three sides are plotted as points.
- **Multiaxial:** Reference frame comprises a network of evenly spaced lines radiating from a centre; in vector graphs, the radiating lines are drawn at true azimuth. In star graphs or circular graphs, lines radiate at uniform angles. Attribute values are plotted along the radiating lines and successive points are joined.
- **Special:** *Relative Temperature curve, Water Balance graph, Ombrothermic graph, Hydrograph, Rating curve, Lorenz curve, Rank-Size graph, Ergograph* and *Composite graph*

Note: All presentations must have a title, legend (where necessary), grid lines, scaling of the axis, labelling of the attribute and attribute values with proper units, and other contextual items.

BIAXIAL GRAPH

Biaxial graphs are drawn on a reference frame of two mutually orthogonal axes. They are generally known as *line graphs* and are of two types—*open* and *closed*.

Open Line Graph

In these the points are plotted by means of rectangular coordinates with reference to a point of origin at which the x-axis (horizontal line) and y-axis (vertical line) intersect at right angles. The positive values are plotted to the right of the origin (zero) of the x-axis and above the origin (zero) of the y-axis, while the negative values are placed to the left of the origin (zero) of the x-axis and below the origin (zero) of the y-axis.

Table 5.2 Types of Graphs

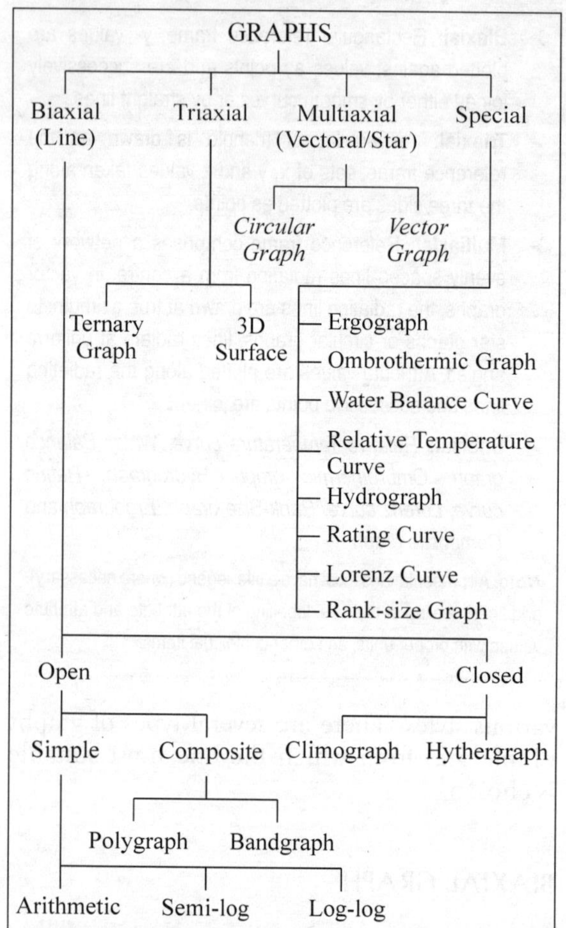

Both the positive and negative values of the axes increase away from the origin.

Normally, *period data*, i.e., data on time frame and *point data*, i.e., spatial or vector data are represented in these graphs. The chronological units are shown on the x - axis and the attributes of the variable on the y - axis. While plotting a curve, the vertical scale along the y - axis must be graduated from zero. However, scale-breaks can be used in demanding situations. The curves must stand out clearly from the background. Therefore, it is ruled more distinctively than the axes and guidelines. When several curves are drawn on the same frame each should be clearly identified and distinguished by line, style or colour. A line graph is normally a visual aid to understand the overall changes in value. As it rarely represents the true locus, points should be joined by a series of short straight lines. However, smoothening may be done by applying the *statistical methods of curve fitting*.

A *simple line graph* represents only a single series of values connected by a curve (Fig. 5.2). A *polygraph* is a mutliple line graph in which several sets of values are represented by distinctive lines for direct comparison (Fig. 5.3). A *band graph* is simply an aggregate or a compound line graph in which the trends of values in both the total and its constituents are shown by a series of lines on the same frame. The area between two successive lines is distinctively shaded for clarity (Fig. 5.4).

Line graphs are normally drawn on an *arithmetic scale,* but when the range and

Fig. 5.2 Simple Line Graph

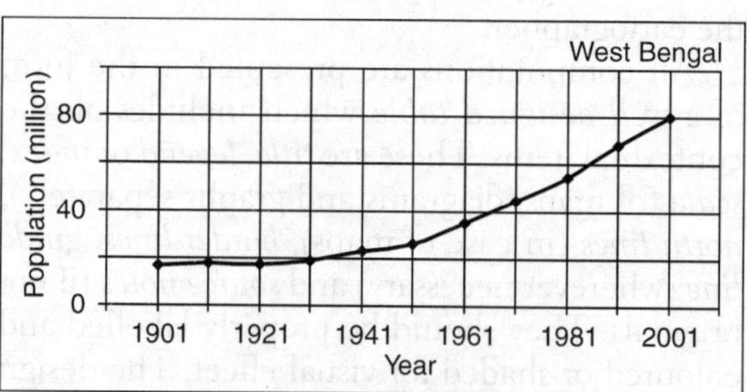

Fig. 5.3 Polygraph

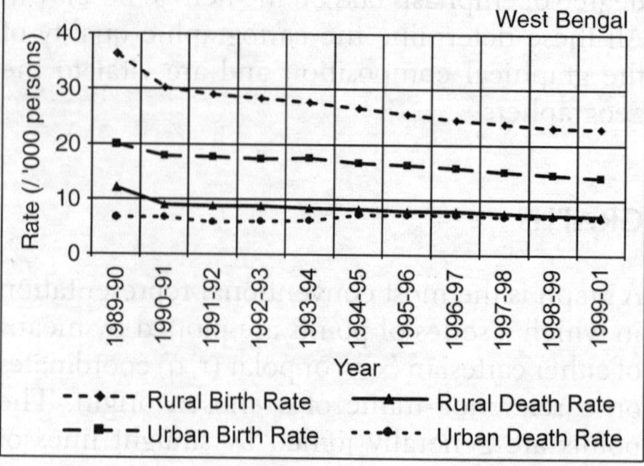

dispersion of the data are very high, a *logarithmic scale* is taken. Such graphs are called *log graphs*. When logarithmic scales are taken along both the x - axis and the y - axis, the graphs are called *log graphs* (Fig. 5.5). When a logarithmic scale is taken along the y - axis and an arithmetic scale along the x - axis, they are called *semi-log graphs* (Fig. 5.6). These are particularly used when the variable is characterised by a very large range of values and a roughly constant ratio of increase or decrease. In *arithmetic scale*, such data would produce a *curve* (up-concave or up-convex or sloping upward or sloping downward) which is difficult to interpret while in *log graphs* it would produce a straight line.

Fig. 5.4 Bandgraph

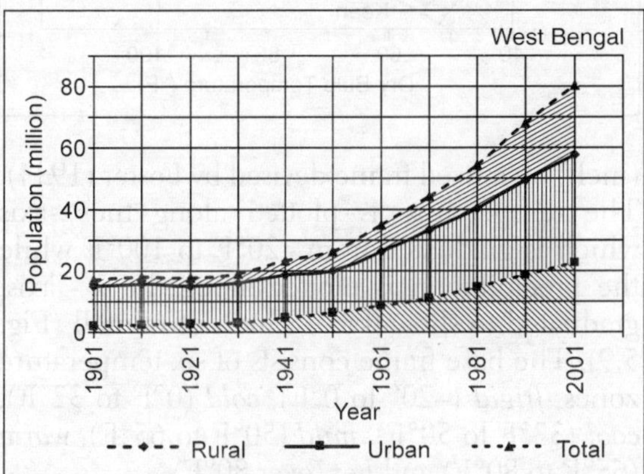

Fig. 5.5 Structure of Log-scale

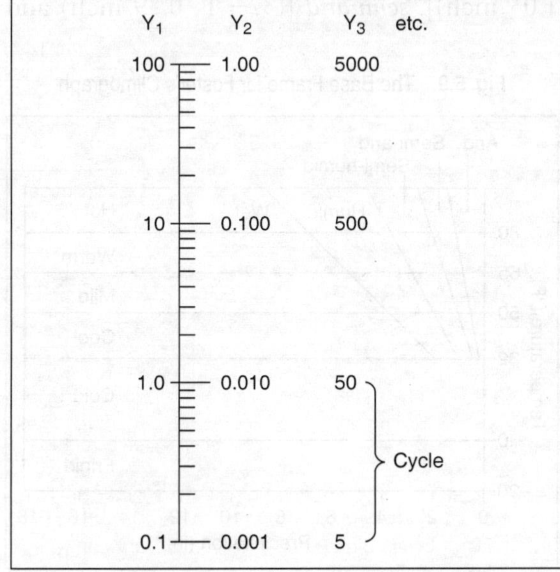

Fig. 5.6 Semi-log Graph

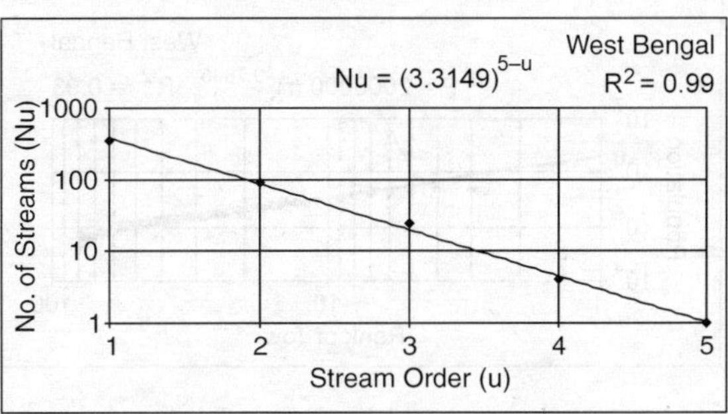

A log scale is based on the principle that the *increase* or *decrease of values occurs in terms of a common ratio*. Thus equal distances measured along a log scale represents equal ratios. In log graph papers, the vertical axis is divided into parts of equal lengths, called *cycles*. A single cycle accommodates a tenfold increase while two cycles make provision for a hundredfold increase. This is so because a common logarithm is the power to which 10 must be raised to produce a given number. Thus, $100 = 10^2$, and the logarithm of 100 is 2.00 and so on. The log scale can never begin with zero, as log (zero) = ∞ (infinity). Therefore any positive value of the user's choice may be taken as the origin. The figure at the top of the first cycle is ten times of that at the bottom of it and the figure at the top of the second cycle is ten times of that at the bottom of it (i.e., the top of the first cycle) and so on. Thus, the vertical scale values are spaced in proportion to the differences between their logarithms (Fig. 5.7).

Once the frame is ready, the position of x within a cycle with top and bottom figures as t and b respectively can be evaluated as:

$$y = \frac{\log x - \log b}{\log t - \log b} \times d$$

$= (\log x - \log b) \times d$ (since, $\log t - \log b = 1$)
where, d = length of a cycle on graph.

Fig. 5.7 Log-log Graph

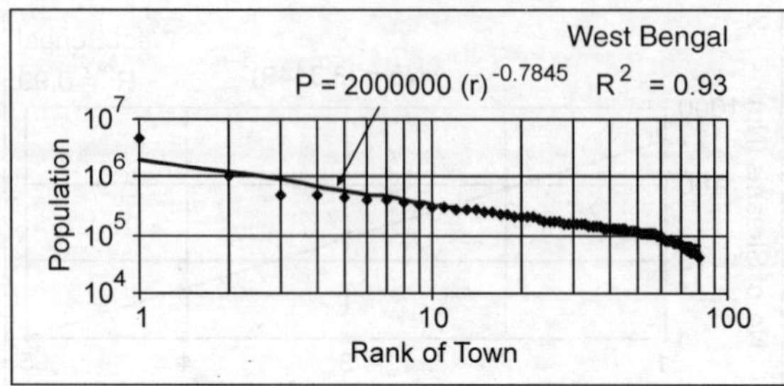

In case no suitable logarithmic paper is available, it can easily be constructed by applying the above principle with a chosen cycle length. Enough rulings should be drawn to avoid any confusion.

Closed Line Graphs

Climograph

A climograph (or climogram) is a 12-sided closed polygon required for plotting the mean-monthly values of selected climatic elements of a particular station against one another. The shape and position of the resultant graph on the framework provides an index to the general climatic character of a place. The base frame of climographs was first conceived by Ball (1910) and later expanded and improved by Leighly (1926), USDA (1941), Foster (1944) and Taylor (1949).

a. The USDA–type

In this the mean-monthly values of wet-bulb temperatures (°F) are plotted against those of the dry-bulb temperatures (°F) on a fixed frame devised by the United States Department of Agriculture (1914). Each month is marked by a letter symbol and the points are then successively joined to form a closed 12 - sided polygon. It is normally used to describe climatic conditions in terms of human physiological comfort (Fig. 5.8).

b. The Foster–type

In this the mean-monthly values of temperatures (°F) are plotted against those of precipitation

Fig. 5.8 The Base Frame for USDA Climograph

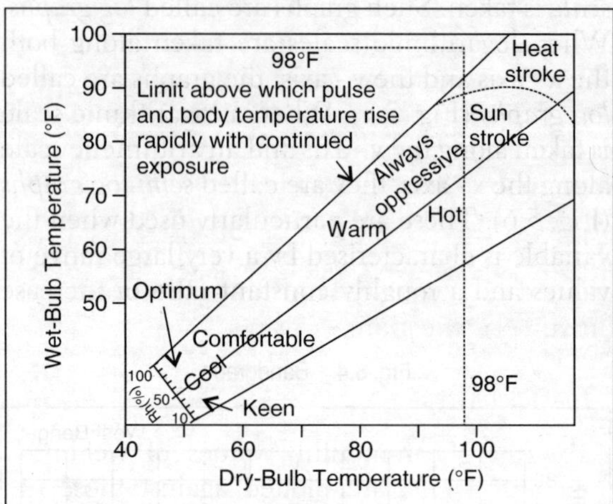

(inch) on a fixed frame devised by Foster (1944). The temperature is plotted along the y-axis which is graduated from –20°F to 100°F while the precipitation is plotted along the x - axis, graduated to show 0 to 18 inch of rainfall (Fig. 5.9). The base frame consists of six temperature zones, *frigid* (–20° to 0°F), *cold* (0°F to 32°F), *cool* (32°F to 50°F), *mild* (50°F to 65°F), *warm* (65°F to 80°F) and *hot* (over 80°F).

The last four zones are further subdivided into *arid* [(32.4°F, 0.32 inch) and (83.2°F, 1.03 inch)], *semi-arid* [(32.4°F, 0.59 inch) and

Fig. 5.9 The Base Frame for Foster's Climograph

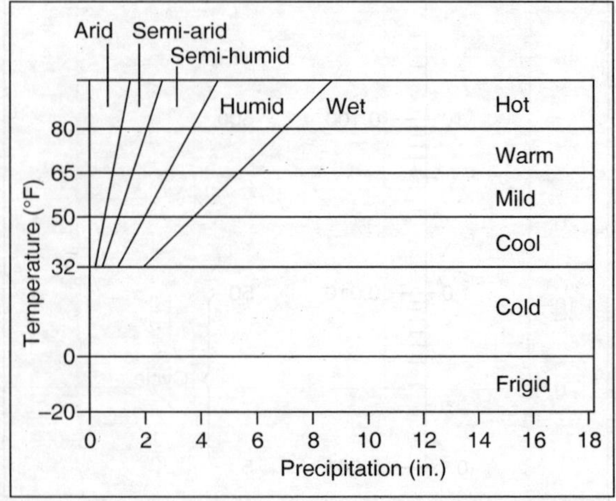

(83.2°F, 1.93 inch)], *subhumid* [(32.4°F, 1.10 inch) and (83.2°F, 3.60 inch)], *humid* [(32.4°F, 2.05 inch) and (83.2°F, 6.73 inch)] and *wet* [(32.4°F, more than 2.05 inch) and (83.2°F, more than 6.73 inch)]. The subdivisions of the last four zones have the temperature and rainfall given in brackets. Each month is marked by a letter symbol and is successively joined to form the polygonal climograph. It is basically designed to illustrate Thornthwaite's system of climatic classification.

c. The Taylor–type

In this the mean-monthly values of wet-bulb temperatures (°F) are plotted against those of relative humidity (%) on a fixed frame devised by Taylor (1949) to show the physiological effects of climate on man. The wet-bulb temperature is plotted along the y-axis, graduated from –10°F to 90°F while the relative humidity is plotted along the x-axis, graduated from 20% to 100%.

The four corners are marked as *raw* (SE), *muggy* (NE), *scorching* (NW) and *keen* (SW). Raw denotes low wet-bulb (below 40°F) and high relative humidity (over 70%), muggy denotes high wet-bulb (over 60°F) and high relative humidity (over 70%), scorching denotes high wet-bulb (over 60°F) and low relative humidity (below 40%) and keen denotes low wet-bulb (below 40°F) and low relative humidity (below 40%).

Fig. 5.10 Climograph after Taylor

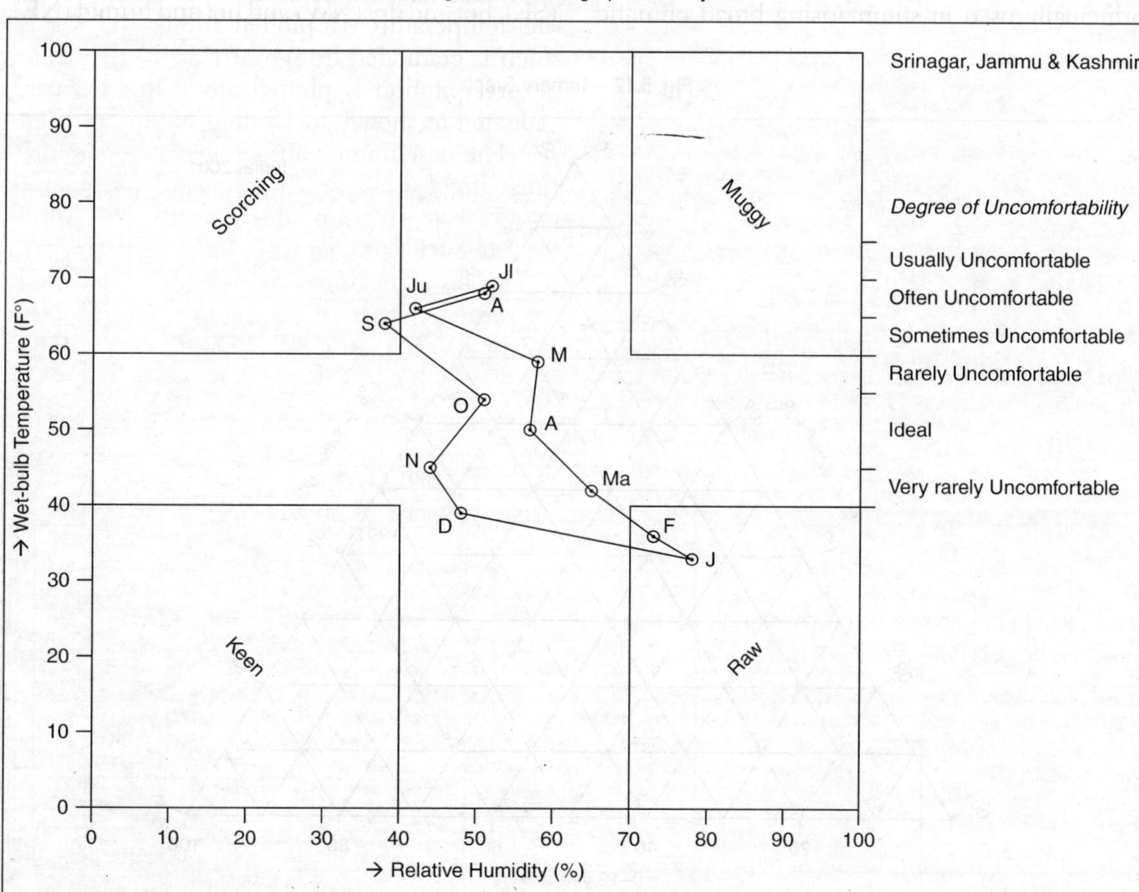

A *scale of discomfort* is shown on the right of the rectangular frame as: *very rarely uncomfortable* (below (45°F), *ideal* (45°F – 55°F), *rarely uncomfortable* (55° – 60°F), *sometimes uncomfortable* (60° – 65°F), *often uncomfortable* (65° – 70°F) and *usually uncomfortable* (above 70°F). Each month is marked by a letter symbol and is successively joined by straight lines to form the polygonal climograph (Fig. 5.10).

Fig. 5.11 Hythergraph

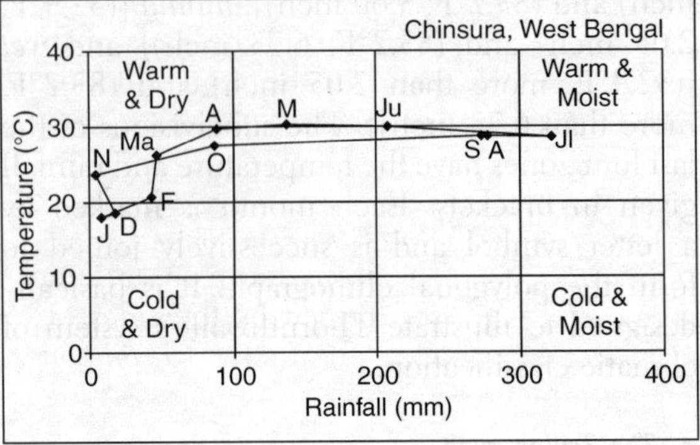

Hythergraph

The hythergraph is another type of climograph devised by Taylor (1916). In this mean-monthly temperature values are plotted as ordinates and mean-monthly rainfall values as abscissae. This is drawn as a 12-sided polygon in just the same way as that of the climograph (Fig. 5.11). These are principally used in summarising broad climatic differences in relation to human activity, more precisely in the context of settlement. The four quarters naturally denote four distinctive climatic conditions – cold and dry (SW), cold and wet (SE), hot and dry (NW) and hot and humid (NE).

Fig. 5.12 Ternary Graph

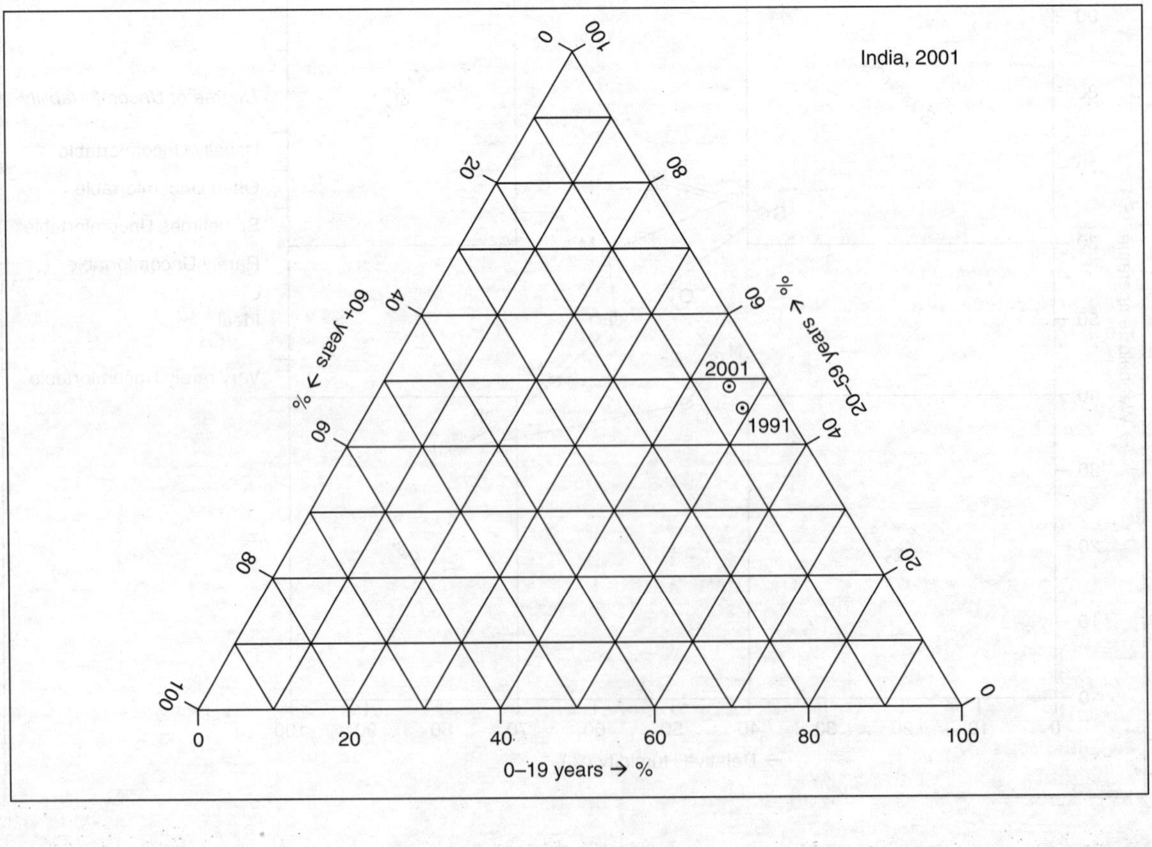

TRIAXIAL GRAPH

Triaxial graphs are drawn on the basis of three axes. They are useful in the representation of three interrelated variables.

Ternary Graph

These are triangular graphs drawn on a reference frame of three axes formed by the sides of an equilateral triangle. In these *tri-component* data are plotted in the form of points (Vincent 1947). Each component is first expressed as a percentage and is arranged uniformly from 0 to 100% along the three sides of an equilateral triangle (Fig. 5.12). Since the component values all add up to 100 percent, all three values are plotted on the graph by a single point. The location of this point within the triangle gives an immediate impression of the compositional character of that sample relating to a certain place or time. Therefore, these are extremely useful in taxonomic problems involving a large number of samples. Each point is marked by a letter or a number corresponding to the respective samples. The most commonly used data for this graph are: *age composition of population* (young, adult and old); *occupational structure of population* (primary, secondary and tertiary); *textural composition of soil* (sand, silt and clay); *composition of mass movement processes* (flow, slide and heave); etc.

The 3D-surface

These are highly sophisticated statistical graphs very useful in revealing the relationship among three variables taken for analysis. Normally there must be one dependent (z) and two independent (x and y) variables plotted along three mutually orthogonal axes such that the resultant surface be z = f (x, y). Variation of any geographical element (z) on space (x, y) or with space (x) and time (y) or variation of an element (z) in relation to any two factors (x and y) may be best represented by this graph (Fig. 5.3b and Fig. 5.13).

Fig. 5.13 3D Surface of Relative Relief

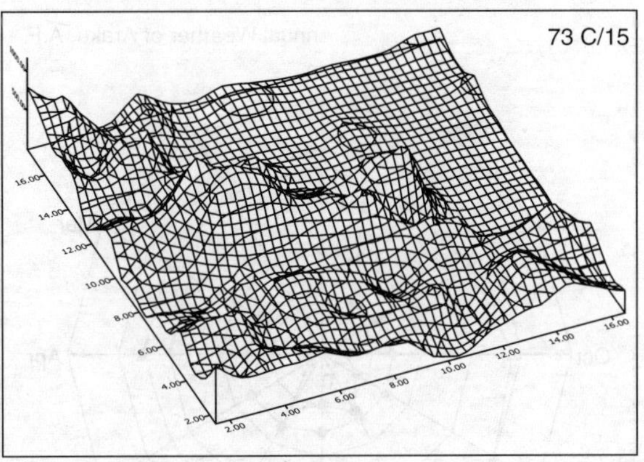

MULTIAXIAL GRAPHS

As the name denotes, such graphs are drawn on a frame of many axes.

Star Graph

In these, values are plotted as radii from a point of origin with the help of polar coordinates: radius vector (*distance from the point of origin*) and vectoral angle (*direction*). Obviously, stargraphs are especially useful where vector values are involved, as in wind roses. These are particularly drawn to represent the continuous rhythm (*diurnal or annual*) of certain geographic elements concerning cyclic variations, for example, *hourly data* for a day of sunshine, soil temperature, air temperature, etc., *monthly data* for a year of temperature, rainfall, humidity, wind, etc. and *wind frequency data* corresponding to direction and velocity, etc.

A scale is first chosen to represent the monthly, hourly or directional data and concentric circles are drawn at regular intervals. From the centre (i.e., the point of origin) the required number of radius vectors of corresponding lengths are drawn as straight lines. Hence, an hourly graph is represented by a *24-sided polygon*, a monthly graph by a *12-sided polygon* (Fig. 5.14) and a wind rose (directional graph) commonly by an *octagonal polygon* (Fig. 5.15). Stargraphs are also called polar

Fig. 5.14 Circular Graph

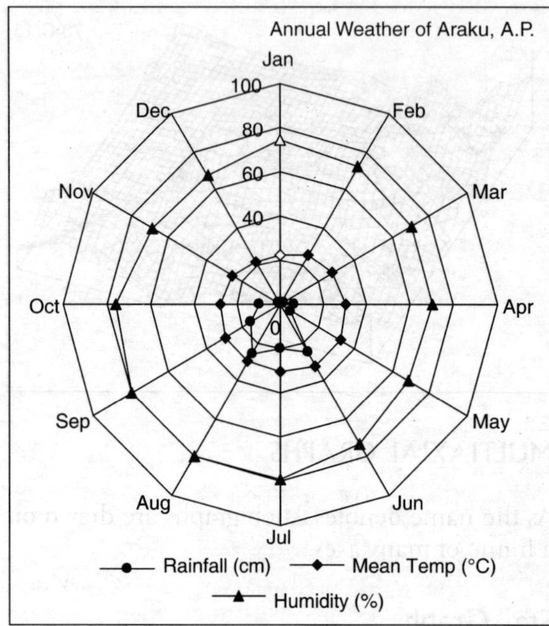

graphs (due to like-shape and like-form), *clock graphs* (as rhythmic timely data are plotted) and vector graphs (as vector data are commonly used).

Fig. 5.15 Vector Graph

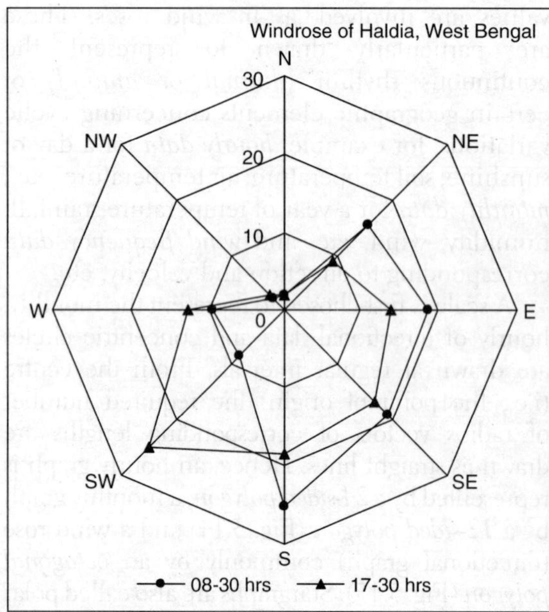

SPECIAL GRAPHS

Hypsometric Curve

A hypsometric curve describes the area-height relationship of a drainage basin dimensionlessly (Strahler 1952). In this, x is the ratio between area (a) and area of the whole drainage basin (A) and y is the ratio of the height (h) between the mouth of the basin and the contour which defines the lower limit of a basin, and the relative relief of the basin (H). The hypsometric curve relating x and y must pass through (1,0) and (1,0) points, but its position within the square graph is a measure of the state of the erosion of the basin concerned (Fig. 5.16).

The shape of a hypsometric curve is an indicator of dominant geomorphic processes at work in a watershed (diffusive or fluvial). A convex curve implies that more of the watershed's area (or volume of rock and soil) is held relatively high in the watershed. The diffusive hillslope processes such as landsliding, rainsplash, inter-rill erosion, soil creep, etc. play a larger role. Similarly, a concave curve indicates that bulk of the basin's area (or volume of rock and soil) remains at relatively low elevation; more material has been removed from higher areas and either transported to lower areas or completely advected out of the basin by channelized /linear /fluvial /alluvial processes.

The simple way to characterize the shape of hypsometric curve for a given drainage basin is to calculate its hypsometric integral (HI). The integral is defined as the area under the hypsometric curve, that provide valuable information not only on the erosional stage of the basin, but also on the tectonic, climatic and lithological factors controlling it. It can be computed in the following two ways—

a) when spot heights of a large number of points well-distributed in the basin are known (or from a DEM), the HI is found from the following equation:

$$HI = \frac{\bar{Z} - Z_0}{Z_x - Z_0}$$

Fig. 5.16 Hypsometric Curve

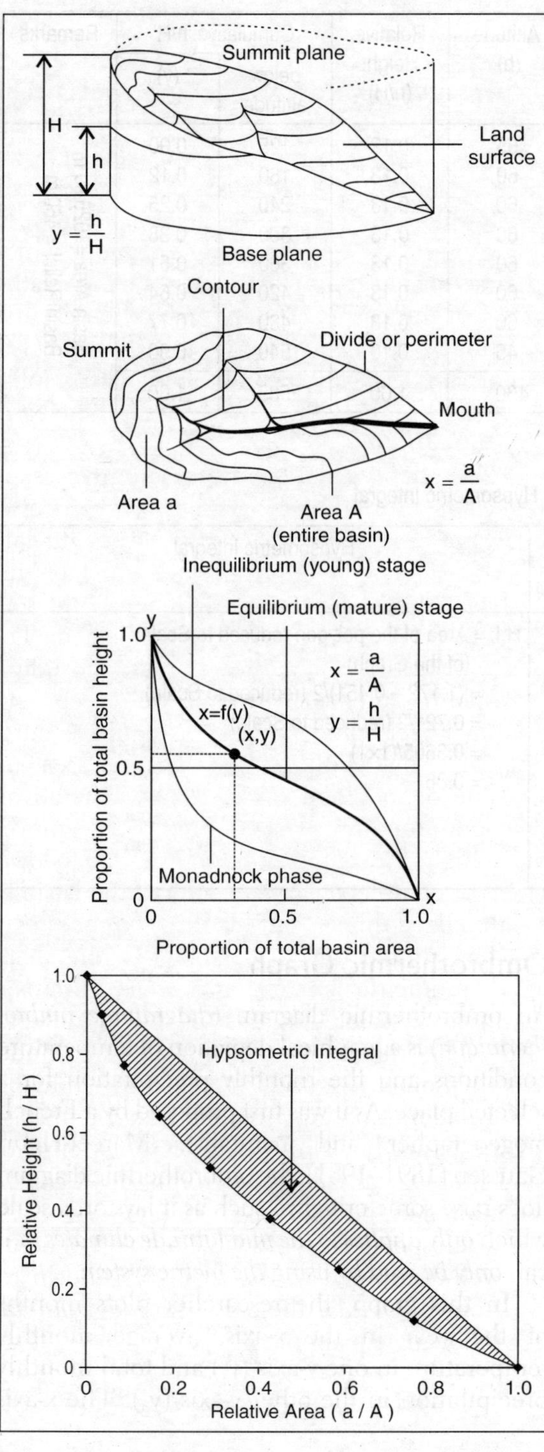

where, Z_x = highest elevation of the basin, Z_0 = lowest elevation of the basin, $\overline{Z}$ = mean elevation of the basin

b) when co-ordinates of the points defining the hypsometric curve (x,y: where x = a/A and y = h/H) (Fig. 5.16), the HI is found from the following equation:

$$HI = \frac{|\Sigma x_i.y_{i+1} - \Sigma x_{i+1}.y_i|}{2}$$

Theoretically, value of hypsometric integral ranges between 0 and 1. Low HI indicates old and more eroded areas and evenly dissected drainage basins. High HI indicates that most of the topography is high relative to the mean, such as a smooth upland surface cut by deeply incised streams indicating young and less eroded areas. HI values of < 0.30 describe "tectonically stable", "denuded", "mature" basins (Davis, 1929). HI values > 0.60 indicate "unstable", "actively uplifting", "young" basins.

Willgoose & Hancock (1998) consider (a) HI > 0.5 dominated by diffusive processes (mainly hillslope processes), (b) HI < 0.5 are considered dominated by fluvial erosion (channel processes play a larger role), and (c) HI ≈ 0.50 represents more "balanced", flattened S-shaped, or straight hypsometric curves that suggest a relatively stable, but still developing landscape.

However, one should be cautious: *it is possible for basins with very different geomorphic histories to have identical HI values.* Similar-looking curves can be produced by complex interactions of climate, tectonism, sedimentation, rock resistance (Bishop et al., 2002).

Relative Temperature Graph

Conrad (1942) used Koppen's concept of relative temperatures to produce a temperature curve in which the variations due to differences in average temperatures and amplitudes at any two stations are eliminated. The amounts by which the consecutive monthly means at any station exceed the temperature of the coldest month are expressed as proportions of the differences in the average temperatures

Table 5.3 Worksheet of Hypsometric Curve Analysis

Altitudinal Class (m)	Area (a)	Relative area (a/A)	Cumulative 'a/A' above altitude	(x)	Altitude (h)	Relative Height (h/H)	Cumulative 'h/H' below altitude	(y)	Remarks
125–180	245	0.239	125	1.000	55	0.12	125	0.00	Basin Area =1025 sq m Basin Relief = 460 m
180–240	180	0.176	180	0.761	60	0.13	180	0.12	
240–300	164	0.160	240	0.585	60	0.13	240	0.25	
300–360	145	0.141	300	0.425	60	0.13	300	0.38	
360–420	120	0.117	360	0.284	60	0.13	360	0.51	
420–480	82	0.080	420	0.167	60	0.13	420	0.64	
480–540	54	0.053	480	0.087	60	0.13	480	0.77	
540–585	35	0.034	540	0.034	45	0.10	540	0.90	
	1025	1.000	585	0.000	460	1.00	585	1.00	

Table 5.4 Computation of Hypsometric Integral

Coordinates x	Coordinates y	Cross Products $x_i \cdot y_{i+1}$	Cross Products $x_{i+1} \cdot y_i$	Hypsometric Integral
1.000	0.000	0.120	0.000	H.I. = Area of the polygon reduced to Scale (of the Graph) = (1.172 – 0.451)/2 (reduced to Scale) = 0.721/2 (reduced to Scale) = 0.3605/(1×1) = 0.36
0.761	0.120	0.190	0.070	
0.585	0.250	0.222	0.106	
0.425	0.380	0.217	0.108	
0.284	0.510	0.182	0.085	
0.167	0.640	0.129	0.056	
0.087	0.770	0.078	0.026	
0.034	0.900	0.034	0.000	
0.000	1.000	Sum = 1.172	Sum = 0.451	

between the coldest and the warmest months. Thus the relative temperature of a month is given by:

$$R_T = (T–Tn)/(Tx–Tn)$$

where, R_T = relative temperature of a month, T = mean temperature of a month, Tn = mean temperature of the coldest month, and Tx = mean temperature of the warmest month.

These values are then plotted as ordinates and the months as abscissa (Fig. 5.17). The horizontal scale is so graduated as to enable the relative temperatures for the six months of the year to be compared directly with those for the last six.

Ombrothermic Graph

An ombrothermic diagram (*diagramme ombrothermique*) is a graphical depiction of temperature conditions and the monthly precipitation for a selected place. As it was first proposed by a French biogeographer and naturalist, Marcel-Henri Gaussen (1891–1981), the ombrothermic diagram does raise some criticism such as it lays on a scale which *only applies to the mid-latitude climates* or it can *only be read by using the metric system*.

In this graph, the researcher plots months of the year in the x-axis, average monthly temperature in one y-axis (y_1) and total monthly precipitation in the other y-axis (y_2). The x-axis

Fig. 5.17 Relative Temperature Graph

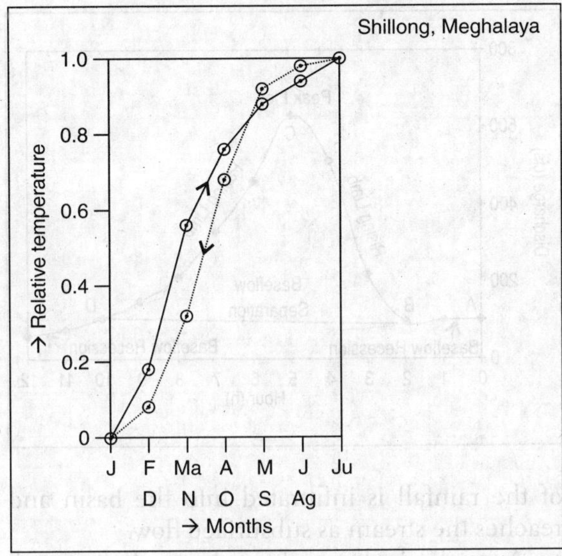

Fig. 5.18 Ombrothermic Graph

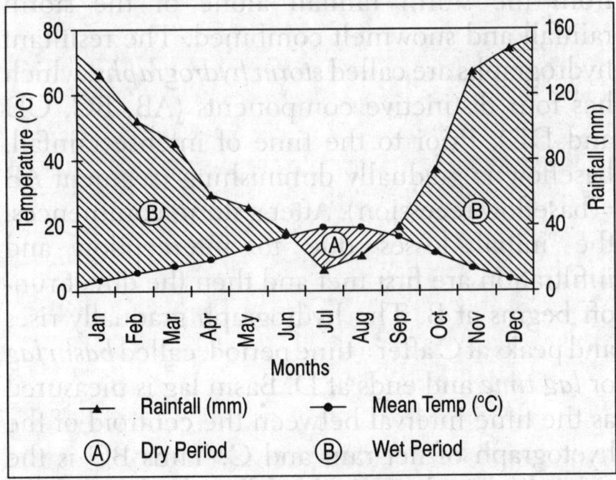

should start with the coldest month; for places in the Northern hemisphere the x-axis should start with January, while those for the Southern hemisphere it should start with July. Temperature should be expressed in degrees centigrade (°C) and precipitation should be expressed in millimetres (mm). It is <u>VERY</u> important to observe the scales of the axes because they <u>MUST</u> respect the following relationship:

Precipitation (P) = Temperature (T) × 2.

As for instance, a 10°C average temperature in the y_1 - axis must be equal to a 20mm of monthly precipitation in the y_2 - axis). Usually, the precipitation curve is represented in blue and the temperature curve in red.

The relationship between the scales of the axes is based on the Gaussen-Bagnouls aridity index, which states that—

(a) $P > 3T$ implies wet period
(b) $3T > P > 2T$ implies semi-wet period
(c) $P < 2T$ implies arid period

This means that when the curve of precipitation falls below the temperature curve there is a *dry period (xeric period)*, when the precipitation curve runs above the temperature curve there is a *wet period*, and when the precipitation curve exceeds 100 mm there is an *excess water period*.

Water Balance Graph

These are line graphs showing the superimposition of the mean-monthly evapo-transpiration and the mean-monthly rainfall. Both are plotted on the same scale (Fig. 5.19). Thus they constitute a valuable means of analysing local climate in terms of seasonal moisture surplus and rainfall efficiency. The portions enclosed by

Fig. 5.19 Water Balance Graph

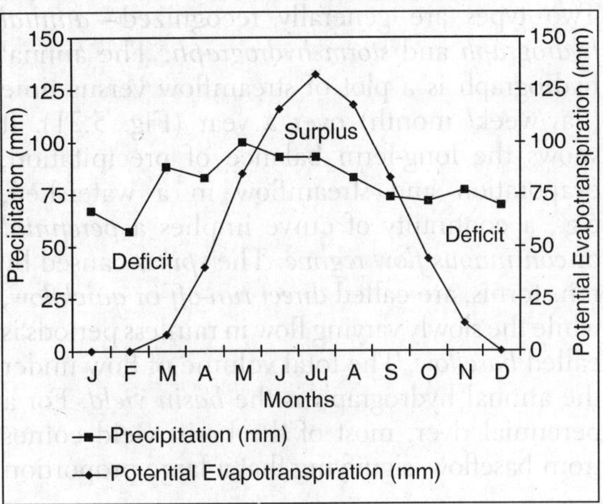

Fig. 5.20 Annual Hydrograph

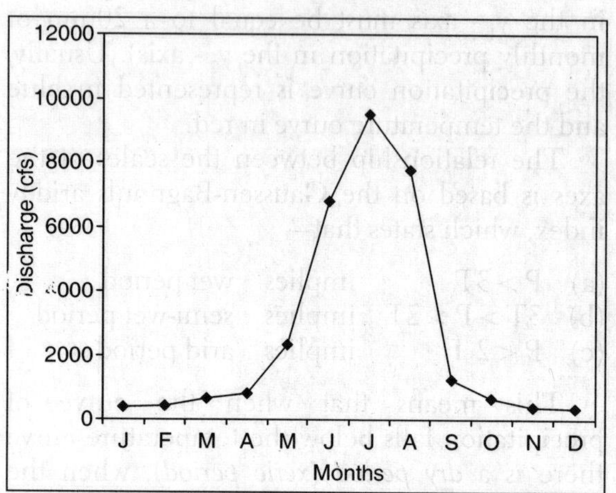

Fig. 5.21 Streamflow Hydrograph

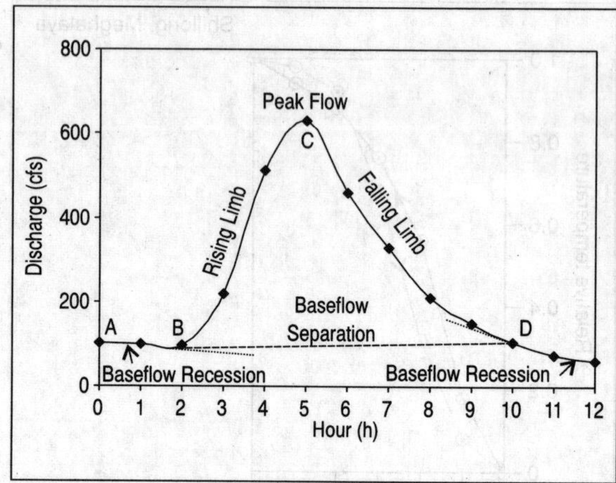

the precipitation curve at the top and potential evapo-transpiration curve at the bottom define the *period of water surplus* and that enclosed by the potential evapo-transpiration curve at the top and precipitation curve at the bottom define the *period of water deficiency*.

Hydrograph

These show the flow rate of streams as a function of time at a given location on the stream. Thus a hydrograph is an integral expression of the physiographic and climatic characteristics that govern the relations between rainfall and run-off of a particular drainage basin (Chow 1959). Two types are generally recognized—*annual hydrograph* and *storm hydrograph*. The annual hydrograph is a plot of streamflow versus time (day/week/ month) over a year (Fig. 5.21). It shows the long-term balance of precipitation, evaporation and streamflow in a watershed, e.g., a continuity of curve implies a *perennial* or *continuous flow regime*. The *spikes* caused by rainstorms, are called *direct run-off* or *quickflow*, while the slowly varying flow in rainless periods is called *baseflow*. The total volume of flow under the annual hydrograph is the *basin yield*. For a perennial river, most of the basin yield comes from baseflow, signifying that a large proportion of the rainfall is infiltrated into the basin and reaches the stream as subsurface flow.

Annual hydrographs show that peak streamflow occurs quite infrequently as it results from the storm rainfall alone or the storm rainfall and snowmelt combined. The resultant hydrographs are called *storm hydrographs*, which has four distinctive components (AB, BC, CD and DE). Prior to the time of intense rainfall, baseflow is gradually diminishing (segment AB = baseflow recession). After a storm commences, the initial losses due to interception and infiltration are first met and then the direct run-off begins at B. The hydrograph gradually rises and peaks at C after a time period, called *basin lag* or *lag time* and ends at D. Basin lag is measured as the time interval between the centroid of the hyetograph of net rain and C. Thus BC is the *rising limb* and CD is the *falling limb*, in which X is the *point of inflection* marked by a change of slope. Actually, up to X there has been inflow of rain and henceforth there is gradual withdrawal of catchment storage. By this time, the GWT is built up by the infiltrating and percolating water and now the ground water contributes more into the stream flow than at the beginning of the storm. Thereafter the GWT declines and the hydrograph again goes on depleting in the exponential form called the *ground water depletion curve* or the *normal baseflow recession curve*. Between B and D, the direct run-off and baseflow can easily be separated and estimated for further analysis.

Rating Curve

For channels, the stage-discharge curve is called a *rating curve* (Fig. 5.22). The principle is: *discharge depends only on flow depth* (stage). The rating curve is developed using a set of measurements of discharge and gauge height in the stream. These measurements are made over a period of months or years so as to obtain an accurate relationship between the stream flow rate or discharge, and the gauge height at the gauging site. Rating curves usually have a *break point*, which is around the stage at which the river spreads out of its banks, or it could be at a lower stage if the river bed cross section changes dramatically. Above that stage, the river does not rise as fast, provided other conditions remain constant. This is illustrated by a change in slope in the rating curve. When a rating curve is unavailable, one can be estimated using Manning's equation.

In rating curves, discharge (cusec or cumec) is plotted on the x-axis and gauge height (ft or m) on the y-axis. Plot scale may be taken in two ways: (a) both the variables in an *arithmetic scale* and (b) both the variables in a *log scale*.

Fig. 5.22 Rating Curve

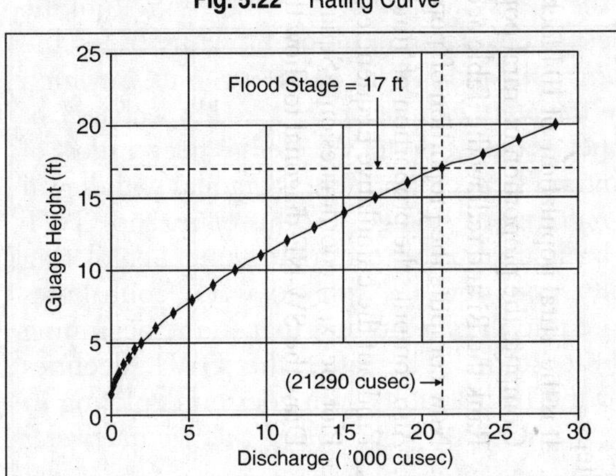

In the first case, the rating curve will appear as *a parabolic curve* (convex), while in the second case it will appear as *a straight line*. Such curves, along with the duration of storm rainfall and hydrological parameters, may be monitored and analysed together to judge the changes in control as well as the propensity of stream flooding in particular. Besides, stream flow is not directly recorded. Instead water level is recorded and stream flow is then deduced by means of a rating curve (Riggs 1985).

Lorenz Curve

A Lorenz curve is a fairly widely used simple graphical method of comparing distributions on 2-dimensional surfaces. Basically, it uses a square graph with x-axis and y-axis having a comparable scale, e.g., percent units (Fig. 5.23). Appropriate data are collected for subdivisions of the total area being considered. If the density of the objects distributed over the area is uniform, a straight-line curve results. It is called the *line of equal distribution* (LED) or 45° with the following characteristics: (a) it is simply the SW-NE diagonal of the square graph, (b) it is obtained by joining the origin (0, 0) of the graph

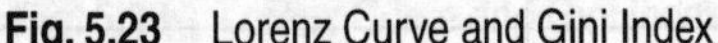
Fig. 5.23 Lorenz Curve and Gini Index

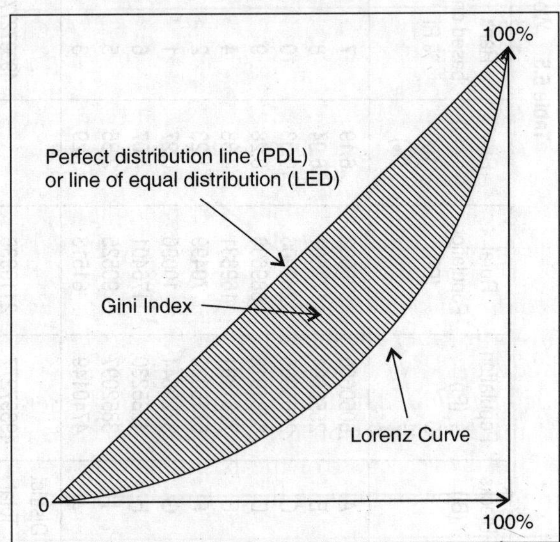

Table 5.5 Worksheet for Lorenz Curve of Rural Population of 10 Blocks

Blocks (Bi)	Population (Pi)	Rural Population (Ri)	Ri = (Ri / Pi) *100 (%)	Rank based on % Ri	Redrawn Table			BPi = (Pi / DPi) *100 (%)	Cumulative BPi (x : %)	BRi = (Ri / DRi) *100 (%)	Cumulative BRi (y : %)
					Rank (ascending order)	Population (Pi)	Rural Population (Ri)				
A	6050605	375033	6.19	7	1	3729644	10090	7.55	7.55	0.48	0.48
B	2555664	177501	6.94	8	2	5715030	70499	11.57	19.12	3.34	3.82
C	2805065	289906	10.33	10	3	4740149	61513	9.60	28.72	2.91	6.73
D	8331912	689636	8.28	9	4	7281881	169831	14.74	43.46	8.04	14.77
E	7281881	169831	2.33	4	5	3852097	90525	7.80	51.26	4.29	19.06
F	5715030	70499	1.23	2	6	4335230	176401	8.77	60.03	8.36	27.42
G	3729644	10090	0.27	1	7	6050605	375033	12.25	72.28	17.77	45.19
H	4335230	176401	4.07	6	8	2555664	177501	5.17	77.45	8.41	53.60
I	3852097	90525	2.35	5	9	8331912	689636	16.87	94.32	32.67	86.27
J	4740149	61513	1.29	3	10	2805065	289906	5.68	100.00	13.73	100.00
District Total	49397277	2110935		49397277	2110935	100.00		100.00			

Steps:

1. It is the 'rural population' of 10 blocks (Bi) of a district (D) for which Lorenz Curve is to be drawn.
2. Compute the proportion of 'rural population' (Ri) of a block with respect to individual block total.
3. Rank the Ri and redraw the Table according to the ascending rank order.
4. Now compute the proportion of 'population' of a block with respect to the district total (BPi).
5. Also compute the proportion of 'rural population' of a block with respect to the district total (BRi).
6. Find out the cumulative frequencies of the BPi (x) and BRi (y) and enter in separate columns.
7. Plot the (x, y) coordinates on a Square Graph and join these by smooth curves to obtain the Lorenz Curve.
8. Draw the SW-NE diagonal to show the line of equal distribution.

Table 5.6 Computation of Gini Coefficient

Blocks	X_i	Y_i	$X_i.Y_{i+1}$	$X_{i+1}.Y_i$	Gini Coefficient
G	7.55	0.48	28.841	9.1776	
F	19.12	3.82	128.6776	109.7104	G = 1 – Area of the polygon
J	28.72	6.73	424.1944	292.4858	reduced to Scale
E	43.46	14.77	828.3476	757.1102	(of the Graph)
I	51.26	19.06	1405.549	1144.172	= 1 – (25516.19 – 21477.09)/2
H	60.03	27.42	2712.756	1981.918	(reduced to Scale)
A	72.28	45.19	3874.208	3499.966	= 1 – 4039.10/2
B	77.45	53.6	6681.612	5055.552	(reduced to Scale)
D	94.32	86.27	9432	8627	= 1 – 2019.5/(100×100)
C	100	100	Sum = 25516.19	Sum = 21477.09	= 0.40

and the point (100%, 100%), and (c) the slope of LED is 45° and (d) inequality of a distribution is directly proportional to the degree of concavity of the curve. Hence, the more the concavity, the more the inequality.

The Gini coefficient (G) is a measure of the inequality of a distribution (Gini 1912). It is defined as a ratio with boundary conditions, $0 \leq G \leq 1$. The *numerator* is the area between the Lorenz curve of the distribution and the uniform (perfect) distribution line (LED), while the *denominator* is the area under the uniform distribution the (LED). Thus, G = 0 corresponds to perfect equality (i.e., each unit has the same quantity) and G = 1 corresponds to perfect inequality (i.e., only a single unit has all the quantity, while everyone else has zero). The Gini coefficient satisfies four important principles: *anonymity*, *scale independence*, *population independence* and *transfer principle*. As a measure of inequality, G has several advantages: (a) it is a measure of inequality by means of a ratio analysis, (b) it can be used to compare distributions across different populations, (c) it is sufficiently simple and can be compared across countries and be easily interpreted, (d) it can be used to indicate how the distribution has changed within a country over a period of time and (e) to see if inequality is increasing or decreasing. G can be calculated from the following formula:

Fig. 5.24 Lorenz Curve

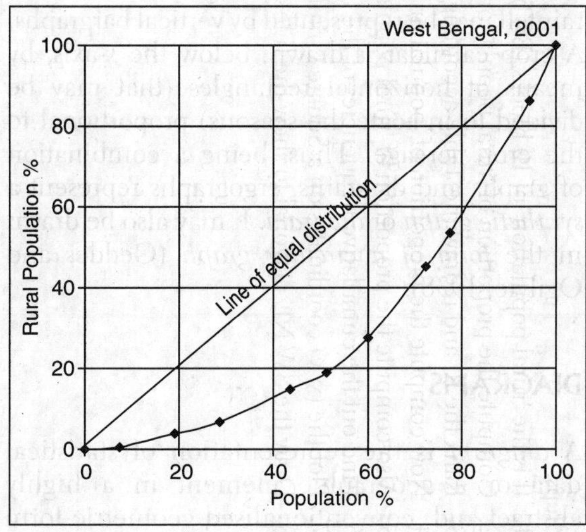

$$G = 1 - \frac{\sum x_i.y_{i+1} - \sum x_{i+1}.y_i}{2}$$

(reduced to scale)

where, i = 1, 2, 3, …n, (x, y) are coordinates of the points plotted, and 10000 referents the factor of reduction to scale.

Rank-Size Curve

When size is plotted against the rank for every town/city, the relationship (on a logarithmic scale) is shown by a downward sloping straight line (Fig. 5.25 and Fig. 5.26). This means that

Fig. 5.25 Rank-Size Graph: Log-log Scale

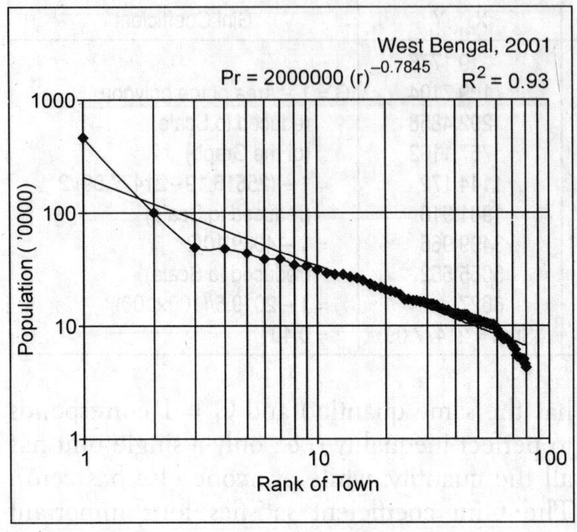

Fig. 5.26 Rank-Size Graph: Arithmetic Scale

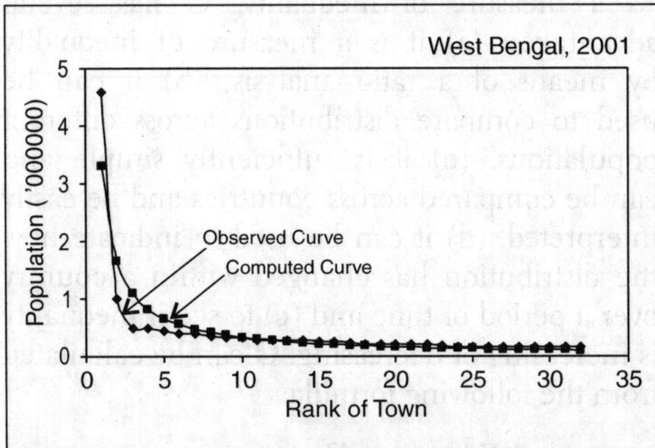

city size is inversely proportional to its rank. In other words, the product of the city size and its rank is constant, being equal to the size of the leading city in the system of cities (Auerbach 1913, Zipf 1949). If all the towns/cities of an area are ranked in descending order of population, the population of the nth ranked town will be $\frac{1}{n}$ th of the largest one.

Thus, $P_r = P_1.(r)^{-q}$,

where r = rank of a city, P_r = population of a city of rank, r, P_1 = the estimated population of the largest city and q = an exponent which generally has a value close to 1.

The tendency of the largest city to be 'excessively' big with stunting effects on cities of nearby rank is the *primate distribution* case (Jefferson 1939, Linksy 1965, Harris 1971). According to Richardson (1973) the rank-size distribution may be interpreted as a very general model according to the value of the exponent; q = 1 implies the *rank-size distribution*, q>1 represents *metropolitan dominance*, and q<1 stands for an urban system in which *intermediate cities are relatively large*. The limiting cases are: q = infinity (only one city) and q = 0 (all cities are of the same size). The rank-size distribution matches well with the allometric growth model and Pareto distribution (Parr 1970, Berry 1961, Beckman 1958, Nordbeck 1971).

Ergograph

These are composite graphs drawn specially to show the relationship between cropping patterns and the annual rhythm or pattern of climatic elements. The principle is based on the fact that *the cycle of plant growth* (sowing, growing, flowering, maturing and harvesting) corresponds with the annual cycle of weather (Fig. 5.27). It is a multivariable graph: for each month, elements of climate like rainfall, temperature, humidity, sunshine, soil moisture, etc. are plotted along the y-axis in the form of polygraphs. However, rainfall may be represented by vertical bargraphs. A crop calendar is drawn, below the x-axis, by means of horizontal rectangles (that may be divided to indicate the seasons) proportional to the crop acreage. Thus, being a combination of graphs and diagrams, ergographs represent a *synthetic graph* or *diagram*. It may also be drawn in the *form of a circular graph* (Geddes and Ogilvie, 1938).

DIAGRAMS

A *diagram* is the representation of statistical data or a geographic element in a highly abstract and conventionalised geometric form

Fig. 5.27 Ergograph

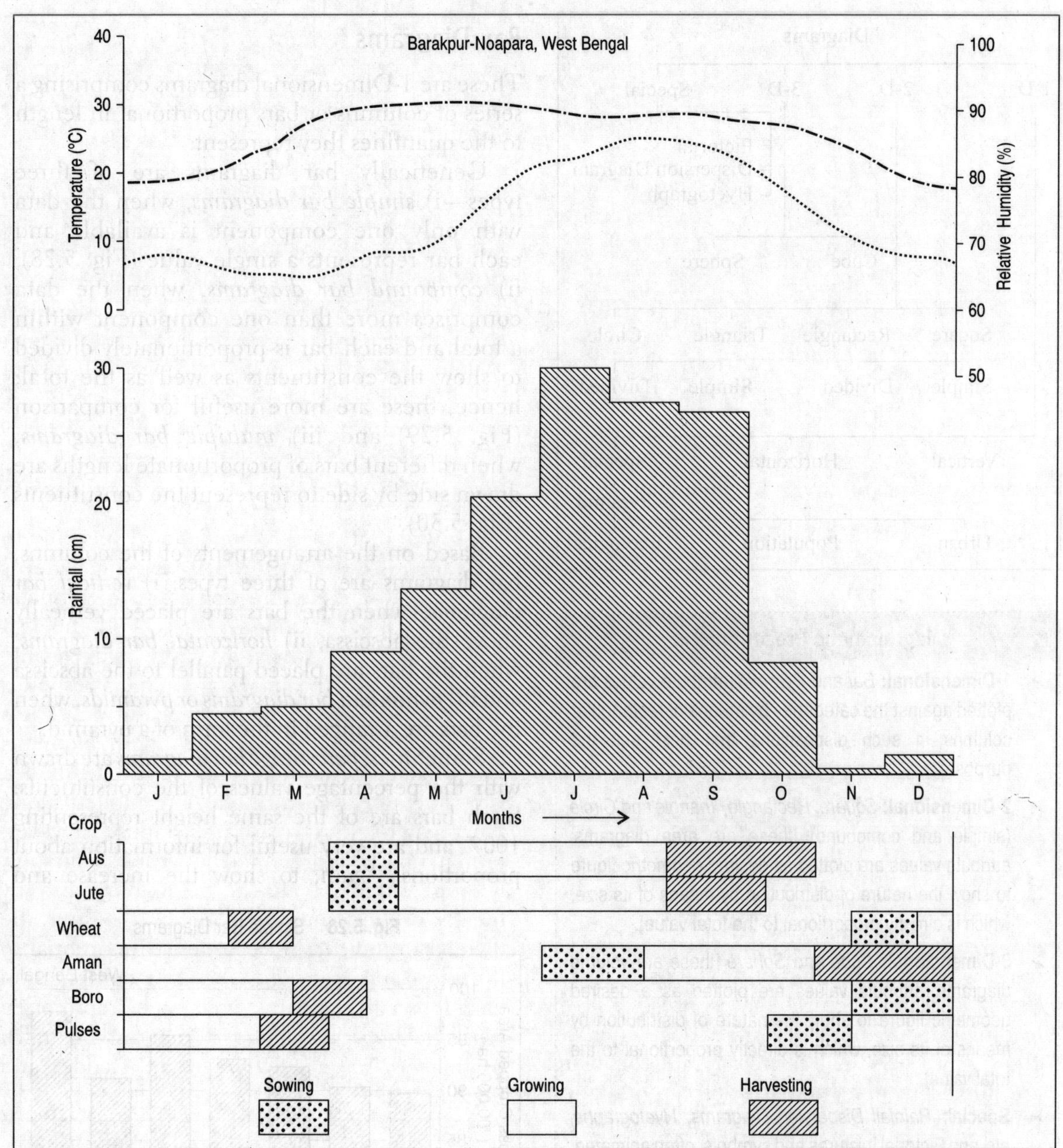

with emphasis on one selected element. Hence, three technical classes of diagrams based on the nature of data may be distinguished—*purely statistical, purely geographical* and *statistical-geographical*. Based on the geometry of the figures, there may be a number of diagrams (Table 5.3) from which the most suitable one is drawn.

Table: 5.37 Types of Diagrams

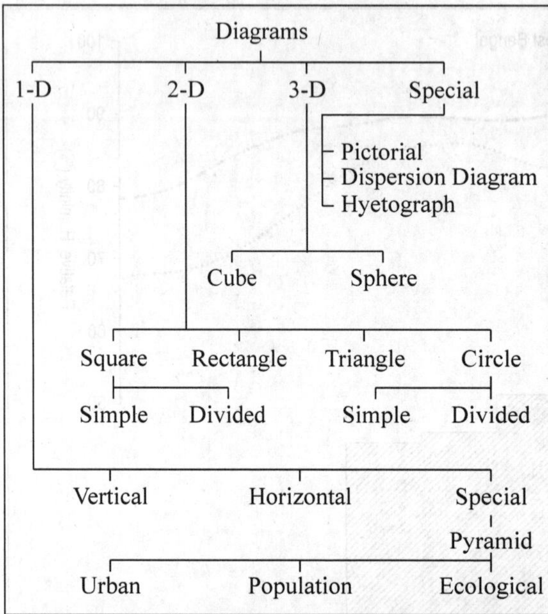

Diagrammatic Presentations

- **1-Dimensional:** *Bar* and *Pyramid* [attribute values are plotted against the categories by means of rectangular columns in such dispositions that best suit the purpose].
- **2-Dimensional:** *Square, Rectangle, Triangle* and *Circle* (simple and compound) [these are area diagrams; attribute values are plotted as desired geometric figure to show the nature of distribution by means of its size, which is directly proportional to the total value].
- **3-Dimensional:** *Cube* and *Sphere* [these are volume diagrams; attribute values are plotted as a desired geometric figure to show the nature of distribution by means of its size, which is directly proportional to the total value].
- **Special:** *Rainfall Dispersion Diagrams, Hyetographs,* etc. and Pictorial [pictures and symbols, often animated, are used and drawn as proportional to the total value].

Note: All presentations must have a title, a legend, a proportional scale, labelling of the attributes and attribute values with proper units, and other contextual items.

1-DIMENSIONAL DIAGRAMS

Bar Diagrams

These are 1-Dimensional diagrams comprising a series of columns or bars proportional in length to the quantities they represent.

Genetically bar diagrams are of three types—i) *simple bar diagrams,* when the data with only one component is available and each bar represents a single value (Fig. 5.28); ii) *compound bar diagrams,* when the data comprises more than one component within a total and each bar is proportionately divided to show the constituents as well as the total; hence, these are more useful for comparison (Fig. 5.29) and iii) *multiple bar diagrams,* when different bars of proportionate lengths are drawn side by side to represent the constituents (Fig. 5.30).

Based on the arrangements of the columns, bar diagrams are of three types: i) *vertical bar diagrams,* when the bars are placed vertically above the abscissa, ii) *horizontal bar diagrams,* when the bars are placed parallel to the abscissa and iii) *pyramidal bar diagrams or pyramids,* when the bars are arranged in the form of a pyramid.

Sometimes, *percentage bar diagrams* are drawn with the percentage values of the constituents. Such bars are of the same height representing 100% and are very useful for information about proportions. Again, to show the increase and

Fig. 5.28 Simple Bar Diagrams

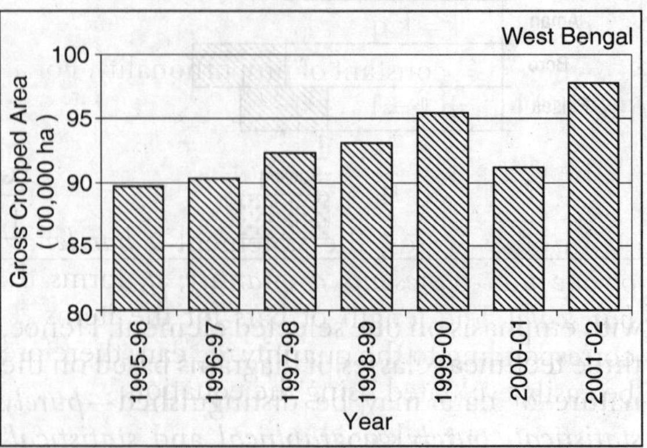

Fig. 5.29 Compound Bar Diagrams

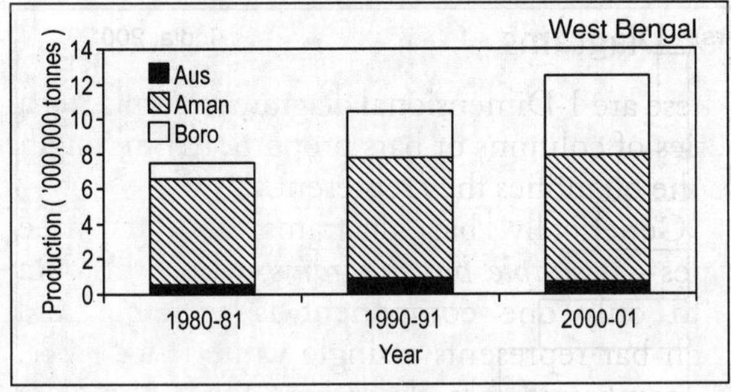

Fig. 5.30 Multiple Bar Diagrams

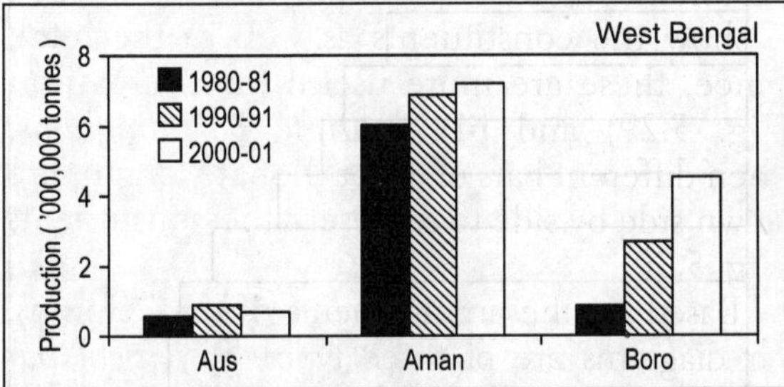

decrease or profit and loss or surplus and deficit or positive and negative deviations, a *two directional bar diagram* or 'deviation bar diagram' is drawn. Here, vertical bars are drawn above and below the horizontal zero line.

The basic principle of bar diagrams is that the length of a bar (ℓ) for an item (i) is directly proportional to the quantity of the item (q) it represents. Empirically,

$$\ell \propto q$$

or, $\ell = k.q$

where, k = constant of proportionality. For an unit bar length $\ell = 1$

hence, $k = \frac{1}{q} \Rightarrow \frac{1}{q}$

In other words, the statement *a bar length of one unit represents a quantity, q,* forms the *bar scale*. The length of bars for the items (ℓ_i) corresponding to the quantity, q_i, can, therefore, be easily evaluated using the equation,

$$\ell_i = k.q_i$$

While *choosing* a bar scale, the ease of computation, construction and comprehensibility are of prime concern. Individual bars should be *neither* exceedingly short and wide *nor* very long and narrow. A *worksheet* must be drawn to show the computation in detail. An important point to remember is that the *origin* of the bar scale must be *zero*. However, space may be saved by using a *scale break* in some situations where there is a large range of values. Bars are separated by spaces not less than half the width of a bar or more than the width of a bar. Individual bars are essentially *equal in width* and *uniformly spaced*. However, in rainfall diagrams and in some special cases, no intervening spaces are left giving them the appearance of a battleship. *Guide lines* are helpful for understanding the diagram and may be extended throughout the diagram. *Colours or shades* may be applied for a better visual appeal.

Pyramids

These are special kinds of bar diagrams where bars are arranged typically in a pyramidal form. These are extensively used in *urban, population* and *ecological* studies. In case of *urban pyramids*, absolute or percentage distribution of towns or cities corresponding to different size-classes is shown. Class frequencies are plotted on a selected scale in the form of horizontal bars placed symmetrically on both sides of the size-class column. The proportion of population living in different size-classes of towns may also be represented by an urban pyramid. This kind of pyramid has been intensively used by the urban geographers to identify the structural properties of the urban system of a country.

In *population studies*, pyramids are very useful to portray the age-sex composition. The simplest method is to construct the pyramid by building bars to represent age groups, males on one side and females on the other. The base represents the youngest group and the apex the oldest. Normally uniform age groups are taken and bars of uniform width are drawn. However, in case of unequal age groups, the bar widths are

Fig. 5.31 Age-Sex Pyramid of Population

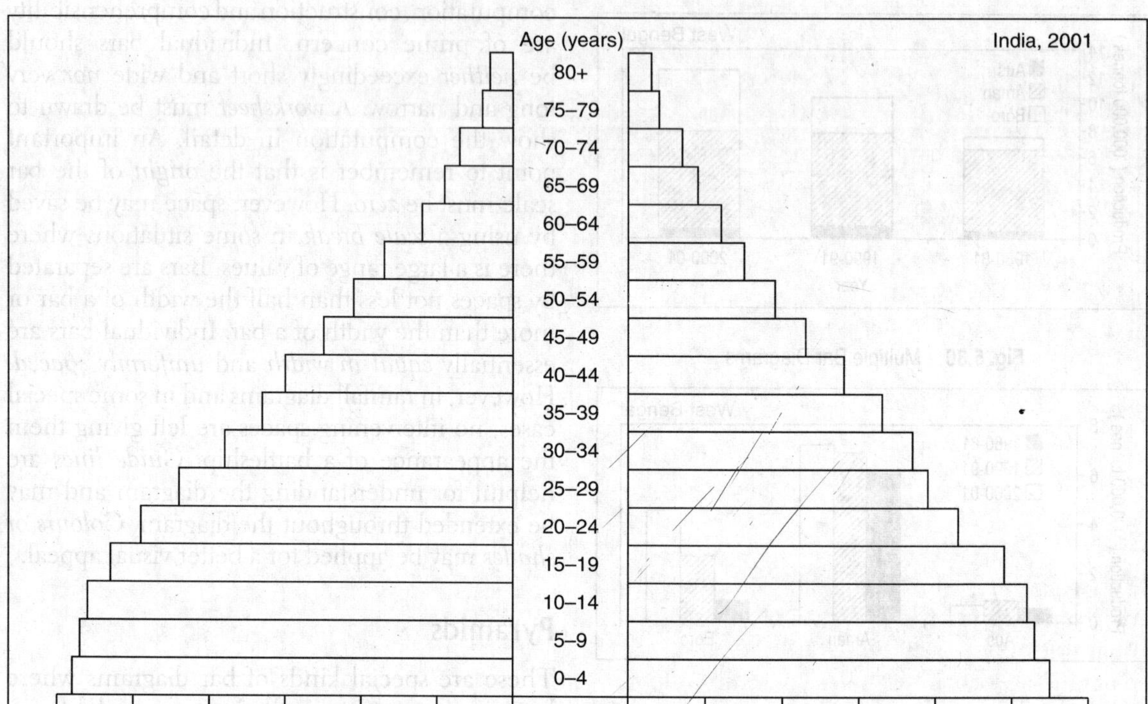

non-uniform. Age group frequencies are plotted either with absolute data or with percentage values and are accordingly called *absolute* and *comparable* graphs.

In case of proportions, there are two possible ways of drawing a pyramid: (a) each male age group may be expressed as percentages of the total male population and similarly each female age group may be expressed as percentages of the total female population in a country and (b) each male and female group may be expressed as percentages of the total population in a country. The second method is more useful as it gives a more realistic view of the age-sex composition of population of a country. Similarly, the agewise working status of a population can be shown by pyramids.

In ecological studies, *trophic structure* and *functions* are shown by means of *ecological pyramids*. Here, the first or primary producers form the base and successive producers the tiers which make up the apex. The three general types of such pyramids are *pyramid of number* (that concerns the number of organisms), *pyramid of biomass* (that concerns the total dry weight, caloric value, etc.) and *pyramid of energy* (that concerns the rate of energy flow and/on productivity at successive trophic levels).

By nature, pyramids are of three types: *simple, compound* and *superimposed.* In the *simple* types, bars represent a single item (Fig. 5.31). When two simple pyramids are drawn on the same frame for comparison, a *superimposed* pyramid is produced. In *compound* types, each bar is subdivided to show different categories. These pyramids are best drawn with a space left between each tier, across which lines are drawn to join the component subdivisions (Fig. 5.32).

Fig. 5.32 Urban Pyramid

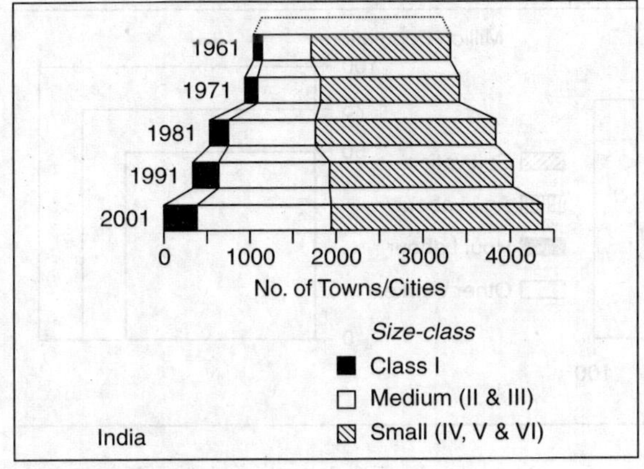

2-DIMENSIONAL DIAGRAMS

Square Diagrams

Squares are 2-dimensional diagrams consisting of a series of squares proportional in size to the quantities they represent. Genetically, square diagrams are of two types: i) *simple square diagrams* where data with only one component is available and each square represents a single value (Fig. 5.33) and ii) *compound square diagrams* where the data comprise more for one component and each square is proportionately divided into rectangular segments to show its constituent parts.

The construction of square diagrams is based on the principle that *the area of a square* (a) *for an item* (i) *is directly proportional to the quantity of the item* (q) *it represents*. Empirically,

$$a \propto q$$

$$\text{or,} \quad a = k.q,$$

where k = constant of proportionality.

For a square with unit area, a =1,

$$\text{hence,} \quad k = \frac{a}{q} \Rightarrow \frac{1}{q}$$

In others words, the statement, *a square of one unit area represents a quantity, q,* forms the *area scale* for the squares. If the side of a square with area a is x,

$$x^2 = a$$

$$\text{or,} \quad x^2 = k.q.$$

$$\therefore \; x = \sqrt{k.q}$$

Here, the sides of the squares for the items (x_i) corresponding to the quantity, q_i, can therefore be easily evaluated using the equation:

$$x_i = \sqrt{k.q_i}$$

The scale for the square diagrams should be chosen in such a way that individual squares are not too large or too small for the base map. A *proportional scale* is diagrammatically represented with at least three squares corresponding roughly to the *largest*, *smallest* and *median* values of the distribution.

A worksheet is usually drawn showing the computations in detail (Table 5.8). When maps are not available these are drawn on the *same baseline* with *uniform* intervening *space*. Colours or shades may be applied for visual appeal. In compound square diagrams, percentage values instead of absolute values may be evaluated and used to subdivide the squares. The squares afford ready information concerning proportions and facilitate arithmetic comparisons (Fig. 5.34).

Fig. 5.33 Proportional Square Diagrams: Area under Rice, West Bengal, 2004–05

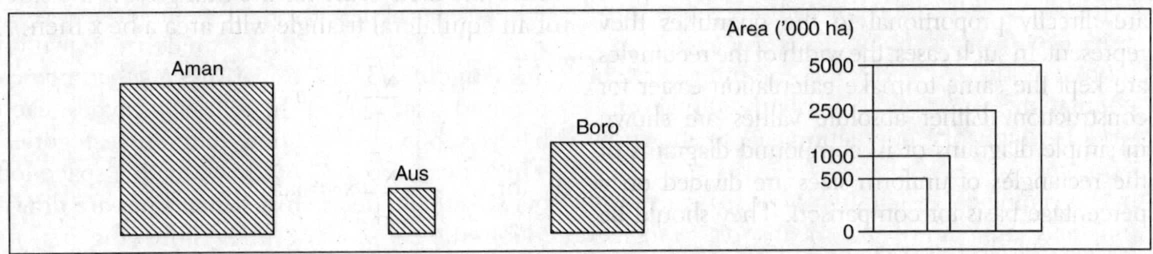

Fig. 5.34 Compound Square Diagrams: Occupational Structure of Population, West Bengal, 2001

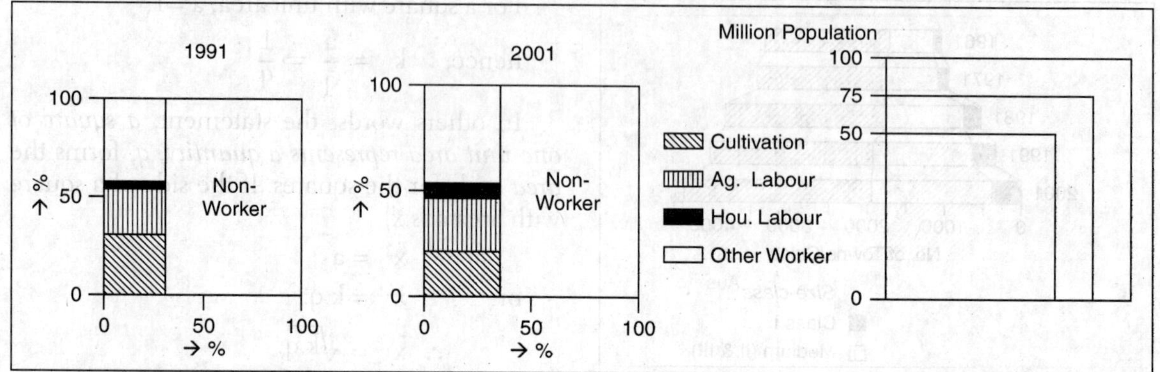

Table 5.8 Worksheet for Square Diagrams

Rice Crop	Area ('000 ha) 2004–05 (A_i)	$x_i = \sqrt{A_i}$	Scale (s)	Side of Square (x_i/ s) cm
Aman	4088.4	2021.98	1 cm to 1000 units	2.02
Aus	320.8	566.39		0.56
Boro	1376.4	1173.20		1.17
for	5000	2236.07		2.23
propor-	2500	1581.14		1.58
tional	1000	1000.00		1.00
scale	500	707.11		0.71

Rectangle Diagrams

These are similar to *compound bar diagrams* as their lengths are *directly proportional to the values* they represent but they are not plotted in series. When the scale is linear, they are often deliberately widened in order to allow names and figures to be placed within the constituent divisions. The scale, on the other hand, may be areal, i.e., the area of both the rectangle and its constituent parts are directly proportional to the quantities they represent. In such cases, the width of the rectangles are kept the same to make calculation easier for construction. Either absolute values are shown in simple diagrams or in compound diagrams or the rectangles of uniform sizes are divided on a percentage basis for comparison. They should be uniformly spaced on a horizontal baseline when maps are not available. Suitable colours or shades may be applied for visual toning.

Triangle Diagrams

Triangle diagrams consist of a series of equilateral triangles proportional in size to the quantities they represent (Fig. 5.35). These are essentially 2-dimensional diagrams based on the principle that the *area of an equilateral triangle* (a) *for an item* (i) *is directly proportional to the quantity of the item* (q) *it represents*. Empirically,

$$a \propto q$$

or, $$a = k.q$$

where, k = constant of proportionality.

For an equilateral triangle with unit area, a = 1;

hence, $$k = \frac{a}{q} \Rightarrow \frac{1}{q}$$

In other words, the statement, *an equilateral triangle of unit area represents a quantity, q*, forms the *area scale* for the triangles. If the side of an equilateral triangle with area a be x then,

$$\frac{\sqrt{3}}{4} x^2 \propto a$$

or, $$\frac{\sqrt{3}}{4} x^2 = k.a$$

Fig. 5.35 Proportional Triangle Diagrams: Area under Rice, West Bengal, 2004–05

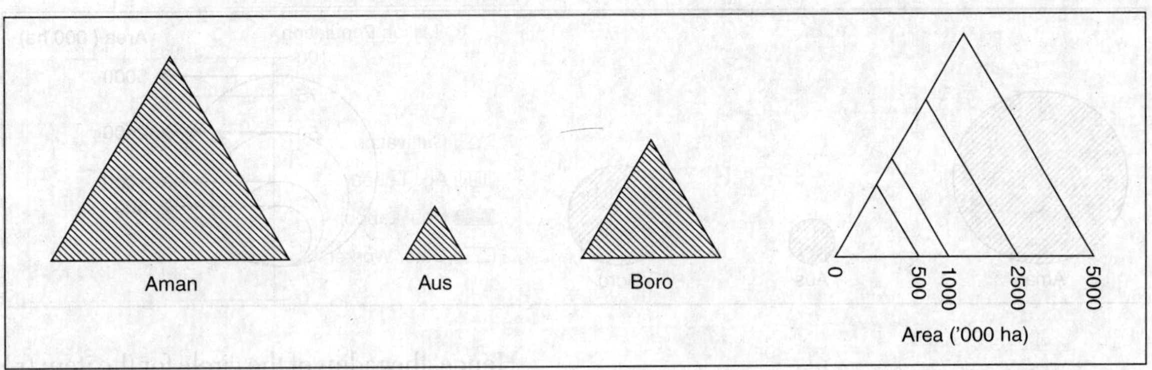

$$\text{or,} \quad x^2 = \frac{4k.q}{\sqrt{3}}$$

$$\therefore \quad x = \sqrt{\frac{4kq_i}{\sqrt{3}}}$$

Hence, the sides of the equilateral triangles for the items (x_i) corresponding to the quantity, q_i, can therefore be easily evaluated using the equation:

$$x_i = \sqrt{\frac{4q_i}{\sqrt{3}}}$$

The scale for the triangle diagrams should be carefully chosen so that individual triangles do not become too large or too small for the base map. A *proportional scale* must be diagrammatically represented with at least three equilateral triangles corresponding roughly to the *largest*, *smallest* and *median* values of the distribution. A *worksheet* should neatly be drawn showing the calculation in detail. When maps are not available, they are drawn on the same baseline with uniform intervening spaces. Colours or shades may be applied for better *visual* appeal (Table 5.9)

Circle Diagrams

Circles are 2-dimensional diagrams consisting of a series of circles proportional in size to the quantities they represent. Genetically, circular diagrams are of two types: i) *simple circle diagrams* when the data with only one component is available and each circle represents a single value and ii) *compound circle diagrams or wheel diagrams or pie diagrams or pie charts* when the data comprises more than one component within a circle and is proportionately divided into angular segments to show its constituent parts (Fig. 5.36).

Table 5.9 Worksheet for Triangle Diagrams

Rice Crop	Area ('000 ha) 2004–05 (A)	$x_i = \sqrt{\frac{4A_i}{\sqrt{3}}}$	Scale (s)	Side of Triangle (x_i/s) cm
Aman	4088.4	3072.74	1 cm to 1000 units	3.07
Aus	320.8	860.73		0.86
Boro	1376.4	1782.88		1.78
for	5000	3398.09		3.39
Propor-	2500	2402.81		2.40
tional	1000	1519.67		1.52
Scale	500	1074.57		1.07

The construction of circle diagrams is based on the principle that the *area of a circle* (a) *for a representable item* (i) *is directly proportional to the quantity of item* (q) *it represents*. Empirically,

$$a \propto q$$

$$\text{or,} \quad a = k.q$$

where k = constant of proportionality.

For an circle of unit area, a =1;

Fig. 5.36 Proportional Circle Diagrams: Area under Rice, West Bengal, 2004–05

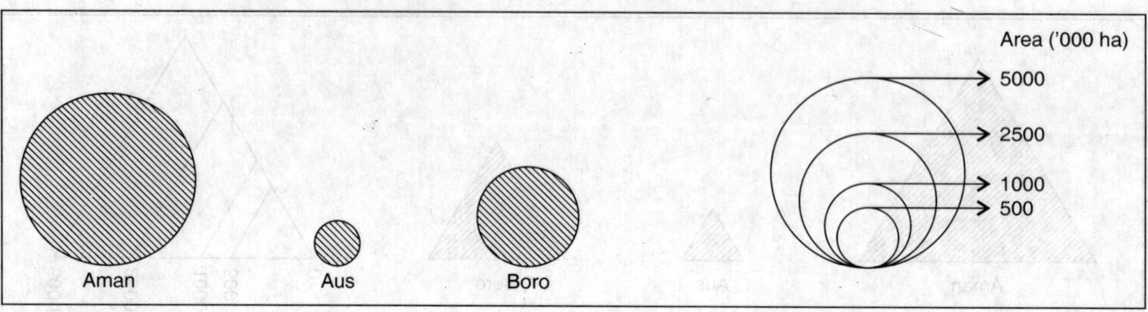

hence, $k = \frac{a}{q} \Rightarrow \frac{1}{q}$

or, $k = \frac{1}{q}$

In other words, the statement, *a circle of one unit area represents a quantity, q*, forms the *area scale* for the circles. If the radius of a circle with area, a, be r,

$$\pi r^2 = a$$

or, $$\pi r^2 = k.q$$

or, $$r^2 = k.\frac{q}{\pi}$$

$$\therefore\ r = \sqrt{k.\frac{q}{\pi}}$$

Hence, the radius of the circle for the item (r_i) corresponding to the quantity, q_i, can therefore be evaluated from the equation:

$$r_i = \sqrt{k.\frac{q_i}{\pi}}$$

The scale for the circle diagrams are carefully selected so that individual circles are not too large or too small for the base map. A proportional scale must be diagrammatically represented with at least three circles corresponding roughly to the largest, smallest and median values of the distribution. A worksheet is drawn showing the computations in detail. Colours or shades may be applied for an effective visual toning. When maps are not available, they should be drawn touching the

Fig. 5.37 Proportional Pie Diagrams: Occupational Structure of Population, West Bengal

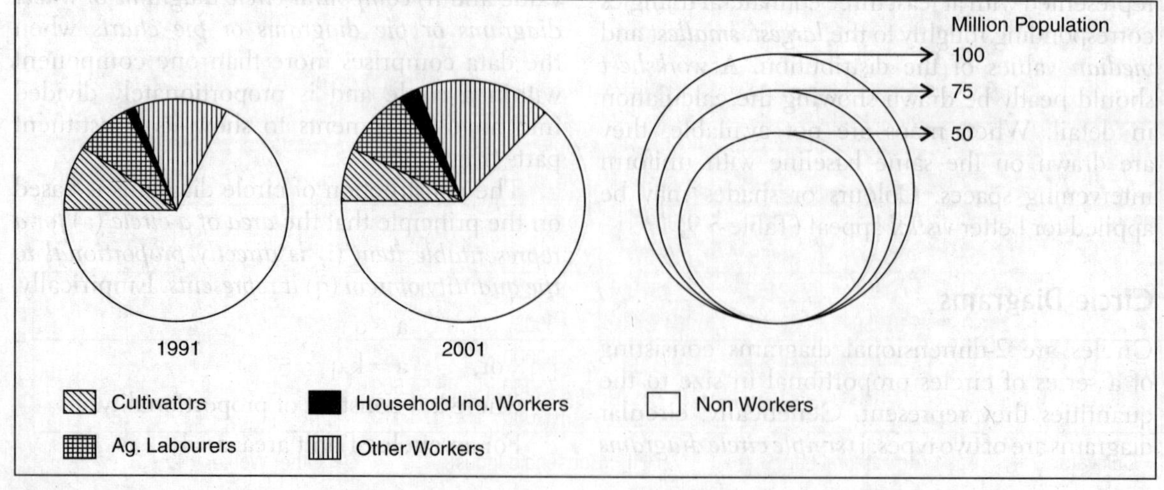

same baseline with uniform intervening spaces or they must be centred at the particular place or at the areal centre of the administrative unit.

In case of *pie diagrams* the angular segments (s) for the constituent items (i = 1, 2, 3, ...) are found by the formula:

$$r_{i_1} = \frac{360°}{q} \times i_i$$

so that, $i_1 + i_2 + i_3 + \ldots + i_n = q$

and $s_1 + s_2 + s_3 + \ldots + s_n = 360°$

A separate worksheet is drawn to show the angular segments for each subdivision of the items and their total must be checked in a separate column. The drawing of the angular segments for each circle must start from a fixed line (either the radius drawn due west or north) and the respective order of drawing should be maintained throughout. A well planned legend of colours or shades must be drawn for the divisions of the main item. To afford ready information about proportions, percentage values corresponding to the angular segments, may be written on the outside or inside of its boundary as well (Table 5.10).

Table 5.10 Worksheet for Circle Diagrams

Rice Crop	Area ('000 ha) 2004–05 (A_i)	$x_i = \sqrt{\frac{A_i}{\pi}}$	Scale (s)	Radius of Circle (x_i/s) cm
Aman	4088.4	1140.78	1 cm to 1000 units	1.14
Aus	320.8	319.55		0.32
Boro	1376.4	661.91		0.66
for	5000	1261.57		1.26
propor-	2500	892.06		0.89
tional	1000	564.19		0.56
scale	500	398.94		0.39

3-DIMENSIONAL DIAGRAMS

Cube Diagrams

Cube diagrams consist of a series of cubes proportional in size to the quantities they represent. These are 3-dimensional diagrams involving volumes (Fig. 5.38). The construction of cube diagrams is based on the principle that the *volume of a cube* (v) *for a representable item* (i) *is directly proportional to the quantity of the item* (q) *it represents*. Empirically,

$$v \propto q$$
$$v = k.q$$

where k = constant of proportionality.

For an unit cube, v = 1;

$$\text{hence,} \quad k = \frac{v}{q} \Rightarrow \frac{1}{q}$$

In other words, the statement a *cube of one unit volume represents a quantity*, q, forms the *volume scale* for the cubes. If the side of a cube with volume, v, is x,

Table 5.11 Worksheet for Proportional Pie Diagrams

Year	Categories of Workers				Total (T_i)	$r_i = \sqrt{T_i/\pi}$	Scale (s)	Radii of Circles (r/s : cm)
	C	AgL	HHI	O				
1991	6407	5481	929	9098	68078	147.207	1 cm to 100 units	1.47
	33°53′	28°59′	4°55′	48°07′	360°			
2001	5613	7351	2153	14378	80222	159.798		1.59
	25°11′	32°59′	9°40′	64°34′	360°			
Note: $S_i = \frac{360}{T_i} \times i$	For proportional scale				50000	126.156		1.26
					75000	154.509		1.54
					100000	178.412		1.78

C = Cultivator AgL = Agricultural Labourer HHI = Household Industry O = Other Worker

Fig. 5.38 Proportional Cube Diagrams: Area under Rice, West Bengal, 2004–05

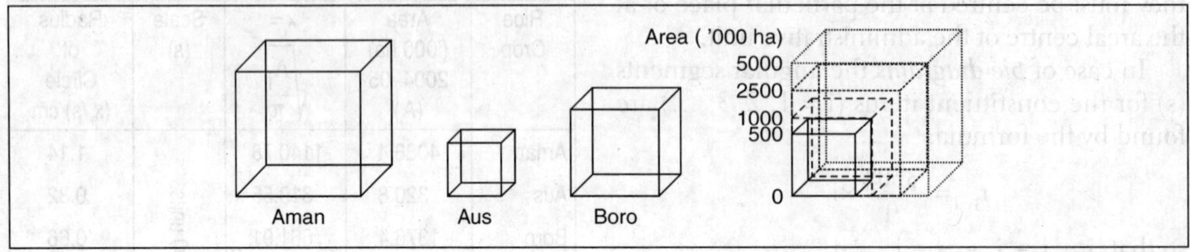

$$x^3 \propto v$$

or, $$x^3 = k.q$$

$$\therefore\ x_i = \sqrt[3]{k.q_i}$$

Hence, the sides of the cube for the items (x_i) corresponding to the quantity (q_i) can therefore be easily evaluated from the equation:

$$x_i = \sqrt[3]{k.q_i}$$

The scale for the cube diagrams should be carefully chosen so that individual cubes are not too small or too large for the base map. When maps are not available they are drawn on the same baseline with uniform intervening spaces. While drawing the cubes, an isometric perspective with 30° – 45° inclination is normally applied. A *worksheet* is neatly drawn to show the computations in detail. A *proportional scale* is diagrammatically represented with at least three cubes corresponding roughly to the *largest*, *smallest* and *median* values of the distribution. Shades may be applied for visual appeal (Table 5.12).

Table 5.12 Worksheet for Cube Diagrams

Rice Crop	Area ('000 ha) 2004–05 (A_i)	$x_i = \sqrt[3]{A_i}$	Scale (s)	Side of Cube (x_i/s) cm
Aman	4088.4	159.90	1 cm to 1000 units	1.59
Aus	320.8	68.46		0.68
Boro	1376.4	111.24		1.11
for	5000	170.99		1.71
propor-	2500	135.72		1.36
tional	1000	100.00		1.00
scale	500	79.37		0.79

Sphere Diagrams

Spheres are 3-dimensional diagrams comprising a series of spheres proportional in size to the quantities they represent (Fig. 5.8b). The construction of sphere diagrams is based on the principle that *the volume of a sphere* (v) *for a representable item* (i) *is directly proportional to the quantity of the item* (q) *it represents*. Empirically,

$$v \propto q$$

or, $$v = k.q,$$

where, k = constant of proportionality.

For a sphere with unit volume, v = 1;

hence, $$k = \frac{v}{q} \Rightarrow \frac{1}{q}$$

In other words, the statement, *a sphere of one unit volume represents a quantity*, q, forms the *volume scale* for the spheres. If the radius of a sphere with volume, v, is r,

then $$\frac{4}{3}\pi r^3 = v \text{ or,}$$

$$\frac{4\pi r^3}{3} = k.q$$

or, $$r^3 = \frac{3kq}{4\pi}$$

$$\therefore\ r = \sqrt[3]{\frac{3kq}{4\pi}}$$

Fig. 5.39 Proportional Sphere Diagrams: Area under Rice, West Bengal, 2004–05

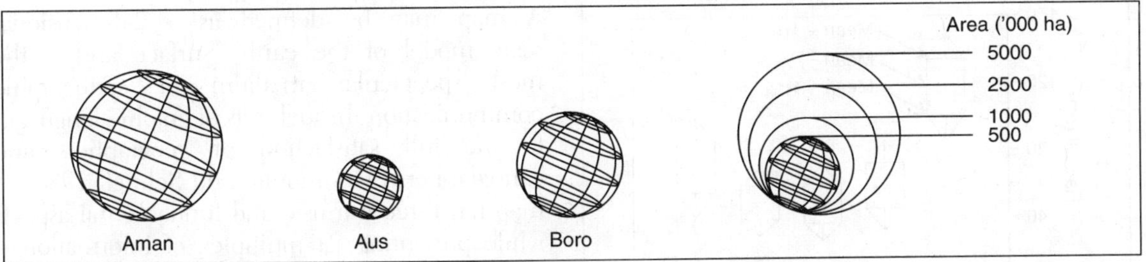

Hence, the radii of the spheres for the items (r_i) corresponding to the quantity, q_i, can, therefore, be evaluated from the equation, $r_i = \sqrt[3]{\frac{3q_i}{4\pi}}$.

The scale for the sphere diagrams should be carefully chosen so that individual spheres are not too small or too large for the base map. When maps are not available they should be drawn touching the same baseline with uniform intervening spaces; otherwise they must be centred at the particular place or at the areal centre of an administrative unit. A proportional scale must be diagrammatically represented with at least three spheres corresponding roughly to the largest, smallest and median values of the distribution. A worksheet must be neatly drawn to show the computations in detail. *Graticules* should be drawn carefully on the surface of the spheres to produce a *3-dimensional effect* (Table 5.10).

Table 5.13 Worksheet for Sphere Diagrams

Rice Crop	Area ('000 ha) 2004–05 (A_i)	$x_i = \sqrt[3]{\frac{3A_i}{4\pi}}$	Scale ... (s)	Radius of Sphere (x_i/s) cm
Aman	4088.4	99.19	1 cm to 1000 units	0.99
Aus	320.8	42.47		0.42
Boro	1376.4	69.01		0.69
for	5000	106.08		1.06
propor-	2500	84.19		0.84
tional	1000	62.04		0.62
scale	500	49.24		0.49

Rainfall Dispersion Diagram

These are statistical diagrams useful for analysing the spread of a series of rainfall values around its median. Rainfall is plotted as 1-Dimensional columnar diagrams, in which, within a range of distribution, quartiles are marked and shaded to emphasise the nature of its dispersion. Naturally, it gives a more rational estimation of rainfall variability and rainfall regimes (Crowe 1933). These are drawn using both the mean monthly rainfall data as well as the mean annual rainfall data and are really useful when drawn with at least 30–35 years of rainfall data (Mathews 1936).

While drawing, months or years are taken on the x - axis and are represented by columns of uniform width. For monthly dispersion diagrams, columns are placed with no gap in between but for annual dispersion diagrams, columns are

Fig. 5.40 Rainfall Dispersion Diagram

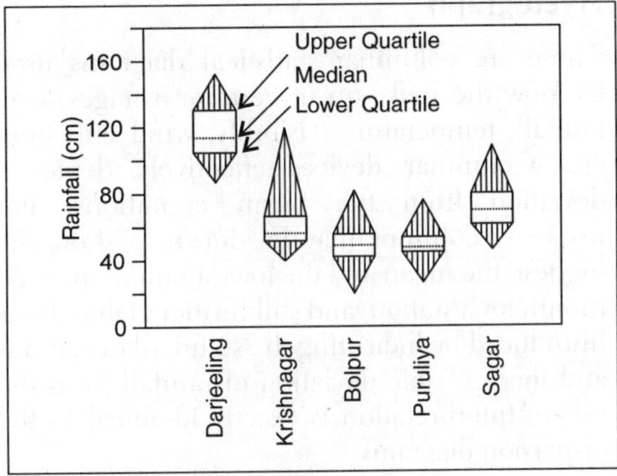

Fig. 5.41 Hyetograph

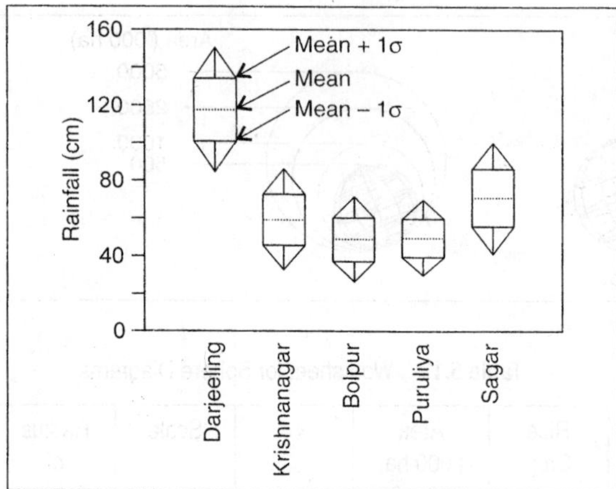

uniformly placed on the x-axis. Figures of rainfall are plotted as dots of uniform size and in case of recurring rainfall figures, such dots are placed sidewise. In each column, horizontal lines are drawn at the minimum, lower quartile, median, upper quartile and maximum value marks (Lackey 1937). For each column the portions beyond the range are erased off and the spaces in between the above marker lines are suitably shaded to emphasise the nature of rainfall dispersion around the median. The shorter the columns, the more the reliability of median rainfall and the longer the columns, the less the consistency of rainfall around the median.

Hyetograph

These are columnar statistical diagrams used to show the make-up of certain averages, e.g., rainfall, temperature, visibility, wind direction, etc. Columnar devices effectively depict a deviation from the mean conditions. For instance, columns may be drawn to show the highest, the mean and the lowest rainfall in each month for a station, and still further elaborations introduced by indicating the standard deviations and the probable deviation of rainfall from the mean. Interpretation is exactly identical to the dispersion diagrams.

MAP

A map may be defined as a 2-dimensional scale model of the earth's surface and is the most spectacular medium of cartographic communication through sets of graphic languages to the full satisfaction of geographers and cartographers (Monmonier and Schnell 1988). A map has three intrinsic and fundamental aspects while presenting a multiplex of information—*location, attributes* at locations and their *spatial relations*. Therefore, maps serve as useful devices of spatial analysis and function as scientific tools or aids to the development of geographical hypothesis. In this age of ISC (I = information, S = satellite and C = computer), maps are defined as *tangible graphic* representations of the cultural and physical environment of the earth's surface (Dent 1985). The *art* and *science* of map making is based on the following seven principles (Raisz 1962):

- Maps are drawn on a *predetermined scale*. Each feature is placed exactly in the proper direction from other points at a horizontal distance proportionate to the scale of the map.
- Maps are *selective*. Only those features that are relevant and important for the purpose of the map are shown.
- Maps emphasise *certain* of the selected features.
- Maps are *symbolised*. All features are shown by standardised symbols that are characteristically

Features of a Good Map Layout Design

- ***Visual balance*** with other map components (e.g., title, legend, scale bar, etc.) to facilitate map reading.
- ***Visual clarity*** by resolving spatial conflicts between map elements generalisation.
- ***Visual hierarchy*** by the appropriate use of font and symbol size.
- ***Visual contrast*** by the appropriate use of colour and shade patterns.
- ***Provision of suitable*** context or reference information.
- ***Appropriate positioning*** of text and annotations.

Conditions for Map Generalisation

- ***Congestion:*** when too many features are packed into a limited geographic space.
- ***Coalescence:*** when features touch overlap one another due to inadequate resolution or symbolisation.
- ***Conflict:*** when the spatial representation of a feature is incomplete with its background.
- ***Complication:*** results from using data from different sources, at different scales, or automated with different tolerance levels.
- ***Inconsistency:*** when a set of generalisation decisions are applied non-uniformly across a map.
- ***Imperceptibility:*** occurs when a feature falls below the minimum display size on a map.

meant for the portrayal of nominal, ordinal and interval or ratio data (Fig. 5.42).

- Maps are *generalised*. Intricate detail is simplified, particularly on small-scale maps.
- Maps are usually *lettered*, *titled* and *labelled*.
- Maps are usually related to a *system of parallels* and *meridians*.
- Visibility and tangibility are definitely the two essential and obvious characteristics of the traditional map products. Based on the attributes of visual and tangible reality, Moellering (1980) proposes four classes of maps:
- A *real map* is any cartographic product that is directly viewable and permanent (e.g., conventional sheet map, globe, orthophoto map, machine-drawn map, plastic relief map and block diagrams),
- A *virtual map* (VM-I) is directly viewable as a cartographic image but has only a transient tangible reality; maps are displayed on a CRT fall in this category (e.g., CRT map image: refresh, storage tube, plasma panel, cognitive map or 2-dimensional image),
- A *virtual map* (VM-II) has a permanent reality but cannot be directly viewed as a cartographic object. Spatial data recorded on a hard copy medium like paper but not as a cartographic image fall in this category (e.g., gazetteer, anaglyph, traditional field data, stored hologram, stored fourier transform and laser disk data), and finally
- A virtual map (VM-III) has neither visual nor tangible reality. Digital images on magnetic disks or tapes fall in this category (e.g., digital memory data, magnetic disk or tape data, video animation, digital terrain model, and cognitive map with relational geographic data).

Operations of Map Generalisation

- ***Simplification:*** of a subset of the original coordinate pairs.
- ***Smoothing:*** relocating or shifting coordinate pairs in order to plane away any small perturbations.
- ***Aggregation:*** grouping point features into higher order classes.
- ***Amalgamation:*** joining small features into a larger map element.
- ***Merging:*** drawing a single line for parallel line features.
- ***Collapse:*** decomposing area or line features into point features.
- ***Refinement:*** discarding smaller features from among a cluster of features.
- ***Typification:*** similar to refinement but uses a representative pattern of features or symbols.
- ***Exaggeration:*** amplifying the shape or size of features to meet the specific need of a map.
- ***Displacement:*** shifting the position of features to gain clarity.
- ***Classification:*** grouping together objects into categories of features sharing identical or similar characteristics.

Real maps are certainly a combination of analogue and symbolic models. For example, a traditional thematic map consists of two important components: *a geographic base map* representing the spatial attributes and *a thematic overlay* representing some nonspatial attributes of the features being mapped. In such a map, the geographic base map is an analogue model because the cartographic forms are similar to the real world patterns but different in scale. The thematic overlay is a symbolic model because an abstract language such as, *shape patterns* or *shade patterns* or *line patterns*, represents this information.

An important characteristic of a map is its *function* performed through a variety of media as *iconic models, analogue models and symbolic models*. To be valid, a map must communicate information to the user of the map. Information instructs the reader to distinguish something new or different from something that is ordinary (Peucker 1972). While producing a map, the cartographer must sort the available data or pieces of facts in order to glean information from redundant facts. Thus maps are certainly the *cartographic abstractions* attained through the processes of *selection, classification, simplification* and *symbolisation* of the information about the environment (Dent 1985). Information is *selected* based on the objectivity or the purpose of the map, while through *classification*, objects with identical or similar attributes are placed in groups. *Simplification* eliminates the unnecessary details and *symbolisation* replaces the form of real world objects with a cartographic representation. In the most general sense, a map is a data model: *a general description of specific groups of entities and the relationships between these entities*. Nystuen (1963) has identified *five* fundamental spatial concepts that are properties of spatial distributions intrinsic to a map:

- *direction*, or the orientation associated with the line of sight from a point,
- *distance* or the physical separation between points in space, which is very important in geographical analysis because spatial interaction usually declines with distance,
- *connectiveness* or relative location/contiguity/ adjacency that exists among cartographic objects,
- *a neighbourhood* that exists around any object containing other elements that are in some way connected to it; it is also very important in spatial analysis because events at one object often influence events at neighbouring objects, and
- *absolute location* or the physical position of a point defined by a metric irrespective of the location of any other point.

Use of Visual Variables for Symbolisation

Feature Dimension	Level of Measurement	
	Nominal (Qualitative)	Ordinal/Interval/ Ratio (Quantitative)
Point	hue (colour) shape orientation	size value (colour) chroma (colour)
Line	hue (colour) shape orientation	size value (colour) chroma (colour)
Area	hue (colour) shape pattern orientation	value (colour) chroma (colour) size
Volume	hue (colour) shape pattern orientation	value (colour) chroma (colour) size

A map is a medium for the expression of the above spatial concepts, and as such, it represents a spatial language. By this, maps serve as useful devices of spatial analysis and function as scientific tools or aids to the development of geographical hypothesis. Thus, maps have a parallelism with geometry, the technical language of space (Harvey 1969). Therefore, a map is a set of cartographic features that have some physical or cultural manifestation on the surface, subsurface or suprasurface of a celestial sphere. Based on the scale, maps are of six types: *wall maps, atlas maps, medium-scale maps, topographical maps, cadastral maps* and *large-scale plans* (Table 5.14).

Based on the objectivity, two distinct classes of maps are usually recognised. These are— *multi-purpose maps* and *single-purpose maps* (Table 5.15). Topographical maps depict as much of

Table 5.14 Classification of Maps based on Scale

Type	Sub-type	Scale	Extent
Wall Map		1:5M – 1:100M	Country, Continent, World
Atlas Map		1:50M – 1:100M	Country, Continent, World
Medium-scale Map		1:5M – 1:10M	District, State, Country
	Million Sheet	1:1,000,000 or 1:1M	4° x 4°
	Degree Sheet	1:250,000	1° x 1°
Topographical Map	Quadrant Sheet	1:100,000	30' x 30'
	Inch Sheet or 15' Sheet	1:50,000	15' x 15'
	Special Sheet	1:25,000 or 1:10,000	5' x 7'30" and 7'30" x 7'30"
Cadastral Map	Small	1:3960	Large Mouza
	Large	1:1980	Small Mouza
Large-scale Plan		1:100	Building Plan

Table 5.15 Classification of Maps based on the Nature of Data and Symbolisation Techniques

Main Class	Sub-class	Maps	Symbolisation Process	Example
I. General		General	Letter, Point, Line and Area Symbols, Colour-Patterns and Conventional Symbols	Topographical Maps
	A. Purely Qualitative (Nominal Data)	1. Chorochromatic	Line Symbols and Colour-Patterns	Landuse Maps
	B. Semi-quantitative (Nominal and Ordinal Data)	2. Choroschematic	Letter, Point, Line and Area Symbols, Colour-Patterns, Conventional Symbols and Pictographs	Administrative or Political Maps
		3. Qualitative Dot	Point Symbol, and Colour-Patterns	Settlement Maps Distribution Maps
II. Thematic	C. Purely Quantitative (Interval and Ratio Data)	4. Quantitative Dot	Point and Area Symbols and Colour-Patterns	Rural and Urban Population Distribution
		5. Flowline	Line Symbols and Colour-Patterns	Traffic Flow or Interaction Maps
		6. Isopleth	Line Symbols	Contour Maps
		7. Choropleth	Line Symbols, Line and Colour-Patterns	Population Density Maps
		8. Diagrammatic	Graphs and Diagrams	Distribution Maps
		9. Cartogram	Shape-Patterns, and Colour-Patterns	World Distribution Maps
	D. Special (Digital Raster and Vector Data)	10. Landform	Oblique Photographs	Virtual Maps
		11. Animated	Movie	Virtual Maps

the surface features of a particular area as is possible within the scale constraints (Lawrence 1971). Hence, these are multi-purpose maps portraying the spatial association among the diverse geographical objects/phenomena of the earth's surface (Board 1967). These maps are also called *general maps* or *general reference maps* in which the contextual items are shown as marginal notations and a standard legend fixed by the ICA (International Cartographic Association) is rigidly followed. The single-purpose maps are known as *thematic maps* as they concentrate on the spatial distribution of a *single phenomenon* or *the relationship between phenomena* (Star and Estes 1990). They are used to explore the *structure of a given distribution*, i.e., the character of the whole as consisting of the interrelationships of the parts. Thematic mapping concerns a goal for an arrangement of mappable items, that achieves the maximum

Fig. 5.42 Mapping Symbols for Qualitative and Quantitative Data

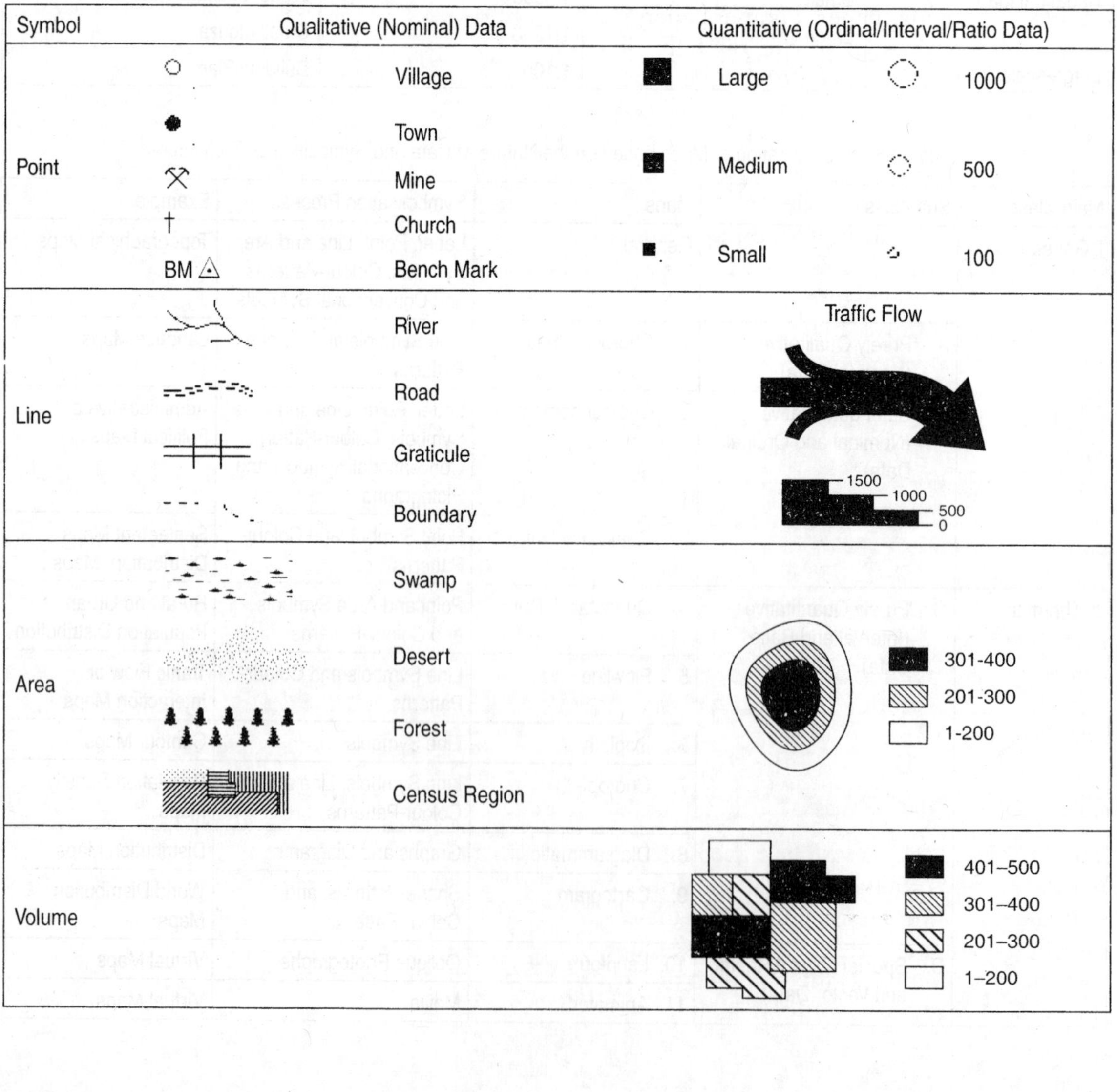

Symbol	Qualitative (Nominal) Data	Quantitative (Ordinal/Interval/Ratio Data)
Point	Village Town Mine Church BM Bench Mark	Large 1000 Medium 500 Small 100
Line	River Road Graticule Boundary	Traffic Flow 1500 1000 500 0
Area	Swamp Desert Forest Census Region	301-400 201-300 1-200
Volume		401–500 301–400 201–300 1–200

clarity, legibility and relative contrast of the detailed items. It involves two stages—*first*, structuring the major elements and *second*, graphic outlining, i.e., the position of detail within each major element. On such maps it is customary to place explanatory items and contextual notes such as title, legends and inset within the map frame itself. Maps were and still are drawn by manual means using a variety of hand equipment, e.g., drawing board with clips, set squares, compass, dividers, protractors, rulers, diagonal scale, french curves, pencils, erasers, rotring pens, templates, stencils, transparencies, art papers, tracing papers, graph papers, etc. However, nowadays maps are digitally drawn on computers and hard copies can be extracted of as many copies as desired from the plotters and printers. For this, cartographers use various graphics and mapping softwares, such as, SYMAP, LINMAP, COLMAP, MAPIT, Grass, Idrisi, Gram++, Ilwis, Mapinfo, ArcView, Geomatica, Arc/Info, ArcGis, ERDAS, etc.

Based on the nature of data and the cartographic controls of mapping, thematic maps are of four types: *qualitative, semi-quantitative, quantitative* and *special* (Table 5.14). Qualitative maps involve nominal data (e.g., *chorochromatic maps*), semi-quantitative maps involve symbolisation with mainly ordinal data (e.g., *qualitative dot maps and choroschematic maps*), quantitative maps concern interval data and ratio data (e.g., *quantitative dot maps, diagrammatic maps, choropleth maps, isopleth maps, flowline maps and cartograms*) and special categories are mainly *virtual maps* generated digitally using computers, e.g., *landform maps and animated maps*.

Chorochromatic Maps

These are also known as *colour-patch maps* (choro = area, chromos = colour). The basic principle is that tints or colours are used to display the spatial distributions on maps (Fig. 5.43). These maps are essentially qualitative as the mapping procedure does not consider any numerical information. On maps, the spatial patterns are first outlined and then coloured. Tints of the same colour for single-distribution or different colours for multi-distribution are normally used. Tint variation may be effectively used to emphasise variations in the spatial characteristics of a single element. However, the choice of colours showing the areal distribution of different elements should be as far as practicable, realistically conventional and iconic. A chorochromatic map must contain a well-made legend of colours reflecting geographical impressionism. The mosaic of colours helps to readily identify the nature of spatial pattern of a distribution. *Landuse components, soil groups, vegetation zones, lithological units or rock groups, crop regions*, etc., form the characteristic information layout for ideal chorochromatic mapping.

Choroschematic Maps

This kind of map shows spatial distribution with the help of pictorial symbols, letter symbols, point symbols, line symbols, area symbols and volume symbols (Fig. 5.44). Symbols for each of the elements to be shown are first selected and then drawn carefully with uniform size and boldness. To make the map more useful, a mapping scale is selected such that *a symbol represents a specific quantity*. With this, the number of symbols to be drawn may easily be evaluated with reference to the quantity it represents. The greatest advantage of this method of mapping is that many elements can be shown together on a single map.

Qualitative Dot Maps

These are maps in which dots are used to denote a position, the location of a feature, the intensity at a place, or a representative location for spatial summary data. Examples are a coordinate location, a radio tower, a spot height, the centroid of some distribution, or a conceptual volume at a place, such as the population of a settlement. Naturally, mapping is done with nominal data as well as ordinal data. Even though the marks (i.e.,

Fig. 5.43 Chorochromatic Map

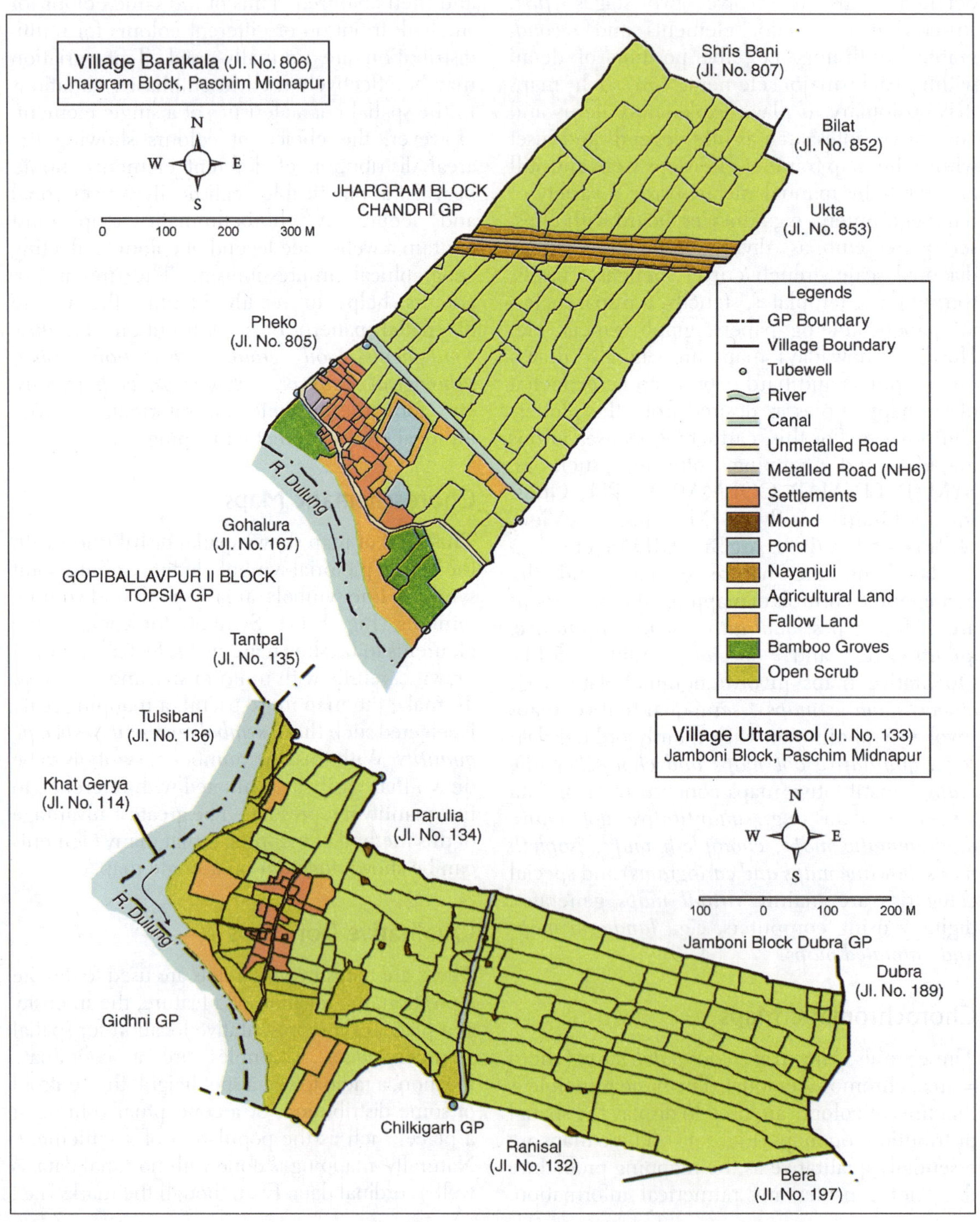

Fig. 5.44 Choroschematic Map

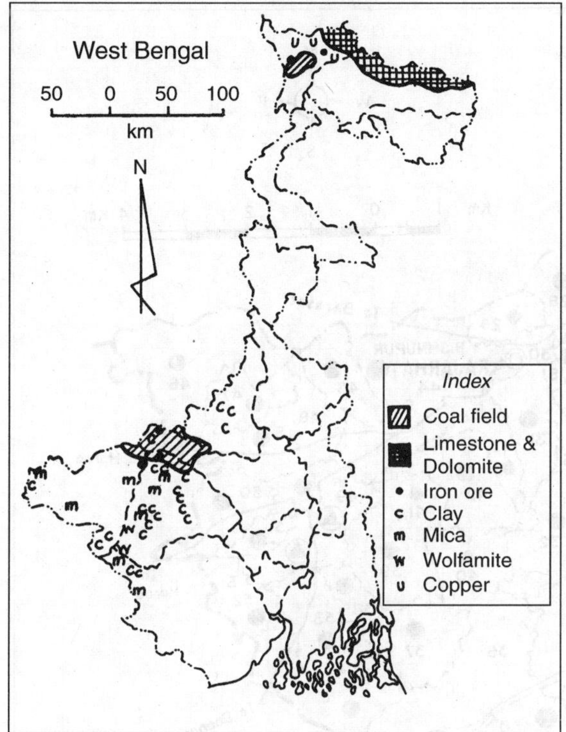

the dots drawn in these maps as circles of different sizes) may cover some map space, they are point symbols as they conceptually refer to a location. The principles are (a) *dot sizes are subjectively scaled, directly varying with the ascending order of data values* and (b) *dots are positioned and centered exactly at the map locations.* Visual variables of hue, orientation and shape can be applied to enhance the quality of map presentations as desired by the cartographer (Fig. 5.45).

Quantitative Dot Maps

In this kind of distribution maps, quantities or values are represented by dots of uniform sizes, each dot having a specific value (Winterbotham 1934). Dot maps are especially useful when the values are unevenly and sporadically distributed. The dots are inserted within the particular administrative units for which data is available. The smaller the units, the more accurate the map (Fig. 5.46).

While constructing a dot map the first step is to examine the range of quantities involved. From this, a value represented by each dot is selected. Basically, the success of a dot map depends absolutely on the choice of this value, called *dot scale*. It is chosen, keeping in mind the size of the administrative unit so that dots are neither numerous nor few.

According to the principle, *the number of dots* (n) *corresponding to an administrative unit* (i) *is directly proportional to the quantity* (q). Empirically,

$$n \propto q$$

or, $n = k.q$

where, k = constant of proportionality.

For a single dot, $n = 1$; therefore, $k = \frac{n}{q} \Rightarrow \frac{1}{q}$.

In other words, the statement, *one dot represents* k *quantity* forms the *dot scale*. Hence, the number of dots for any administrative unit can be found from the equation,

$$n_i = k.q_i \Rightarrow \frac{q_i}{q}$$

While plotting, the dots should be placed evenly and uniformly within each unit. The boundaries of the units are often erased after the insertion of dots. Dots should never be placed in straight rows and columns; rather the vertices of a small equilateral triangle should be assumed and dots should be placed on these. The precision of a dot map can be enhanced by consulting a physical map showing negative areas. The size of the dots depends on the scale of the base map and on the number of dots to be inserted. To avoid the effects of coarseness or blurring and to obtain a finer visual tone, a nomograph may be consulted (Mackay 1949).

Dot maps have three *variants—percentage dot maps, mille dot maps and multiple dot maps.* In the first two categories, the percentage values are mapped. In percentage dot maps, *each dot*

Fig. 5.45 Qualitative Dot Map

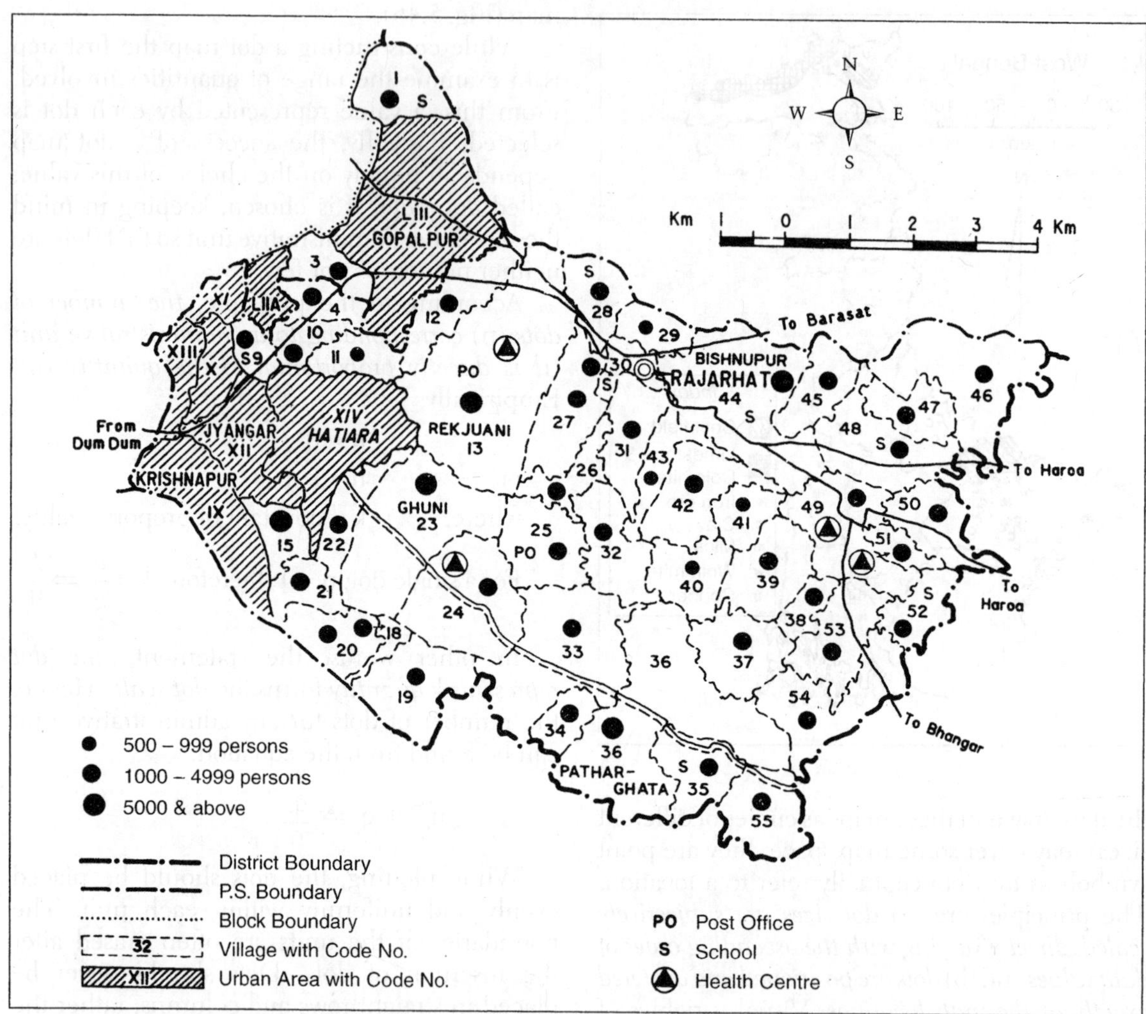

represents 1% while in mille dot maps *each dot represents* 0.1%. Such maps facilitate arithmetical comparisons and give ready information about the fractional distributions and proportions. In multiple dot maps, the distribution of several elements are normally shown using dots of differing colours and sizes.

Flowline Maps

These are the most important *forms of dynamic maps* in which movements of goods, information and people *between and among* places are *shown*. Obviously, the *actual route* of movement on a map is considered and the *quantitative impression is conveyed by the width of the line*. In order to make the map easily commensurable, *a line of uniform thickness may be taken as a unit* and the required number of lines may be drawn closely parallel to each other to represent different amounts of traffic movement along the respective routes (Fig. 5.47). However, this may also be represented by *single lines of variable*

Fig. 5.46 Quantitative Dot Map

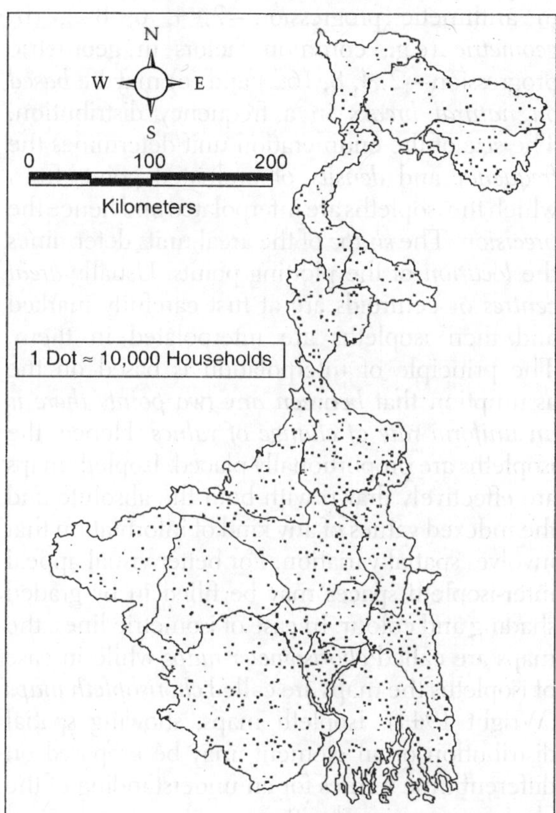

thicknesses with a scale of traffic flows. This form certainly appears like the arterial system of a living organism through which the vital energy continuously flows (Green 1984, Ormsby 1950, Taylor 1952, Christensen 1960).

A variant of flowline map is one in which *qualitative information about flow is represented by means of straight lines or rays drawn directly between two points.* The actual route of flow or the amount of flow is immaterial in such maps (Fig. 5.47). These are used to identify the sphere of influence of towns or cities, where areas served by functions are shown by rays radiating from the town to the rural units. The most important advantage of this is that it allows overlaps of service areas to be mapped. Thus, a great variety of affinities may be expressed by such maps in a most simple and effective manner.

Fig. 5.47 Flowline Map

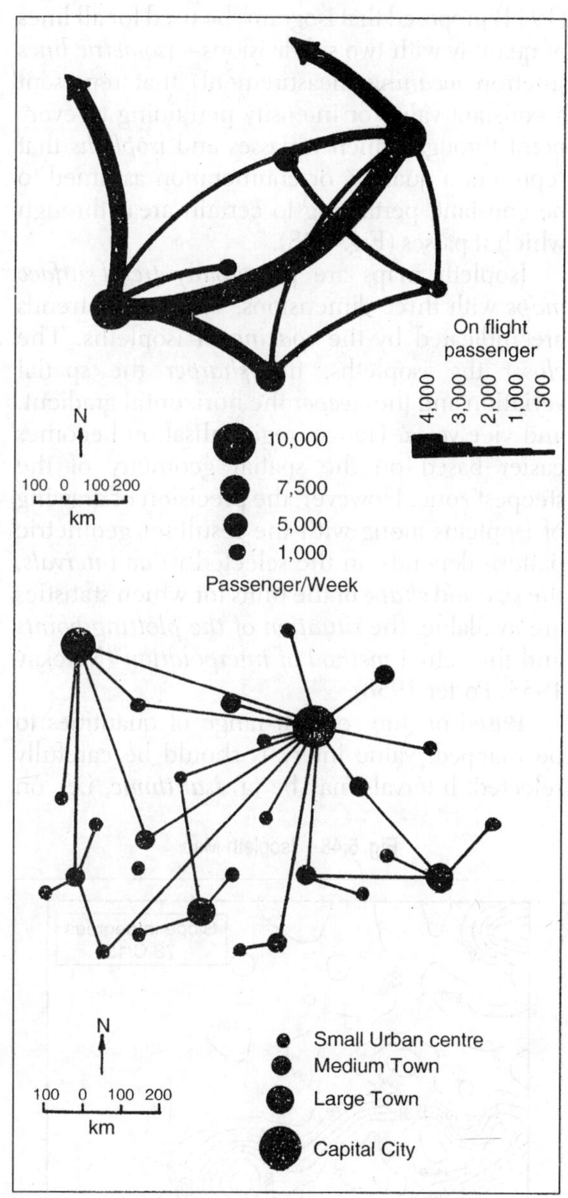

Isopleth Maps

These are *quantitative areal maps* where quantities are indicated by lines of equal value known by a multiplicity of such terms as *isopleth* (iso = equal, plethos = a multitude or crowd), *isarithm, isoline, isometric lines, isontic lines and isogram.* Wright (1944) proposed that isograms be used for all lines of quantity with two subdivisions—*isometric lines* (metron *meaning* measurement) that represent a constant value or intensity pertaining to every point through which it passes and *isopleths* that represent a quantity or enumeration assumed to be constant, pertaining to certain areas through which it passes (Fig. 5.48).

Isopleth maps are principally *trend-surface maps* with three dimensions. The spatial trends are indicated by the *spacing* of isopleths. The *closer* the isopleths, the *sharper* the spatial variation and the *steeper* the horizontal gradient, and vice versa. Hence, regionalisation becomes easier based on the spatial geometry of the steepest zone. However, the precision of drawing of isopleths along with the resultant geometric pattern depends on the selected *value intervals*, the *size* and *shape* of the units for which statistics are available, the *situation of the plotting points* and the actual *method of interpolation* (Mackay 1953, Porter 1958).

Based on the overall range of quantities to be mapped, value intervals should be carefully selected. Intervals may be (a) *isarithmic*, i.e., on a rhythmic interval basis (e.g., common intervals in arithmetic progression—2, 4, 6, 8...), (b) *geometric* (e.g., common factors in geometric progression—2, 4, 8, 16...) and (c) may be *based on natural breaks* in a frequency distribution. The *size* of the enumeration unit determines the *frequency* and *density* of evaluated points from which the isopleths are interpolated and hence the *precision*. The *shape* of the areal units determines the *location* of the plotting points. Usually *areal centres* or centroids are at first carefully marked and then isopleths are interpolated in them. The principle of interpolation is based on the assumption that *between any two points there is an uniform rate of change of values*. Hence, the isopleths are proportionally placed. Isopleth maps are effectively drawn with both the absolute and the indexed values of any kind of information that involves spatial variation. For better visual appeal inter-isopleth spaces may be filled in by graded shading and colour. In case of isometric lines, the maps are called *chorisometer maps* while in case of isopleths the maps are called *chorisopleth maps* (Wright 1944). Isopleth maps, showing spatial distribution of an element, may be prepared on different time frames for an understanding of the changes in regional pattern.

Fig. 5.48 Isopleth Map

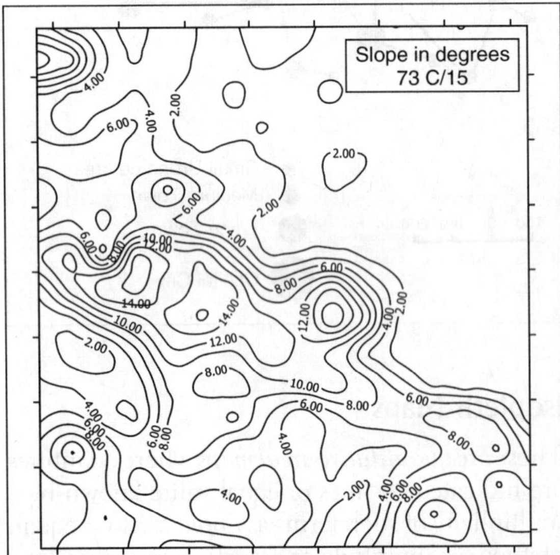

Choropleth Maps

Choropleth maps are technically *quantitative areal* maps that show the spatial distribution of the intensity or density of an element with the help of a system of graded shading or colour, drawn following the boundaries of the administrative units (Fig. 5.49). The basic principle is that the *intensity of shading is directly proportional to the density of elements.* These density maps, related as they are to the administrative units, display only average distributions. Hence, the grouping of a number of units under one average value implies distributional uniformity. This may be far from the real world picture for the broad average may mask a vast range of local variations. Obviously, the more expansive the areal units, the more sweeping the generalisation presented in the map form.

Fig. 5.49 Choropleth Map

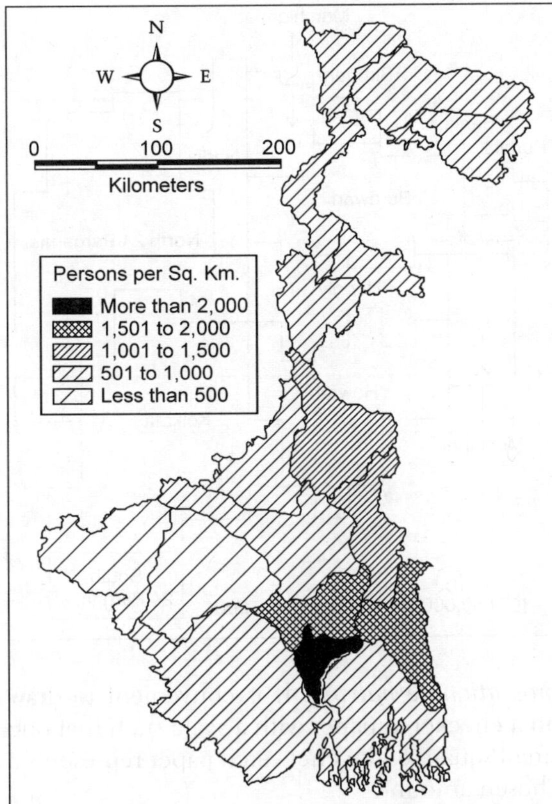

The construction of a choropleth map is a *3-step process* as follows:

- The *first* step is the drawing of a worksheet with four columns—name of the unit (column 1), area (column 2), absolute value of an element (column 3) and the density obtained by dividing the absolute value by the area (column 4) and rows equalling the count of the administrative units.
- The *second* step is the construction of a *choropleth table* showing the columns of the density classes, shading system, administrative units and remarks. The choice of the scale of densities may be based on arithmetical progression with uniform class interval, or geometrical progression with rapid increasing intervals, or quartile deviation or mean deviation or standard deviation of the dataset or any other criteria chosen by the researcher that seems to better suit his/her purpose.
- The *third step* involves the meticulous drawing of shades (of either line or colour) following the administrative boundaries as per the choropleth table.

Choropleth maps are the basic tools of human geographers. The smaller the administrative units, the more the map precision. *Intervals* should be wisely chosen depending more on *experience* rather than on the theoretical character of distribution.

Diagrammatic Maps

These maps, as the name suggests, show the *representation of statistical data over the map*

Fig. 5.50 Diagrammatic Map

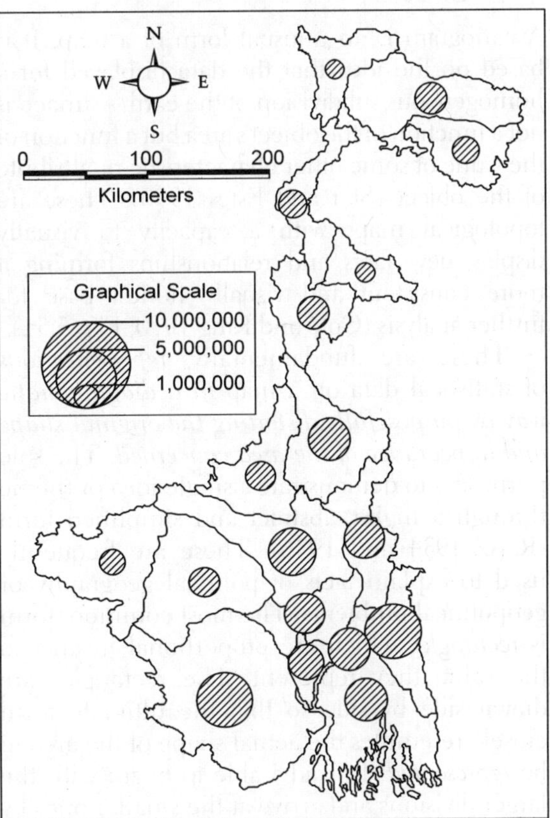

by means of suitable graphs and diagrams (Fig. 5.50). The principles and procedures of such mapping are: i) the *centres* of circles and spheres *must correspond* to the *exact site* of the place or to the *areal centres* of the administrative units, ii) the *base* of any graph or of a bar, rectangle, square, triangle or cube should be placed at the *exact centre* of the region or at any point to be so selected that it may lie, as far as practicable, within the limits of the area concerned. For this purpose the scale of representation should be thoughtfully chosen and iii) if, in spite of all considerations, diagrams of contiguous administrative units appear to overlap, they should be drawn *sequentially for places in the ascending order of size*. The smaller ones must be clearly visible while the larger ones may be allowed to be partly consumed or eclipsed.

Cartogram

A cartogram is an unusual form of a map. It is based on the idea that the data displayed for a homogeneous subdivision of the earth's surface is not a function of the object's area but a function of the value of some other characteristic or attribute of the object (Star and Estes 1990). These are topological maps with a capacity to visually display new facts and relationships forming a more consistent and visually honest base for further analysis (Cole and King 1970; Fig. 5.51).

These are fundamentally *representations* of statistical data on a map *in a diagrammatic way by purposefully distorting the original shape and appearance of the area concerned*. The sole purpose is to demonstrate a single idea or theme through a highly abstract and simplified form (Raisz 1934 and 1936). These are frequently used to explain facts of political geography or geopolitical problems. The most common form is *rectangles* which are proportional in area to the value they represent. The rectangles are drawn side by side so that the ultimate form closely resembles the actual shape of the area to be *represented*. It is advisable to begin with the larger divisions and arrive at the smaller ones by *proportionate halving*. It is convenient to draw on a checkered paper with a scale such that one small square of the checkered paper represents a chosen amount.

Fig. 5.51 Cartogram

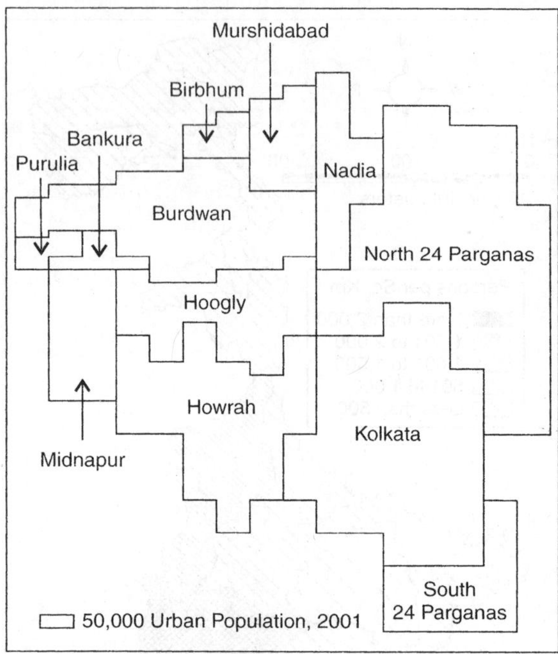

Landform Maps

These are special maps that depict the earth's surface as if it were viewed *from an oblique areal point of view* (Monmonier and Schnell 1988).

Fig. 5.52 Landform Map

Such a synthetic 3-dimensional representation requires a significant amount of effect, skill and time. These are most commonly generated with the help of computer programmes (Fig. 5.52).

Animated Maps

These are more common and popular nowadays, as computer graphic techniques make them easier to create. Animation makes it possible to easily display the sequences through time. These methods of data display are sometimes called movie loops. An ideal example would be the growth of a city as its population and area increases through time. Hence, it serves as an excellent summary tool, since patterns occurring over years of change may efficiently be presented to the audience in a minute or two.

EXERCISE

1. Why does statistical data need to be represented?
2. What are the ways in which statistical data is presented?
3. Discuss in detail the 'tabular representation' of data.
4. In what way is graphical presentation of data superior to other forms of presentation?
5. What are the contextual items of a graphical presentation?
6. What is a graph? Classify it. Explain the quality of data suitable for various graphs.
7. What are biaxial graphs? Classify them. State the situations in which each is used.
8. What are semi-log and log-log graphs? Explain their construction and state the situations in which each is used.
9. What are climographs? Explain their principles of construction and interpretation.
10. What are hythergraphs? Explain their principles of construction and interpretation.
11. What are ternary graphs? Explain their principles of construction. Discuss the situations in which they are used.
12. What are multiaxial graphs? Classify them. Discuss the situations in which each is used.
13. What are vector graphs? Which type of data is best represented by it?
14. What are circular graphs? What type of data is best represented by it?
15. What are dispersion graphs? Explain their principles of construction and interpretation.
16. What are ergographs? Explain their principles of construction and interpretation.
17. Project: Collect suitable data from Census, Development Reports, Statistical Abstracts, etc. and illustrate the representation of statistical data with the help of graphs.
18. What is a diagram? Classify it. Explain the quality of data suitable for various diagrams.
19. What are 1-dimensonal diagrams? Classify them. State the situations in which each is used.
20. What are pyramidal diagrams? Explain the principles of construction of urban, population and ecological pyramids.
21. What are 2-dimensional diagrams? Classify them and explain their principle of construction.
22. What are pie diagrams? Explain their principles of construction and use.
23. What are compound square diagrams? Explain their principles of construction and situations in which each is used.
24. What are 3-dimensional diagrams? Classify them. Discuss the situations in which each is used.
25. What are sphere diagrams? Which type of data is best represented by it?
26. Project: Collect suitable data from Census, Development Reports, Statistical Abstracts, etc. and illustrate the representation of statistical data with the help of diagrams.
27. What is a map? What are its properties? What are its characteristics? What are its functions?

28. What are the principles of map making?
29. Classify maps based on visual and tangible reality.
30. Distinguish between real maps and virtual maps.
31. What are the components of spatial analysis on a map?
32. Explain: Map is a data model.
33. Explain: Map is the medium of cartographic communication.
34. Explain the terms: generalisation of data, symbolisation of data, map design, map layout, digital maps, map typography, map colour-pattern, map reproduction.
35. Classify maps based on scale.
36. Distinguish between general maps and thematic maps.
37. Classify thematic maps based on the nature of mapable data.
38. What are isopleth maps? Explain their principles of construction, use and interpretation.
39. What are choropleth maps? Explain their principles of construction, use and interpretation.
40. What are diagrammatic maps? Explain their principles of construction, use and interpretation.
41. What are quantitative dot maps? Explain their principles of construction, use and interpretation.
42. What are flowline maps? Explain their principles of construction, use and interpretation.
43. What are cartograms? Explain their principles of construction, use and interpretation.
44. What are choroschematic maps? Explain their principles of construction, use and interpretation.
45. What are qualitative dot maps? Explain their principles of construction, use and interpretation.
46. What are chorochromatic maps? Explain their principles of construction, use and interpretation.
47. What are landform maps? Explain their principles of construction, use and interpretation.
48. What are animated maps? Explain their principles of construction, use and interpretation.
49. Project: Collect suitable data from Census, Development Reports, Statistical Abstracts, etc. and illustrate the representation of statistical data with the help of maps.
50. Compare the characteristics of manually produced and digitally generated maps.

Unit III

Map Interpretation

6

Geological Map

HIGHLIGHTS

- Map Reading
- Map Analysis
- Interpretation of Geological Maps
- Example 1
- Example 2
- Example 3
- Example 4

Geological maps show the geographical pattern of the composition of the earth's surface by means of lithology, structure and succession of geological formations. These are useful to students of geography (particularly, geomorphology), geology and other allied disciplines. Being a function of structure, process and stage, landforms evolve as a synthesis of lithology, structure, tectonic history and subaerial processes that are finely tuned and well synchronised in operation. Thus, reading and interpreting a geological map necessitates a deeper understanding of the geomorphological processes, lithology and structure of the constituent rocks along with the chronology of formations. It brings out the sequence of both the geological and geomorphological evolution of a region on the earth's surface. A complete work out of a geological map therefore needs a-priori knowledge of geology and geomorphology and basic skills in map reading, data acquisition from maps and its subsequent manipulation.

A *geological map* represents the outcrops of the different rock strata of a particular region superimposed upon its topography. *Lithology* of the geological formations are shown by conventional colours or in black and white by conventional symbols inserted accurately according to the precise location and extent of the different outcrops. The *succession* is shown in columns in accurate chronological sequence, i.e., the oldest bed occupying the bottom and the youngest the top. The discontinuity in the succession column denotes an *unconformity*. The attitude of the rock-beds or the *structure* is shown by the relations between outcrop pattern and the contour layout. The latter forms the topographic mosaic of the area concerned.

The intersection of a rock body with the topography is called its outcrop (Fig. 6.1). The sedimentary rocks are formed in layers, strata or beds. The plane that separates two successive beds is called a *bedding plane*. For all practical purposes, it is treated as a *geometric plane*. Thus, a bed is bounded by two bedding planes, namely

Fig. 6.1 Bedding Planes and Outcrops

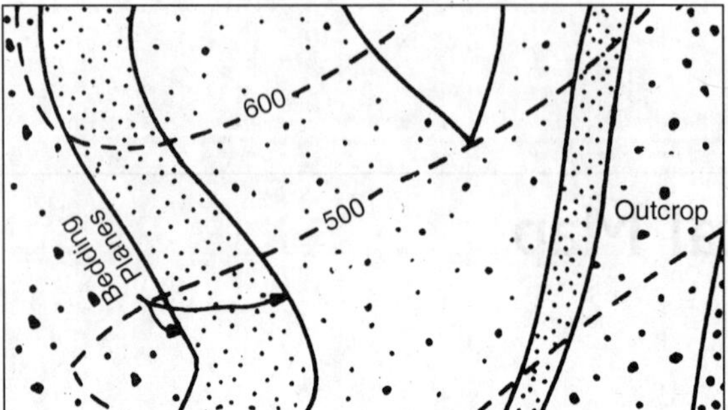

the *upper bedding plane* which represents the interface between the bed and the overlying younger bed and the *lower bedding plane* which separates the bed from the underlying older bed (Fig. 6.2). The bedding planes may be *horizontal, tilted, folded* or *faulted* in any direction. The inclination of a bedding plane with respect to the horizontal plane is called its *dip*. Dip measured along the direction of maximum inclination is called the *true dip*. The *geographical* direction in which the beds dip is called the *dip direction*. As dip is a vector quantity, the attitude of a bed is shown both by its *magnitude* and *direction*. The direction perpendicular to the direction of true dip is called the *strike* and a line drawn in this direction is called a *strike line*. It is essentially the strike line along which the upper and lower bedding planes of a strata intersect a horizontal plane. Naturally, strike is the direction along

Fig. 6.2 Strike and Dip of a Bed

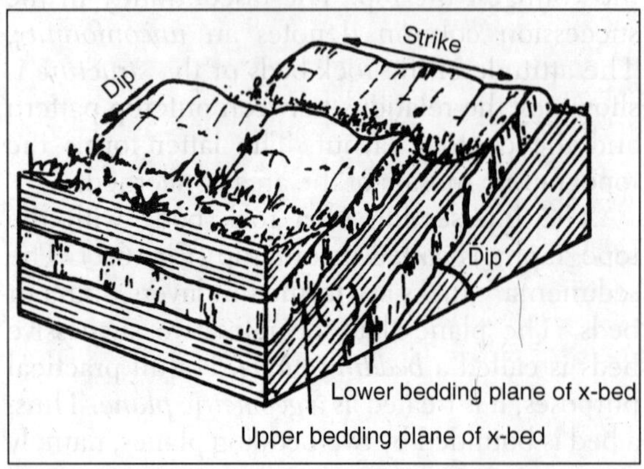

Fig. 6.3 True and Apparent Dip

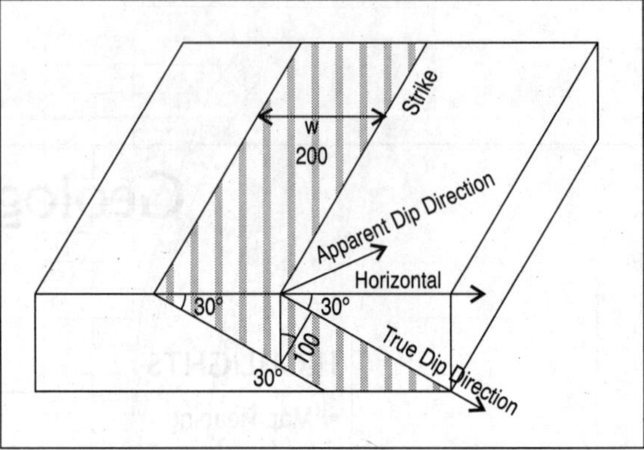

which a bed appears to be horizontal. Thus, an infinite number of strike lines can be drawn on a bedding plane at varying elevations. Strike lines are essentially parallel to one another as each one of them represents a level surface at different elevations. Hence, the strike lines are also known as *stratum contours* (Fig. 6.3).

MAP READING

Identification of Unconformities

The beds of a single series of formation are conformable to one other. But in case of two or more series of formations, they are mutually unconformable. This is so because they are formed in different geological ages and contain beds that differ in attitude, lithology and succession. This discordance between any two series of formations is called an *unconformity* (Fig. 6.4). On a geological map it can easily be identified from the following:

i. The beds of the younger series appear to cover the beds of the older series.
ii. The bedding planes of the older series appear to be truncated or to end abruptly.
iii. The dips of the beds in two formations differ.
iv. The strike lines of the different formations usually intersect.
v. A basal conglomerate bed is often associated with an unconformity.

Fig. 6.4 Unconformity Separating the Upper and the Lower Series of Formations

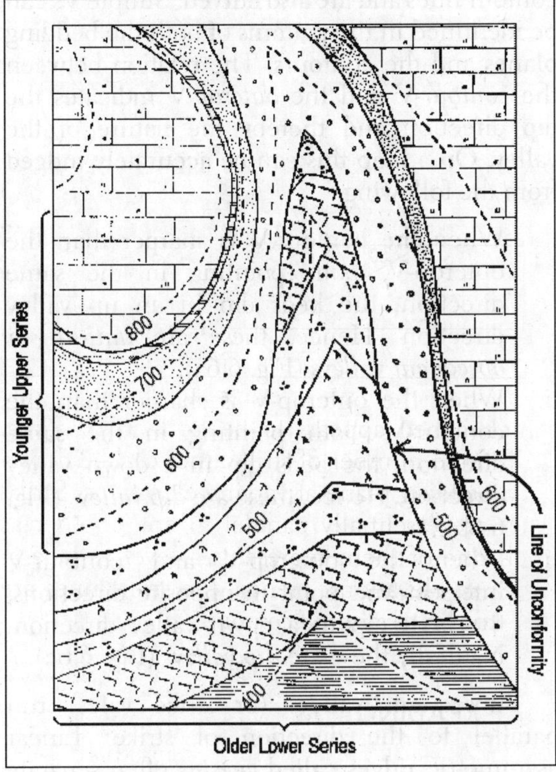

The surface or plane that separates any two series of formations forms the *surface or plane of unconformity*. Its intersection with the topography forms the *line of unconformity*. The dip of the plane of unconformity conforms to the dip of the beds of the upper younger series.

Identification of Horizontal and Vertical Structures

When the bedding planes run parallel to the contours, the outcrop pattern appears to be entirely guided by the contour pattern. In such situations, it is not possible to draw any strike line and the dip of the beds become zero (i.e., $\alpha = 0$, $\theta = 0$) or, in other wards, the beds are horizontal. On a map, it is shown by a + sign. Younger beds occur at higher elevations and older beds at lower elevations. The contour patterns represent *tablelands, mesas* and *buttes* separated by valleys and well formed spurs. Hence, the region represents, in most cases, a *tableland* or *a spur* and *valley topography*. Sometimes an outlier is formed when subaerial denudation exposes younger rocks surrounded by older rocks.

When the outcrops are all straight, irrespective of the contour pattern, and the bedding planes cut across the contours and run straight, the beds are *vertically* disposed. In such situations, strike lines at different elevations coincide with one another so that inter-strike intercept vanishes to zero. Hence, the dip of the beds becomes 90°. On a map it is shown by a † sign (Fig.6.5).

Identification of Uniclinal Structures

In a series, when all the beds dip uniformly in a particular direction, a *uniclinal structure* is formed. In such situations, the outcrops of the

Fig. 6.5 Horizontal and Vertical Beds

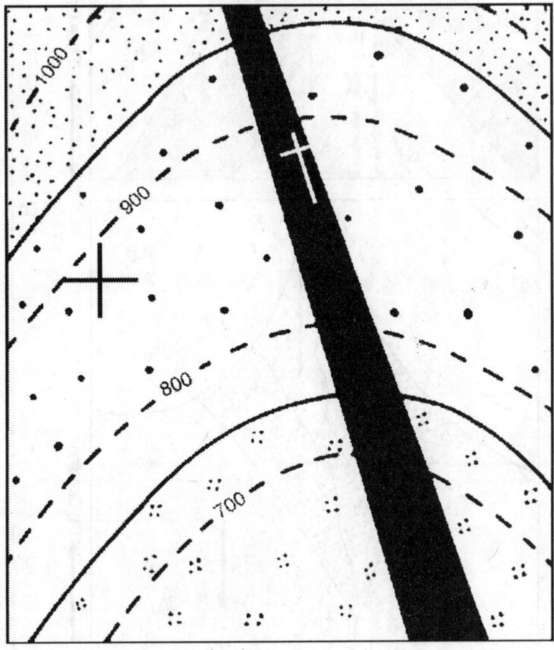

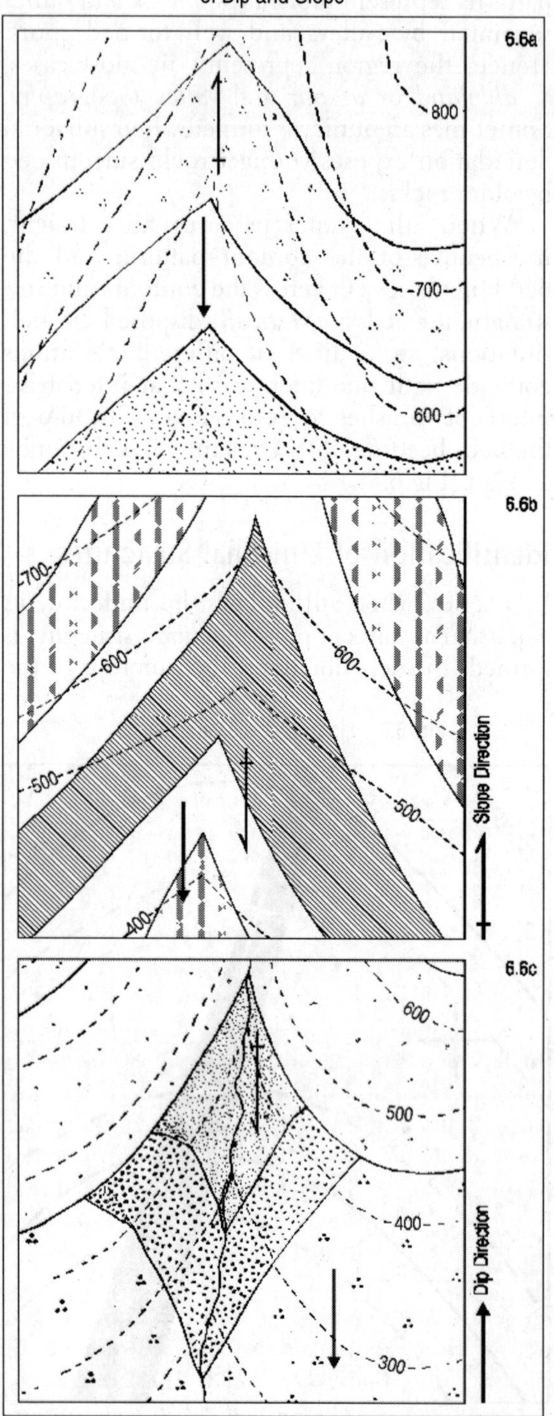

Fig. 6.6 Application of V-rules Showing the Directions of Dip and Slope

beds are curved and run more or less parallel to one other. The bedding planes cut across the contour lines and are also curved. Simple Vs can be identified in the patterns of both the bedding planes and the contours. The relation between the *contour*-V and the *outcrop*-V indicates the dip direction and thereby the nature of the valley. On a map this can be accurately judged from the following:

i. When the contour-V is sharper than the outcrop-V, both pointing in the same direction, the beds dip in an up valley direction. Hence, these are *antidip* or *obsequent valleys* (Fig. 6.6a).
ii. When the outcrop-V is sharper than the contour-V, both pointing in the same direction, the beds dip in a down valley direction. Hence, these are *dip valleys* (Fig. 6.6b).
iii. When the outcrop-V and contour-V intersect and point in opposite directions, the beds dip in a down valley direction. Naturally these are *dip valleys* (Fig. 6.6c).

Subsequent valleys (or strike vales) run parallel to the direction of strike. Linear asymmetric ridges called *cuestas* often separate these valleys. Obsequent valleys are shorter in length and drain the steeper antidip or escarpment slopes of a cuesta, while longer secondary consequent valleys drain its gentle dip slopes. Thus, the network of valleys developed on an uniclinal structure forms a rectangular trellis pattern. Subaerial denudation, especially valley deepening, often exposes younger rocks along it, forming valley inliers. Similarly, outliers also develop particularly where, as a result of huge subaerial erosion, younger rocks surround the older country rocks. Thus the topography developed on the uniclinal structure is unique to it and is called a scarpland landscape.

Identification of Folded Structures

In case of *folding*, a fold may either be an *anticline* or a *syncline* or a combination of the two. The criteria for identification of such structures are (Fig. 6.7):

i. The outcrop of a bed generally repeats itself.
ii. If, along with the first criteria, the beds are found to dip in opposite directions in any two successive outcrops, it is definitely a case of folding.
iii. There occurs essentially some beds, on the two sides of which another bed repeats itself. The former bed is definitely the *core bed* of either an anticline or a syncline.
iv. In most cases, the Vs of the outcrops combine to form either an 'X' or an 'H' pattern.
v. In case of a syncline, younger beds form the core while older beds occur progressively outward on either side.
vi. In case of an anticline, the core is formed by older beds. The younger beds occur progressively on either side of the older beds in the direction of the limbs.

Usually on maps, valleys occupy the anticlinal folds and hills occupy the synclinal folds. Thus *inversion of relief* is common. The landscape represents a *ridge and valley* topography. Usually

Fig. 6.7 Faults and Vertical Intrusion

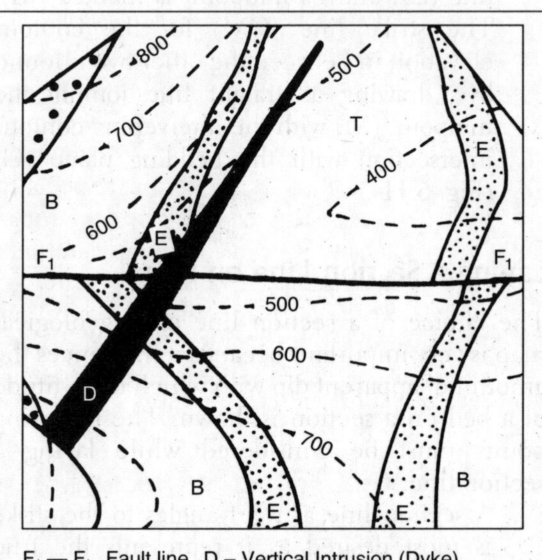

longitudinal subsequent valleys run parallel to the strike lines while relatively short obsequent valleys follow the dip or antidip directions. The anticlinal valleys are often bounded by *inward facing escarpments*. Outliers are found in regions of denuded synclines while inliers in areas of denuded anticlines.

Identification of Faults

On a geological map, the *fault trace* or the *line of fault* is marked as F–F or f–f. It is indicated by a thick line. A straight fault trace indicates a *vertical fault* while a curved fault trace indicates an *inclined fault*. Normally, there occurs a shifting in the position of the same bed on either side of the fault (Fig. 6.8 and Fig. 6.9). The vertical displacement is called *throw,* while the horizontal displacement is called *heave*. The angle which the fault plane makes with the normal plane is called *hade,* while the angle between the horizontal plane and the fault plane is called fault angle or dip of the fault (β). For a more or less flat and level surface,

$$\text{Throw} = \text{Heave} \times \tan\beta$$

Normally, on the downthrown side, the younger bed occurs against an older bed on the corresponding upthrown side. The throw can be measured from the map by comparing the strike lines of a bedding plane on one side of the fault with that of the same bedding plane on the other side. The difference between the two values of the same strike line gives the amount of throw.

The *apparent displacement* of the beds on the surface depends on the relation between the fault trace and the strike of the beds. On the basis of this, faults can be classified into *strike, dip* and *oblique faults*. Strike faults run parallel or subparallel to the strike, dip faults run parallel or subparallel to the dip, while oblique faults run in a direction intermediate between these two extremes.

Faults are essentially *younger* in age than the beds which they cut across but are older than the bedding planes at which they terminate abruptly. *Fault guided valleys* are common features and sometimes *fault scarps, fault line scarps, rift valleys, horst,* etc. characterise the landscape. This landscape is generally termed *blockfaulted topography* or *basin and range topography*.

Identification of Intrusions

Usually igneous intrusions are younger than the beds in which they occur. *Dykes* cut across the formations in the map and are normally straight but they may be curved as well (Fig. 6.8). *Sills* are embedded intrusions and run parallel to the bedding planes of the sedimentary rocks in which they occur (Fig. 6.9). *Batholiths* have very irregular boundaries on the map, quite inconsistent with the topography. They show no regular dip and appear to have eaten away the beds which were present where such bodies now occur.

Fig. 6.8 Faults and Horizontal Intrusion

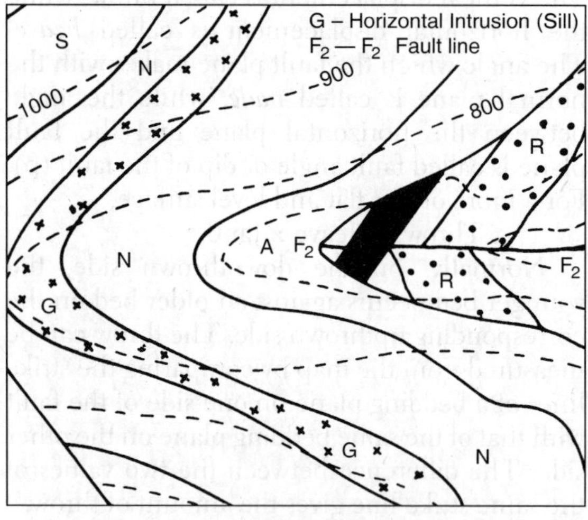

Fig. 6.9 Construction of Strike Lines

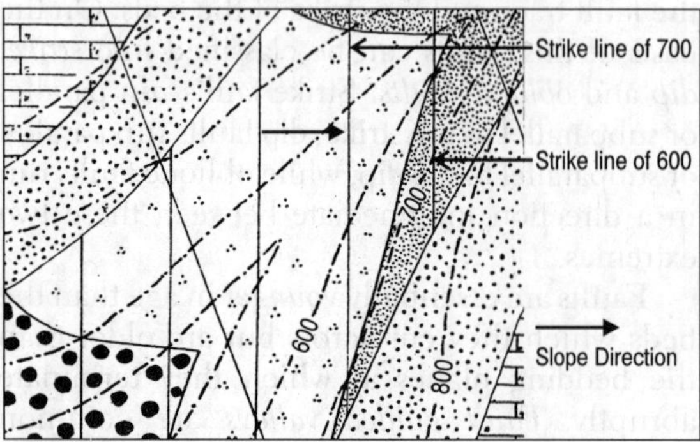

MAP ANALYSIS

Drawing a Strike Line

When the dip direction is given on the map, the strike direction is found by drawing a straight line at right angles to the dip direction. In cases where the dip direction is not shown, it is drawn with the help of contours, as follows:

i. When a contour intersects the same bedding plane more than once, a straight line drawn through these points of intersection gives the strike line for that bedding plane at that contour elevation (Fig. 6.10).

ii. Sometimes a bedding plane intersects more than two contours but each contour situation intersects at a single point only. In such situations, two alternate contour intersections with the same bedding plane at suitable locations are joined by a straight line (ab) and its midpoint is marked (M). The strike line (CM) for the contour elevation in between the other two is found by drawing a straight line joining the midpoint (M) with the intervening contour intersection with the bedding plane (C) (Fig. 6.11).

Laying a Section Line on Map

The choice of a section line on a geological map is very important because it influences the amount of apparent dip with which the attitude of a bed on a section is drawn. The following points must be considered while laying a section line:

i. A section line at right angles to the strike is most desired as it represents the true attitude of the beds.

Visual Identification of Structures

Unconformity

- Bedding planes of the older series appear to end abruptly, covered by the beds of younger series
- Strike lines on its two sides intersect and beds have different attitudes

Horizontal Structure

- Contours run parallel to the bedding planes (usually denoted by a plus sign)

Vertical Structure

- Bedding planes cut across the contours such that the inter strike distance is zero (usually denoted by a cross)

Uniclinal Structure

- Contour-Vs and outcrop-Vs intersect one another

Folded Structure

- Outcrops repeat on either side of the core beds
- The Vs of the outcrops combine to form either X or H pattern
- If younger beds form the core, it is a syncline, and if older beds form the core, it is an anticline

Faulted Structure

- A fault line is indicated by a thicker line on the either side of which bedding planes are displaced
- A straight fault trace indicates a vertical fault and a curved fault trace indicates an inclined fault

Intrusion

- Conventional symbols for igneous rocks are used to mark these on maps
- Dykes (= vertical and near vertical intrusions) cut across the formations
- Sills (= horizontal intrusions) are embedded and run parallel to the bedding planes
- Batholiths have irregular boundaries and no bedding planes

ii. In case of two or more series of formations, a section line is chosen along or parallel to the angular bisector of the strike intersection.

iii. The section line is normally laid in such a way that it can represent the maximum number of beds.

Determining the Dip of a Bed

The dip of a bed is the angle at which the bed is inclined with respect to the horizontal plane. It varies with the direction along which it is measured. It is *maximum* along the true dip direction and is *minimum* or nil along the strike direction. When the dip is measured along the direction perpendicular to the strike, it is called the *true dip* and the direction is called the *true dip direction*. When the dip is measured in any other direction (i.e., between true dip and strike) it is called the *apparent dip* and the direction of measurement is referred to as the *apparent dip direction* (Fig. 6.10).

To find the true dip direction at least two strike lines for two successive contours are drawn on the same bedding plane. A perpendicular line is then drawn between these two strike lines at any point. The direction of this perpendicular line from the higher strike level to the lower strike level specifies the direction of the true dip (Fig. 6.11).

To find the true dip (q), the following measurements are taken:

i. The perpendicular intercept (XY) between two successive strike lines drawn at different elevations (S_h and S_l) on the same bedding plane on a map.

ii. The level difference between the strike lines = (h–l), where h = higher contour and

Fig. 6.10 Construction of Strike Lines (Special Case)

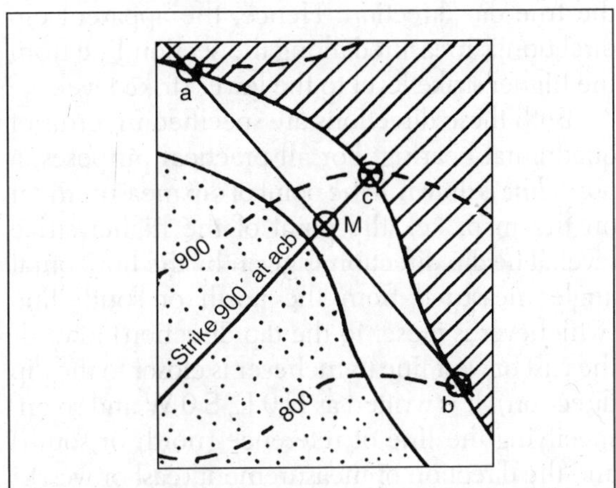

Fig. 6.11 Determination of Direction and Amount of Dip of a Bed

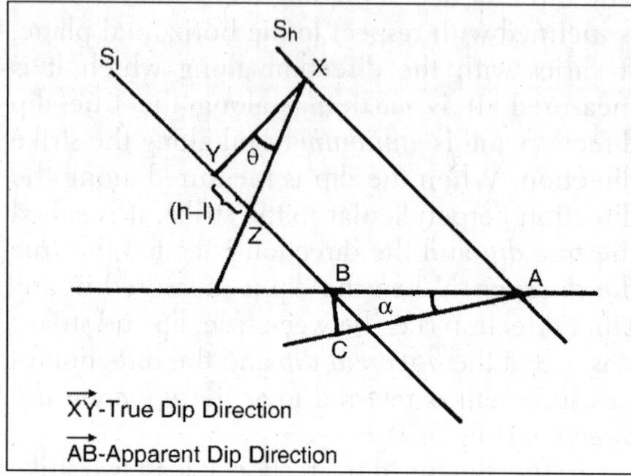

l = lower contour. This is reduced on the given map scale (say, YZ) (Fig. 6.12).

Now a right angled triangle is drawn by plotting its base (XY), perpendicular (YZ) and hypotenuse (ZX). ∠ZXY describes the true dip. It may be measured either graphically with a protractor or solved from the formulae

$$\tan \angle ZXY = \frac{YZ}{XY}$$

$$\text{or,} \quad \angle ZXY = \tan^{-1} (YZ/XY)$$

$$\therefore \theta = \tan^{-1}\left[\frac{\text{Strike difference}}{\text{Perpendicular Strike Intercept}}\right]$$

On a map containing more than one series of formation, the section line does not conform to the true dip direction. Hence, the apparent dip direction is measured along the section line from the higher strike level to the lower strike level.

Both these directions are specified in terms of quadrantal bearing. For all practical purposes, a *north line* is drawn at the point of dip measurement on the map, i.e., the point of the higher strike level. The dip direction is given by the horizontal angle measured from the north or south line (whichever is closer to the dip direction) towards the east or west line (whichever is closer to the dip direction). It is written as N θ E, S θ W and so on, specifying the line of reference (north or south) and the direction of measurement (east or west).

Fig. 6.12 Mapwork to Find the Direction and Amount of True and Apparent Dip

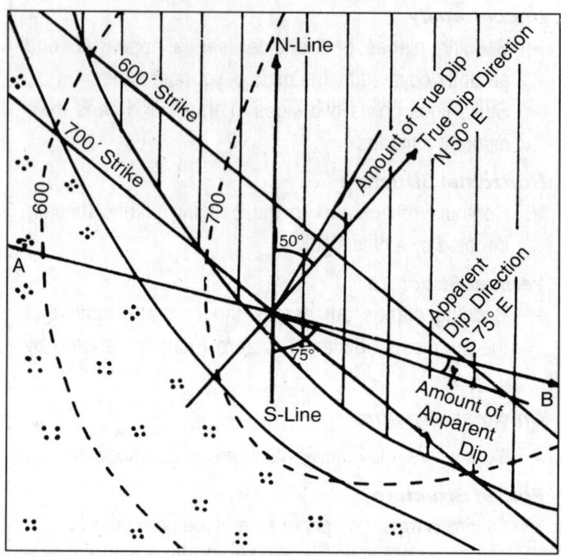

In determining the apparent dip (α), the same procedure is followed, but instead of taking the perpendicular inter strike intercept (XY), the strike intercept (AB) on the section line is taken. Let the strike level difference reduced to map scale be BC (Fig. 6.11). Thus, ∠BAC gives the apparent dip

$$\tan \angle BCA = \frac{BC}{AB} \text{ [from rt}\angle\Delta ABC]$$

$$\text{or,} \quad \angle BCA = \tan^{-1} (BC/AB)$$

$$\therefore \alpha = \tan^{-1}\left[\frac{\text{Strike difference}}{\text{Strike Interception Section Line}}\right]$$

Since the true dip (q) involves the shortest possible distance (XY) between two parallel strike lines, it is always larger than the apparent dip, i.e., θ > α.

Construction of a Geological Section

After going through the map in detail, the strike lines are first drawn and the section line is carefully laid. In order to construct the geological section, the steps to be followed are explained below.

Determination of the Thickness of Beds

The thickness of a bed is defined by the distance between its upper and lower bedding planes. It is of two types—*true* and *apparent*. The true thickness (T_t) is the perpendicular distance between the bedding planes. The vertical distance between the bedding planes gives the apparent or vertical thickness (V_t). These can easily be measured from the section with the help of a divider and a diagonal scale. The reading then needs to be reduced to map scale to find the actual ground equivalent.

The vertical thickness can also be determined from a map by finding the difference between the strike lines of its upper bedding plane from those of the lower bedding plane.

The thickness of a bed is related to the dip of the bed (θ) and width of the outcrop (w). The width of a outcrop means the distance between the two bedding planes on a map measured at right angles to the strike direction. For a flat terrain, these parameters are related by the following equations:

$$T_t = V_t \cdot \cos\theta$$

$$V_t = w. \tan\theta$$

and $$T_t = w.\sin\theta$$

Construction of a Geological section

***First**, read the geological map very carefully and identify*

- ➢ The unconformity (if any) and mark it
- ➢ The younger and older series of formations

***Second**, taking one series of formation at a time*

- ➢ Draw the topographic section along the given line on the given map scale.
- ➢ Draw at least two consecutive strike lines on a bedding plane.
- ➢ Take the inter strike intercept on the base of the topographic section.
- ➢ Drop a perpendicular (= strike level difference, reduced to map scale) in the dip direction and draw a line joining the two points to find the amount of dip.
- ➢ Project the bedding planes vertically on the topographic section and draw lines parallel to their respective lines of dip.
- ➢ Shade the beds properly and draw up a legend to show the chronological sequence/ succession of beds.

INTERPRETATION OF GEOLOGICAL MAPS

Procedures of Interpretation

After the completion of exercises on preliminary map reading and drawing of geological sections, interpretation is done under the following heads.

Interpreting a Geological Map through Visual Study/with Sections

- ➢ Identify the **unconformity** and series of formations and mark with colour pencils on the map.
- ➢ Identify the **Contour-Vs** and **Outcrop-Vs**.
- ➢ If the **Contour-Vs** are **sharper** than the Outcrop-Vs (both pointing in the same direction), the beds **dip** in an **upvalley** direction, i.e., these are anti-dip valleys, likely to drain the steeper escarpments.
- ➢ If the Outcrop-Vs are **sharper** than the Contour-Vs (both pointing in the same direction), the beds **dip** in a **downvalley** direction, i.e., these are dip valleys, likely to drain gentler dip slopes.
- ➢ If the Contour-Vs and Outcrop-Vs **intersect** and point in **opposite directions**, the beds **dip** in a **downvalley** direction, i.e., these are dip valleys, likely to drain gentler dip slopes.
- ➢ Identify the chronological sequence of the formation of the beds and construct the geological history of evolution.
- ➢ Draw the river valleys on the map, identify the drainage pattern and also their genetic class.
- ➢ From the contour pattern, identify the landforms and the landscape.
- ➢ Interpret the landforms of the landscape, produced as a result of fluvial processes on the geological structure.

Introduction

The map number and the scale of the map should be mentioned. The *length* and *direction* of the section line need to be clearly stated.

Topography

The *maximum* and *minimum* elevation along with the *direction* of the regional slope, as indicated by the contour pattern, are to be stated. The *pattern* of relief variation should be discussed in terms of *landform units*, namely hills, ridges, valleys, spurs, cols, saddles, etc. The nature of the *landscape* formed by the assemblage of landform units needs to be clearly stated, e.g., ridge and valley topography, spur and valley topography, basin and range topography, scarpland landscape, etc.

General Geology

In this, the following items are to be clearly stated—numbers and names of formations, numbers and names of each rock beds, numbers of unconformities, the nature of intrusions and faults and the general nature of the outcrop as a whole.

Structure

The attitude of the beds is described and analysed here. The true and apparent dips with the directions of the beds of different series or both limbs of a fold or on both sides of a fault need to be stated first. The type of structure formed by the beds of a series, i.e., horizontal, uniclinal, folded or faulted, are to be mentioned. The *occurrence* of unconformities, igneous bodies, superficial deposits like alluvium, loess, raised beach, etc. are also to be noted.

If there are *folds*, the number of anticlines and synclines, the direction of fold trace, the inclination of the axial plane, the number of core beds, the type of folding, e.g., close or open, symmetrical or asymmetrical, isoclinal or recumbent, plunging or non-plunging, etc. are to be stated. Fold traces of the anticlines and synclines are also to be clearly indicated on the map.

If there is a *fault*, it is to be mentioned whether it is vertical or inclined. If the fault is inclined, identify whether it is a normal or a reverse fault. The upthrown and downthrown sides have to be mentioned together with the amount of throw, heave, head, dip of the fault, age of faulting, length and direction of fault trace and whether it is a dip or strike or oblique fault.

If igneous bodies or superficial deposits are present, their nature, type, mode of occurrence, extent, geographic location and age of formation are to be discussed in detail. Finally, the structural parameters of the formations need to be arranged in a tabular form:

Table Structural Characteristics

Beds	*Age of Deposition*	*Dip with Direction*		*Remarks*
		Apparent	*True*	

Succession or Sequence of Beds

The sequence of beds, i.e., the order of superposition from the oldest to the youngest is determined and arranged in respective order. The thickness of the beds are found and tabulated against the beds as below:

Table Geological Succession

Beds	*Age of Deposition*	*Thickness*		*Remarks*
		Vertical	*True*	

Geological History

In this, the order in which the beds are formed is emphasised. In chronological order, the *sequence of events* such as bed formation, tilting, folding, faulting, intrusion, subaerial denudation, upheaval and submergence of land, etc. are precisely pointed out to reveal the geological evolution.

Topography in Relation to Structure

Once the lithology, structure, geological history and landform units are identified and ascertained, the geomorphological class may be specified as:

i. Whether the valley is a dip valley or an antidip valley or a strike valley or a superimposed valley or an anticlinal valley or a subsequent valley or a fault-guided valley.
ii. Whether there is any tableland, outlier, spur (*horizontal structure*); hogback, cuesta, dip slope, escarpment, valley inliers, cap rock (*uniclinal structure*); anticlinal valley, synclinal hill, inward facing escarpment, homoclinal ridge, outliers, inliers (*folded structure*); fault scarp, fault line scarp, block mountain, rift valley (*faulted structure*); hill massif, weathering escarpment (*intrusions*).
iii. Whether the pattern of drainage represents a trellis or part of it, a rectangle or part of it, a fish-bone pattern and also whether these are concordant or discordant to structure.
iv. Whether there is any correlation between the width of the outcrop and the dip of the bed, between the escarpment and the dip and between the slope and the dip of the bed. Comparisons may be made in case of a large area with a number of formations.

Conclusions may be drawn about the degree of adjustment between the development of valleys and drainage patterns and the underlying lithology and structure. This may be illustrated with the help of a block diagram.

EXAMPLE 1

Introduction: Pure Uniclinal

The geological map (No. 1, Scale: 1 inch to 1000 ft) has been analyzed and interpreted with the help of a geological section drawn along the line CD (length = 550 ft; direction = due East or N90°E).

Topography

Topographically, the region is part of a plateau terrain with elevation ranging between a little below 400 ft to a little over 1200ft, thereby producing a relative relief of the order of about 800 ft plus. The region is drained by seven valleys that together form a typical rectangular pattern. Subaerial denudation has produced a ridge-and-valley landscape with a variety of landforms like a typical V-shaped valley, escarpment, spurs, saddles and ridges (Fig. 6.12).

Succession or Sequence of Beds

The geological succession or the chronological sequence of formation of the rock beds has been shown in the following Tables 6.1 and 6.2

Drainage and Landforms on Geological Structure

Structure	*Drainage Elements*	*Landform Elements*
Horizontal	Consequents	Spur, Mesa, Butte, Tablelands
Uniclinal	Consequents, Subsequents, Obsequents, Resequents, Rectangular Trellis	Strike Valley, Dip Slope, Escarpment, Saddle, Valley-inlier, Cuesta, Hogback, Monoclinal Ridge, Salient, etc.
Folded	Longitudinal and Lateral Subsequents, Obsequents	Dip Slope, Saddle, Inward-facing Escarpment, Homoclinal Ridge, Anticlinal Valley, Synclinal Hill, Cuesta, Hogback, Inversion of Relief, etc.

Fig. 6.13 Geological Map No. I and Geological Section along CD

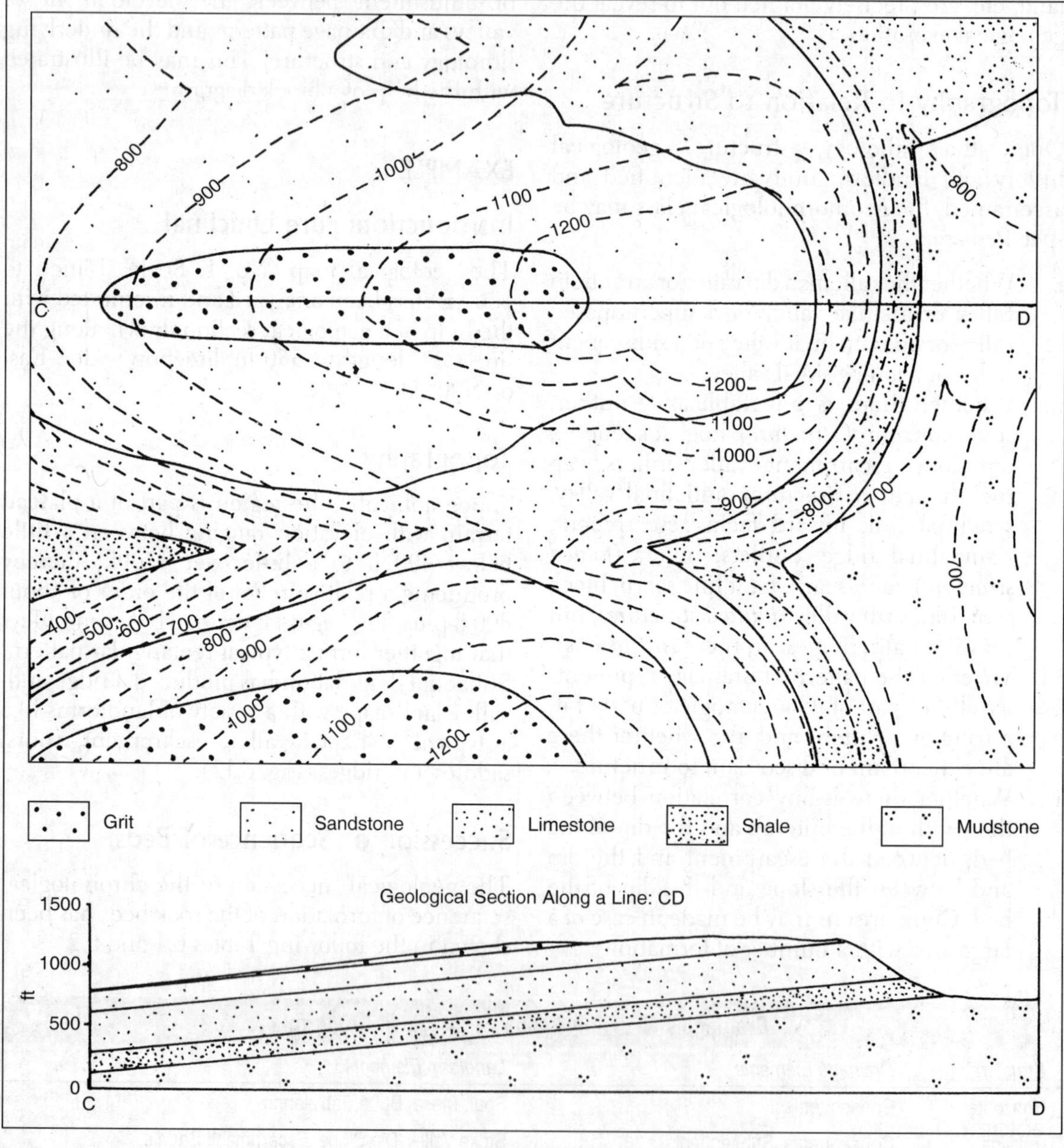

Structure

The region shows a single series formation that comprises a set of five sedimentary beds, viz., Grit, Sandstone, Limestone, Shale and Mudstone respectively, of which Grit is the youngest bed and Mudstone is the oldest bed. All these beds are conformable to one another and dip uniformly from D to C (S90°W) at an angle of 6°00′ thereby forming a uniclinal structure, the true dip and direction being 7°30′, S84°30′W.

Table 6.1 Sequence of Beds

Beds	*Age of Deposition*	*Thickness (ft)*		*Remarks*
		Vertical	*True*	
Grit		> 50	> 50	Youngest Bed
Sandstone		150	120	↑
Limestone	Single Series Formation	245	225	
Shale		140	120	
Mudstone		> 550	> 550	Oldest Bed

Table 6.2 Parameters of Geological Structure

Beds	*Age of Deposition*	*Dip with Direction*		*Remarks*
		Apparent	*True*	
Grit Sandstone Limestone Shale Mudstone	Single Series Formation	6º00′ S90ºW	7º30′ S84º30′W	Conformable Beds forming Uniclinal Structure

Fig. 6.14 Topography in Relation to Structure (Map No. 1)

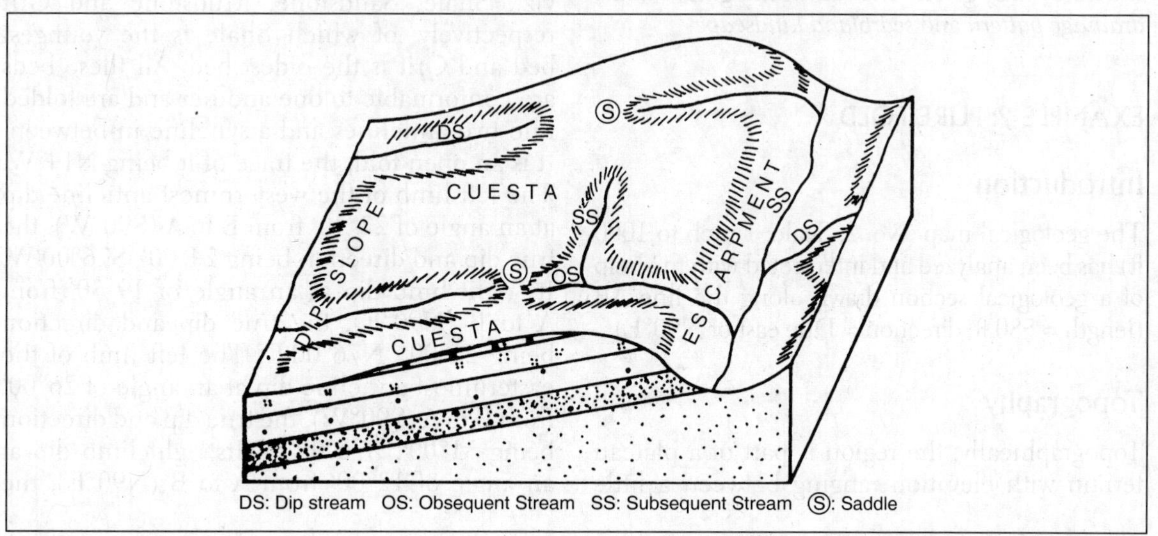

Geological History

The geological evolution of the region may be traced out in the following sequence of events—

1. Formation of the sedimentary beds under marine condition in the order Mudstone, Shale, Limestone, Sandstone, and Grit, respectively.
2. Upliftment associated with tilting followed by subaerial denudation giving rise to the present topography.

Topography in Relation to Structure

Two parallel valleys run westward with a broad spur in between them. Both of these are *dip valleys* as they drain the dip slope, i.e., run towards the dip direction. On the eastern half, there is a north – south flowing major valley disposed parallel to the structural strike and is therefore a *strike valley* or *subsequent valley*. It mainly occupies the beds of softer Mudstone and Shale. It is joined by one left bank and two right bank and tributaries. The left bank tributary is again a *dip valley* while the right bank tributaries drain the steeper escarpment and are *anti dip* or *obsequent valleys*. The southern-most *obsequent valley* is fed by a small headwater sub-tributary that originates from the flank of Sandstone. The watershed of the easterly and westerly drainage is marked by *saddles* – one formed in the northern half, and the other in the southern half. These two separates two asymmetrical ridges, called *cuestas*. Ideally, the underlying *uniclinal structure* has given rise to *rectangular trellis drainage pattern* and *scarpland landscape*.

EXAMPLE 2: PURE FOLD

Introduction

The geological map (No. 2, Scale: 1 inch to 1000 ft) has been analyzed and interpreted with the help of a geological section drawn along the line AB (length = 580 ft; direction = Due east or N90°E).

Topography

Topographically, the region is part of a plateau terrain with elevation ranging between a little below 400 ft to a little over 900ft, thereby producing a relative relief of the order of about 500 ft plus. The region is drained by two parallel major valley systems, running in opposite directions in the western (N – S) and eastern (S – N) parts with a dissected ridge in between. Subaerial denudation has produced a ridge-and-valley landscape with a variety of landforms like V-shaped valley, ridge, spurs, saddles, inward-facing escarpment and ridges (Fig. 6.12).

Succession or Sequence of Beds

The geological succession or the chronological sequence of formation of the rock beds has been shown in the following Tables 6.3 and 6.4.

Structure

The region shows a single series formation that comprises a set of four sedimentary beds, viz., Shale, Sandstone, Mudstone and Grit respectively, of which Shale is the youngest bed and Grit is the oldest bed. All these beds are conformable to one another and are folded into two anticlines and a syncline in between. It is an open fold, the trace of it being N14°W. The left limb of the westernmost anticline dip at an angle of 22°30′ from B to A (S90°W), the true dip and direction being 24°00′, S76°00′W. Its right limb dip at an angle of 19°30′ from A to B (N90°E), the true dip and direction being 21°00′, N76°00′E. The left limb of the easternmost anticline dip at an angle of 26°00′ from B to A (S90°W), the true dip and direction being 31°00′, S76°00′W. Its right limb dip at an angle of 12°00′ from A to B (N90°E), the

Table 6.3 Sequence of Beds

Beds	*Age of Deposition*	*Thickness (ft)*		*Remarks*
		Vertical	*True*	
Shale		> 250	> 250	Youngest Bed
Sandstone	Single Series	120	100	↑
Mudstone	Formation	300	275	
Grit		> 700	> 700	Oldest Bed

Fig. 6.15 Geological Map No. 2 and Geological Section along AB

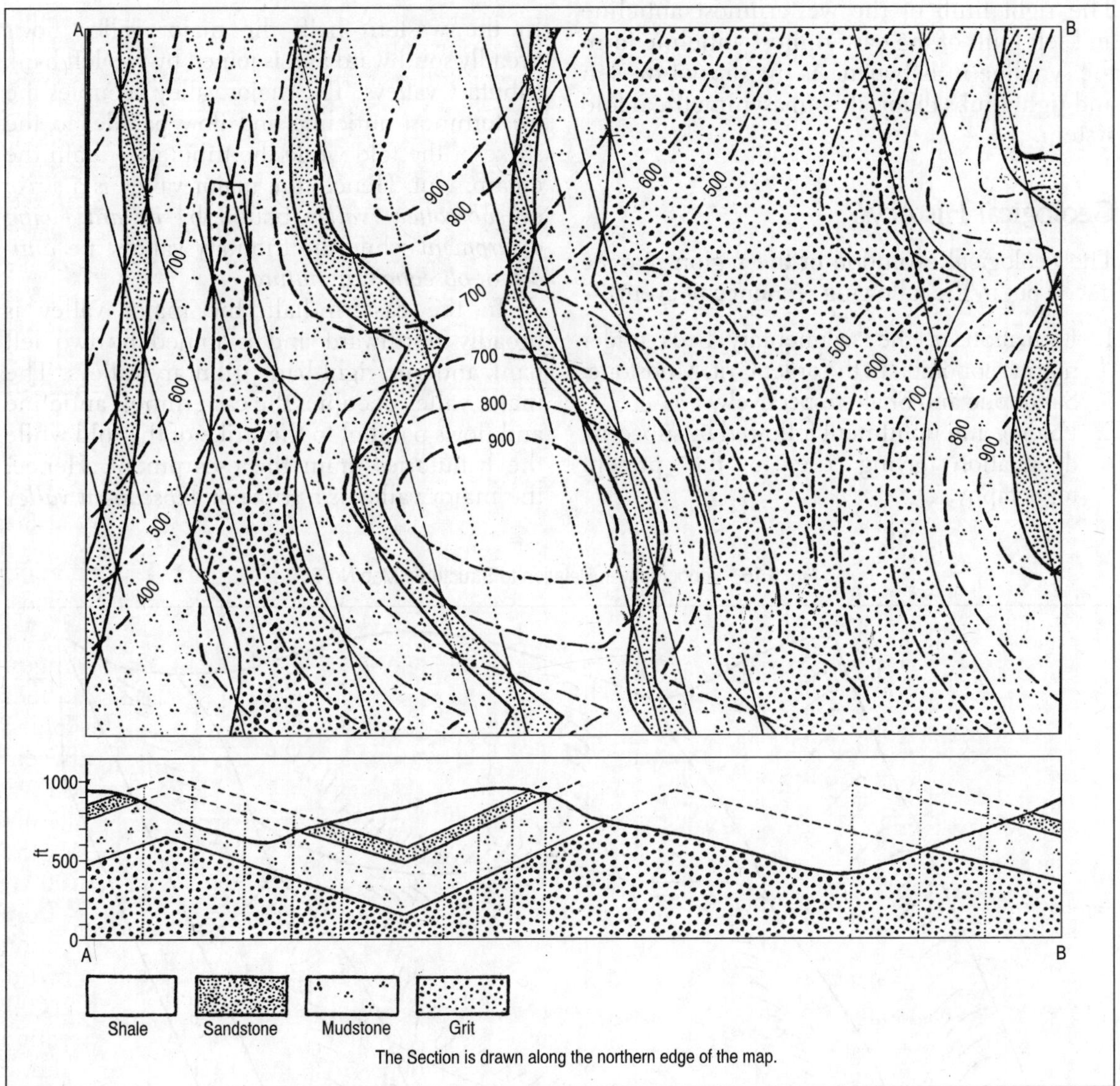

The Section is drawn along the northern edge of the map.

Table 6.4 Parameters of Geological Structure

Beds	*Age of Deposition*	*Dip with Direction*		*Remarks*
		Apparent	*True*	
Shale Sandstone Mudstone Grit	Single Series Formation	1) L: 22°30′, S90°W 1) R: 19°00′, N90°E 2) L: 26°00′, S90°W 2) R: 12°00′, N90°E	1) L: 24°00′, S76°W 1) R: 21°00′, N76°E 2) L: 31°00′, S76°W 2) R: 13°00′, N76°E	Conformable Beds forming Folded Structure

true dip and direction being 13°00′, N76°00′E. The right limb of the westernmost anticline and left limb of the easternmost anticline form the syncline in between. As the dips of the left and right limbs differ, it is an asymmetrical fold system.

Geological History

The geological evolution of the region may be traced out in the following sequence of events—

1. Formation of the sedimentary beds under marine condition in the order Grit, Mudstone, Sandstone and Shale respectively.
2. Folding and upliftment followed by subaerial denudation giving rise to the present topography.

Topography in Relation to Structure

In the western half, the major valley flows broadly southward and is joined by two left bank tributary valleys. The major valley occupies the westernmost anticline and flow parallel to the strike of the fold while the tributaries drain the escarpment. Hence, the major valley is a *strike* or *subsequent valley* bound by *inward facing escarpment* while the tributary valleys are *anti-dip* or *obsequent in nature*.

In the eastern half, the major valley is broadly northward and is joined by two left bank and one right bank tributary valleys. The major valley occupies the easternmost anticline and flows parallel to the strike of the fold while the tributaries drain the escarpment. Hence, the major valley is a *strike* or *subsequent valley*

Fig. 6.16 Topography in Relation to Structure: (Map No. 2)

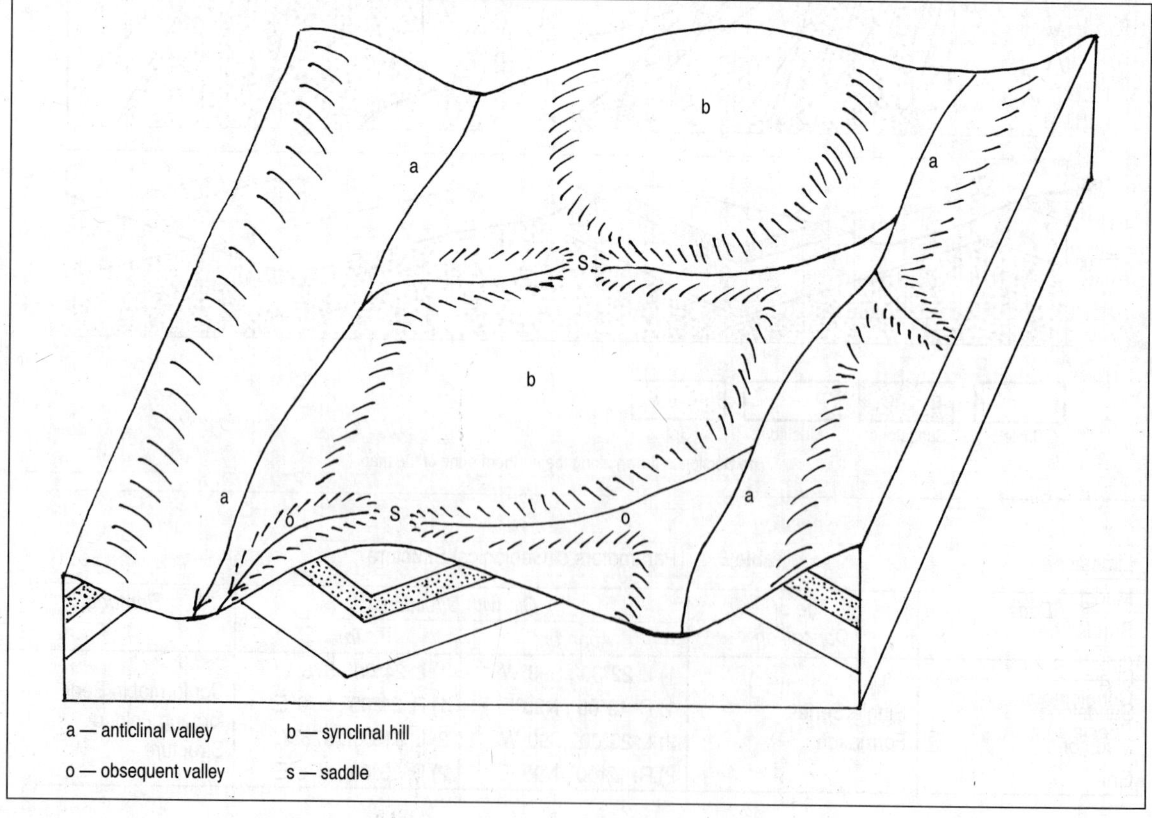

bound by *inward facing escarpment* while the tributary valleys are *anti-dip* or *obsequent in nature*.

The syncline is occupied by a *dissected ridge* over which two distinct *saddles* have been formed. Ideal *spurs* have been formed in between the tributary valleys. Thus, *synformal topography* (valley) on *anticlinal structure* and *antiformal topography* (ridge) on *synclinal structure* have developed in the course of drainage development on folded structure giving rise to *inversion of relief*.

EXAMPLE 3: UNICLINAL + FOLD WITH FAULT AND INTRUSION

Introduction

The geological map (No. 3, Scale: 1 inch to 1000 ft) has been analyzed and interpreted with the help of a geological section drawn along the line AB (length = 600 ft; direction = due East or N90°E).

Topography

Topographically, the region is part of a plateau terrain with elevation ranging between a little below 200 ft to a little over 1000ft, thereby producing a relative relief of the order of about 800 ft plus. The region is drained by two major valleys that run roughly parallel to one another in a NE – SW direction with a broad spur in between them. The one draining the NW part of the region is a relatively smaller one while the other flowing almost diagonally across the region is a larger one with a small right bank tributary that flows in a NNE – SSW direction and joins the trunk stream near the SW part of the region.

Succession or Sequence of Beds

The geological succession or the chronological sequence of formation of the rock beds has been shown in the following Tables 6.5 and 6.6

Structure

There are two formations here – Carboniferous and Ordovician, of which Carboniferous is the upper/younger series. The Carboniferous series consists of only Limestone bed that dips from A to B (N90°E) at an angle of 3°05′, its true dip with direction being 8°08′, N22°E. It lies unconformably over the Ordovician series.

The older Ordovician series consists of four sedimentary beds, viz., conglomerate, grit, sandstone, and mudstone respectively in order of formation. All these beds are folded into an asymmetrical syncline, the trace / trend of which is N10°W. The beds of the left limb dips from A to B (N90°E) at an angle of 13°05′, its true dip with direction being 14°02′, N80°E. The beds of the right limb dips from B to A (N90°W) at

Table 6.5 Sequence of Beds

Beds	*Age of Deposition*	*Thickness (ft)*		*Remarks*
		Vertical	*True*	
Limestone	Carboniferous	> 750	> 750	Young Series: Only Bed
Mudstone	Ordovician	> 200	> 200	Older Series: Youngest Bed
Sandstone		125	100	↑
Grit		225	200	
Conglomerate		> 500	> 500	Oldest Bed
Special Outcrops	Granite Dolerite Quartz Porphyry			Boss: Intrusion Dyke: Vertical Intrusion Sill: Horizontal Intrusion

an angle of 17°05′, its true dip with direction being 18°26′, S80°W. The age of folding is post-Ordovician or pre-Carboniferous.

The area is crossed by two faults—(a) the one in the north west is a normal vertical fault (N48°E), the vertical displacement being about 200 ft towards left; the age of faulting is post-intrusion of quartz porphyry; and (b) the other in the east is also a normal vertical fault (N22°E), the vertical displacement being about 200 ft towards right; the age of faulting is post-intrusion of quartz porphyry. Thus, the region between the

Fig. 6.17 Geological Map No. 3 and Geological Section along the Line AB

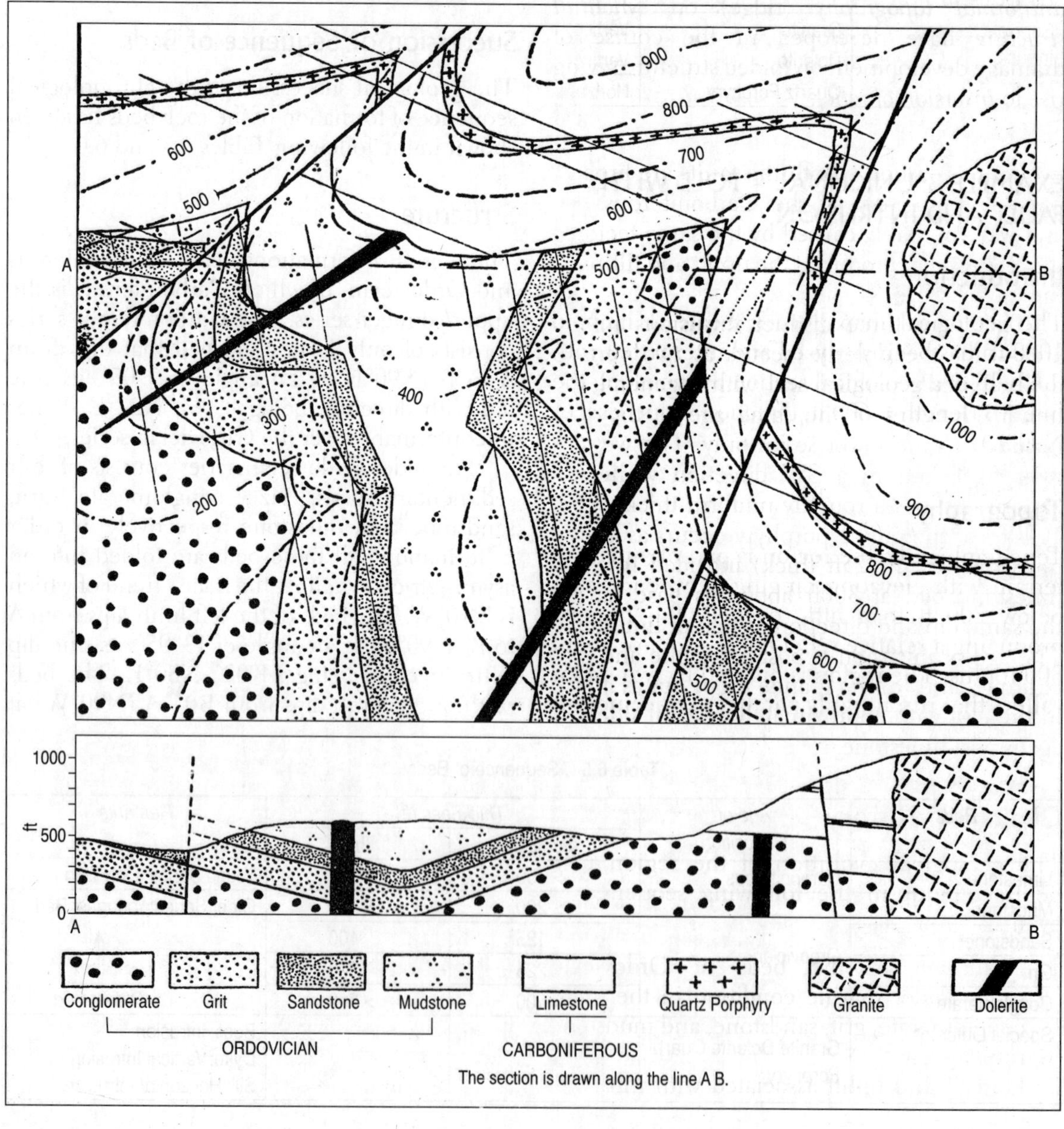

The section is drawn along the line A B.

Table 6.6 Parameters of Geological Structure

Beds	*Series*	*Dip with Direction*		*Remarks*
		Apparent	*True*	
Limestone	Carboniferous	3°05′, N90°E	8°08′, N22°E	Uniclinal Structure
Mudstone Sandstone Grit Conglomerate	Ordovician	L. Limb: 13°05′, N90°E R. Limb: 17°05′, N90°E	L. Limb: 14°02′, N80°E R. Limb: 18°26′, S80°W	Folded Structure associated with faults and intrusions
Special Outcrops: Intrusions	Granite	Vertical		Boss (≈ 900 ft)
	Dolerite	Vertical		Dyke (100ft, 150 ft)
	Quartz Porphyry	Horizontal		Sill (30 ft)

two faults is a horst. Both the faults are broadly parallel to the dip of the Carboniferous bed and were probably caused by the same tectonic movement that produced the uplift and tilting of the upper series.

There are three distinct intrusions in the area— (a) the dolerite occurs as two vertical dykes, broadly parallel and with a general NE trend. They are of the same age as they are intruded in the lower series only. The western dyke thins out in a SW direction while the eastern one has a roughly uniform thickness of 100 ft, (b) the quartz porphyry occurs as a non-transgressive sill (25ft thick) intruded into the limestone bed at 300ft above its base having the same dip and direction as that of limestone. The age of intrusion is post-limestone and pre-faulting., and (c) the granite crops out as a boss in the east with approximately vertical sides; its age is post-limestone.

Geological History

The geological evolution of the region may be traced out in the following sequence of events—

1. Formation of the beds of Ordovician series under marine condition in the order conglomerate, grit, sandstone, and mudstone respectively.
2. Folding and uplift associated with intrusion of dolerite dykes followed by subaerial erosion and then submergence.
3. Formation of the beds of Carboniferous series, i.e., limestone unconformably over the eroded Ordovician series followed by intrusion of the quartz porphyry and the granite.
4. Upliftment associated with tilting and faulting followed by subaerial denudation giving rise to the present topography.

Topography in Relation to Structure

Two major valleys run roughly parallel to one another in a NE – SW direction with a broad spur in between them. Both of them run opposite to the dip direction and are therefore anti-dip or obsequent valleys. As they maintain their direction while crossing the beds of the lower series, the drainage occupied by these beds is superimposed and is naturally discordant in nature. The western valley is a relatively shorter one and is a fault-guided one in the lower series. The eastern valley runs almost diagonally and has a short right bank tributary (only over the lower series).

An antiformal topography may be identified over the syncline, thereby giving an impression of 'relief inversion'. The higher land in the area is occupied by the granite boss that has a relatively flat top. The other intrusions do not make a definite physical feature.

EXAMPLE 4: UNICLINAL + FOLD WITH FAULT AND INTRUSION

Introduction

The geological map (No. 4, Scale: 1 inch to 1000 ft) has been analyzed and interpreted with the help of a geological section drawn along the line CD (length = 475 ft; direction = N88°30′E).

Topography

Topographically, the region is part of a plateau terrain with elevation ranging between a little below 400 ft to a little over 1000ft, thereby producing a relative relief of the order of about 600 ft plus. The region is drained by two parallel valleys running northwest to southeast with spurs in between.

Succession or Sequence of Beds

The geological succession or the chronological sequence of formation of the rock beds has been shown in the following Tables 6.7 and 6.8.

Structure

There are two formations here—Carboniferous and Ordovician, of which Carboniferous is the upper/younger series. The Carboniferous series consists of a set of three sedimentary beds, viz., Mudstone, Grit and Sandstone respectively, of which Mudstone is the youngest bed and Sandstone is the oldest bed. All these beds dip uniformly at an angle of 6°00′ from D to C (S88°30′W), thereby forming a uniclinal structure, their true dip with direction being 11°30′, N72°30′W. It lies unconformably over the Ordovician series.

The older Ordovician series consists of five sedimentary beds, viz., Shale, Grit, Conglomerate, Sandstone and Limestone respectively in order of formation. All these beds are folded into an asymmetrical anticline, the trace/trend of which is N37°30′E. The beds of the left limb dips from D to C (S88°30′W) at an angle of 15°00′, its true dip with direction being 17°30′, N52°30′W. The beds of the right limb dips from C to D (N88°30′E) at an angle of 22°30′, its true dip with direction being 25°30′, S52°30′E. The age of folding is post-Ordovician or pre-Carboniferous.

The area is crossed by two faults of post-Carboniferous age—(a) the one in the north east is a normal vertical fault (F_1: S54°′E), the vertical displacement being about 125 ft towards right, and (b) the other in the south central part

Table 6.7 Sequence of Beds

Beds	Age of Deposition	Thickness (ft)				Remarks
		Vertical		True		
Mudstone	Carboniferous	> 50		> 50		Younger Series
Grit		275		270		
Sandstone		60		60		
Shale	Ordovician	> 200	(a)	> 200	(a)	Older Series
Grit		175	(a)	150	(a)	
Conglomerate		275	275	250	250	
Sandstone		75	75	60	60	
Limestone		> 500	(b)	> 500	(b)	
Special Outcrops	Vertical Intrusion Horizontal Intrusion					Dyke Sill

Table 6.8 Parameters of Geological Structure

Beds	*Age of Deposition*	*Dip with Direction*		*Remarks*
		Apparent	*True*	
Mudstone		6°,	11°30′,	
Grit	Carboniferous	S88°30′W	S88°30′W	Conformable Beds forming Uniclinal Structure
Sandstone				
Shale		LL: 15°,	LL: 17°30′,	
Grit		S88°30′W	N52°30′W	
Conglomerate	Ordovician			Conformable Beds forming Folded Structure associated with intrusions and faults
Sandstone		RL: 22°30′,	RL: 25°30′,	
Limestone		S88°30′W	S52°30′E	
Special Outcrops	Vertical Intrusion Horizontal Intrusion			Dyke (≈ 450 ft) Sill (≈ 50 ft)

Fig. 6.18 Geological Map No. 4 and Geological Section along CD

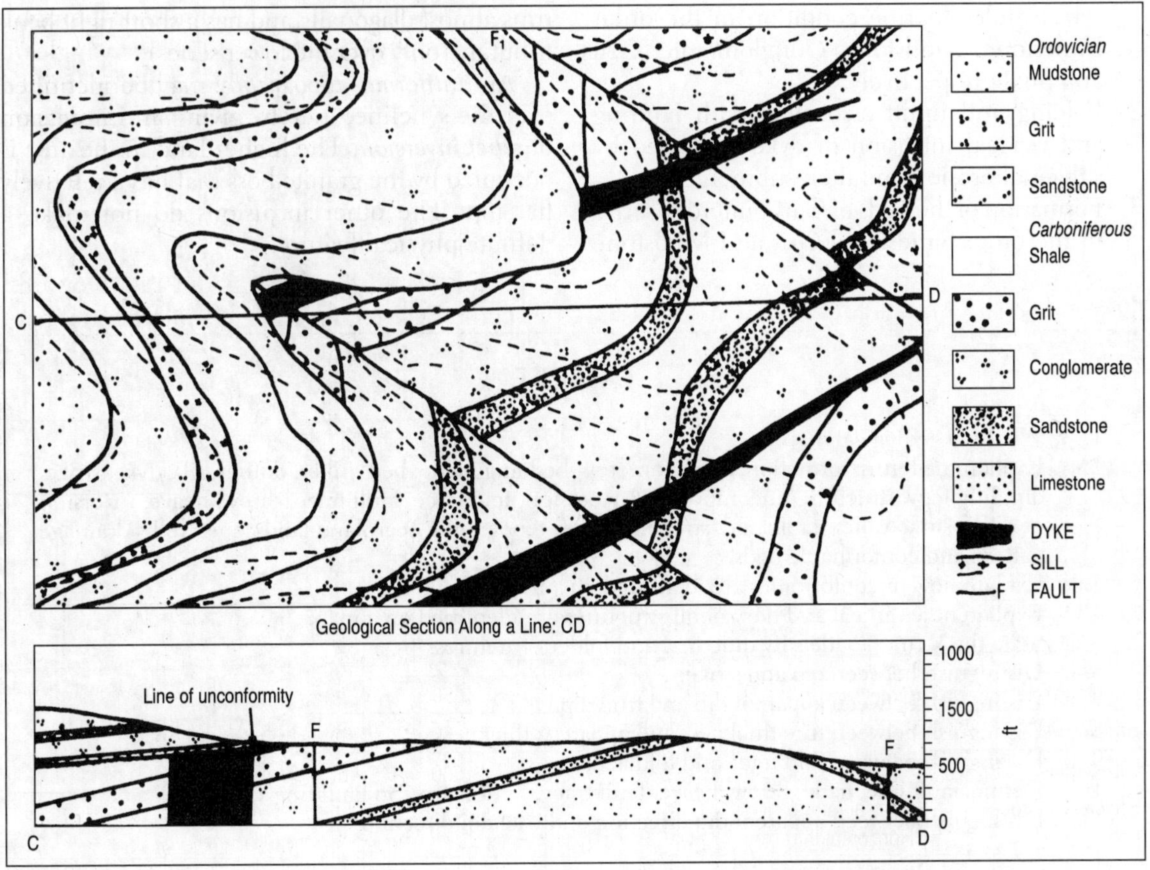

is also a normal vertical fault ((F_2: S54°E),), the vertical displacement being about 50 ft towards right. Thus, the region is a step-faulted one.

There are two distinct intrusions in the area— (a) the vertical intrusions occur as two dykes, broadly parallel and with a general ENE trend. They are of same age as they are intruded in the lower series only. Both the dykes thin out in a ENE direction and (b) the horizontal intrusion occurs as a sill (50ft thick) intruded into the Grit bed at 150ft above its base having the same dip and direction as that of Grit. The age of intrusion is post-Shale and pre-faulting.

Geological History

The geological evolution of the region may be traced out in the following sequence of events—

1. Formation of the beds of Ordovician series under marine condition in the order Limestone, Sandstone, Conglomerate, Grit and Shale respectively.
2. Folding and uplift associated with faulting and vertical intrusion of dykes followed by subaerial erosion and then submergence.
3. Formation of the beds of Carboniferous series in the order Sandstone, Grit and Mudstone unconformably over the eroded Ordovician series followed by the horizontal intrusion of the sill.
4. Upliftment associated with tilting followed by subaerial denudation giving rise to the present topography.

Topography in Relation to Structure

Two major valleys run roughly parallel to one another in a NW – SE direction with a broad spur in between them. Both of them run opposite to the dip direction and are therefore *anti-dip* or *obsequent valleys*. As they maintain their direction while crossing the beds of the lower series, the drainage occupied by these beds is *superimposed* and is naturally *discordant* in nature. The western valley is a relatively shorter one and is a *fault-guided* one in the lower series. The eastern valley runs almost diagonally and has a short right bank tributary (only over the lower series).

An *antiformal topography* may be identified over the syncline, thereby giving an impression of *relief inversion*. The higher land in the area is occupied by the granite boss that has a relatively flat top. The other intrusions do not make a definite physical feature.

EXERCISE

1. What is a geological map?
2. Explain the terms: formation, series, outcrop, bedding plane, bed, inlier, outlier, sill, dyke, strike, dip, lithology, structure, unconformity, succession, fold trace, fault trace, throw, heave, inversion of relief, strike valley, tableland, spur and valley topography, scarpland landscape, trellis drainage pattern and conformable beds.
3. Explain how unconformities are identified on a map?
4. Explain how vertical and horizontal structures are identified on a map?
5. State the V-rules to identify uniclinal and folded structures?
6. Distinguish between dip and strike.
7. Distinguish between apparent dip and true dip.
8. Distinguish between true thickness and apparent thickness.
9. Distinguish between fold trace and fold access.
10. Define: fault line, hade, throw, heave, fault trace, fault scarp and fault line scarp.
11. Distinguish between apparent dip direction and true dip direction.

12. Explain how the dips and strikes of the beds of a formation are measured on a map.
13. Draw all necessary strike lines for each formation on the geological map and find the strike directions.
14. Mark the line(s) of unconformity and identify the older and younger formations in the map.
15. State the structure of the formation by applying the V-rules in the map.
16. Find the apparent and true dip directions of the beds shown in the map.
17. Find the magnitudes of the apparent and true dips of the beds shown in the map.
18. Find the apparent or vertical and true thickness of the beds shown in the map.
19. Find the trace of the fold and also its direction in the map.
20. Find the inclination of the fold-axis and give the nature of folds.
21. Find the trace of the fault and also its direction in the map.
22. Find the inclination of the fault plane in the map.
23. Find the magnitude and direction of throw and state the nature of fault in the map.
24. Study the pattern of intrusion and give its nature.
25. Identify the genetic type of valleys and the associated landforms in the map.
26. Draw a neat geological section along the best suited line on the map.
27. Describe the topography of the area covered by the map.
28. Based on the geological map, prepare the successive table in an appropriate order and explain it.
29. Based on the the geological map, find the structural parameters of the beds and explain it.
30. Based on the geological map, trace the geological history of the formations.
31. Interpret the geological map under the head: structure in relation to topography.

Note: Q. 13–21 relate to Maps 1–6.

Map 1

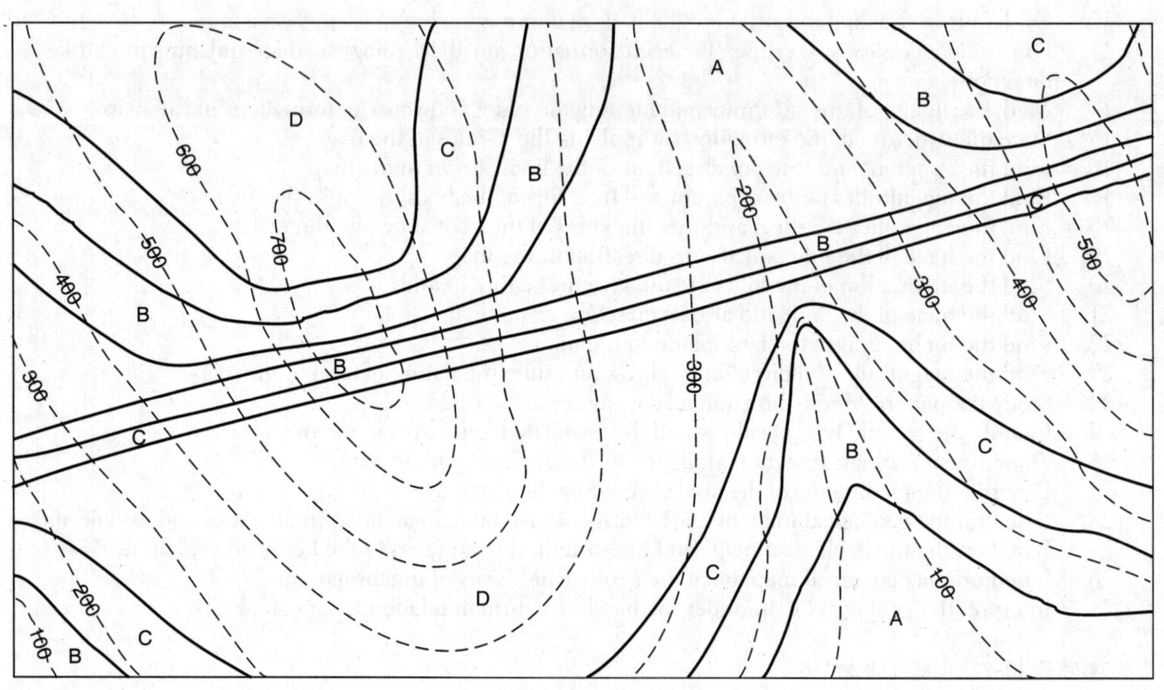
A
B
C
D
C
B
C
B
B
B
C
B
C
C
D
A
C
B
100
200
300
400
500
600
700
800
300
200
300
400
500
100

Map 2

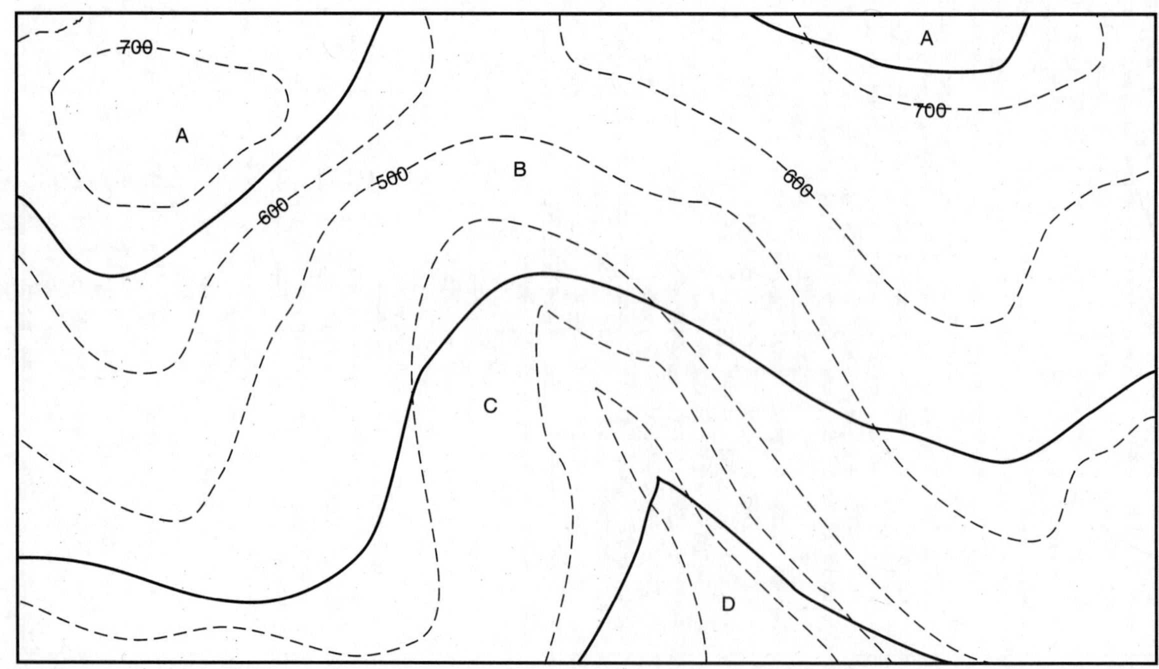
700
A
A
700
B
500
600
600
C
D

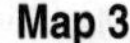

Map 3

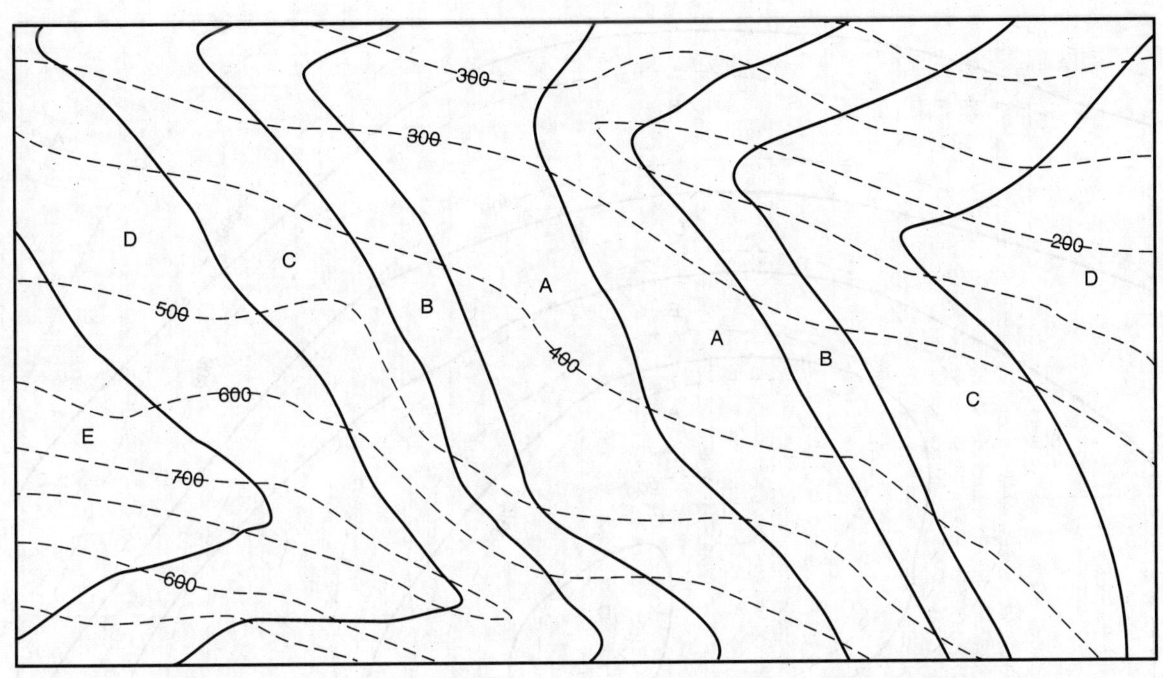
300
300
200
D
C
B
A
A
B
C
D
500
400
600
E
700
600

Map 4

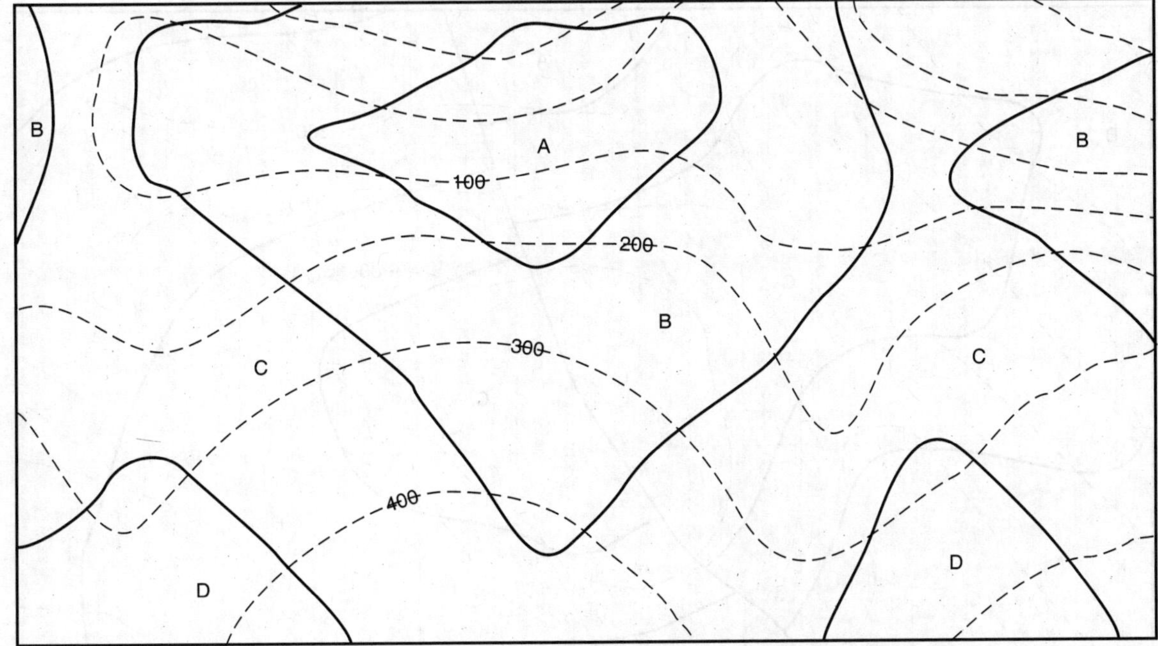
B
A
B
100
200
B
300
C
C
400
D
D

Map 5

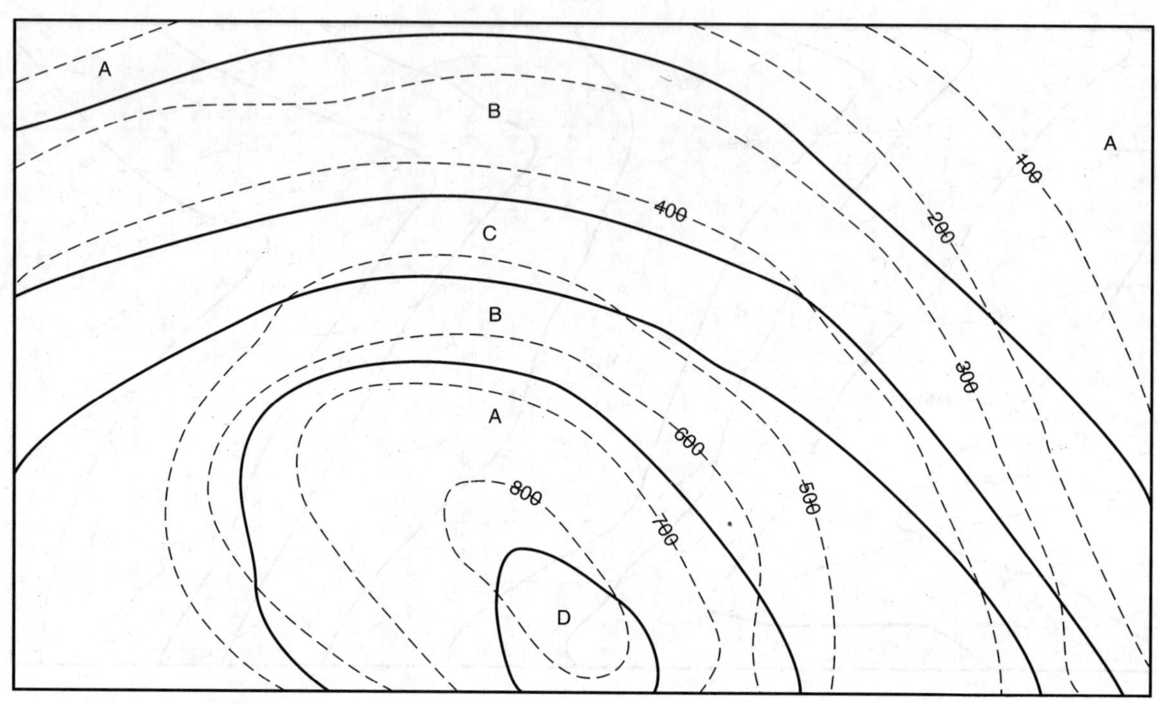

Map 6

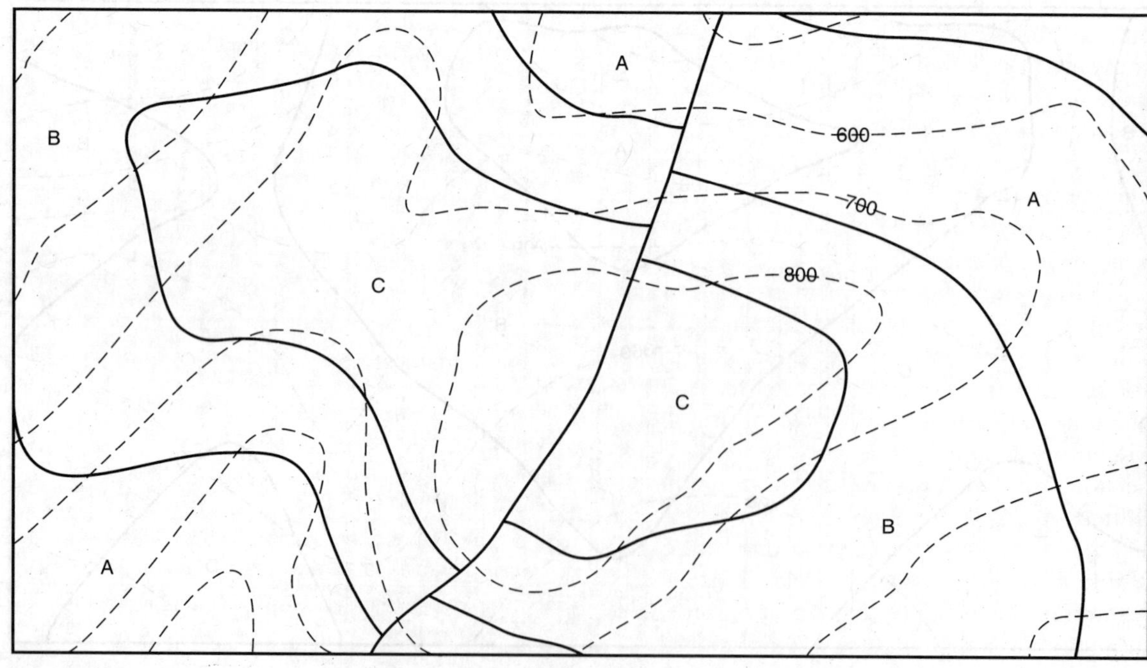

7

Weather Map

HIGHLIGHTS

- Weather Elements on a Map
- Salient Features of Indian Seasons
- Procedures of Interpretation
- Example: Indian Daily Weather Report, 9 August 1965

The weather of a place is defined as the sum total of the atmospheric conditions as described by the weather elements at a particular instant of time. A *weather map* is, therefore, one that portrays weather elements recorded at a particular instant of time for the whole world or part of it. Normally, the weather elements are shown either by conventional symbols or by absolute numerical values or by codes. The weather elements usually recorded and shown are *temperature* (maximum and minimum), *pressure* (continuous), *wind* (direction and velocity), *cloud* (type and coverage), *precipitation* (form and amount), *other atmospheric phenomena* and *sea conditions*.

In India, the Meteorological Department was set up in 1864 with its headquarters in Simla. Till 1949, weather maps provided information about only three elements—pressure, wind and precipitation. After World War I, the Indian Meteorological Department expanded and became more efficient and its headquarters was shifted to Pune. In India there are about 350 weather stations belonging to 5 different classes (Table 7.1). In the I, II and III class stations, data regarding each weather element is recorded and subsequently transmitted by WMO codes to the forecasting stations. These are supplemented by the observations received from the IV and V class stations and from ships. At the headquarters, these are immediately compiled and mapped. The upper air data are separately collected and plotted at the Class I stations. Nowadays, satellite pictures provide accurate information about the cloud- structure, thermo-humid potential, association, rate and direction of movement and other behaviour from which forecasting has become more easy, precise and accurate.

Table 7.1 Categories of Observing Stations

Class	*Facilities*	*Message Transmitted*
I	Eye-reading and self-recording instruments	Twice daily
II	Only eye-reading instrument	Twice daily
III	Only eye-reading instrument	Once a day
IV	Only temperature and rainfall data are collected	Once a week
V	Only rainfall data are collected	Once a week

WEATHER ELEMENTS ON A MAP

Pressure

It is defined as the *weight of an unit column of air* at any place. It varies with altitude, latitude and temperature. It is measured with the help of a *barometer* (simple Mercury type, Fortin's or Aneroid) or a *barograph* (Fig. 7.1, Fig. 7.2 and Fig. 7.3). It is expressed in millibars or inch/cm/mm of mercury column. At the mean sea level, the *normal pressure* is equivalent to 760 mm Hg or 76 cm Hg. This is taken as an unit of pressure, called one *standard atmosphere* (1 atm). One millibar is equivalent to a pressure of 1,000 dynes per square cm (1 dyne = the force that, while acting upon a mass of 1 gram, will impart to it an acceleration of 1 cm/sq sec). A pressure of 1,000 mb is called 1 bar.

$$\begin{aligned}\text{Thus } 1 \text{ atm} &= 76 \text{ cm Hg} \\ &= \frac{76.\rho.g.10^4}{10^3} \text{ N/m}^2 \ (\rho = 13.6 \text{ gm/cc}, \ g = 981 \text{ cm/s}^2) \\ &= 101396.16 \text{ N/m}^2 \\ &= 1013.9616 \text{ millibar} \\ &= 1.0139616 \text{ bar}\end{aligned}$$

On a map, pressure is shown by *isopleths* of bars (= pressure) or *isobars* which are imaginary lines joining places with equal magnitude of surface air pressure. The pattern of isobars, their spacing, trends and values together indicate the pressure conditions of a region.

Wind

Horizontally moving air is called wind. It is measured as a vector parameter as it has both direction and velocity. The direction of flow is measured with the help of a *wind vane* while the velocity is measured with the help of an *anemometer* (Fig. 7.4 and Fig. 7.5). On a map, this is shown by an arrow. The layout of the arrows indicates the direction of the wind while the number of hands or quills denote its

Fig. 7.1 A Fortin's Barometer

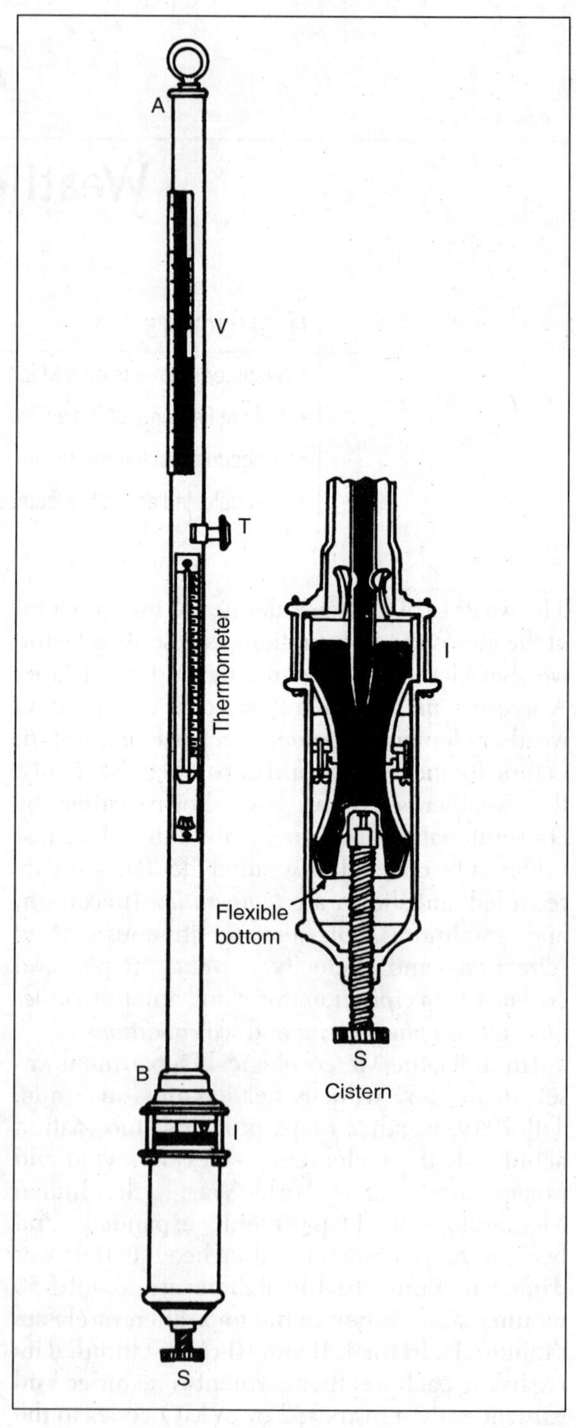

Fig. 7.2 An Aneroid Barometer

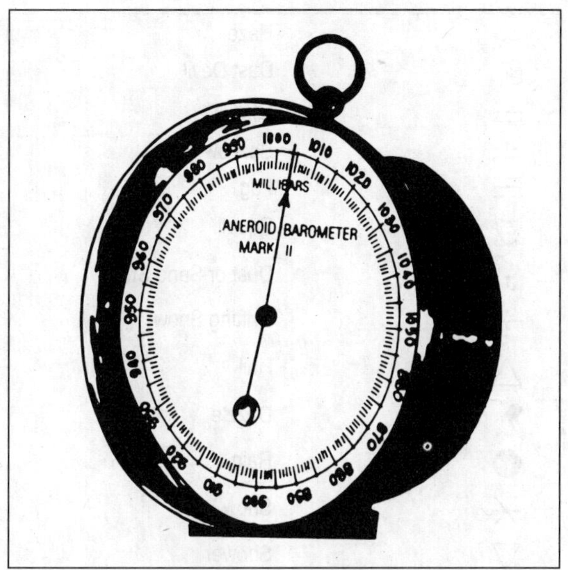

Fig. 7.3 A Barograph

Fig. 7.4 A Wind Vane

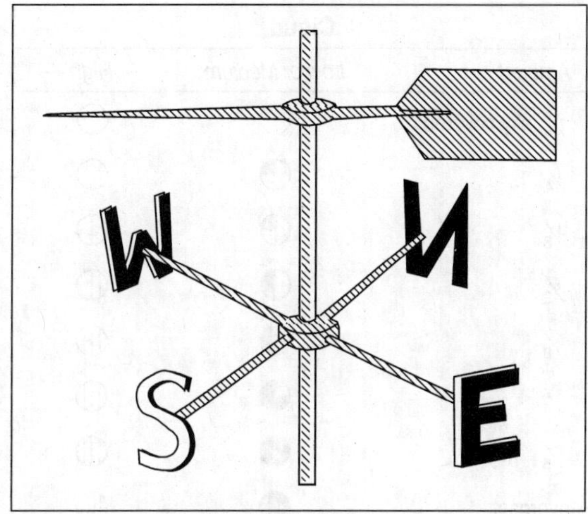

Fig. 7.5 An Anemometer

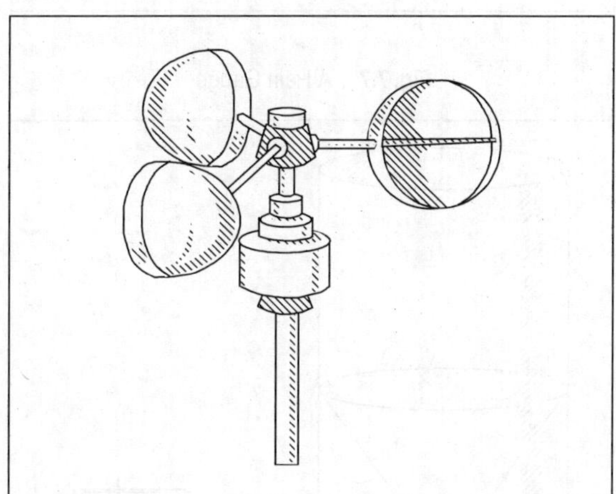

velocity. On weather maps, velocity and wind direction usually refer to the observations taken at 33 ft or 10 m above the ground (Table 7.2).

Cloudiness

This represents the amount of sky covered by clouds. The observer imaginarily divides the sky into 4 equal quadrants and then subdivides each of these quadrants into two equal parts. The estimate of the cloud cover is then taken as 1/8th part or 12.5% of the sky. The observations are recorded as tiny circles with different conventional symbols (Fig. 7.6).

Rainfall

The amount of rainfall is recorded with the help of a *rain gauge* (Fig. 7.7). This is represented by the exact numerals (equivalent to the amount recorded) in suitable units, either inch or cm or mm. Usually the rainfall data relates to the past 24 hours.

Fig. 7.6 Map Symbols for Cloudiness

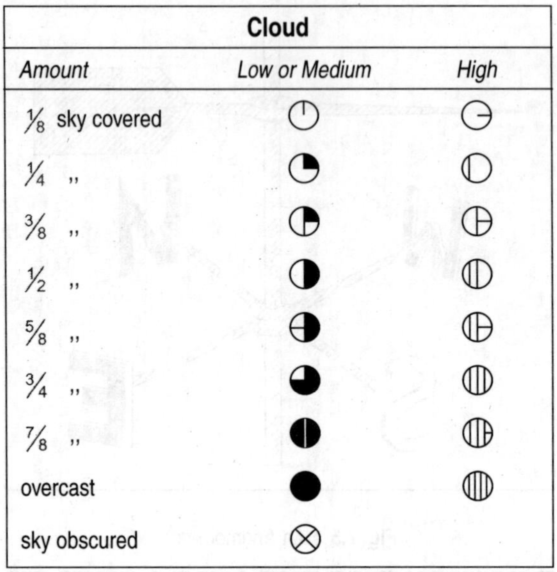

Cloud		
Amount	*Low or Medium*	*High*
⅛ sky covered		
¼ ,,		
⅜ ,,		
½ ,,		
⅝ ,,		
¾ ,,		
⅞ ,,		
overcast		
sky obscured		

Fig. 7.7 A Rain Gauge

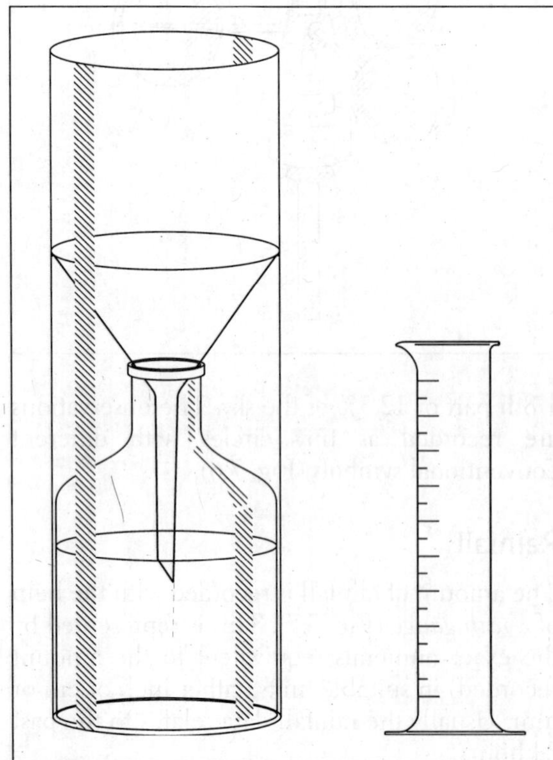

Fig. 7.8 Map Symbols for Atmospheric Phenomena

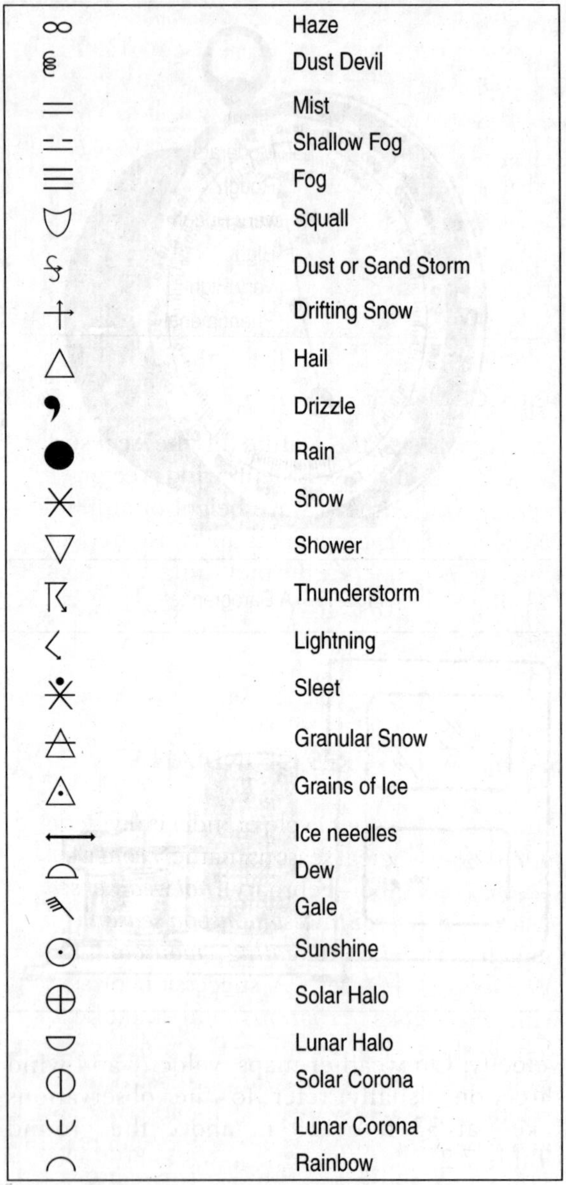

Symbol	Phenomenon
	Haze
	Dust Devil
	Mist
	Shallow Fog
	Fog
	Squall
	Dust or Sand Storm
	Drifting Snow
	Hail
	Drizzle
	Rain
	Snow
	Shower
	Thunderstorm
	Lightning
	Sleet
	Granular Snow
	Grains of Ice
	Ice needles
	Dew
	Gale
	Sunshine
	Solar Halo
	Lunar Halo
	Solar Corona
	Lunar Corona
	Rainbow

Atmospheric Phenomena

Haze, mist, fog, shallow fog, ground / fog, drizzle, snow, sleet, hail, thunderstorm, dust or sandstorm, dust devil, dew, depression, sky obscured are recorded. These are represented by conventional symbols (Fig. 7.8).

Fig. 7.9 Map Symbols for Sea Condition

Sea Conditions	
Cm	Calm
Sm	Smooth
Sl	Slight
Mod	Moderate
Ro	Rough
V. Ro	Very Rough
Hi	High
V. Hi	Very High
Ph	Phenomenal

Sea Conditions

This represents the nature of the sea surface as influenced by persistent wind vector. It is measured in terms of wave height or amplitude, wave length, wave steepness and wave frequency, which together define the prevailing wave regime. On the basis of these, *calm, slight, smooth, moderate, rough* and *very rough* seas have been scaled and denoted by letter symbols (Fig. 7.9).

SALIENT FEATURES OF INDIAN SEASONS

The annual weather cycle of India is divided into *four* meteorological seasons, namely *cold weather season* (December–February), *hot weather season* (March–May), *southwest monsoon season* (June–September) and *retreating monsoon season* (October–November). A succession of seasons with numerous variations and modifications defines the Indian weather. The process repeats itself year after year with almost unvarying exactitude. Each season is characterised by its own typical assemblage of weather phenomena represented in the *weather charts*, as below:

Cold Weather Season

- High pressure areas develop over the northern and northwestern parts of the subcontinent while low pressure areas develop over the Indian Ocean and the Bay of Bengal.
- Normally pressure decreases from north to south or from land to sea and accordingly pressure gradients are formed.
- Wedge and cols of high pressure generally form over the northwestern parts and the Indo-Gangetic plains.
- Isobars are moderately spaced out all over India excepting near the high pressure areas where relatively steeper gradients may prevail.
- Usually clear sky and fine weather conditions prevail.
- Calm to light air conditions dominate.
- Due to cold weather disturbances, rainfall may occur over the Punjab plains, vale of Kashmir and even the Ganga Plains.
- Snowfalls are usually recorded over Jammu and Kashmir and other Himalayan terrains.
- Downvalley katabatic winds bring cold wave conditions in the northern plains.
- Haze, mist and fog are commonly recorded near the coastal areas, industrial towns and metropolitan cities.

Hot Weather Season

- Low pressure areas demarcated by enclosed isobars develop over the plateau areas of Chhotanagpur, Malwa, Maharashtra, Karnataka and Telengana.
- High pressure areas develop over the southern seas.
- Magnitude of the total pressure differences is not as high as it would be by the end of the season. Pressure gradient is gentle to moderate but the direction of pressure gradient varies with the location of the thermal lows.
- Moderate breeze to fresh gale conditions may develop near the Assam valley, the lower Ganga plains, the arid to semi-arid tracts of north and northwest India and the Deccan plateau.
- Dust storms and dust devils are common over Rajasthan while northeasterly or northwesterly winds in the form of *kalbaisakhi* may be recorded over the Assam valley and the lower Ganga plains.
- Rain with thunder and lightning may be

recorded over isolated areas in the Assam valley, the lower Ganga plains and the Deccan plateau.

- The last days of the season may experience some pre-monsoon cyclones.
- Slight, calm and smooth seas are usually recorded.

Southwest Monsoon Season

- Intense low pressure areas develop over northwest India and relatively high pressure areas are recorded over the Indian ocean.
- In general, pressure decreases from south to north and from southeast to northwest.
- The axis of monsoon trough usually prevails along a line extending from Punjab to Orissa.
- Near the land-water transition, the transverse side of the monsoon trough and low pressure areas, steeper pressure gradients prevail.
- Isobars show WSW–ENE trends over the Arabian Sea, NW–SE trends over Peninsular India and southeast to northeast trends over the Bay of Bengal.
- Isobars are parallel to one another and are relatively widespread over the Ganga Plains and the extrapeninsular areas.
- Westerly, southwesterly and southerly components of winds are dominant.
- Gentle breeze to moderate gale conditions prevail. The wind speed is particularly high in case of geostrophic and cyclostrophic situations.
- The NLM line is usually marked and within it the sky shows cloudiness with varying intensity.
- This is the period of general rain. Widespread rainfall occurs over the western coast, the northeastern hill states, the sub-Himalayan region, the Andaman and Nicobar islands and the lower Ganga Plains.
- Local thunderstorms may be recorded over the plateau areas.
- Monsoon cyclones and depressions affect the general weather, particularly near the eastern coastal states.
- Rough and moderate sea conditions dominate the coastal seas.

Retreating Monsoon Season

- Relative high pressure develops over the extreme northwestern parts of the subcontinent.
- Relative low pressure areas become scattered

Interpreting Weather Maps

Salient features of pressure, temperature, wind, bad weather and atmospheric phenomena:

Pressure Condition

- *Location of Bar High*: magnitude and location in relation to land and sea, and season
- *Location of Bar Low*: magnitude and location in relation to land and sea, and season
- *Trends of Isobars*: geometry of isobars, and their regional trends in relation to land and sea, and also season
- *Pressure Gradient*: location of the steepest and the gentlest gradients, their regional distribution and variation in relation to thermal condition

Wind Condition

- *Wind Direction*: major directions and their regional variations in relation to seasons, trends of isobars and the location of high and low pressure areas
- *Wind Velocity*: locations of the highest and the lowest velocity and their regional variation in relation to site, season, pressure gradient and surface friction

Sky Condition

- *Nature of Cloud*: occurrences of high, medium and low clouds in relation to wind
- *Cloud Cover*: amount of coverage and regional variation in relation to wind

Precipitation

- *Forms*: nature of precipitation in relation to thermal and wind conditions
- *Distribution*: areas of occurrences in relation to wind and sky condition

Other Atmospheric Phenomena: if any, their probable occurrence and origin

Sea Condition: nature of sea along the off-shore regions in relation to prevailing winds

Weather Forecasts: short-term forecasts based on the prevailing inter-element relations over the Ganga Plains, the Arabian Sea, the Bay of Bengal, the Indian Ocean and the Peninsular India.

- Isobars are moderate to widely spaced.
- Calm to gentle breeze dominates the weather.
- Usually fine weather prevails excepting the eastern coastal states and the Indo-Gangetic plains where sky conditions may be shown as variably cloudy.
- Rainfall may occur over Tamil Nadu, coastal Andhra Pradesh and coastal Orissa, West Bengal, Bihar and at some stations in Uttar Pradesh and the northeastern hill states.
- Offshore winds in the morning hours and onshore winds in the afternoon are usually experienced.
- Coastal seas remain smooth, calm or slight.

PROCEDURES OF INTERPRETATION

The study and interpretation of weather maps are to be arranged in the following order.

Pressure Conditions

Location of Bar High

In this, the *geographical occurrence* of high pressure, its *nature, intensity, demarcating isobaric value* and possible *mode of formation* are to be described and stated. A sketch map showing the high pressure areas may also be prepared.

Location of Bar Low

Similar to **Location of Bar High**.

Trends of Isobars

First, a sketch map showing the main components of the orientation of isobars is to be prepared. This is followed by an analysis of the orientation or trends of isobars over different areas *in relation* to parallels of *latitudes, season, land-sea contrast, pressure cells* and *dynamic factors of surface heating. Axis of low pressure, wedge, col*, etc., are also to be especially stated.

Pressure Gradient

In this, the general nature of the *pressure decreased, direction* and the *magnitude* of the steepest, moderate and gentle pressure gradients both over land and sea are to be indicated. *Representative* isobaric sections with proper labelling must be drawn to emphasise the degree of spatial variation. From the isobaric map, according to a chosen inter-isobar spacing, a sketch is to be drawn to find out the areas of occurrence of different *pressure gradient zones*. Spatial occurrence of these is then analysed in terms of climatic controls.

Wind Conditions

Wind Direction

First, the general direction of wind in relation to seasons and *directions of pressure gradient* are to be mentioned. A *wind rose* may be prepared from a database collected from the map in the form as shown in Table 7.2. This would reveal the *dominant directions* of wind all over the map and explain seasonal surface circulation as controlled by the general climatic factors. From this, a sketch map could be prepared to show the dominant wind direction zones. The spatial distribution of the dominant components are then interpreted accordingly.

Table 7.2 Worksheet for Wind Rose

Wind Direction	*Calm*	*N*	*S*	*E*	*W*	*NE*	*NW*	*SE*	*SW*
Tally Mark									
Frequency									
% Frequency									

Wind Velocity

Under this heading, the amount and place of recording of the *highest velocity*, the general features of *velocity distribution* in terms of pressure gradient, cyclonic circulation and geostrophic effect are to be clearly mentioned.

A *bar diagram* or *frequency histogram* may be prepared from a database collected from the map (Table 7.3: number of column depending on the highest velocity recorded on the map). This would give a vivid idea of the velocity distribution. The modal value and nature of skewness would suggest a lot about the general nature of the distribution. A sketch map can also be prepared to show the wind velocity zones which are then described and analysed in detail.

Sky Conditions

Nature of Cloud

Mainly the places or regions of occurrences of *high*, *low* or *medium* clouds are analysed. Interpretation of this is done in terms of the prevailing winds and the thermo-humid conditions. A sketch map showing the regional pattern may also be prepared.

Cloud Cover

The general *distribution of cloudiness* is discussed and analysed here. A *bar diagram* or *frequency histogram* may be prepared from the database collected from the map in the form of the Table 7.4. This will help find the exact cloud cover over different parts of the map. A sketch map can easily be prepared to show *blue sky*, *less cloudy* (1/8th–1/4th), *moderately cloudy* (3/8th–5/8th), *largely cloudy* (3/4th–7/8th) and *overcast* zones. These could then be analysed in terms of available sunshine hour, back radiation and night temperature, surface comfortability, prevailing winds, etc.

Precipitation

Under this heading, the general *distribution of rainfall* and places of rainfall in descending order of magnitude are to be described. A *sketch map* is then prepared to indicate the areas with *heavy rainfall*, *moderate rainfall*, *light rainfall* and *light drizzle*. A geographical explanation of bad weather conditions is to be given in terms of physical factors, type and mode of instability associated with rainfall, surface features and surface circulation patterns.

Weather Map: Data Acquisition and Display

Pressure Condition

- Sketch map showing the location of high and low pressure centres/areas/belts
- Sketch map showing the trends of isobars
- Isobaric sections along suitable directions to show the gradients of pressure
- Sketch map showing the spatial variations of pressure gradients

Wind Condition

- Database and wind rose showing the prevailing major wind directions
- Sketch map showing the spatial distribution of the prevailing major wind directions
- Database and histogram showing the distribution of wind velocity
- Sketch map showing the spatial distribution of wind velocities

Sky Condition

- Sketch map showing the distribution of high, medium and low clouds
- Database and histogram showing the distribution of cloud cover
- Sketch map showing the spatial distribution of cloud cover

Precipitation

- Sketch map showing the distribution of different forms of precipitation
- Sketch map showing the distribution of rainfall

Atmospheric/Sea Phenomena

- Sketch map showing the distribution of various atmospheric phenomena
- Sketch map showing the nature of sea condition

Pressure–Wind–Cloud–Rainfall Relation

- Transect charts along suitable lines to show the inter-element relations

Table 7.3 Worksheet for Wind Velocity Distribution

Wind Velocity	*Calm*	*Light Air (5 knots)*	*Light Breeze (10 knots)*	*Gentle Breeze (15 knots)*	----	*Storm (55 knots)*	*Hurricane (60 knots)*
Tally Mark							
Frequency							
% Frequency							

Table 7.4 Worksheet for Distribution of Cloudiness

Amount of Cloudiness (Part of Sky)	*Azure Blue Sky*	*1/8th*	*1/4th*	*3/8th*	*1/2*	*5/8th*	*3/4th*	*7/8th*	*Overcast Sky*
Tally Marks									
Frequency of Occurrence									
% Frequency									

Other Atmospheric Phenomena

Places of occurrences of *haze, mist, shallow fog, fog, dust devil, dust storm, drizzle, rain, snow, hail, shower, squall, thunderstorm, lightning, dew, rainbow, obscure sky*, etc. are to be stated in relation to their modes of formation. A sketch map may also be prepared to show their distribution.

Pressure-Wind-Cloudiness-Rainfall Relation

Location of high and low pressure areas and direction of pressure gradient together with latitudinal location, airmass trajectory and surface features control the wind vector of a place. Depending on the trajectory of wind flow, clouds are formed. This in turn influences the rainfall distribution. This can be analysed with the help of a *transect chart* drawn between any two suitable locations.

Pressure Departures from Normal

This can be well judged from the deviation map that is given. Areas of normal pressure and positive and negative departures within different ranges are to be described and analysed.

Temperature Departures from Normal

From the given deviation map, areas of normal, positive and negative departures of the mean maximum or the mean minimum temperature within different ranges are to be described and analysed.

Sea Conditions

Conditions of the sea in terms of wave patterns near or off a particular coast and high seas are to be pointed out in relation to the seasonal surface wind and oceanic circulation pattern.

Weather Forecasting

From an analysis of the weather elements and the deviation maps, usually a *short range forecast* for the next 24 or 48 hours can be predicted in respect of:

i. Wind	Anticipated direction and force
ii. Sky	Amount of cloud cover and nature of rainfall
iii. Temperature	Departure from normal and change in night temperature
iv. Special weather phenomena	Thunderstorm, fog, squalls, heavy shower, etc.
v. General weather conditions	Areas of fair weather

However, the degree of precision in forecasting depends on special skills and on one's specific observations. This can only be very broad and oversimplified.

EXAMPLE: INDIAN DAILY WEATHER REPORT, 9 AUGUST 1965

The weather map shows the weather conditions as recorded at 08.30 hrs IST on 9 August 1965. Obviously the weather experienced represents that of the southwest monsoon season which is precisely the period of general rain in India. The primary characteristics as shown by the map are:

i. Low Pressure	100 mb (northwest India).
ii. High Pressure	1010 mb (over ocean).
iii. Isobaric Pattern	Monsoon trough (Punjab-Orissa coast); only 6 isobars, patchy in outline.
iv. Wind	Westerly and southwesterly winds dominant; gentle to moderate breeze.
v. Cloud	Overcast conditions prevail all over India.
iv. Rainfall	Highest 7 cm, mostly in the North-Eastern Hill States.
vii. Sea	Mainly calm and smooth.

Hence, though it is monsoon in full swing, the lack of well distributed cloudiness and rainfall, the absence of any steep pressure gradient and high velocity winds, and the prevalence of a generally unperturbed sea are typical. Probably it is the phase of a *dry spell* that normally alternates with a *wet spell* in the weather sequence of the monsoon season (Fig. 7.10).

Identification of Seasons

Season	*Diagnostic Features*
Cold Weather Season (December–February)	1. High pressure (Wedge/Col) over land and Low over seas with land–to–sea gradient 2. Clear sky; fair weather; haze, mist and fog along the coastal and industrial areas 3. Western disturbance brings rain over the Indo-Gangetic plain
Hot Weather Season (March–May)	1. Enclosed Lows over the peninsular plateaus and high over the seas 2. Strong winds and rain due to heat thunderstorms (kalboishakhi/mango shower) 3. Dust storms near the desert margins; loo in the north-central part
South–West Monsoon Season (June–September)	1 Intense low over the north-western part and high over the sea 2. Low pressure/Monsoon trough extends from Orissa to Punjab 3. Bay depressions off the Orissa and Andhra coasts with gentle breeze to moderate gale 4. Cloudy sky with widespread heavy/very heavy rain; moderate/rough sea
Retreating Monsoon Season (October–November)	1. Relative high pressure over northwest India and relative lows over the Ganga plains, Chhotanagpur plateau and southern seas 2. Moderate to widely spaced isobars with calm/gentle breeze 3. Sky variably cloudy with rains mainly along the Coromondal coast; calm to slight sea

Fig. 7.10 Indian Daily Weather Report: 08.30 hrs IST, Monday 9 August 1963

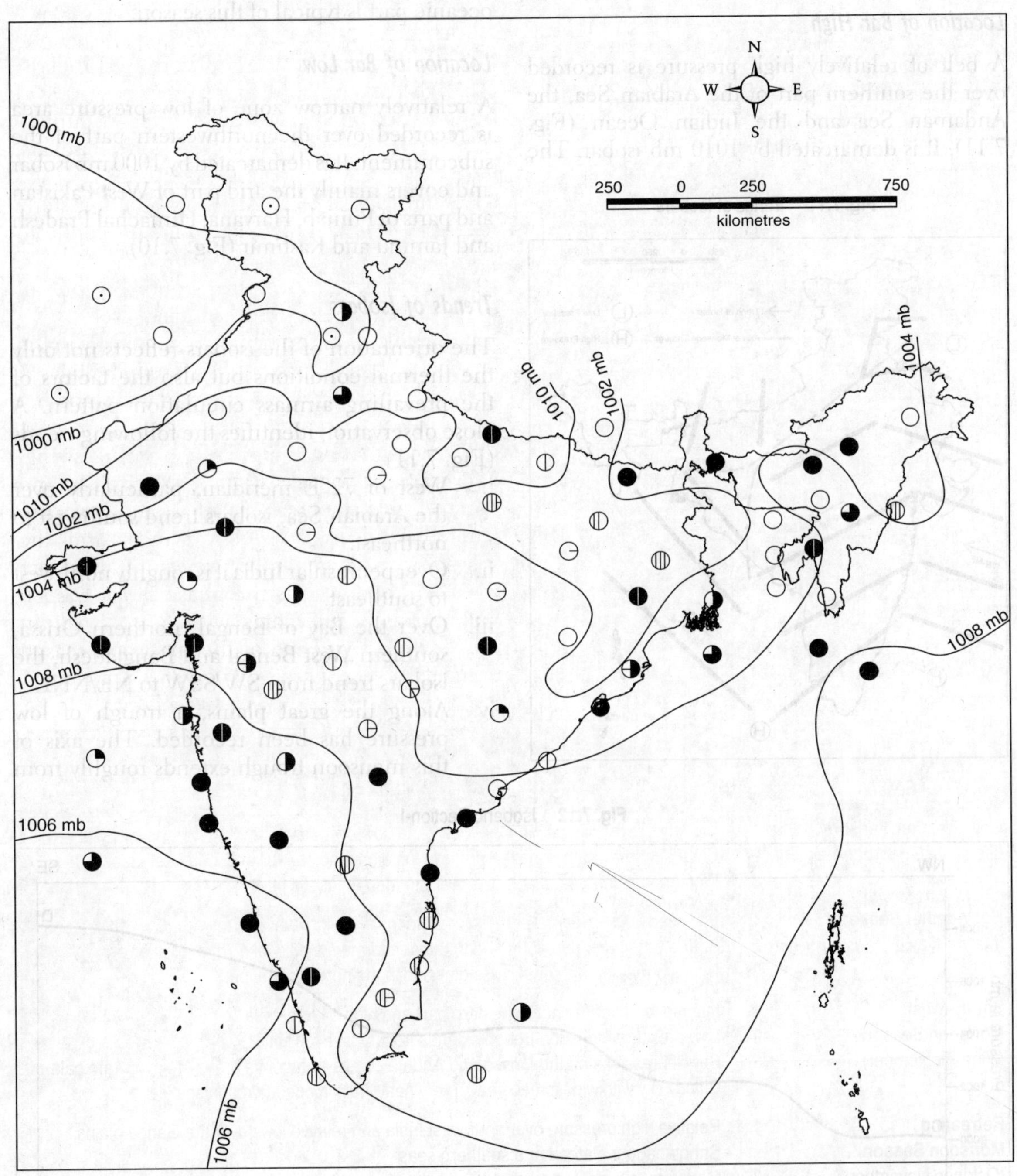

Pressure Conditions

Location of Bar High

A belt of relatively high pressure is recorded over the southern part of the Arabian Sea, the Andaman Sea and the Indian Ocean (Fig. 7.11). It is demarcated by 1010 mb isobar. The broad latitudinal orientation of this belt over the oceanic part is typical of this season.

Fig. 7.11 Trends of Isobars

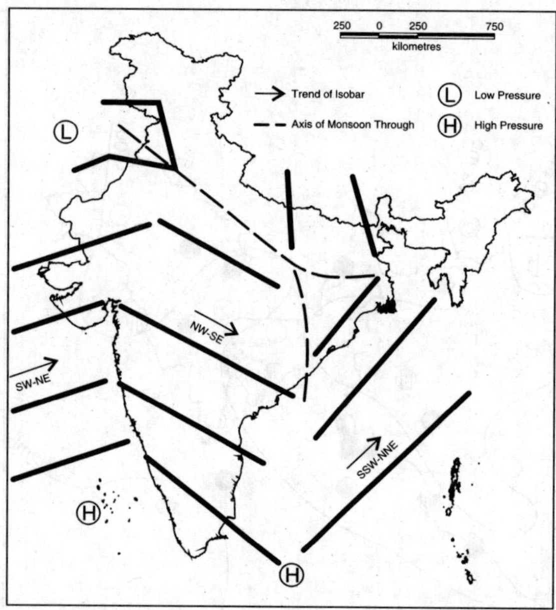

Location of Bar Low

A relatively narrow zone of low pressure area is recorded over the northwestern part of the subcontinent. It is demarcated by 1000 mb isobar and covers mainly the arid part of West Pakistan and parts of Punjab, Haryana, Himachal Pradesh and Jammu and Kashmir (Fig. 7.10).

Trends of Isobars

The orientation of the isobars reflects not only the thermal conditions but also the factors of the prevailing airmass circulation pattern. A close observation identifies the following trends (Fig. 7.11)

i. West of 72°E meridian, particularly over the Arabian Sea, isobars trend southwest to northeast.
ii. Over peninsular India it is roughly northwest to southeast.
iii. Over the Bay of Bengal, northern Orissa, southern West Bengal and Bangladesh, the isobars trend from SW/SSW to NE/NNE.
iv. Along the great plains, a trough of low pressure has been recorded. The axis of this monsoon trough extends roughly from

Fig. 7.12 Isobaric Section-I

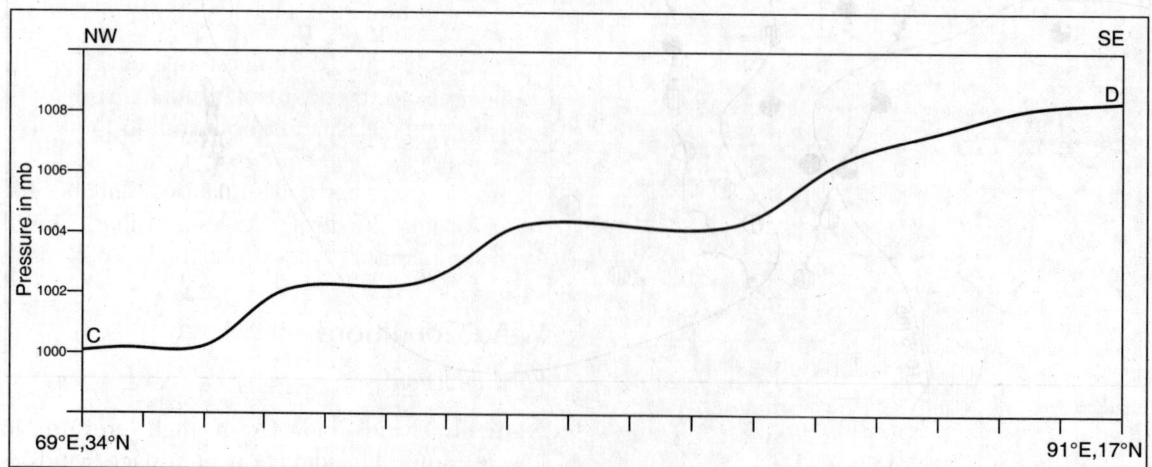

Punjab to Orissa or Bengal in a northwest to southeast fashion.

Pressure Gradient

In general, pressure decreases from south to north over the Arabian sea, from south-southwest to north-northeast over the peninsula, from east to west over the Bay of Bengal and from southeast to northeast over the Indo-Gangetic plains. These directions essentially conform to the directions of the prevailing pressure gradient patterns. A pressure gradient is simply the rate at which pressure falls horizontally per unit distance. Therefore the magnitude of pressure gradient varies inversely with the spacing of isobars. The closer the isobars the steeper the gradient and vice versa. The isobaric sections (Fig. 7.12 and Fig. 7.13) as well as the map shows that pressure gradient is not uniform throughout the subcontinent. Between 1006 mb and 1010 mb isobars, mainly gentle gradient persists. Between 1000 mb and 1006 mb isobars, the gradient is relatively steep in the western part and moderate in the eastern part. The steepest pressure gradient of 4 mb/100 km has been recorded near Bhopal (77°E, 24°N) while the gentlest pressure gradient of 0.14mb/100km has been recorded along the 85°E meridian over the Bay of Bengal.

Fig. 7.13 Isobaric Section-II

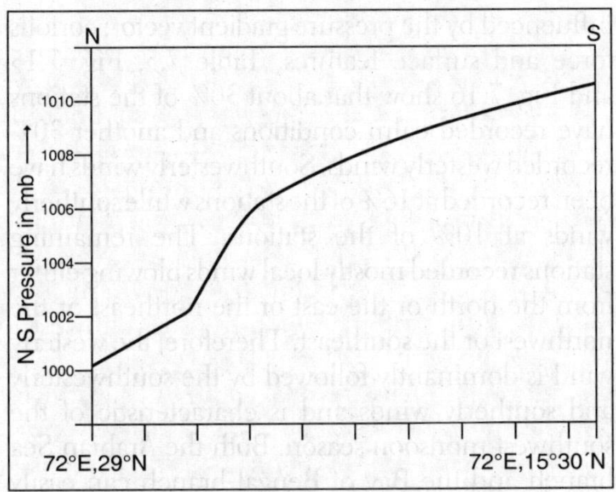

Fig. 7.14 Pressure Gradient Zones

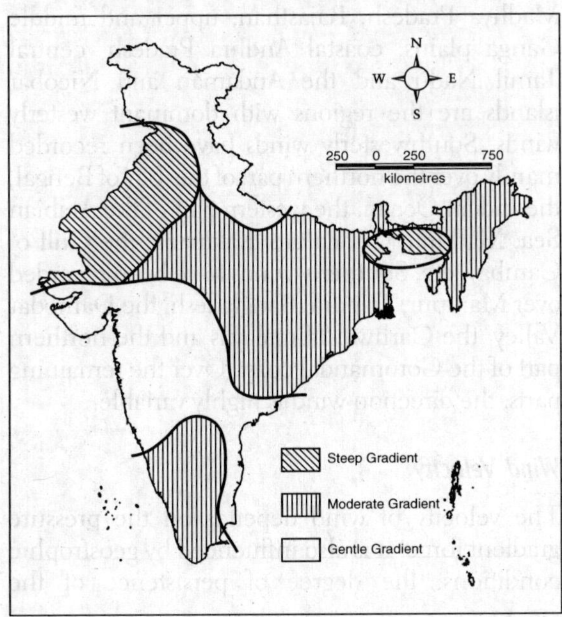

On the basis of the surface pattern of pressure gradient, the subcontinent may be divided into three pressure gradient zones (Fig. 7.14).

i. *Steep gradient zone*
This occurs in two separate areas. The first zone extends over a larger area covering parts of Gujarat, Rajasthan, Madhya Pradesh, Punjab and Pakistan. The second zone is spread over a much smaller area in the northern part of Bangladesh.

ii. *Moderate gradient zone*
This occupies the largest area covering the whole of extra peninsular India, Orissa, parts of Kerala, Karnataka and Andhra Pradesh.

iii. *Gentle gradient zone*
This covers parts of Karnataka, Maharashtra, Gujarat, Madhya Pradesh, Bihar, Tamil Nadu, Andhra Pradesh and the Bay of Bengal.

Wind Conditions

Wind direction

Normally, winds blow from high pressure to low pressure. The direction of surface winds is

Table 7.5 Database for Wind Rose

Wind Direction	*Calm*	*N*	*S*	*E*	*W*	*NE*	*SE*	*NW*	*SW*
Tally Mark	卌 卌 卌 卌 卌 |	|.	卌 |||	|||	卌 卌 卌 卌 卌 |	|.	|.	|||	卌 卌 |||
Frequency	27	2	9	3	27	1	2	4	14
% Frequency	30.3	2.2	10.2	3.4	30.3	1.1	2.2	4.5	15.8

Fig. 7.15 Wind Rose

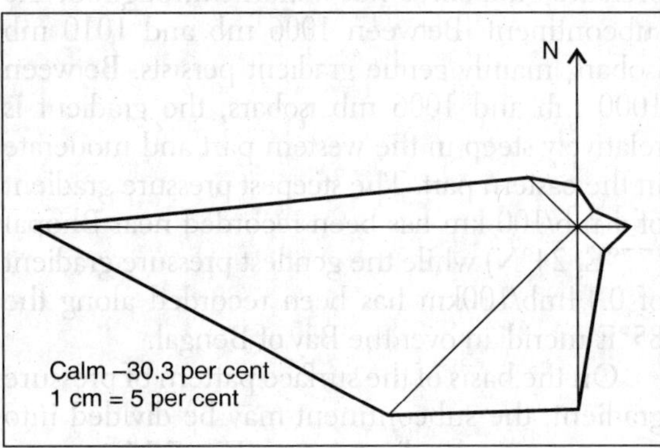

Fig. 7.16 Dominant Wind Direction Zones

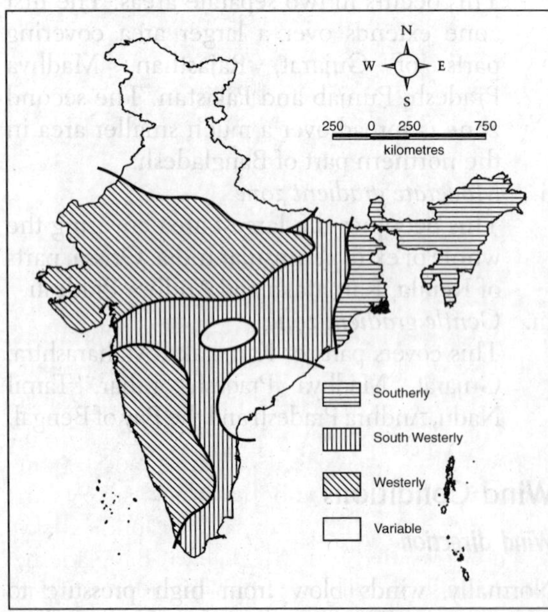

influenced by the pressure gradient vector, coriolis force and surface features. Table 7.5, Fig. 7.15 and Fig. 7.16 show that about 30% of the stations have recorded calm conditions and another 30% recorded westerly winds. Southwesterly winds have been recorded at 16% of the stations while southerly winds at 10% of the stations. The remaining stations recorded mostly local winds blowing either from the north or the east or the northeast or the northwest or the southeast. Therefore, the westerly wind is dominantly followed by the southwesterly and southerly winds and is characteristic of the southwest monsoon season. Both the Arabian Sea branch and the Bay of Bengal branch can easily be identified. Coastal and interior coastal Kerala, coastal Maharashtra, Gujarat, the western parts of Madhya Pradesh, Rajasthan, upper and middle Ganga plains, coastal Andhra Pradesh, central Tamil Nadu and the Andaman and Nicobar islands are the regions with dominant westerly winds. Southwesterly winds have been recorded mainly over the northern part of the Bay of Bengal, the Indian Ocean, the western part of the Arabian Sea, Madurai, Cuddalore, Kakinada, the Gulf of Cambay, etc. Southerly winds have been recorded over Manipur, Tripura, Bangladesh, the Damodar Valley, the Garhwal Himalayas and the northern part of the Coromandal coast. Over the remaining parts, the direction wind is highly variable.

Wind Velocity

The velocity of wind depends on the pressure gradient force. It is also influenced by geostrophic conditions, the degree of persistence of the

Table 7.6 Database for Wind Velocity Distribution

Wind Velocity	*Calm*	*Light Air (5 knots)*	*Light Breeze (10 knots)*	*Gentle Breeze (15 knots)*	*Moderate Breeze (20 knots)*
Tally Mark	卌 卌 卌 卌 卌 卌 卌 I	卌 卌 卌 卌 卌 卌 I	卌 卌 II	卌 II	II
Frequency	36	31	13	7	2
% Frequency	40.4	34.8	14.6	7.9	2.3

prevailing system and also the occurrences of cyclonic components, if any. In this season of southwest monsoon, the pattern of circulation is mainly governed by the flow vector of the transient equatorial monsoon airmass. Table 7.6 and Fig.7.17 show that the velocity distribution is positively skewed with light air to light breeze conditions prevailing over most of the subcontinent. About 40% of the stations experienced calm conditions. The highest velocity of 20 knots has been recorded at two places on ships in the Arabian Sea (65°E, 20°N) and in the Indian Ocean (86°E, 6°N). A gentle breeze of 15 knots has been recorded at New Delhi, Veraval, Tiruchirapalli, Hyderabad, Kanpur and on ships at (66°E, 10°N) and (70°E, 15°E).

Fig. 7.17 Wind Velocity Zones

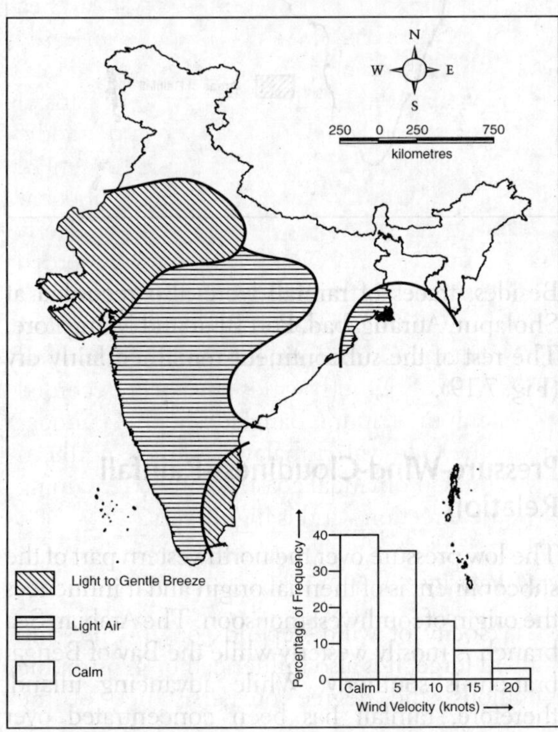

Sky Conditions

Nature of Clouds

This is a season of general rain. But interestingly, the map shows that nearly 29% of the stations recorded clear sky conditions and about 27% recorded medium and high clouds. The remaining areas were covered with low clouds responsible for the observed pattern of rain. Medium and high clouds have been mainly recorded over South Kerala, Tamil Nadu, parts of Andhra Pradesh, Madhya Pradesh, Maharashtra, Uttar Pradesh, Bihar and Orissa.

Cloud Cover

The moist and rainbearing southwesterly monsoon winds have produced clouds of varying intensity depending upon the local situations of atmospheric instability. Table 7.7 and Fig. 7.18 show that nearly 33% of the stations recorded completely overcast sky while about 29% of the stations absolutely clear sky. Overcast conditions were found over the whole of the western coast, parts of Rajasthan, interior Karnataka, coastal Andhra Pradesh, part of north Bihar plain and Bihar plateau, Assam valley and Darjeeling Himalayas. Cloudless sky has been recorded over Meghalaya, Bangladesh, part of Orissa, Madhya Pradesh, Rajasthan, Uttar Pradesh, Punjab, Haryana, Jammu and Kashmir

Table 7.7 Database for Cloud Coverage

Cloudiness	Azure Blue Sky	1/8th – 3/8th	1/2 – 5/8th	3/4 – 7/8th	Overcast
Tally Mark	𝍸 𝍸 𝍸 𝍸 𝍸 ||	𝍸 𝍸 𝍸 ||	||	𝍸 𝍸 |	𝍸 𝍸 𝍸 𝍸 𝍸 ||
Frequency	29	19	4	11	29
% Frequency	29.2	21.3	4.5	12.4	32.6

and the whole of Pakistan. Over the rest of the subcontinent, partial cloudy sky ranging from 1/8th to 7/8th was recorded.

Precipitation

During the past 24 hours, rainfall has been widespread over the northeastern Hill States. Silchar recorded the highest rainfall of 7 cm. Siliguri (5 cm), Vishakhapatnam (4 cm), Ratnagiri (3 cm), Tejpur (2 cm) and Gauhati (1 cm) recorded varying amounts of rainfall.

Fig. 7.18 Intensity of Clouds

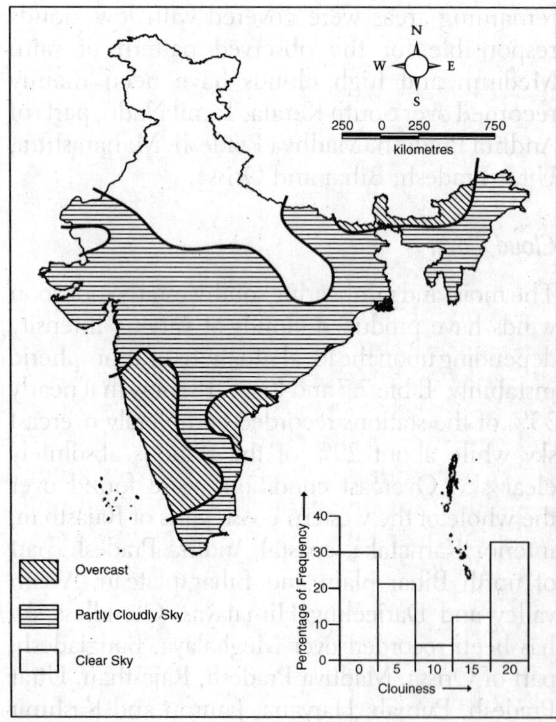

Fig. 7.19 Areas of Rainfall

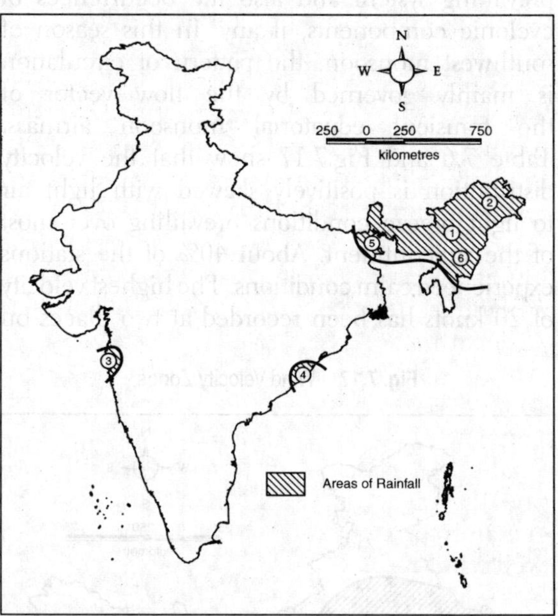

Besides, traces of rainfall were also recorded at Sholapur, Aurangabad, Port Blair and Mangalore. The rest of the subcontinent remained fairly dry (Fig. 7.19).

Pressure-Wind-Cloudiness-Rainfall Relation

The low pressure over the northwestern part of the subcontinent is of thermal origin and it influences the origin of southwest monsoon. The Arabian Sea branch is mostly westerly while the Bay of Bengal branch is southerly. While advancing inland, therefore, rainfall has been concentrated over the windward slopes of the orographic barriers involving the effects of conditional instability. The two branches meet along the axis of monsoon trough along which depressions are likely to move affecting general distribution of rain. Since this is not a period of continuous rain, the observed dryness suggests that a dry spell is ensuing.

Other Atmospheric Phenomena

Nil.

Pressure Departures from Normal

The 08.30 hrs pressure was normal over the southern half of Kerala, Tamil Nadu, Uttar Pradesh, Himalayas and western half of Nepal. Over the rest of the subcontinent it was above normal. A departure of +4 and above covers Gujarat, west and central Madhya Pradesh and adjoining parts of Rajasthan and Maharashtra. A departure of pressure between +2 and +4 range covers the whole of Maharashtra, Orissa, Bihar, West Bengal, Meghalaya, Tripura and parts of Rajasthan, Madhya Pradesh, Uttar Pradesh, Karnataka, Andhra Pradesh, Assam and the North Eastern Hill States. Over the remaining parts of the country, departure was well within +2.

Temperature Departures from Normal

The minimum temperature was below normal over the North Eastern Hill States, most of Bangladesh, Darjeeling Himalayas, Sikkim, Bhutan, Meghalaya, eastern Nepal and the adjoining Siwaliks, Kashmir and Kumaun Himalayas, Punjab, Haryana, Rajasthan, the whole of the eastern coast and in a narrow zone covering parts of Madhya Pradesh, Maharashtra and Kerala. Between these two zones, departure was above normal in a belt extending from North to South. A departure of –2 and more was recorded over western Rajasthan, the western part of Jammu and Kashmir, eastern Nepal, Sikkim, Bhutan, Darjeeling Himalayas and North-Eastern Hill States. A departure of +2 and more has been recorded at four separate narrow zones extending over Madhya Pradesh, Bihar, West Bengal, Andhra Pradesh, Maharashtra, Karnataka and Kerala.

Sea Conditions

The sea conditions fluctuated between calm and slight. The sea was calm in the southern part of the Coromandal coast. Further north, near Madras, it was smooth and off Kakinada slightly rough. The Arabian Sea was smooth at (64°E, 19°N) and (70°E, 17°N) and Minikoy but was slightly rough at (65°E, 9°N) and (70°E, 14°N). The Indian Ocean was slightly rough at (90°E, 5°N) and smooth at (86°E, 5°N).

Weather Forecasting

From the above analysis a short-range forecast for the next 24 to 48 hours may be made as follows:

i. The night temperatures are likely to increase at places with cloud cover of more than 50%, particularly over the peninsula.
ii. The minimum temperature is likely to increase over the semi-arid and arid regions and the Punjab plains.
iii. The pressure is likely to normalise over most of the subcontinent.
iv. Light and scattered rainfall may occur at one or two places in peninsular India, the coastal Orissa, the Gangetic West Bengal, the Darjeeling Himalayas, the Assam valley and the North Eastern Hill States.

EXERCISE

1. What is a weather map? What does it portray? Why it is important to geographers?
2. Name the instruments that record atmospheric temperature, pressure, wind velocity, wind direction, cloudiness, sunshine, rainfall, evaporation, and evapotranspiration.
3. Name the Headquarters of IMO and WMO.
4. Name the different types of weather recording stations in India.
5. Define atmospheric pressure. State the units of its measurement.
6. Name the different climatological seasons along with their duration in India.
7. What are the salient features of weather observed during the cold weather season in India?
8. What are the salient features of weather observed during the hot weather season in India?
9. What are the salient features of weather observed during the southwest monsoon season in India?
10. What are the salient features of weather observed during the retreating monsoon season in India?
11. State the salient characteristics of the weather elements as shown in the given weather map and identify the seasons.
12. Critically study the isobaric pattern in the weather map. Identify the trends of isobars and explain it.
13. Draw a sketch map to show the areas of high pressure and low pressure and the trends of isobars with such features as axis of low pressure, axis of monsoon trough, wedge of high pressure, cyclonic circulation and the general pattern of isobars.
14. What is meant by pressure gradient? How it is measured on a weather map? Find the directions and magnitudes of the steepest, gentlest and average pressure gradient shown on the given weather map.
15. Draw isobaric sections along suitable directions on the given weather map to show the variations of pressure gradient. Explain it.
16. Draw a sketch map to illustrate the distribution of pressure gradient zones and interpret it.
17. For the given weather map draw a wind rose and analyse it to identify the seasons.
18. State the laws relating to wind direction. Draw a sketch map to show the different wind direction zones and interpret it.
19. Draw a statistical diagram to analyse the distribution of wind velocity over a given weather map.
20. Draw a sketch map to show the distribution of different wind velocity zones and interpret it.
21. Draw a statistical diagram to analyse the distribution of cloudiness over a given weather map.
22. Draw a sketch map to show the distribution of zones of different cloudiness and interpret it.
23. Draw a sketch map to show the distribution of rainfall and interpret it.
24. Draw a sketch map to show the distribution of other atmospheric phenomena and interpret it.
25. Draw a sketch map to show the distribution of different sea conditions and interpret it.
26. Draw a transect chart to illustrate the relation between locations of pressure belts and wind directions.
27. Draw a transect chart to illustrate the relation between pressure gradient and wind velocity.
28. Draw a transect chart to illustrate the relation between wind conditions and cloudiness.
29. Draw a transect chart to illustrate the relation between cloudiness and rainfall.
30. With the help of a transect chart, explain the relations between pressure condition, wind condition, cloudiness, rainfall and other atmospheric phenomena.
31. Interpret the weather map under the heads: pressure conditions, wind conditions, cloud cover and rainfall. Comment on the weather conditions that would be most likely in the next 24–48 hours.
32. Critically analyse the weather conditions portrayed in the weather map and identify the seasons.
33. Interpret the weather map and make a short-range forecast of weather.

Map 1

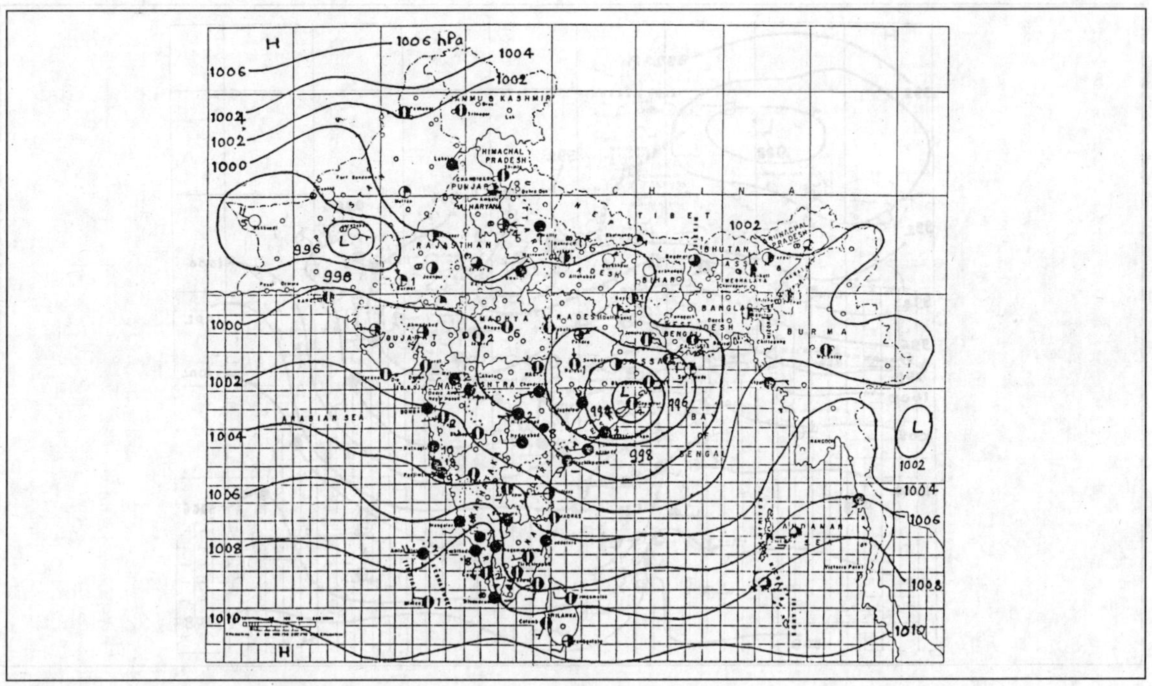
H
1006 hPa
1004
1002
1006
1002
1000
JAMMU & KASHMIR
HIMACHAL PRADESH
PUNJAB
HARYANA
RAJASTHAN
L
996
998
TIBET
1002
BHUTAN
BURMA
BANGLADESH
1000
1002
ARABIAN SEA
1004
L
998
996
BAY OF BENGAL
L
1002
1004
1006
1008
1010
H
1006
1008
1010

Map 2

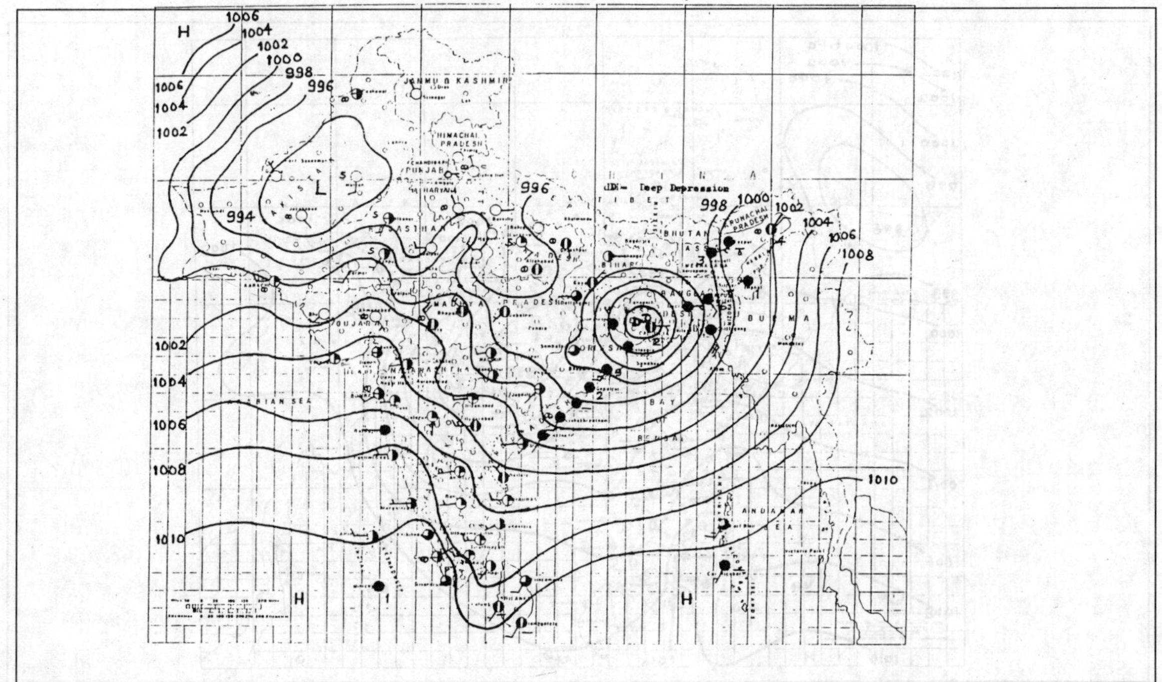
H
1006
1004
1002
1000
998
996
1006
1004
1002
JAMMU & KASHMIR
HIMACHAL PRADESH
PUNJAB
HARYANA
L
994
996
(D) – Deep Depression
998
1000
1002
1004
1006
1008
BHUTAN
BANGLADESH
BURMA
GUJARAT
1002
1004
ARABIAN SEA
1006
1008
1010
1010
H
H

Map 3

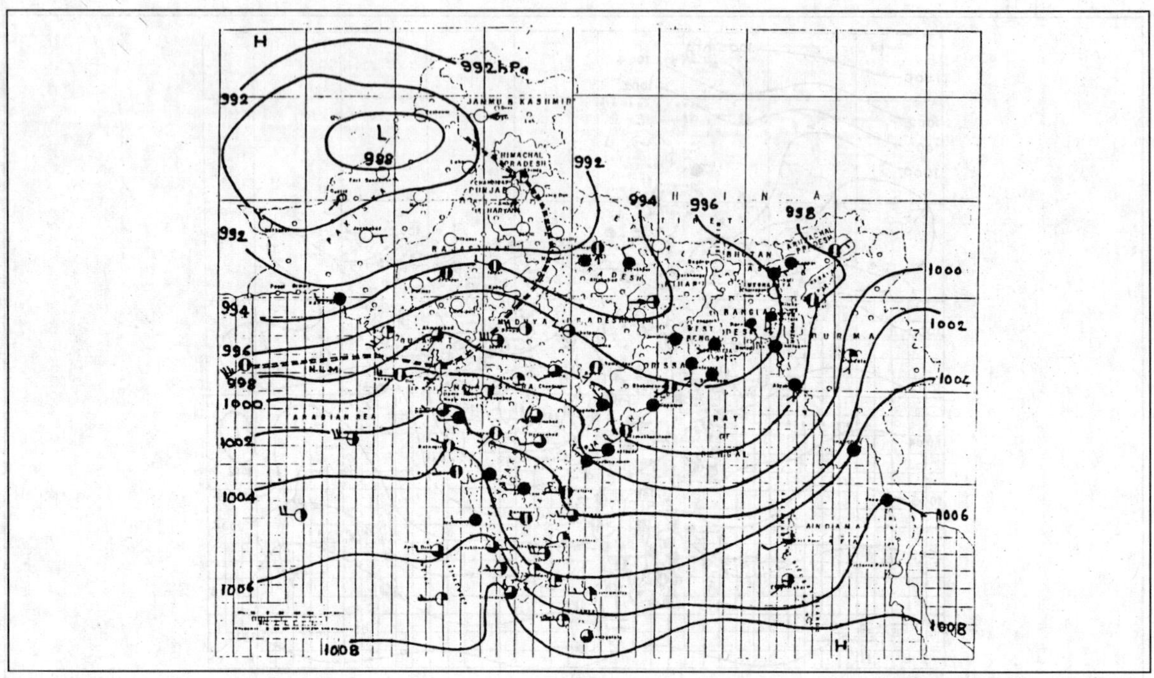
H
992 hPa
JAMMU & KASHMIR
HIMACHAL PRADESH
PUNJAB
HARYANA
992
L
998
994
996
998
1000
1002
1004
1006
1008
N.L.M.
ARABIAN SEA
MADHYA PRADESH
ORISSA
BHUTAN
BURMA
BAY OF BENGAL
H

Map 4

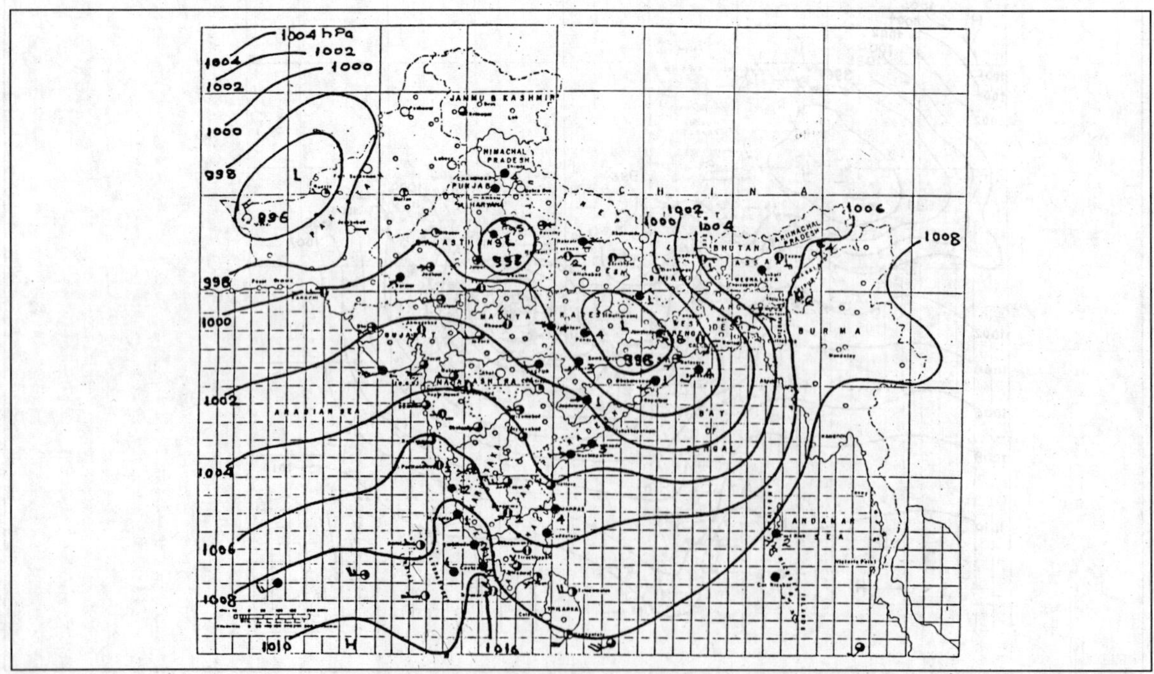
1004 hPa
1002
1000
1004
1002
1000
998
L
996
JAMMU & KASHMIR
HIMACHAL PRADESH
PUNJAB
998
1000
1002
1004
1006
1008
1010
1006
H
C H I N A
BHUTAN
ARUNACHAL PRADESH
BURMA
ARABIAN SEA
BAY OF BENGAL
ANDAMAN SEA
1016

Map 5

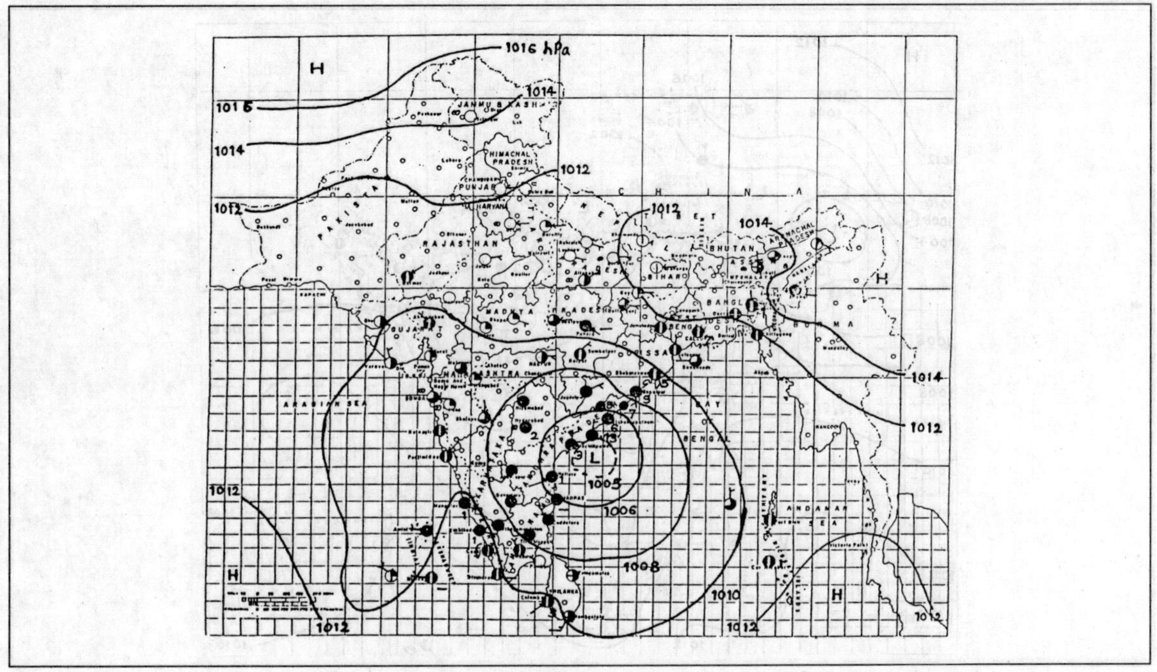
1016 hPa
H
1016
1014
1012
HIMACHAL PRADESH
PUNJAB
HARYANA
RAJASTHAN
CHINA
BHUTAN
ARABIAN SEA
BENGAL
L
1005
1006
1008
1010
ANDAMAN SEA

Map 6

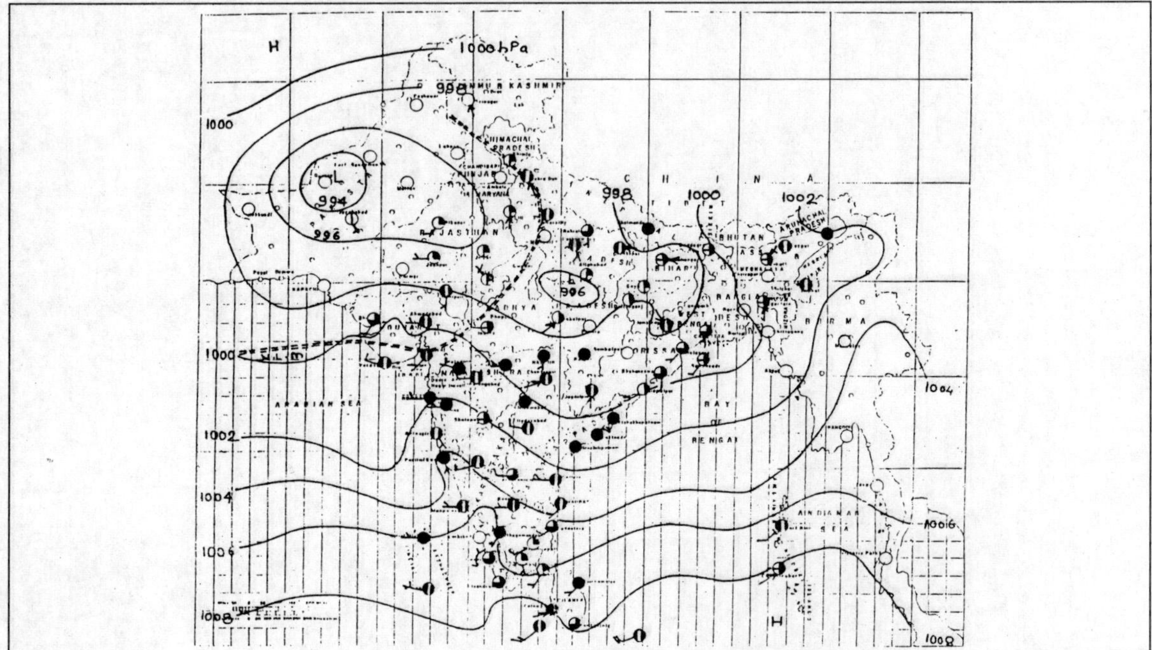
H
1000 hPa
1000
HIMACHAL PRADESH
PUNJAB
HARYANA
L
994
996
998
CHINA
1002
BHUTAN
ARABIAN SEA
1004
1006
1008

Map 7

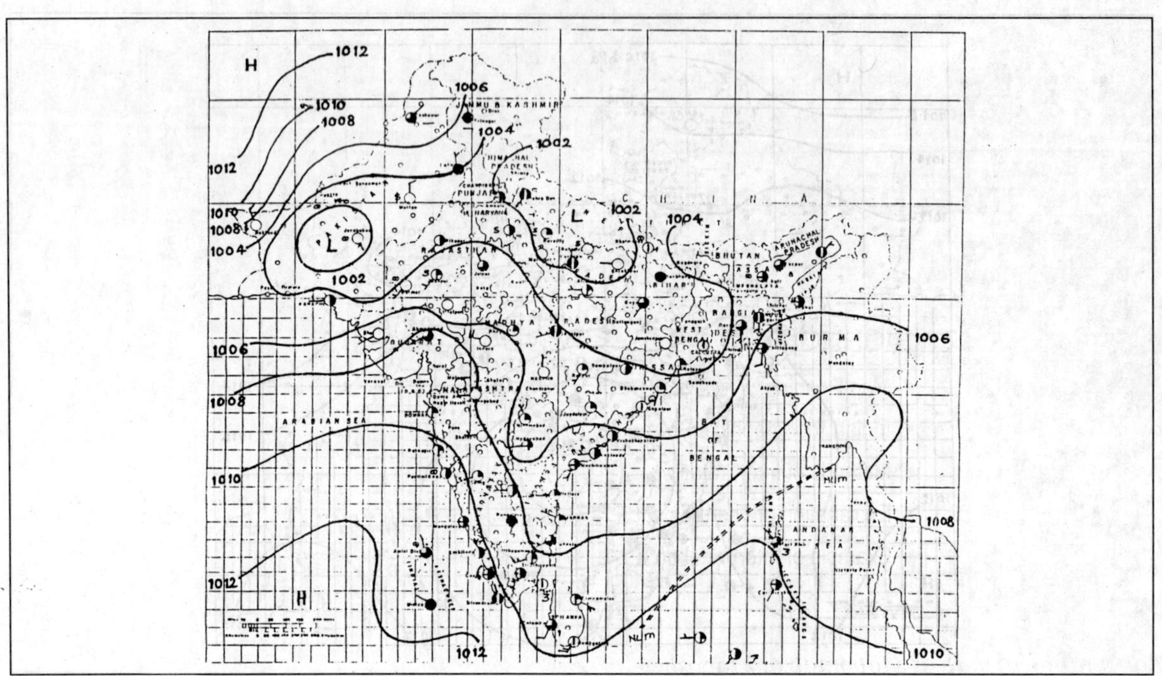
H
1012
1010
1008
1006
1004
1002
L
JAMMU & KASHMIR
PUNJAB
RAJASTHAN
ARABIAN SEA
BHUTAN
ARUNACHAL PRADESH
BURMA
BAY OF BENGAL
ANDAMAN SEA

8

Interpretation of Topographical Map

HIGHLIGHTS

- Map Reading
- Indian Topographical Sheets
- Methods of Interpretation
- Explanation of Features and Patterns
- Example of Interpretation

A *topographical map* or a *topomap* or a *toposheet*, as it is generally called, shows the surface features of the earth as it exists during the period of survey in as much detail as the scale allows by means of *cartographic* and *conventional symbols*. Hence, genetically, it is a *large scale geocoded general reference map* that uses a great variety of *symbolisation*, *generalisation*, *typography* and *colour* to portray in detail both the natural and manmade features for an area, e.g., relief and topography, drainage networks and waterbodies, forests and other vegetation, settlements with functions and utilities, and transport and communication networks. Naturally, it contains a large number of names—both specific (namely hills, rivers, forests, villages, towns, highways and expressways) and descriptive (namely general features, such as, weirs, fords, post offices, mines, etc.).

As it presents the ground in great detail, it is a *storehouse* of all information necessary to understand and comprehend both the physical and human geography of an area. Thus it forms the most important *practical tool* of a geographer. For interpretation, the user always chooses one as per its date of survey, extent of revision and map scale. Currently, information of a particular area acquired from multi-dated topographical maps is being integrated with those from the multi-dated satellite images for detection of changes, its measurement, monitoring, manipulation and modelling.

The ability to interpret a topographical map is fundamental to the understanding of the geography of an area. For this, it is necessary to study the map in detail through three distinct tasks—*first*, identifying the geographical features shown by conventional symbols (point, line, area, letter, colour); *second*, measuring their attributes and spatial patterns; and *third*, comprehending their occurrence and origin. Mapcraft is a special skill by which a mental picture of the ground is constructed from a map and is basic to map reading and interpretation. However, real *mapcraft* lies in the ability to visualise the ground form shown by contours and spot heights.

The geometric pattern of the nature and orientation of the hills, ridges, valleys, watersheds, etc., helps us visualise not only a 3D mental image of the ground but also the underlying geological conditions. Climatic, hydrological and soil conditions can intuitively be identified from the nature and occurrence of valleys and waterbodies, forests and grasslands and human use. Settlement and landuse, along with transport and communication networks, are a manifestation of the geographical systems produced by overlaying a layer of human features on the physical layers. The conventional symbols are indeed almost innumerable and not fully standardised yet. Hence, each map provides detailed marginal information specific to a particular area for the users. Mapcraft is therefore a practical skill that is to be developed by practice only. In the process, experience is acquired and an instinctive feel for maps is developed for comprehending the geography of an area.

MAP READING

While reading a topographical map, the first task is to have a good look at the *information* contained in the *margins* (i.e., the area of paper surrounding the map). This is essential for a full

Fig. 8.1 Part of a Topographical Map

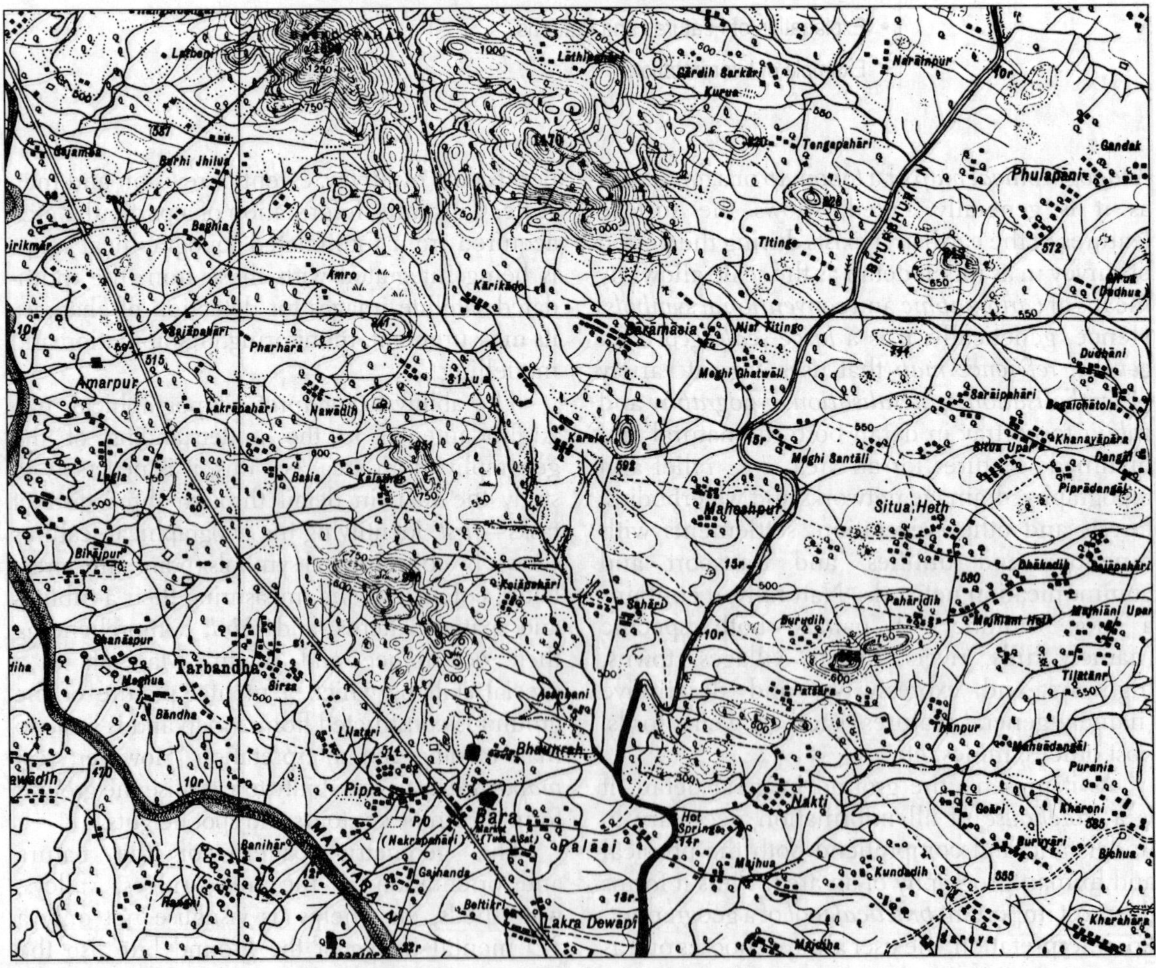

understanding and use of the map and hence deserve more attention than is often paid. The information included is:

On Top Margin

Administrative division(s), year of survey, name of the toposheet, magnetic declination (at a point of time and its rate of increase or decrease) and topographical sheet number.

On Left and Right Margins

Special information about a place (in the form of reference to a particular map).

On Bottom Margin

Symbols (both line, pictorial and letter) especially used in the map to represent lines of transport and communication, spring, mine, well, fort, watch tower, camping ground, deserted village, church, mosque, grave, post and telegraphic office, police station (all in a box format); index to sheets; name of publication directorate; date of publication; date of revision; scale (both in statement, ratio and graphical form); contour interval; notes on symbols and colours used; references to map compilation; administrative index; information about administrative boundaries; surveying stations, units of elevation

Fig. 8.2 Conventional Symbols Used in a Topographical Map

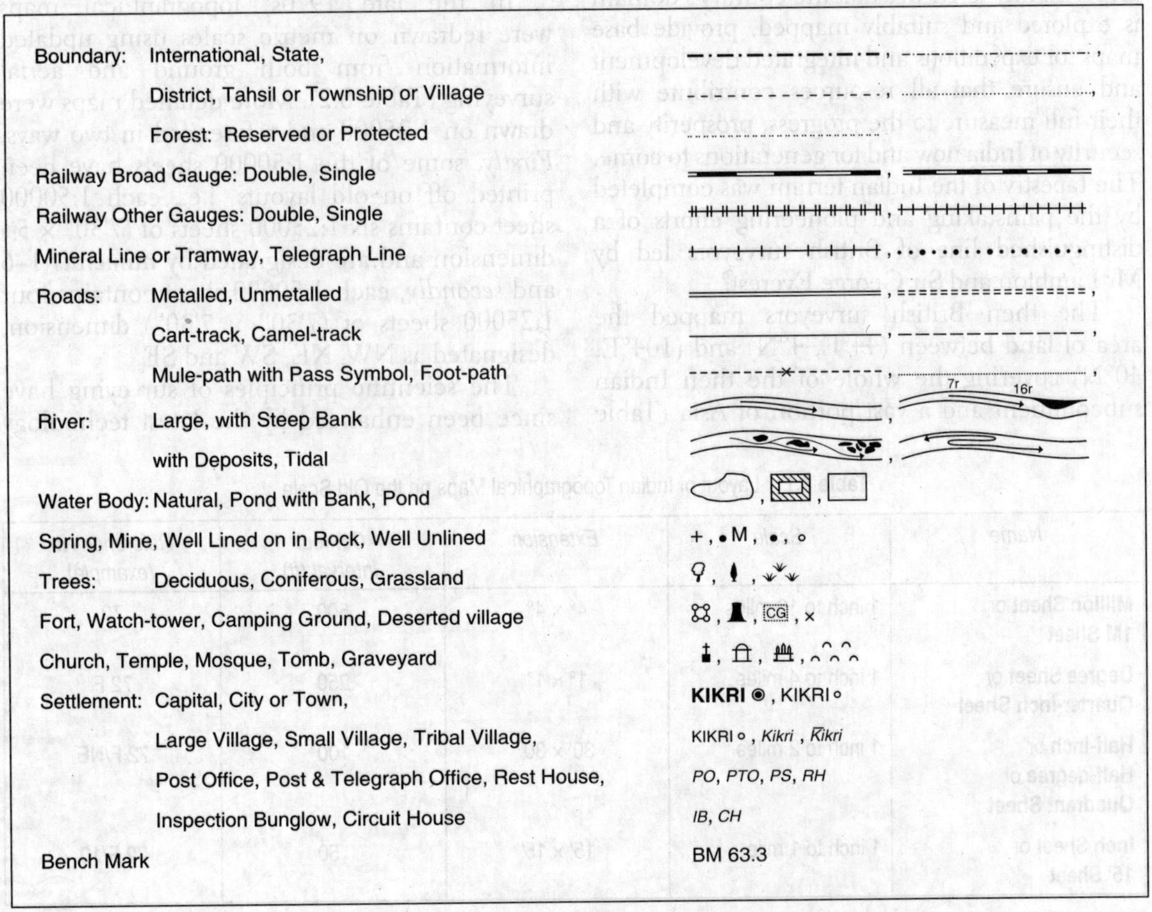

Boundary: International, State,
District, Tahsil or Township or Village
Forest: Reserved or Protected
Railway Broad Gauge: Double, Single
Railway Other Gauges: Double, Single
Mineral Line or Tramway, Telegraph Line
Roads: Metalled, Unmetalled
Cart-track, Camel-track
Mule-path with Pass Symbol, Foot-path
River: Large, with Steep Bank
with Deposits, Tidal
Water Body: Natural, Pond with Bank, Pond
Spring, Mine, Well Lined on in Rock, Well Unlined — + , M , ,
Trees: Deciduous, Coniferous, Grassland
Fort, Watch-tower, Camping Ground, Deserted village — CG , ×
Church, Temple, Mosque, Tomb, Graveyard
Settlement: Capital, City or Town, — **KIKRI** , KIKRI
Large Village, Small Village, Tribal Village, — KIKRI , *Kikri* , *Kikri*
Post Office, Post & Telegraph Office, Rest House, — *PO, PTO, PS, RH*
Inspection Bunglow, Circuit House — *IB, CH*
Bench Mark — BM 63.3

and landmarks (all in a box format); copyright information, and so on (Fig. 8.2).

On each map, the geographical coordinates of the sheet corners are shown in degrees and minutes.

INDIAN TOPOGRAPHICAL SHEETS

The Survey of India (SOI), the National Survey and Mapping Organisation of India under the Department of Science & Technology, is the oldest scientific department of the Government of India. With headquarters at Dehradun, Uttaranchal, it was set up in 1767 and has evolved rich traditions over the years. As the nation's principal mapping agency, it bears a special responsibility to ensure that the country's domain is explored and suitably mapped, provide base maps for expeditions and integrated development and ensure that all resources contribute with their full measure to the progress, prosperity and security of India now and for generations to come. The tapestry of the Indian terrain was completed by the painstaking and pioneering efforts of a distinguished line of British surveyors led by Mr Lambton and Sir George Everest.

The then British surveyors mapped the area of land between (44°E, 4°N) and (104°E, 40°N) covering the whole of the then Indian subcontinent and a vast portion of Asia (Table 8.1). In this layout, the above area was first divided into 106 uniform rectangles of (4° × 4°) dimension and were designated by *numerals*, 1–106 (Fig. 8.3). These were drawn on a 1 inch to 16 miles scale (1:1,000,000) and are known as *million sheets* or 1M Sheets. Each 1M sheet contains 16 *degree sheets* of (1° × 1°) dimension drawn on a 1 inch to 4 miles scale (1:250,000) and are designated by *alphabets* A–P. A degree sheet is then further divided in two ways. *Firstly*, each such sheet contains 4 *quadrant sheets* of (30' × 30') dimension drawn on a 1 inch to 2 miles scale and designated by NW, NE, SW and SE and *secondly*, each degree sheet contains 16 *inch sheets* of (15' × 15') dimension drawn on a 1 inch to a mile scale and are designated by *numerals* 1–16.

In the late 1970s, topographical maps were redrawn on metric scales using updated information from both ground and aerial surveying (Table 8.2). More detailed maps were drawn on 1:25000 and referenced in two ways. *Firstly*, some of the 1:50000 sheets have been printed off on old layouts, i.e., each 1:50000 sheet contains six 1:25000 sheets of (7'30" × 5') dimension and are designated by *numerals* 1–6 and *secondly*, each 1:50000 sheet contains four 1:25000 sheets of (7'30" × 7'30") dimension, designated as NW, NE, SW and SE.

The scientific principles of surveying have since been enhanced by the latest technology

Table 8.1 Layout of Indian Topographical Maps on the Old Scale

Name	*Scale*	*Extension*	*Contour Interval (ft)*	*Reference No. (example)*
Million Sheet or **1M Sheet**	1 inch to 16 miles	4° × 4°	500	72
Degree Sheet or **Quarter-inch Sheet**	1 inch to 4 miles	1° × 1°	250	72 F
Half-inch or **Half-degree** or **Quadrant Sheet**	1 inch to 2 miles	30' × 30'	100	72 F/NE
Inch Sheet or **15' Sheet**	1 inch to 1 mile	15' × 15'	50	72 F/10

Fig. 8.3 Layout of Million Sheets Covering the Indian Subcontinent

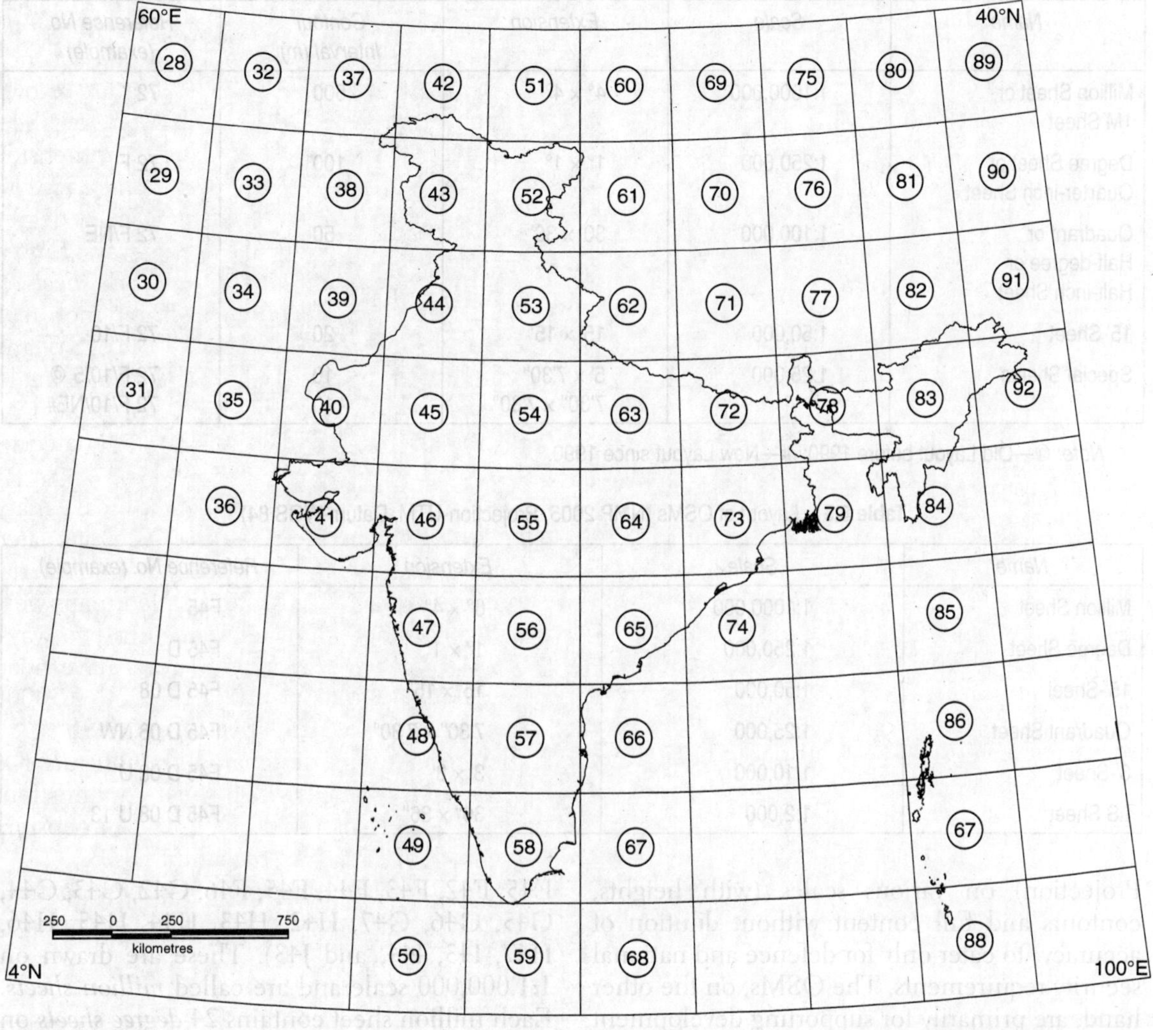

(RS and GIS) to meet the multidisciplinary needs of national security, sustainable national development, and new information markets. The SOI has taken a leadership role in providing the user focused, cost effective, reliable and quality geospatial data, information and intelligence by introducing a new *national map policy* (NMP) on 19 May 2005 that reshapes altogether the long existing frame of mapping layout, scheme of map referencing, scale and dimension of mapping, nature of map product and finally, their mode of access. The SOI promotes an active exchange of information, ideas and technological innovations among the data producers and users across the globe, who get access to such data of the highest possible resolution at an affordable cost in the real-time environment. It will produce two different series of maps, such as, *Defence Series Maps* (DSMs) and *Open Series Maps* (OSMs).

The DSMs are topographical maps (on Everest/ WGS-84 Datum and Polyconic/UTM

Table 8.2 Layout of Indian Topographical Maps on a Metric Scale

Name	*Scale*	*Extension*	*Contour Interval (m)*	*Reference No. (example)*
Million Sheet or 1M Sheet	1:1000,000	4° × 4°	500	72
Degree Sheet or Quarter-inch Sheet	1:250,000	1° × 1°	100	72 F
Quadrant or Half-degree or Half-inch Sheet	1:100,000	30' × 30'	50	72 F/NE
15' Sheet	1:50,000	15' × 15'	20	72 F/10
Special Sheets	1:25,000	5' × 7'30" 7'30" × 7'30"	10 10	72 F/10/5 @ 72 F/10/NE#

Note: @—Old Layout before 1990; #—New Layout since 1990.

Table 8.3 Layout of OSMs (NMP 2005, Projection-UTM, Datum-WGS 84)

Name	*Scale*	*Extension*	*Reference No. (example)*
Million Sheet	1:1000,000	6° × 4°	F45
Degree Sheet	1:250,000	1° × 1°	F45 D
15'-Sheet	1:50,000	15' × 15'	F45 D 08
Quadrant Sheet	1:25,000	7'30" × 7'30"	F45 D 08 NW
3'-Sheet	1:10,000	3' × 3'	F45 D 08 U
LS Sheet	1:2,000	36" × 36"	F45 D 08 U 13

Projection) on various scales (with heights, contours and full content without dilution of accuracy) to cater only for defence and national security requirements. The OSMs, on the other hand, are primarily for supporting development activities in the country. These are drawn on UTM Projection on WGS-84 datum. Each of these OSMs (in both hard copy and digital form) with complete topographical database will become "unrestricted" after obtaining a one-time clearance from the Ministry of Defence. The SOI will ensure that no civil and military vulnerable areas and vulnerable points (VAs/VPs) are shown on the OSMs.

In this layout, India is covered by 32 UTM zones of (6° × 4°) dimension (i.e., B46, C42, C43, C44, C46, D42, D43, D44, D46, E43, E44, E45, F42, F43, F44, F45, F46, G42, G43, G44, G45, G46, G47, H42, H43, H44, H45, H46, H47, I43, I44, and J43). These are drawn on 1:1,000,000 scale and are called *million sheets*. Each million sheet contains 24 *degree sheets* on 1:250,000 scale and are designated by *alphabets*, A–X. Each degree sheet contains 16 sheets of (15' × 15') dimension on 1:50,000 scale and are designated by *numerals* 01–16. A 15' sheet is then divided in *two* ways. *Firstly*, each sheet contains 4 *quadrant sheets* of dimension (7'30" × 7'30") on 1:25,000 scale and are designated as NW, NE, SE and SW. *Secondly*, each sheet contains 25 sheets of dimension (3' × 3') on 1:10000 scale and are designated again by alphabets, A–Y. Each 3'-Sheet contains 25 sheets of (36" × 36") dimension on 1:2000 scale and are designated

Fig. 8.4 Indian Topographical Map: Dimension and Scale

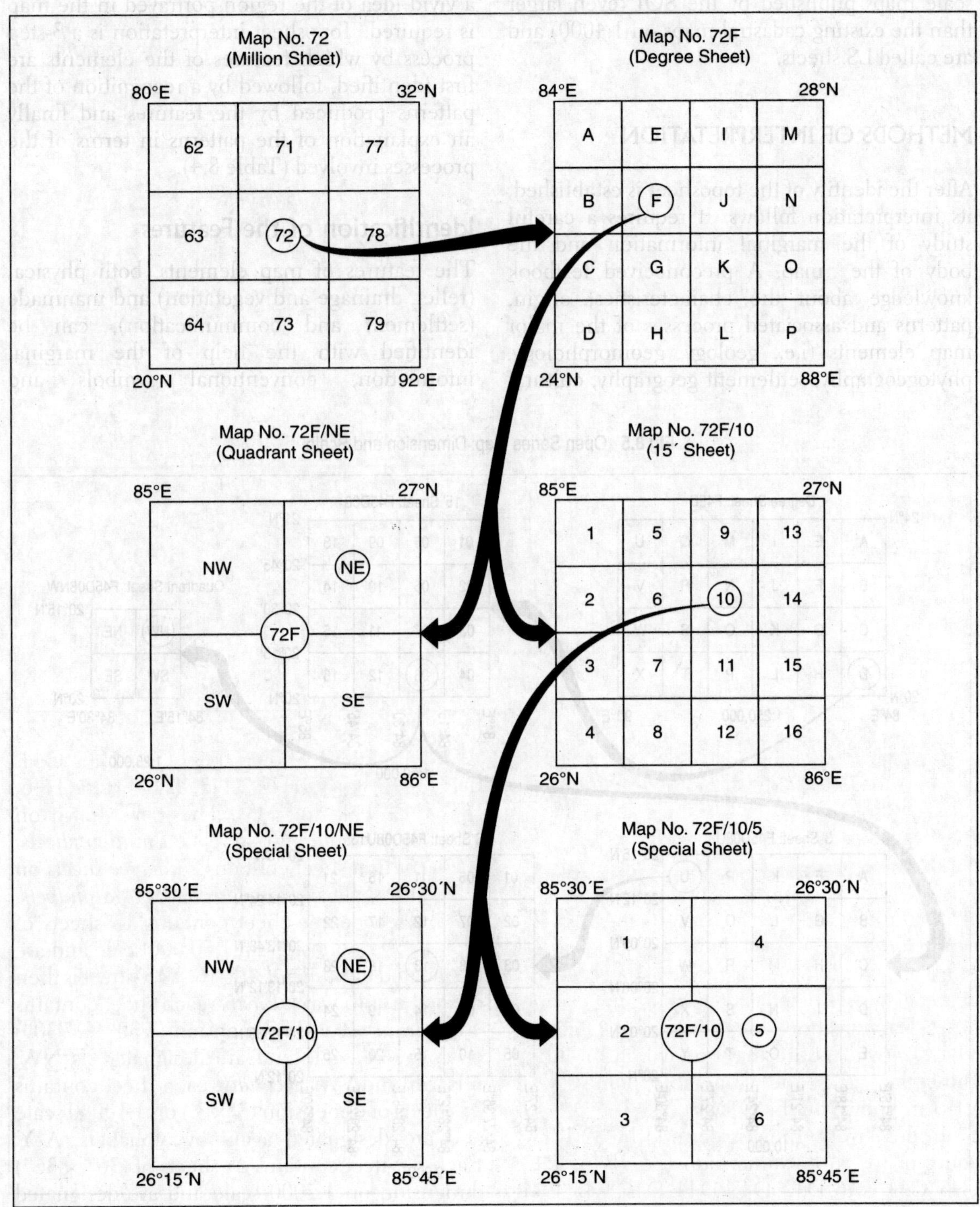

again by numerals, 01–25. These are the largest scale maps published by the SOI (even larger than the existing cadastral maps on 1:4000) and are called LS sheets.

METHODS OF INTERPRETATION

After the identity of the toposheet is established, its interpretation follows. It requires a careful study of the marginal information and the body of the map. A preconceived textbook knowledge about the characteristics, origin, patterns and associated processes of the major map elements (i.e., geology, geomorphology, phytogeography, settlement geography, cultural geography, economic geography, etc.) as well as a vivid idea of the region portrayed in the map is required. Toposheet interpretation is a 3-step process by which features of the elements are first identified, followed by a recognition of the patterns produced by the features and finally an explanation of the patterns in terms of the processes involved (Table 8.4).

Identification of the Features

The features of map elements, both physical (relief, drainage and vegetation) and manmade (settlement and communication), can be identified with the help of the marginal information, conventional symbols and

Fig 8.5 Open Series Map: Dimension and Scale

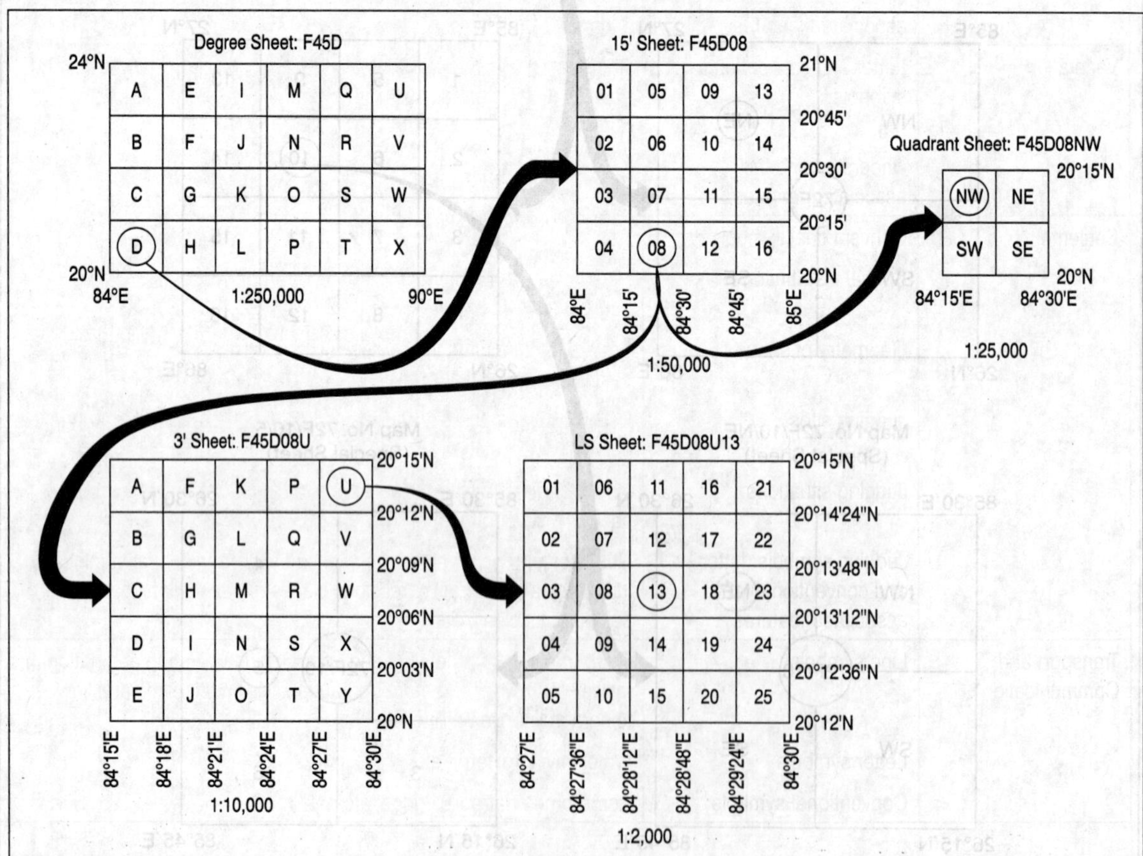

Table 8.4 Identification of Features on Topographical Sheets

Elements	*Symbols/Criteria*	*Features*
Relief	Contour pattern, spacing and values	Hills of various elevation, shape and size; valleys of various shape, size, length and gradient; ridges of various geometric outline; basin; col; saddle; spur; re-entrant; cliff; escarpment; gorge; slope form; water divide; different types of plateaus and plains; etc.
	Conventional symbols	Dune; etc.
	Letter symbols	Rock outcrop; stony waste; sheet rock; etc.
Drainage	Line symbols and associated pattern	Drainage patterns (dendritic, pinnate, parallel, rectangular, radial, annular, angulate, barbed, deranged, contourted, etc); channel patterns (straight, sinuous, meandering, anabranching, braided, etc.); perennial and non-perennial channels; dry bed; canal; abandoned channel; crevasse splay; meander scroll; cut off; chute; etc.
	Conventional symbols	Shoal; island; sand deposit; rocky bed; boulder bed; gully erosion; waterfall; swamp; etc.
	Letter symbols	Weir; barrage; reservoir; ford; ferry; seasonally indundated areas; etc.
	Understanding the associated features	Terrace; river capture; meander core; paleo channel; Alluvial fan; hydrological aspects; etc.
Vegetation	Letter and conventional symbols	Reserve forest; protected forest; scrubland; grassland; jungle type; species type; species density; plantations; orchards and groves; horticulture; silviculture; etc.
	Understanding the associated features	Afforestation; deforestation; etc.
Settlement	Conventional symbols	Deserted village; fort; walled settlement; tribal settlement; etc.
	Spatial distribution	Hamlet; isolated pattern; dispersed pattern; clustered pattern; random pattern
	Geometry of shape	Square; rectangular; circular; semi-circular; linear; A-, L-, T-, Y- shaped settlement; amorphous; etc.
	Judging sites	River bank; wet point; dry point; piedmont; confluence; forest fringe; gap; summit; levee; etc.
	Judging situations	Nodal; agglomerated; nucleated; station; port; cantonment; forest chowki; break-in-transport point; etc.
	Judging symbols (letter and conventional) and associated features	Rural settlement; urban centre; planned town; agricultural, mining, industrial, administrative, religious places or settlements; new settlement; colony; etc.
Transport and Communication	Line symbols	Road (metalled, unmetalled and village); highway (state and national); railway (narrow, broad or metre gauge, division, etc.); mineral line; telegraph line; power line; etc.
	Letter symbols	Waterway; ford; ferry; etc.
	Conventional symbols	Aerodrome or airport; bridge; etc.

Topographical Maps

- Topographical Maps, called Topomaps or Toposheets, are analogous or equivalent to the Ordnance Maps of Great Britain and Europe and the USGS maps in America.
- The British Surveyors prepared 106 Million Sheets in total covering the then Indian Sub-continent and the adjacent countries of Iran, Afghanistan, Tibet and parts of China and South-east Asia.
- Million Sheets (1:1,000,000) are drawn on Lambert's Conical Orthomorphic Projection with Everest Datum.
- Toposheet No. 1 on 1:1,000,000 corresponds to Tabriz in Iran.
- India is covered by 39 Million Sheets: 39– 49, 51–58, 61–66, 72–74, 77–79, 82–84, 86–88, 91 and 92.
- The Inch Sheets (1:63,360) or the 15' Sheets (1:50,000) are drawn on Modified Polyconic or International Projections.
- The estimated number of Million Sheets covering the whole globe is 2222
  a) between 60°N and 60°S of (4° × 6°)
  dimension = 2 × (60°/4°) × (360°/6°)
  = 1800
  b) between 60°N/S and 88°N/S of (4° × 12°)
  dimension = 2 × (28°/4°) × (360°/12°)
  = 420
  c) between 88°N/S and 90°N/S
  covering polar regions = 2 × 1 ⇒ 2
- With advancements in science and technology, Indian topographical sheets have been updated and redrawn with more details on larger scales of various layouts from time to time.
- The open series maps (OSM) on digital formats with a new layout are now available on various scales with fully digitised datalayers since the introduction of the new National Map Policy (NMP) on 19 May 2005.

abbreviations (published by SOI) and a knowledge of contour geometry. There can be no end of the possible list of features and each sheet contains some special features characteristic of that particular region only. It is this *set of special features* that needs primary emphasis in interpretation because these build up the personality of the region. A list of some features commonly found in plateaus and plains of India is given below for the user's facility (Table 8.4).

Recognition of Patterns

It involves *measurement* and *manipulation* of information from maps and their *representation* by means of profiles, graphs, transect charts, plan overlays and thematic maps, The derived representation is of immense value in justifying the spatial analysis of the various characteristics of the elements. For the purpose of thematic mapping the toposheet needs to be divided into grids of practicable sizes depending on map scale and degree of generalisation desired by the users. Data collected for each cell is then represented by means of either isopleths (in which case data is plotted at the cell centre) or choropleths (graded shading based on the cell

Interpretation of a Topographical Map

Identification of Features

- Choosing a data layer
- Identifying the features of the data layer using information given in the margins
- Mapping/Extracting the data layers using map craft and instinctive feel

Recognition of Spatial Patterns

- Qualitative analysis of the themes: points (settlement, etc.), lines (river network, transport network, etc.) and areas (landform, slope zone, forest, etc.)
- Quantitative analysis of the themes: points (e.g., settlement, etc.), lines (e.g., river network, transport network, etc.) and areas (e.g., landform, slope zone, forest, etc.)

Interpretation

- Transect analysis
- Bivariate analysis
- Overlay analysis
- Regionalisation using map algebra

boundary). The plan overlays of typical features of an element may be traced out or neatly redrawn. The contour sections and different graphs are prepared with relevant data generated from the map. Similarly, 2-dimensional sections of two or three map elements (drawn along a line) may be superimposed to produce the transect charts. Table 8.4 summarises the various forms of representations and derivatives of elements with clues for interpretations. The set of technical terms and indices are described in brief.

Contour Profile

A *contour profile* is an outline produced by the plane of a section as intersecting the ground surface. It is obtained by interpolation of contour elevations on a chosen scale and along a chosen line on a map transverse to the grain of the region. Naturally, it depicts the ground form in a *most perceivable* and *precise format*. A series of profiles at regular intervals of time or space are usually drawn to understand the changes in topography. When these profiles are arranged and drawn on a single frame, a *superimposed* profile is formed. This helps us to identify breaks-in-slopes and planation surfaces. The highest parts of the superimposed profiles, as viewed in the horizontal planes of the summit levels from an infinite distance, constitute the *composite* profile. It represents the skyline with a defined perspective. To extract the panoramic effect of the landscape, *projected* profiles are drawn. This is obtained by drawing profiles sequentially, one after the other, starting from the foreground to the distant part and successively erasing the part obscured by the former profile.

Gradient and Slope

Both are measures of *landform geometry* as viewed along a line. Gradient (g) is expressed in the form of a ratio while slope (q) as an angular value. If the ground distance between two points on the map, called the horizontal equivalent, is d and the elevation difference between them, called the vertical interval, is r (both expressed in the same unit), the computation is done by:

$$\textit{Gradient}, \; g \; = 1 : \left(\frac{d}{r}\right)$$

$$\text{and } \textit{Slope}, \; \theta \; = \tan^{-1}\left(\frac{r}{d}\right)$$

Morphometric Indices

Morphometric indices have been derived by the geomorphologists to identify the geometric properties of the land surface. The most common ones are:

- Relative Relief (RR) = $H_x - H_n$ (Smith 1935)
  where, H_x = highest altitude, H_n = lowest altitude
- Dissection Index (DI) = $\dfrac{RR}{H_x}$

  (Dov Nir/Miller 1949)
- Average Slope, $\theta = \tan^{-1}\left(\dfrac{N \times i}{K}\right)$

  (Wentworth 1930)

  where,

  N = number of contour crossings per mile or per kilometre

  i = contour interval

  K = a constant [= 3361 for the mile-grid and 636.6 for the kilometre-grid].
- Ruggedness Number (Rn) = $\dfrac{RR \times Dd}{K}$

  where,

  K = a conversion constant (= 5280 in case of mile-grid when relative relief is expressed in feet and drainage density in miles / sq mile and is 1000 when relative relief is expressed in metres and drainage density in kilometres/square kilometres).

  Dd = drainage density

Table 8.5 Representation of Data Derived from a Topographical Sheet

Elements	*Forms of Representation*	*Derivatives*	*Remarks*
Relief	Profiles	Representative contour profile; superimposed, composite and projected profiles	Used to identify breaks-in-slopes, summit levels, general nature of dissection, etc.
	Graphs	Altimetric frequency curve; frequency graphs of relative relief, dissection index, average slope, etc.; hypsometric curve; etc.	Used to understand the nature of statistical distribution of relief aspect
	Plan overlays	Contour plans of different landform features as identified	Used to identify the features of landform
	Thematic maps	Absolute relief; broad physical divisions; relative relief; dissection index; average slope (Wentworth's); slope zone map (Smith's or Raisz and Henry's)	Used to delineate the spatial properties of relief aspects
Drainage	Profiles	Long and cross profiles of streams; valley forms; etc.	Used to understand fluvial morphology
	Graphs	Frequency graphs of stream frequency, drainage density, constant of channel maintenance, ruggedness number, etc.	Used to explore statistical properties of the constant of channel drainage indices
		Order-number; order-mean length; order-mean basin area; order-mean slope graphs	Used to find out structural properties of the drainage network
		Scatter graphs with relevant pairs of relief and drainage parameters, etc.	Used to test the relation between relief and drainage
	Plan overlays	Different drainage and channel patterns; erosional and depositional features, etc.	Used to understand the geologic and morphometric controls of drainage and prevalent actions of the rivers
	Thematic maps	Drainage basins, divides and networks; stream frequency; drainage density; constant of channel maintenance; ruggedness number; etc.	Used to delineate spatial properties of drainage aspects
	Transect charts	Landform and drainage frequency; Slope and stream dimension (in relation to hydrology, etc.)	Used to identify the interrelation between relief and drainage
Vegetation	Graphs	Frequency graphs of the vegetal cover (%)	Used to understand the statistical properties of spatial distribution
		Scatter graph of the vegetal cover and absolute altitude or relative relief; etc.	Used to test the relationship between relief and vegetation
	Plan Overlays	Species type; species density; plantation; orchards and groves; afforestation; deforestation; badlands; etc.	Used to identify the environmental set-up
	Thematic maps	Vegetal cover; forest cover (reserved and protected); etc.	Used to delineate the spatial pattern of distribution

(Contd on next page)

(Contd from previous page)

Elements	*Forms of Representation*	*Derivatives*	*Remarks*
	Transect charts	Altitude and vegetal cover and species type or species density; etc.	Used to identify the interrelation between relief and vegetation
Communication	Graphs	Network graphs	Used to understand the structural properties of the network
		Scatter graphs of road density and relative relief; road density and forest cover; road density and settlement frequency	Used to identify the relation between transport and relief, transport and forest cover, transport and settlement
	Thematic maps	Connectivity indices	Used to show the distribution of transport connectivity
	Transect charts	Frequency of communication line and altitude or relative relief or forest cover or settlement	Used to describe the relation between transport and communication development frequency, etc. and other parameters
Settlement	Graphs	Distribution of settlement frequency	Used to measure the statistical properties of settlement distribution
		Scatter graphs of settlement frequency and absolute altitude/relative relief/vegetal cover/ frequency of communication lines/connectivity values	Used to identify the relation between settlement frequency and other physical and human features
	Plan overlays	Settlement patterns; sites of settlement; situations of settlement; urban morphology; etc.	Used to identify spatial patterns and settlement morphology in terms of site and situation
	Thematic maps	Settlement frequency; dot map of distribution; etc.	Used to show the spatial distribution of settlement
	Transect charts	Settlement and relief/forest cover/ drainage features/communication	Used to describe the relation between settlement and other physical and man-made features

Slope Forms

Rectilinear slope (contour spacing is regular and uniform on moderate gradient), convex slope (contour spacing is moderate to large towards higher values which rapidly decreases towards lower elevation), concave slope (contour spacing is small to very small towards higher values which rapidly increases towards lower values), cliff (a zone where contours overlap) and escarpment (a zone distinguished by uniformly regular and close contour spacing).

Scales of Slope

A line-scale of slope is a useful graphical device for understanding the surface variation of slope. On a map with scale, 1:63360 (contour interval = 50ft), for a slope of 1 in 20, an horizontal equivalent of 1 inch on map

$$= 1 \text{ mile}/20$$
$$= 264 \text{ ft}$$

Thus a vertical interval of 264 ft is represented by 1 inch on a map. Therefore, a vertical interval of 250 ft will be represented by

= 250/264 inch
= 0.947 inch

A horizontal line (AB) of any convenient length is drawn and divided into 20 equal parts. A perpendicular BC (= 0.947 inch) is dropped on AB. AC is joined by a straight line. From each point on AB perpendiculars are drawn on AB. The lengths of these perpendiculars are then taken as a scale of horizontal equivalents. Slopes between 1 in 1 and 1 in 20 are read from the baseline AB and those between 1 in 20 and 1 in 100 are read from the upper line AC using a contour interval of 50 ft (Fig. 8.5).

Methods of Slope Analysis

Smith (1935) used relative relief as a measure of slope. It proved good only on a maturely dissected plateau of horizontal sedimentary rock structure with uniform slopes and a simple physiographic history (Raisz and Henry 1937). A measure of average slope by Wentworth (1930) is laborious. Raisz and Henry suggested a simple technique in which the map is divided into small regions of uniform contour spacings. For an inch sheet with 50 ft contour interval, they chose six categories of slope: under 50 ft/mile, 50–100 ft/mile, 100–200/mile, 200–300 ft/mile, 300–400 ft/mile, 400–500 ft/mile and above 500 ft/mile. A horizontal scale of standard contour-spacings is drawn first so that the number of contours/mile on the map scale can be checked with a divider and the slope category carefully ascertained. Robinson (1948), on the other hand, used a completely different technique to produce a quantitatively accurate relief map from areal slope data. He divided the map into a network of squares of 0.01 square mile in dimension. The average slope of each square was then evaluated and a quantitative dot map prepared with a scale, 1 dot = 1 degree. The dots were placed with reference to the contours on the map and to the dots in the adjacent squares to produce some appearance of continuity. Miller (1960) adopted the concept of Wood (1942). He examined various mathematical functions of slope and divided the map into slope zones with four-fold slope elements as: *waxing slope* (0°–3°35'), *free face* (3°35'–14°24'), *constant slope* (14°24'–34°14') and *waning slope* (34°14'–90°).

Fig. 8.6 The Gradient of Slope

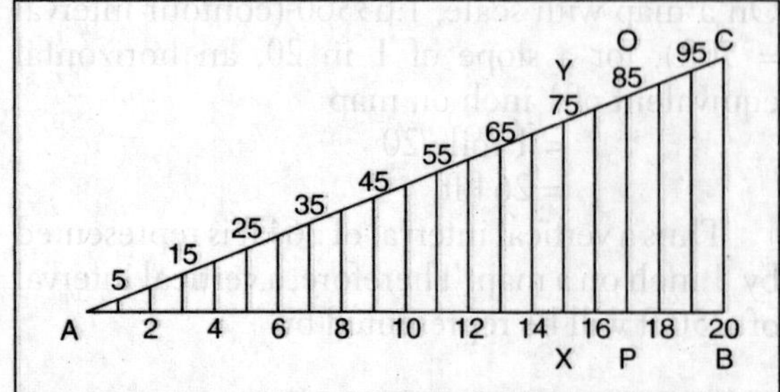

Valleys and Spurs

Valleys usually run along the line joining the Vs of the contour and the Vs point upstream, i.e., towards higher elevation. The segment of arcuate contours between two valleys with tributary relation form a spur.

Hills and Ridges

Hills are marked by a pack of closed contours such that values increase towards the centre. The shape of the contours define the geometry of the hills. The summit may not always be a point or a small plain but the summit area should necessarily be far smaller than the base area. A ridge is formed when a large number of hills occur roughly along a line for a considerable distance. Ridges normally correspond to the regional water divides. On a dissected ridge, *cols* and *saddles* are the typical features.

Breaks-in-Slope

These are specific values of contours that mark a sharp contrast in contour spacing on two sides. On a slope these are marked by points separating two contrasting slope forms.

Accordant Summit Level

Summits of the hills (separated by river valleys and plains) occurring roughly at the same elevation may be taken as geomorphologically accordant.

Altimetric Frequency Curve

This refers to the frequency distribution of spot heights and hill summits. Nature of statistical distribution as can be judged from the shape of the curve, its *skewness*, its *kurtosis*, and number of *modes* form the basis for identifying the existence of plantation surfaces at different altitudinal levels.

Hypsometric Curve

A hypsometric or hypsographic curve is used to show the proportion of the area of the surface at various elevations above or below a given datum. It should be drawn for an area having a physical unity. Here the area between each pair of contours (a) is expressed as a percentage of the total area of the region (A). Similarly, the altitudinal variation (h) between the corresponding pair of contours is expressed as a percentage of the total relative relief (H) of the region. The hypsographic curve is obtained by plotting the (h/H)% on y-axis with respect to (a/A)% on x-axis. The shape and orientation of this curve graphically indicates the stage of the ensuing erosion cycle and the number of cyles already elapsed.

Drainage Basin

It describes the area drained collectively by the network of a river along with its tributaries and subtributaries of various dimensions. It is demarcated by drawing a line along the network boundary or watershed. Its shape reflects the nature of the basin hydrology and can be measured by using the formula:

- Form Factor (F_f) $= A/L^2$
- Circularity Ratio (R_c) $= 4\pi A/P^2$
- Elongation Ratio (R_e) $= P/\pi L$
- Lamniscate Ratio (k) $= \pi L^2/4A$

where, A = basin area, L = maximum basin length, P = basin perimeter

Fluviometric Indices

The most commonly applied indices measuring the fluviometric properties are:

- Drainage Frequency (Df) = number of channels/area
- Drainage Density (Dd) = length of channels/area
- Constant of Channel Maintenance (CCM) $= 1/Dd$
- Drainage Texture = number of contour crenulations/ perimeter of the study unit
- Length of Overland Flow (L_{of}) $= 1/2Dd$

Gully and Bank Erosion

Gully erosion is marked by conventional symbols [⇔⇔] of enclosing and linear patterns of second brackets along the centre of which gullies may be imagined. Bank erosion is marked by open second brackets along the bank of a large shifting channel.

Stream Profiles

The longitudinal profiles of rivers are drawn by plotting the bed elevation (plotted on the y-axis) with the distance from the source (plotted on the x-axis). Cross profiles are similarly drawn across a line transverse to the river. Both profiles may be superimposed for better geomorphic appreciation.

Drainage Patterns

The geometric patterns formed collectively by the drainage lines may be identified from geometric relations as *tree like* (dendritic), *fishbone* (fish bone), *parallel* (parallel), *rectangular* (rectangular), *annular* (circular), *angulate* (rhombic), *radiating* (radial), *converging* (centripetal), *boat-hook* (barbed), *poorly connected* (deranged), *complex* (contorted), etc.

Channel Patterns

Anabranching patterns are formed when the main channel forms an anabranch, i.e., a channel emanating from the main and uniting again after some distance. Braided patterns are marked by wide and shallow channels full of islands or braids and the associated interlace of waterways or mesh of union, deunion and reunion of the main flow. Channel sinuosity index, CSI (the ratio of the channel distance between two points and the straight line distance between the same two points) is a measure of identifying the patterns.

Pattern	Straight	Irregular	Sinuous	Meandering
CSI	1.0	1.0 – 1.3	1.3 – 1.5	>1.5

Obtuse Angular Relation

This is marked by an obtuse angular confluence between two channels in tributary junctions. It represents aberrations in confluence relations denoting a typical fluviological manifestation.

Dry Valley

It is marked by dashed lines (– –) oriented in the form of drainage lines.

Structural Properties of Drainage Network

The properties of network development in a basin may be assessed by evaluating the descriptive measures of statistical distribution of a set of well defined network indices:

- Bifurcation ratio, $R_b = N_u/N_{u+i}$
- Stream length ratio, $R_l = L_u/L_{u+i}$
- Area ratio, $R_a = A_u/A_{u+i}$
- Slope ratio, $R_s = S_u/S_{u+i}$
- Relief ratio, $R_h = H_u/H_{u+i}$

where N_u, L_u, A_u, S_u, H_u, = number, mean length, mean basin area, mean slope and mean basin relief of streams or basins of order, u; the suffix (u+1) denotes the same for next higher order.

The structural regularities of network evolution may also be judged by a verification of the laws of drainage composition:

- Law of Stream Numbers, $N_u = R_b^{s-u}$
- Law of Stream Lengths, $L_u = L_1 R_b^{u-1}$
- Law of Basin Areas, $A_u = A_1 R_a^{u-1}$
- Law of Stream Gradient, $S_u = S_1 R_s^{s-u}$
- Law of Basin Relief, $H_u = H_1 R_H^{u-1}$

where, s = highest order of streams in the basin and L_1, A_1, S_1 and H_1 are empirically determined constants in *geometric* regressions.

Settlement Frequency

It is a measure of settlement density viewed as a point pattern on a plane and is expressed as the number of settlements per unit area.

Settlement Pattern

The spatial pattern of settlements may be delineated with the help of the *nearest neighbour* statistic, R_n. The patterns and the R_n values are:

$$R_n = \frac{\bar{r}_a}{\bar{r}_c}$$

where,

$\bar{r}_a$ = observed mean NND, $\frac{\sum_{i=1}^{n} r_i}{n}$

$\bar{r}_a$ = expected mean NND, $\frac{1}{2}\sqrt{\frac{A}{n}}$

n = number of settlements,

A = area of the study unit,

r_i = distance of ith point to its nearest neighbours.

Pattern	Isolated	Dispersed	Uniform	Clustered
Rn	Very large	>1.0	1.0	<1.0

Settlement Morphology

This indicates the set of features intrinsically related to settlement, namely, size, site, situation and functions. Each of these morphologic elements may easily be identified from a map.

Transport Frequency and Density

Transport frequency refers to the number of transport lines as measured per unit area. Transport density is the length of transport lines as measured per unit area. Both are crude measures of transport development.

Accessibility

This is a measure of the ease of access to a particular place with respect to another place or object or transport lines depending on physical factors. Accessibility is often shown as zones of isopleth of a particular distance from a transport link.

Structure of Transport Network

The connectivity of transport network is usually measured in terms of nodes (n), arcs (a), and subgraphs (p). The common indices are cyclomatic number (μ), alpha index (α), beta index (β) and gamma index (γ). The formulas for calculation are:

$\mu = (a - n + p)$

$\alpha = (a - n + p)/(2n - 5)$

$\beta = \frac{a}{n}$

$\gamma = 2a/n(n - 1)$ for non-planar graphs and

$\gamma = a/3(n - 2)$ for planar graphs

Detour Index

The magnitude of detour one has to make to travel between two different places defines detour index (dT). It is found by dividing the actual distance (AD) along the detour path by the straightline distance (SD) between the places and is expressed as percentage.

Thus, $dT = (AD / SD) \times 100\ \%$ (1)

Again, as a measure of network efficiency, detour index is computed as follows–

$dT = (SD / AD) \times 100\%$ (2)

Example

For, AD = 60 km and SD = 50 km, dT = 120% (Eq. 1) and = 30% (Eqn. 2). This is interpreted in the following two ways–

a) one has to make 120% detour to travel between the two places, and
b) the efficiency of the transport network between the above two places is only 83%.

Transect Chart

The physical and human features distributed along a line on a map are geographically correlated and shown by means of a transect chart constructed by serially drawing the transects of two or more map elements and placing them one above the other in the respective order of relief, drainage, vegetation, settlement and communication. Relief is shown by means of a contour profile with delineations of relief units. Drainage and communication lines are shown as vertical straight lines following the exact map symbols. Vegetation and settlements are shown as rectangular blocks and their boundaries are delineated on the basis of their areal patterns as observed.

Plan Overlays

These are the exact replica of the features of a map drawn on a tracing paper.

EXPLANATION OF FEATURES AND PATTERNS

Once the features and patterns of the elements have been identified and recognised, the act of explanation begins. This is not an easy task. The interpreter must be skilled in mapcraft. He must be well versed with the geographical matrix of the region, especially the underlying lithological composition, structure of the rock beds, climatic regimes, soil characteristics, general economy, people and the local environmental or natural hazards. These would be of immense help in explaining the features and patterns of the map elements.

To any researcher in our country, the toposheet is still the only cheap and reliable source of information on a large scale scenario. Thus, this information system is very important in the collection of data (measurement and manipulation), and in the processing of data (database management, analysis and representation of derived features by graphs, diagrams and maps) to be used in the planning and decision making processes. The most recently edited map is always desirable. However, a series of maps surveyed and edited at different points of time are very useful to identify the changes in landscape in terms of the patterns and processes operating in varied scales and dimensions.

To beginners, the most important thing is the organisation of the information derived from a map while explaining the features in association. The approach must be scientific with deductions following the principles of geographical logic. As an aid to this, an orderly sequence of broad heads with specific contents to be discussed for each map element is shown below.

Relief

Characteristic Features

These include the general physiographic pattern, the proportion of land under mountains or hills, plateaus and plains, and their relative locations, the direction and magnitude of regional slope, orientation of major divides, maximum altitude, minimum altitude, the regional relief and gross dissection index, the general nature of topography in terms of stages of evolution, etc.

Landform Features

Dimensions and the probable origins of features like flat topped hills, low residual isolated hills, conical hills, rectilinear, waxing and waning slopes, ridges, cols, saddles, spurs and valleys, salients and re-entrants, cliffs and escarpments, rock out-crops, boulder deposits, stony wastes, sheet rocks, sand deposits, terraces, breaks-in-slopes, etc.

Morphometric Divisions

Broad physical divisions (based on breaks-in-slopes), relief units (based on relative relief), divisions based on dissection index and average slope along with their geomorphic evolution; interpretation of measures of statistical distribution of morphometric elements; etc.

Special Features

Accordant summit levels, bad lands, lithological and climatic controls on weathering and erosion, interruptions in erosion cycles, etc.

Drainage

Characteristic Features

Degree of drainage development, major rivers (direction and characteristics of flow with major left and right bank tributaries) and drainage basins, etc.

Fluvial Features

Valley forms, long profiles, channel patterns (straight, irregular, sinuous, meandering, braided, anabranching, anastomosing and reticulate); flood plains with associated features (meander neck, cut-off, meander scroll, back swamp, levee, spill channels, chutes, etc.); waterfall, channel deposits (fine grained or coarse grained), shoals, islands, bars (longitudinal and transverse), boulder deposits, dry bed, wind gap, terrace, channel incision, gully erosion, rapids or cascades or cataracts, etc.

Drainage Patterns

Patterns formed collectively by drainage lines (dendritic, fishbone, pinnate, parallel, radial, centripetal, rectangular, barbed, annular, angulate, deranged, contorted, etc.) as well as their morphometric and structural controls are to be analysed.

Fluviometric Properties

Spatial distribution of drainage frequency, drainage density, ruggedness index, length of overland flow, constant of channel maintenance, etc. are to be analysed along with the descriptive measures of statistical distribution.

Special Features

Distribution, perenniality and dimensions of other water features (ponds, lakes, reservoirs and canals); structural properties of drainage network (order-number, order-mean length, order-mean area, order-slope, etc.); relationships; obtuse-angular confluential relation; defunct channels; dendritic gully; bank erosion; dams and barrages; etc.

Physical Relations

Drainage pattern and relief; slope and drainage; relative relief and drainage; drainage and underlying geology; etc.

Vegetation

Characteristic Features

Forest cover (as percentage of total area, nature of forest, dominant species, average density, etc.); environmental factors of vegetative growth and distribution.

Species Type

Distribution in terms of local factors.

Areal Pattern

Distribution (spatial and statistical) and density pattern in terms of reserved and protected forests and jungle types in different physical units.

Special Features

Comments to be made on plantation, orchards and groves; afforestation (social forestry); deforestation (human interference); badlands (loss of land and soil); forest fire (fire line); forest-based economy of the immediate localities (if any); swamps and marshes (water logging); grasslands (pasture, pulp making); etc.

Physical and Human Relation

Relationships between forest (type and density) and terrain character (slope, altitude, relative relief, site) as well as settlement (distribution and expansion) and transport development.

Settlements

Characteristic Features

Settlement coverage (proportion of total area, nature of settlement, average density, etc.); general factors of settlement growth and distribution.

Areal Pattern

Patterns (clustered, uniform, dispersed and isolated) as per the nearest-neighbour parameters are to be analysed in terms of geographical factors.

Morphological Type

Size and shape (square, rectangle, circular, semi-circular, linear, amorphous, A-, L-, T-, Y- shape) to be explained in terms of site and situation.

Sites and Situations

Location components of the map settlements (both rural and urban) are to be discussed separately. Wet point, dry point, river bank, levee, river valley, piedmont, terrace, wind gap, summit plain, river confluence, estuary, forest fringe, etc. (site elements) nodal, nucleated, agglomerations, stations, ports, forts, fords, forest chowkis, mines, power generation points, breaks-in-points, markets, etc. (elements of situation).

Special Features

Deserted village, walled village, tribal localities, social segregation, urban landuse, hill stations, hot springs, recreation centres, new settlements, colonies, etc. are to be discussed and interpreted.

Environmental Relations

Relationships between settlement (site, situation, frequency, density and type) and altitude, relief patterns, slopes, forests, fluvial aspects and transport development.

Transport and Communication

Characteristic Features

Types (roads: metalled and unmetalled; railways: broad gauge and metre gauge; waterways: seasonal or round the year) the general degree of development, and the environmental factors affecting the development of transport networks.

Areal Pattern

Development of transport in terms of environmental constraints in different physical units; geographical interpretation of network connectivity.

Special Features

Mineral lines, telecommunication lines, power lines, intricate mesh, grid patterns, nautilus, highways, expressways, air connection, ropeways, fire lines, etc.

Physical and Human Relations

Relationships between transport and communication (type, frequency and connectivity) and relief pattern, altitude, slope, vegetation, drainage frequency and settlement (type and frequency).

EXAMPLE OF INTERPRETATION

Identity of the Toposheet

The Topographical Map No. 73C/15, first edition, is a gridded *inch sheet* drawn for a dimension of 15'×15' (latitudinal and longitudinal) on scale 1:63,360 (i.e., one inch to a mile) with a contour interval of 50 ft. The area is defined by the parallels, 21°15'N and 21°30'N and the meridians, 84°45'E and 85°00'E. It was surveyed in 1932–33 and the map was published under the direction of the then Surveyor General of India in 1933. Mean Grid North is 1°35' East of True North and magnetic variation from True North is about 1°30' West in 1946 (annual change negligible). The area covered by the sheet is about 278 sq miles including mainly the districts of Bumra and parts of the districts of Pal Lahara, Rairakhol and Talcher of Orissa, India.

Geographically, the region appears to be covered with dense forests. The northwest to southeast diagonal roughly divides the region into two distinct topographic units. The northeastern one is a rolling plain with low relative relief through which the Brahmani river and its tributaries sluggishly flow south-eastward and southward while the southwestern portion represents a dissected plateau landscape with well-formed ridges and valleys. The settlements and communication are both poorly developed and hence dispersed rural settlements predominate with very low frequency and poor connectivity (Fig. 8.21).

Relief

Physiographically, the region is a highly dissected plateau terrain drained by Brahmani river and its tributaries. The line formed by the rivers Jaraikela-Gohira-Brahmani practically divides the region into two distinct topographic units. On the northeastern part lies a terrain with low absolute altitude (819 ft), low relative relief (maximum of 519 ft), southward regional slope, less frequent patchy contours and some isolated low residual hills. Drained by the Jaraikela, Balam, Motuali, Churakhai, Usthali, Brahmani and a number of other tributaries, this forms the erosional rolling plains and flood plains. On the southwestern part lies the proper plateau terrain with high absolute altitude (2,095 ft), very high relative relief (maximum of 1,795 ft), dissection index of 86%, average slope of 10%–15%, south-eastward regional slope, dense and patchy contours and rugged landscape. There are at least five watershed units lying parallely in northwest-southeast fashion in the form of dissected ridges separated by the valleys of Hinjali, Chilanti, Choardhara, Garda and Gohira rivers. Sinduria peak (L463962), in the main central watershed, forms the highest point (2,095 ft) of the region, while the 300 ft contour near the exit of the Brahmani river at L670860 forms the lowest point (Fig. 8.8 and Table 8.5).

The distribution of absolute altitude is uneven. The contour sections drawn across the watersheds and superimposed on the same frame show accordant summits at two different levels of approximately 750 ft and 1,500 ft.

Table 8.6 Morphometric Attributes of the Broad Physiographic Units

Physiographic Units / *Morpho-metric Parameters*	*Plateau Proper*	*Plateau Rim*	*Erosional Plain*
Occurrence	In a narrow strip along the NW-SE diagonal	Flanking the Plateau proper	On the NE, along valleys and flood plains
Approximate area (% of total)	10	30	60
Highest elevation (ft)	2095	1500	750
Lower elevation (ft)	1500	750	300
Relative relief (ft)	595	750	450
Dissection index (ft)	28.40	50.00	60.00
Maximum slope	1:1.5	1:2.0	1:50
Minimum slope	1:80	1:100	1:500
Nature of contour	Dense, parallel, oval, patchy and full of kinks	Same as plateau proper but with less frequency	Few contours, sparse and patchy
Terrain roughness	Very high	Moderately high	Very little
Drainage frequency	Moderate	High	Very low
Nature of drainage	Lower order channels predominate	Large tributaries and subtributaries dominate	Large rivers, ponds and canals dominant
Settlement and communication	Very few or lacking	Isolated but not rare	Dominate the landscape
Landform elements	Ridges, hills of different sizes and shapes, flat summits, 'I'-shaped valleys, cliffs, etc.	Ridges, hills of different sizes and shapes, narrow valleys, cols and saddles, wind gaps, slopes of different forms	Isolated low residual hills, outcrops, gullies badlands, dry beds, silted beds, discontinuous water courses, steep banks, flood plains, point bars, cultivated areas, ponds, cross dams, etc.

Frequency distribution of spot heights and summit elevations show a bimodal curve, modes broadly corresponding to the aforesaid levels. On the basis of these two observations, the region may be divided into three distinct morphometric units: plateau proper (above 1,500 ft), plateau rim (between 750 ft and 1,500 ft) and erosional plain (below 750 ft).

The distinctive morphometric attributes of these are tabulated below (Fig. 8.9 to 8.23).

Plateau Proper

This represents the highest levels of the plateau found at an elevation of 1,500 ft. Sinduria with

Fig. 8.7 Serial Profiles

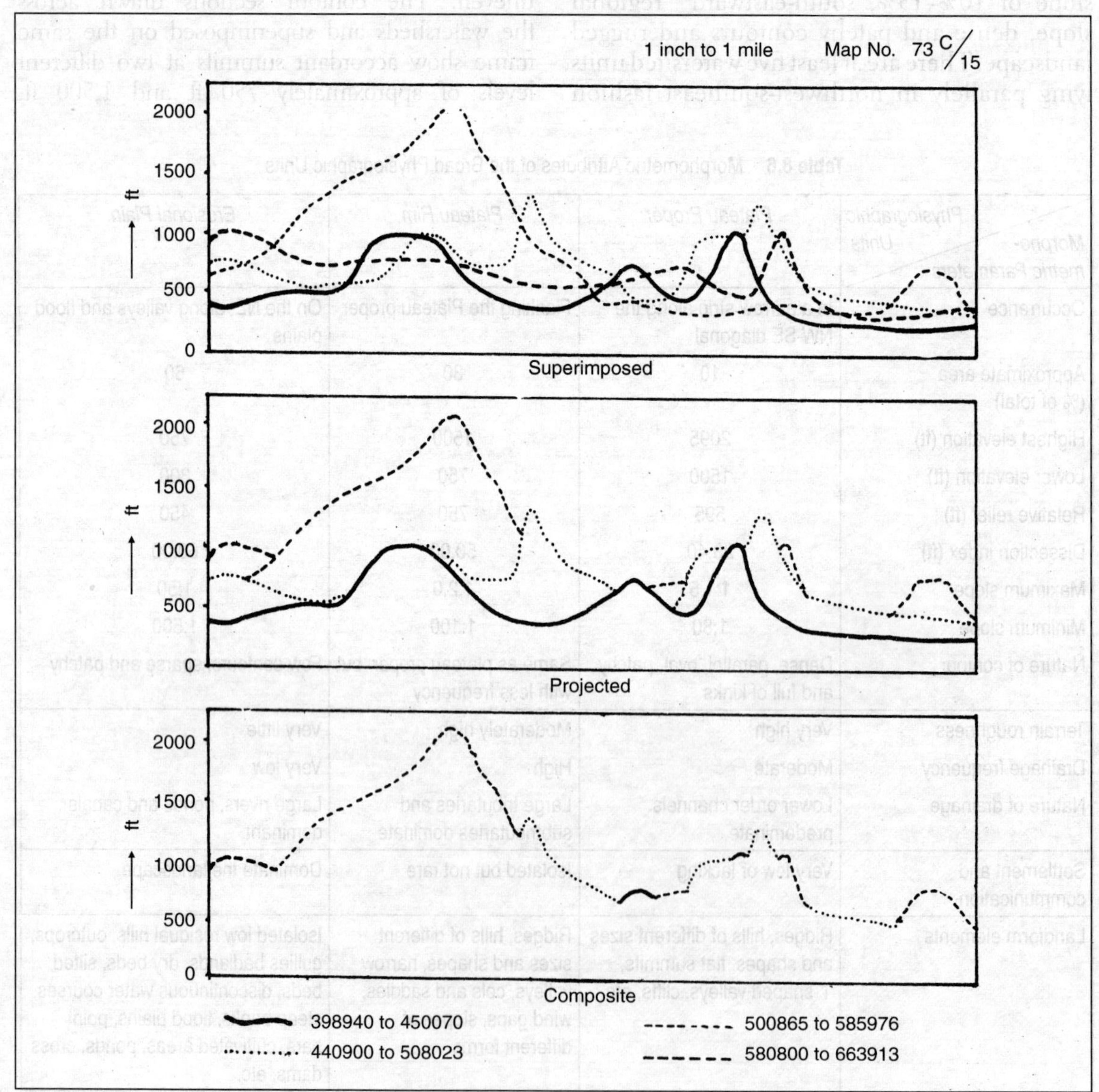

2,095 ft elevation is the highest point (L463942). It is a peak of a flat summit at 2,000 ft level. Other such hills are at L446956 (1787 ft), L479929 (1789 ft), L524917 (1584 ft), etc. The hills are separated by valleys. They appear to be a part of a large ridge that was once continuous in a northwest-southeast fashion in the central part of the map. Hills are commonly large and oval shaped. Summits are occasionally followed by steep slopes. Lower basal slopes are concave signifying progressive denudation while the middle slopes are of convex form. This implies wasting of materials by means of slow denudational processes. Steep cliffs due to slope failures are produced probably due to lithologic strains during rejuvenation. The latter are noted along some lower order channels around L470920, L470960, L445966 and L470940. Being the highest zone, the regional drainage is naturally of radial pattern. The region is generally inaccessible, occupied by the dense mixed jungle of the Gogwa State forest.

Fig. 8.8 Broad Physiographic Divisions: 73 $\frac{C}{15}$

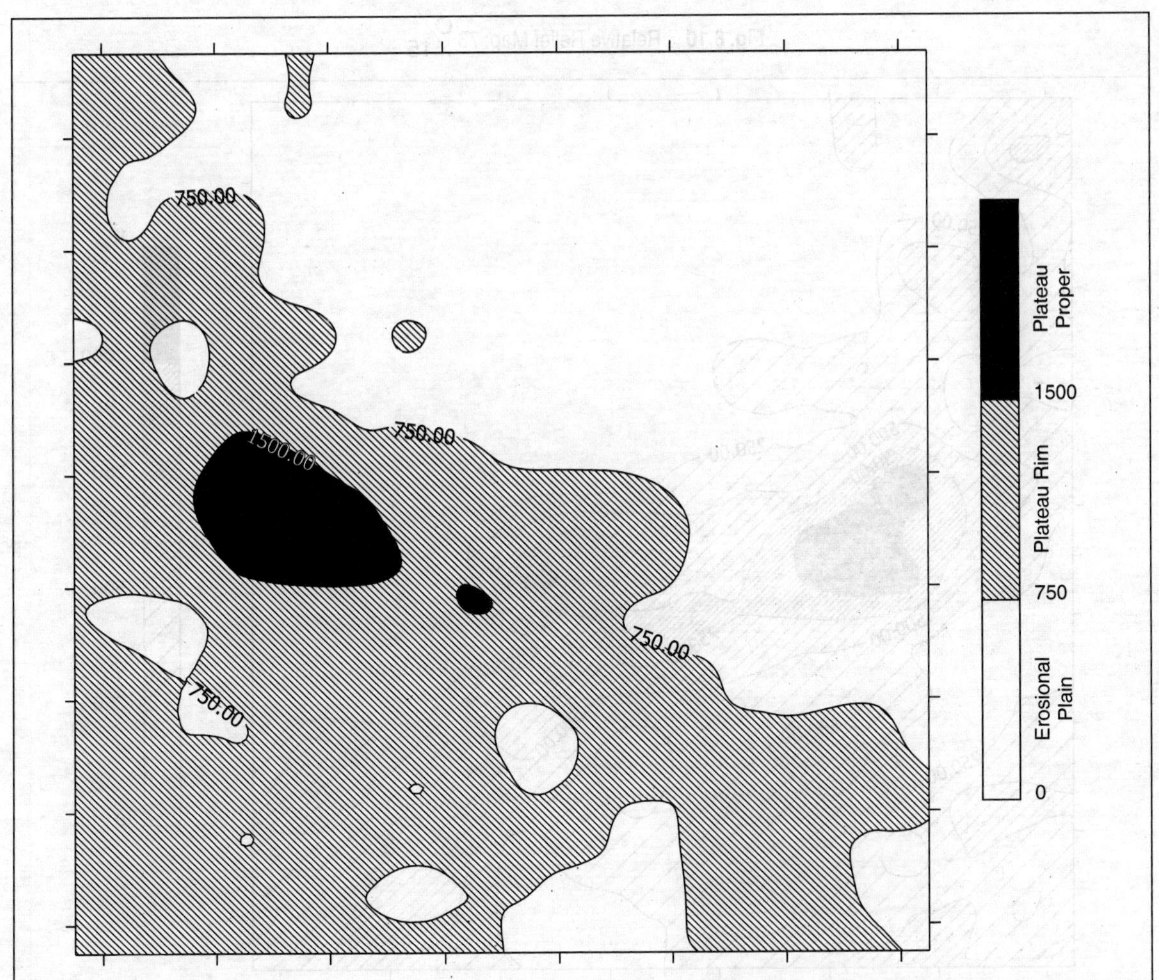

Fig. 8.9 Altimetric Frequency Histrogram

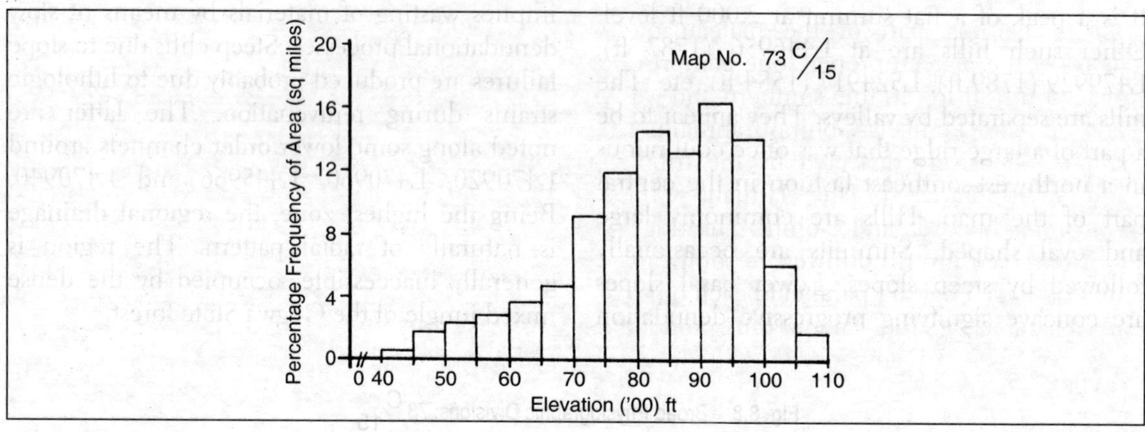

Fig. 8.10 Relative Relief Map: 73 C/15

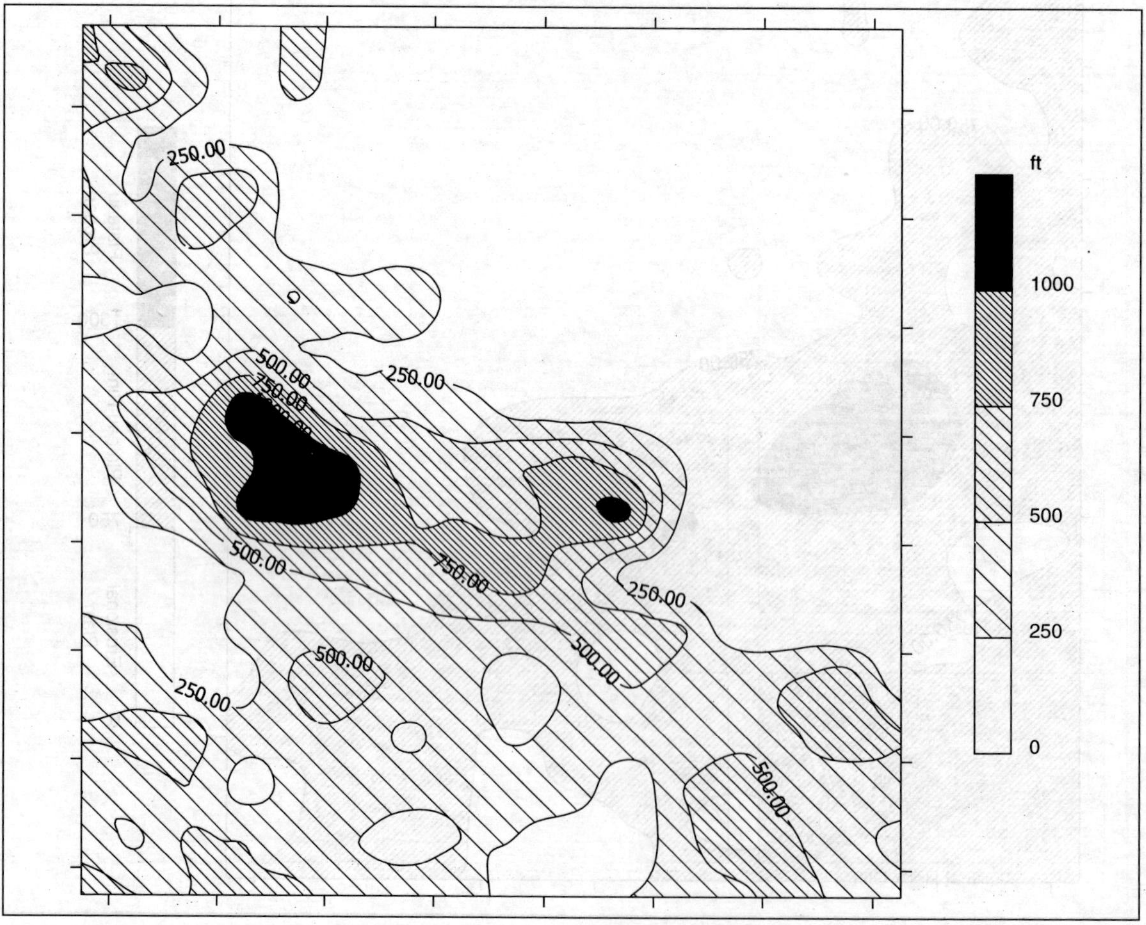

Plateau Rim

The remaining part of the plateau on the flanks of the plateau proper zone comprises the plateau rim. It lies in the southern and western side of the Jaraikela-Gohira-Brahmani rivers and is extensively dissected by the tributaries and subtributaries of the Hinjali, Chilanti, Choardhara, Garda and Gohira. The ridges are well dissected, from northwest to southeast and are oriented. They are parallel to each other and are separated by large valleys, thus producing a *ridge and valley* landscape. Rambhadevi ridge (1411 ft, F409099) has been completely dissected by Gohira river and its tributaries and thus a chain of elongated, oval as well as flat-topped, discontinuous but closely spaced hills have been formed. These are mostly over 1000 ft and lie in a roughly west-east fashion in between the Gohira and Garda rivers. These are Dhubulipathar (1010 ft), Suila (1082 ft), Pandrisila (1055 ft), San Tongri (1051 ft) and Raijhari (852 ft). In between the Garda-Gohira-Brahmani rivers and the headwaters of the Hinjali, Chilanti and Choardhara rivers lies the largest and the longest continuous block of plateau rim extending right from L386970 to L670847 for about 18 miles in the WNW-ESE trend. Chhata (1,499 ft, L561933) is the highest peak. Hills in this segment are mostly over 1,250 ft in elevation. The Amgothu-Baghdhala ridge with an average summit level of 1,000 ft lies between the rivers Choardhara and Chilanti. The Chilanti-Hinjali watershed appears to be absolutely dissected. The small residual hills mostly around 1,000 ft of altitude lie as separate but closely spaced

Fig. 8.11 Typical Landforms: 73 C/15

landform units. On the extreme southwest corner, the Baniakeha-Jhannu watershed lies on the west of river Chilanti in a northwest-southeast orientation. It is also dissected and the hills are mostly 1,200 ft to 1,250 ft in elevation.

Dissection of the ridges has produced some typical landforms, e.g., saddles (L480855, L494883, L442822, L641815, etc); peaked hills like Buriha (L633872), Khamra (L650803), Rambhadevi, Chhata, etc; hills with steeper eastern slope and gentler convex western slope (L573813); vertical rock cliffs (L615813, L590840, L490882, etc.); uniformly sloping hills (L418816, L509878, etc.); arcuate sub-ridges, e.g., Chhata (L561993), Baniakeha (L400880); typical escarpments, e.g., the southwestern slope of the hill at L 485906 etc; the wind gap at Mahasindhu (L397989); cols (F446002, F485003, etc); asymmetric narrow valleys (L400883); asymmetric open valleys (L440800); typical 'V'-shaped valleys (L530903); and so on. All over this region rounded hills are frequently observed. These are demarcated by regular, dense and close contours. This implies that the underlying lithological matrix is made up of hard and crystalline igneous and metamorphic rocks.

Fig. 8.12 Dissection Index Map: 73 C/15

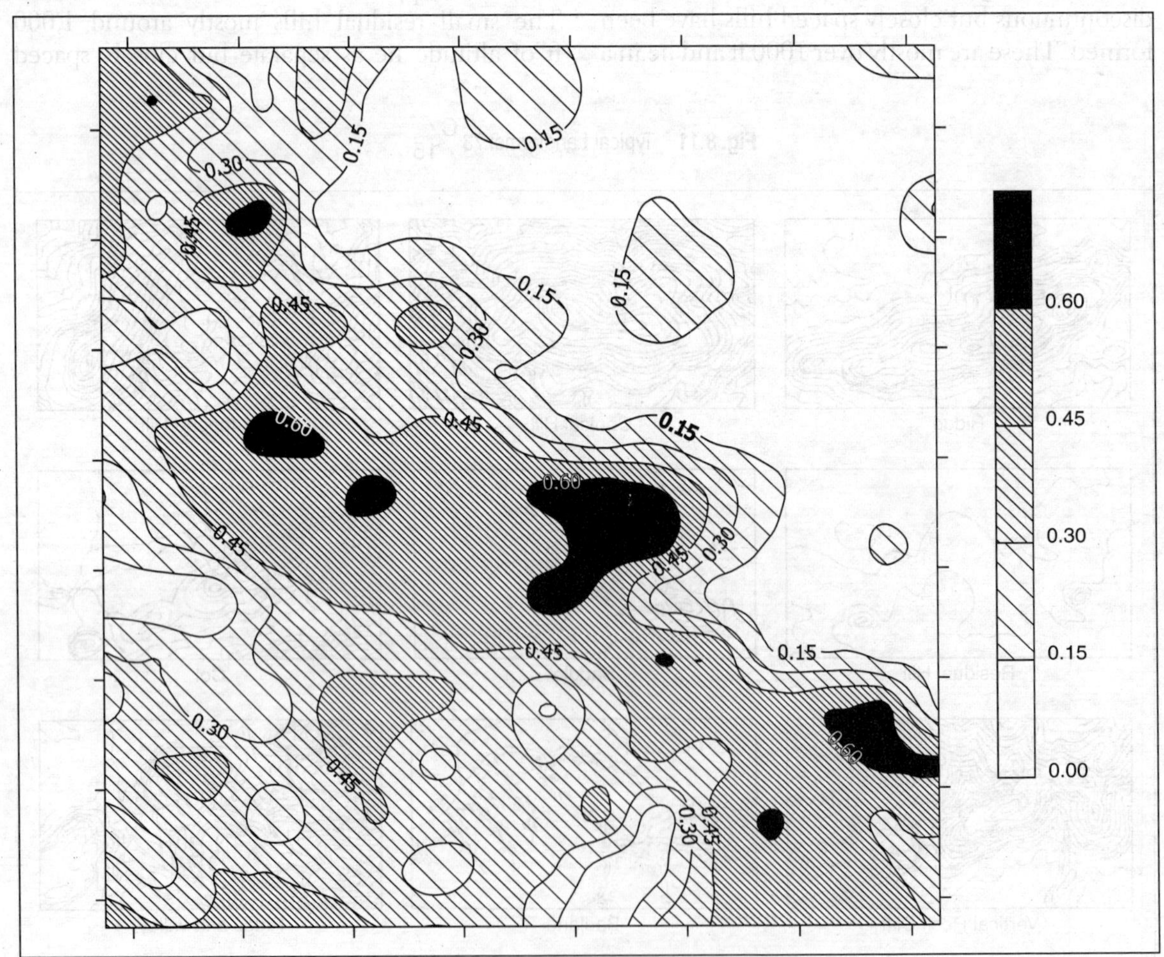

Erosional Plain

This physiographic unit is confined along the large tributary valleys in the south, centre and west, the flood plains of the Brahmani river and the interfluves of Jaraikela-Balam, Balam-Motuali, Motuali-Churakhai, Churakhai-Usthali and Usthali-Brahmani rivers. In the north and east, the regional slope is directed southward while it is southeastward in the southwest. It is a typical rolling plain formed absolutely by fluvial actions. Contours are few and far between. Some low residual hills, e.g., at F520081 (605 ft), F573020 (550 ft), F 627098 (550 ft), F 524050 (500 ft) and F478097 (580 ft) jut out in a disperse fashion breaking the general monotony of the otherwise plain landscape. Clayfree coarse soil and occasional high intensity monsoon downpour coupled with the factors of human interference in the form the jungle clearings have given rise to selective gullying. Hence, extensive badlands have formed particularly along the right bank tributaries of the lower Hinjali river, in headwater regions of the rivers Chilanti and Garda, on the piedmont

Fig. 8.13 Average Slope Map (after Wentworth): 73 C/15

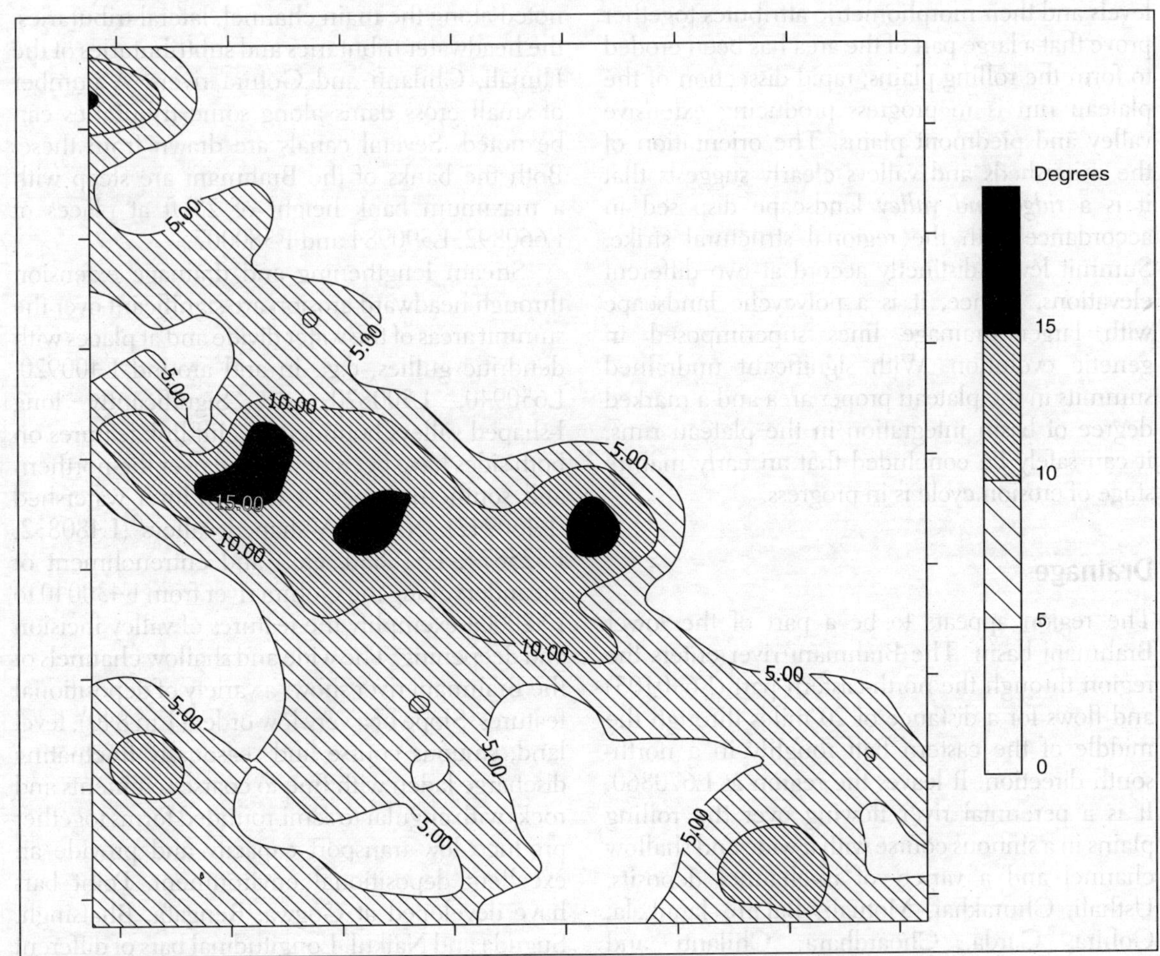

plains of the western parts of the Sinduria ridge, the northern peidmont plains of the Chhata ridge and on the immediate west of the Bamra canal in the Brahmani flood plains.

In general, absolute relief decreases from the top of the watersheds in all directions, the rate being higher in transverse directions. The highest absolute relief (above 1,500 ft) is represented by the plateau proper while the lowest (below 500 ft) by the erosional plains. In between these two lies the plateau rim with moderate to high absolute relief. The relative relief broadly corresponds to the above pattern such that it is highest over the plateau and lowest over the plains. The distribution of the dissection index and the average slope also shows an identical pattern. The topographic levels and their morphometric attributes together prove that a large part of the area has been eroded to form the rolling plains; rapid dissection of the plateau rim is in progress producing extensive valley and piedmont plains. The orientation of the watersheds and valleys clearly suggests that it is a *ridge and valley* landscape disposed in accordance with the regional structural strike. Summit levels distinctly accord at two different elevations. Hence, it is a polycyclic landscape with larger drainage lines superimposed in genetic evolution. With significant undrained summits in the plateau proper area and a marked degree of basin integration in the plateau rims, it can safely be concluded that an early mature stage of erosion cycle is in progress.

Drainage

The region appears to be a part of the lower Brahmani basin. The Brahmani river enters the region through the north-eastern part (F640105) and flows for a distance of 20 miles through the middle of the eastern half roughly in a north-south direction. It leaves the region at L670860. It is a perennial river flowing over the rolling plains in a sinuous course with a wide and shallow channel and a variety of in-channel deposits. Usthali, Churakhai, Motuali, Balam, Jaraikela, Gohira, Garda, Choardhara, Chilanti and Hinjali are the major tributaries (all right bank). All these and their innumerable tributaries, subtributaries and headwater courses drain the area. Interestingly, the four tributaries—Hinjali, Chilanti, Choardhara and Gohira—together compose a regional parallel pattern flowing northwest to southeast with dissected ridges and high watersheds in between. The lowest order channels originate from these and drain both the slopes to intensify the network (Fig. 8.14 to 8.19).

Most of the drainage is non-perennial being either rainfed or springed. Inter-stream areas are defined by narrow and well-formed water divides. Discontinuous small water courses and dry beds can be observed over the Brahmani river and the piedmont plains. Significant gully erosion is noted along the main channel, lateral tributaries, the headwater tributaries and subtributaries of the Hinjali, Chilanti and Gohira rivers. A number of small cross dams along some tributaries can be noted. Several canals are drawn from these. Both the banks of the Brahmani are steep with a maximum bank height of 25 ft at places of L660892, L590984 and F580007.

Stream lengthening and drainage extension through headward erosion are significant over the summit areas of the water divide and at places with dendritic gullies, e.g., in and around L400920, L650940, L500820, etc. Significantly long I-shaped valleys bounded by cliff-link features on both sides have been developed over the northern and southern slopes of the Sinduria watershed (L460942). Narrow V-shaped valleys (L480832, L525874, L518880, etc.) and entrenchment of meanders along the Gohira river from F430040 to F520004 exemplify the features of valley incision and deepening. The wide and shallow channels of the Brahmani river shows a variety of depositional features. Slope of a very low order (1 in 840), level land, sinuous course and seasonally fluctuating discharge laden with fine to coarse sediments and rocks with angular to semi-rounded forms together produce low transport capacity and provide an excellent depositional environment. Point bars have developed at Gogwa, Rengata, Bhatsingh, Suguda and Naikul. Longitudinal bars of different

Fig. 8.14 Drainage Map: 73 C/15

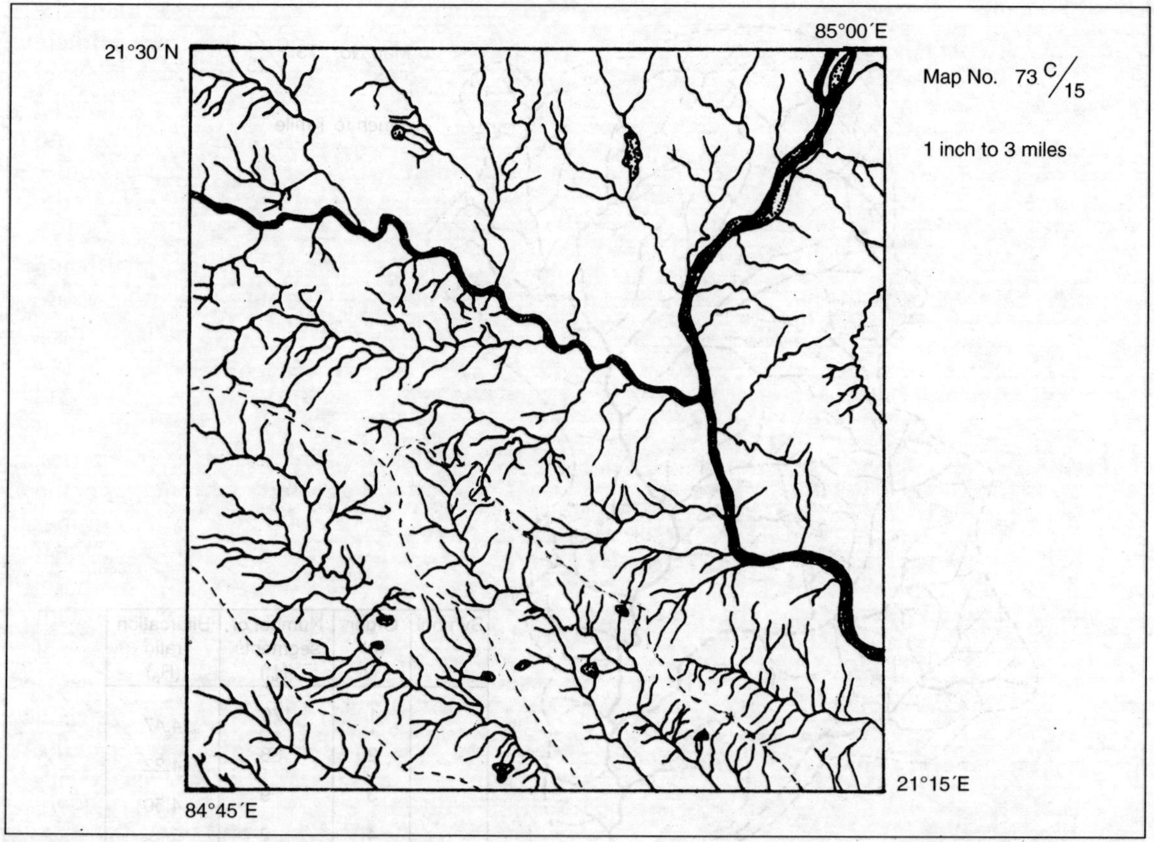

sizes are noted at F620060, F594037, F587025, F583010, L610913 and L665865. Transverse bars can be seen at F610050 and F588028. Large in-channel shoals and islands formed both of river borne sediments and rock bodies lie at F632080, F620060, L665865, etc. At the confluences of Usthali, Churakhai and Gohira rivers, fan-shaped depositions are prominantly developed in the main channel of the Brahmani river.

Drainage frequency has been evaluated to be highest (above 15/sq mile) over the slopes and summits of the watersheds and lowest (below 5/sq mile) over the erosional plains. The same sense of pattern is also observed for drainage density; the highest and lowest figures are respectively above 7 miles/sq mile and below 1 mile/sq mile. The cross profiles are usually V-shaped, that widens out towards the channel mouth. Long profiles are very much up-concave towards the source particularly above the junction of the plateau levels, thus separating the upper and the lower courses. Drainage networks are commonly structurally irregular with a mean bifurcation ratio of 3.79 (and variance 1.09). The huge number of first and second order channels implies that the networks are still developing and being elaborated (Fig. 8.14 and 8.15).

The varying underlying geological structure, surface morphometry and stage of evolution together have given rise to a variety of drainage and channel patterns. The most common of these is the dendritic pattern in which the

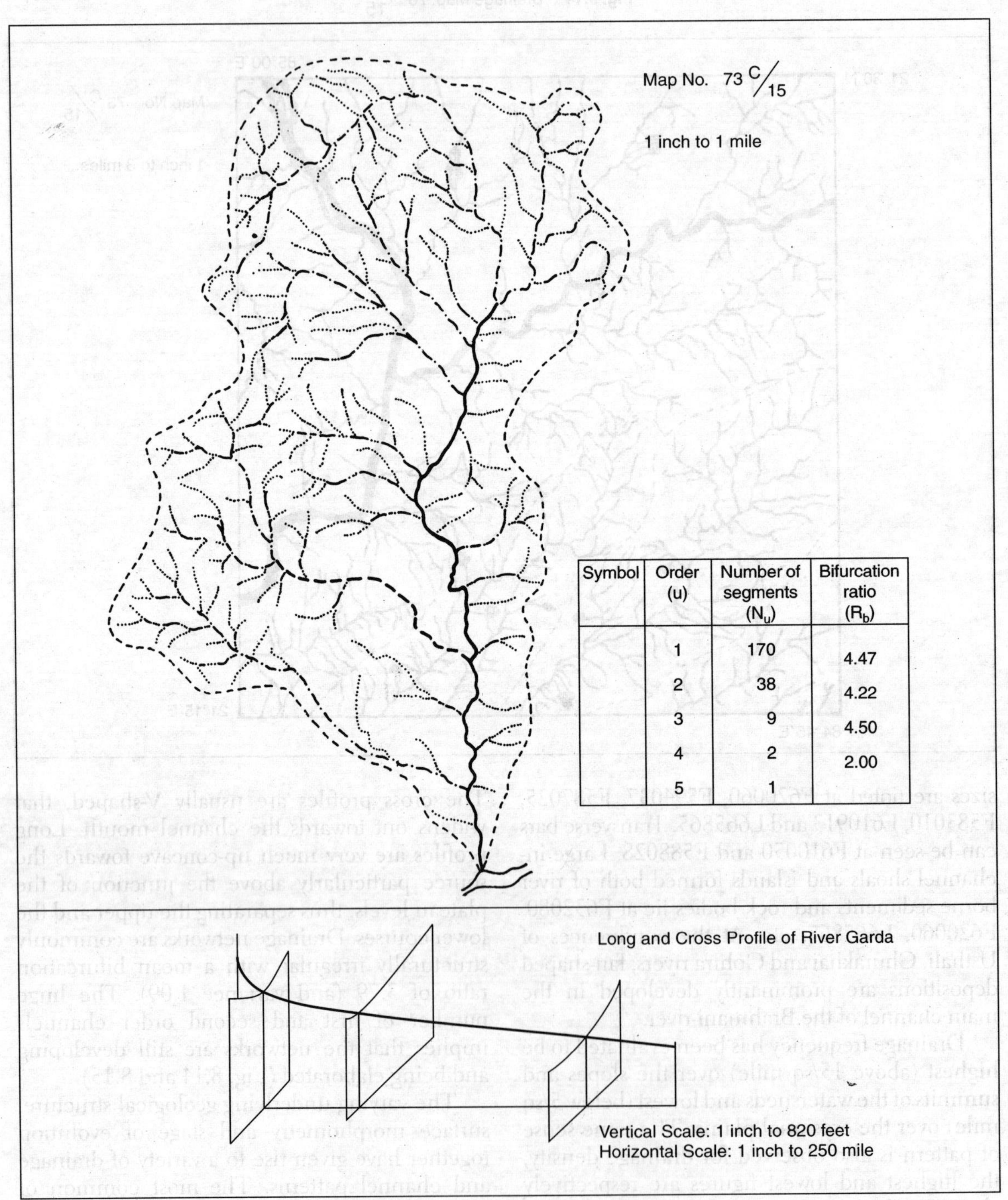

Symbol	Order (u)	Number of segments (N_u)	Bifurcation ratio (R_b)
	1	170	4.47
	2	38	4.22
	3	9	4.50
	4	2	2.00
	5	1	

Fig. 8.15 Garda River Basin

Fig. 8.16 Drainage Frequency Map: 73 C/15

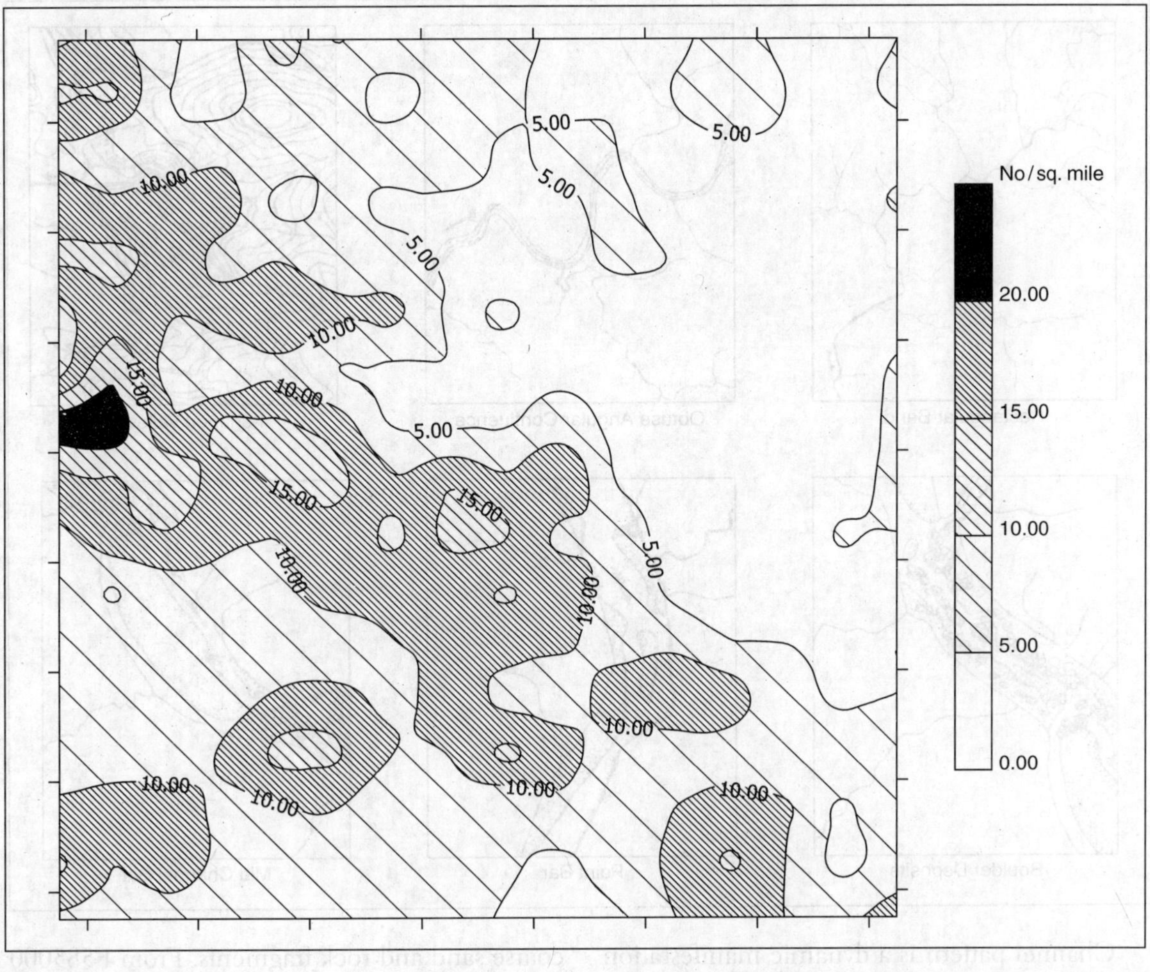

alignments of the tributaries and subtributaries are analogous to the branching arrangement of a tree. The confluence angles are typically acute. This pattern is characteristic of regions with uniform lithology, best represented in the central part of the region. Hinjali, Chilanti, Choardhara, Gohira, Balam, Churakhai and the upper course of Usthali—all flow from northwest to southeast and have thus produced a parallel pattern on the regional scale. However, locally, in areas with broadly identical slope conditions, the network of tributaries have formed parallel patterns. Excellent examples of this can be seen on both slopes of the Rambhadevi ridge (around F400070). Over the plateau proper, drainage lines appear to radiate from around the summits of the high hills. The terrain around Sinduria (L463942), Amgothu (L605829), Raijhari (F496003), Chhata (L561933), Suila (F434028), etc. provide typical examples. The headwater channel of Garda shows a typical boat-hook bend. A closer look along a line between L398982 and F386007 suggests that it is a wind gap. Possibly the headwater of Juanbhanga had been captured by a more potent tributary of Garda to form the existing boat-hook and hence the barbed pattern.

Fig. 8.17 Typical Features of Drainage: 73 C/15

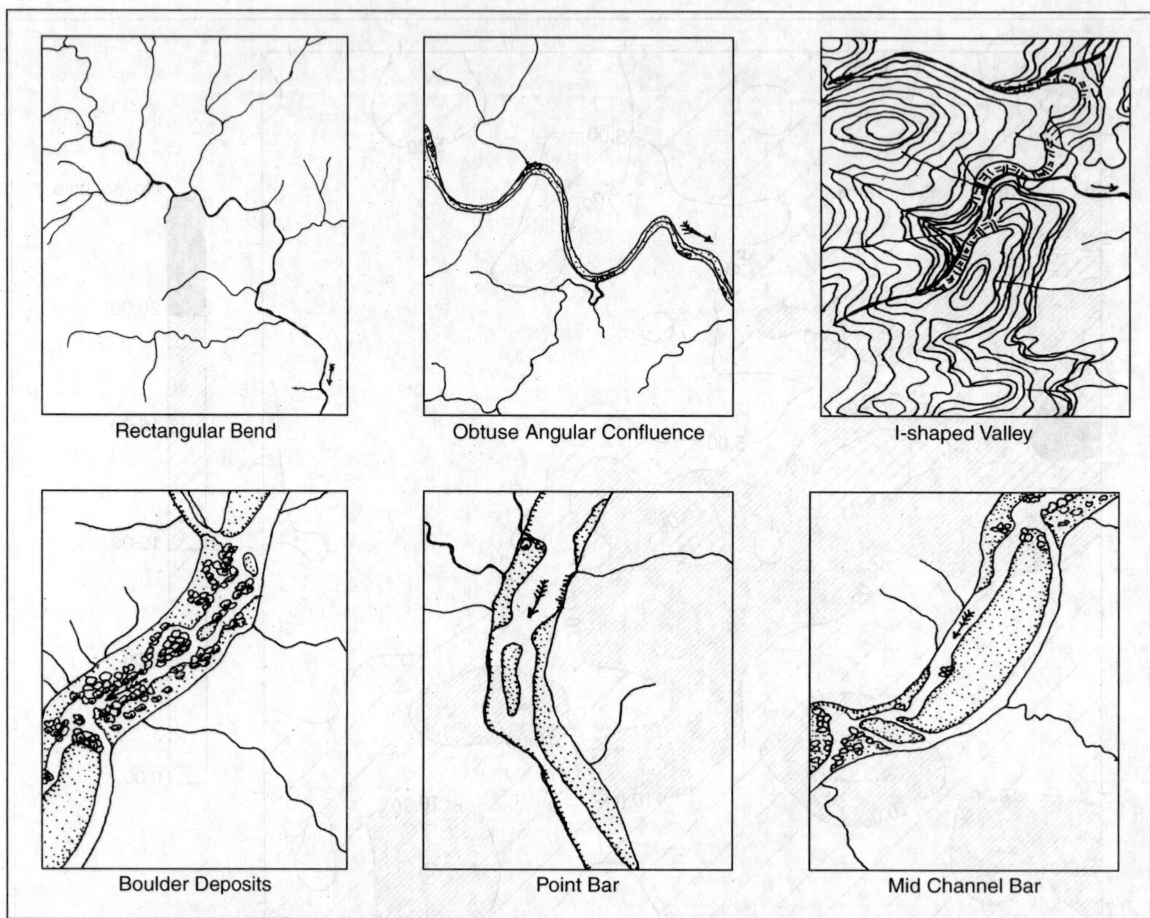

Channel pattern is a dynamic manifestation of bed slope, discharge, load, velocity and cross-section geometry. In general the channels show a sinuous pattern. The index value lies between 1.15 and 1.30. Between F430040 and L552990, with an index value ranging between 1.35 and 1.5, the channel of Gohira presents some beautiful meanders. The main channel of the Brahmani river, however, exemplifies some model features of fluviological interests. From F640103 to F605050 the channel shows braiding with several in-channel sandbars and rock islands. The width-depth ratio is large. The general incompetency arises from a huge load of coarse sand and rock fragments. From F585000 to L610900, the channel is exceptionally straight bounded by steep banks. The remaining segments are sinuous. The longitudinal bars are migratory in nature. The several transverse bars imply the existence of secondary transverse circulations and general incompetency of the channel (Fig. 8.14).

The common hydrological regime is defined by the tropical monsoon climate with alternate dry and wet seasons, hot and cool weather, concentration of 90% of the total annual rainfall within three months (mid-June to mid-September) and the spatiality of rain being

Fig. 8.18 Channel Patterns and Features: 73 C/15

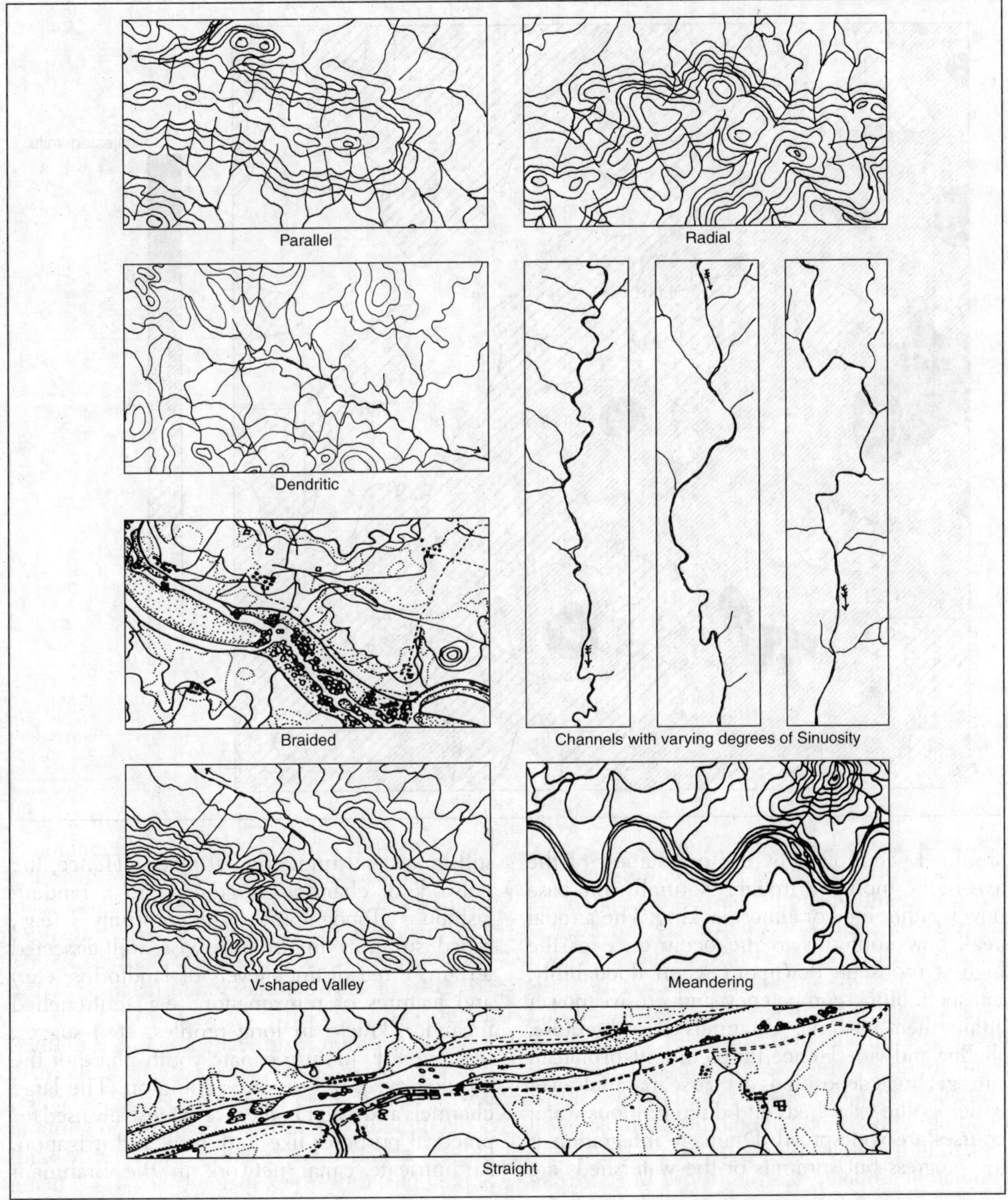

Fig. 8.19 Drainage Density Map: 73 C/15

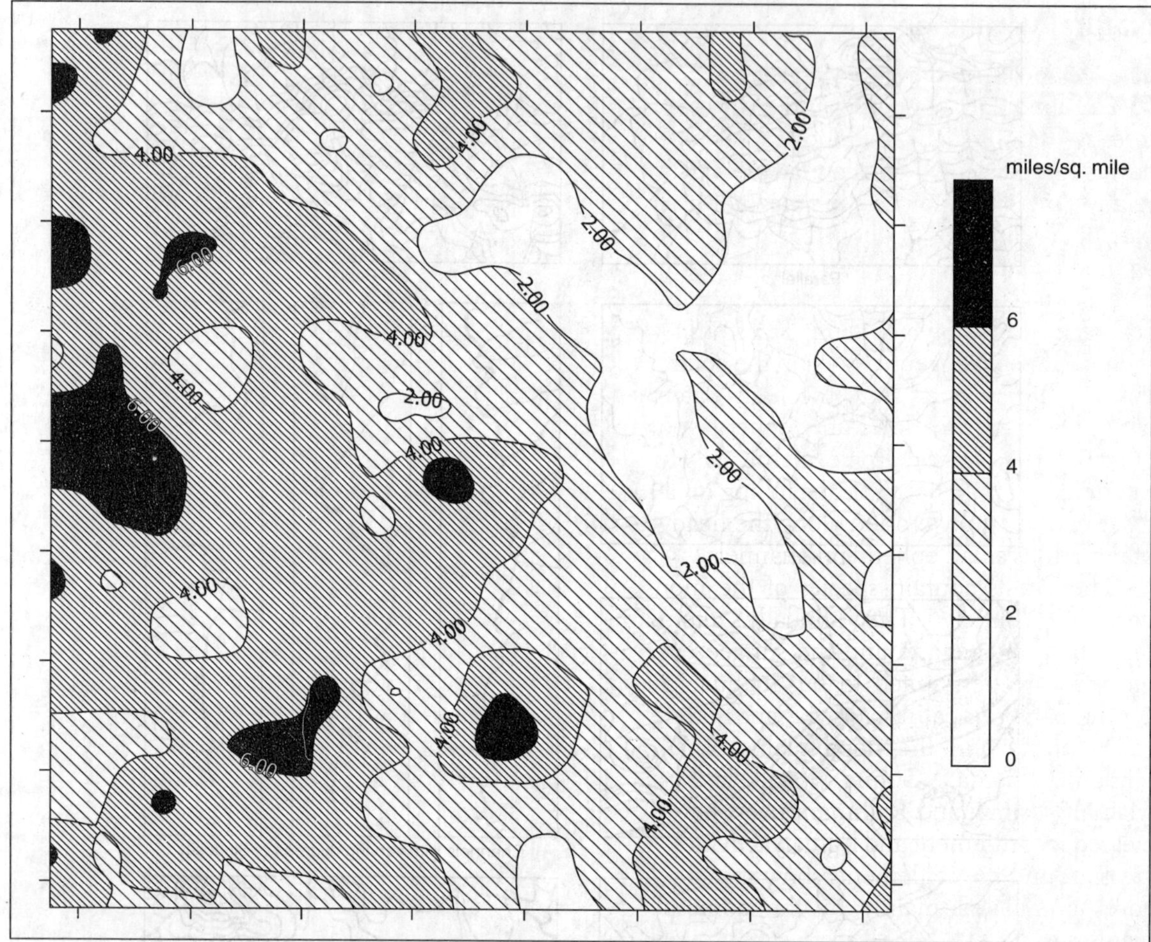

absolutely controlled by the orientation of the axis of the monsoon trough. Naturally, the base flow is either little or almost lacking. The annual peak flow conforms to the occurrence of the highest monsoon downpour when floodability, transportability, competency and erosive power attain their maxima. Channels are, therefore, shallow and wide. Hence, the piedmonts of plateau rims get huge depositions of coarse materials over which gullies, dry beds and discontinuous water courses are widespread. Drainage integration is in progress but summits of the watersheds are still broad and undrained at places. Hence, the network of channels is growing in a random fashion. Topographic unconformity (e.g., broad summits in an otherwise well-dissected terrain, three distinct levels of landforms, etc.) and features of rejuvenation (e.g., entrenched meander, knicks in long profiles, etc.) suggest that an early mature or late youth phase of the ensuing cycle of erosion is going on. The large channels and water bodies are extensively used for practical purposes like ford, ferry and irrigation. An intricate canal network in the Brahmani plains suggests the nature and importance of local agricultural economy.

Natural Vegetation

The region has dense vegetal cover. Nearly 75% of the total area comes under the two state forests—Rambhadebi and Gogwa. The former covers the dissected plateau areas with lower relative relief condition in the northwestern part. It has an area of about 85 sq miles and is typified by fairly dense mixed jungle, mainly sal. The Gogwa State Forest is much larger in areal coverage with an area of about 125 sq miles and is characteristically represented in the central and entire southern part. Here the high watersheds with large variations in relative relief favour the growth of dense mixed jungle, mainly sal. Human interference in the form of tree felling on a local scale induces gully erosion along the headwaters of numerous small ephemeral channels.

The forests contain species of the monsoon deciduous variety. The annually alternating thermo-humid regime and the surface cover of red soils are best suited to sal, mahua, palash, kendu, bamboo, palm, palmyra, and so on. The valley plains and the flood plains of Hinjali, Chilanti, Garda, Gohira, Jaraikela, Balam, Matuali, Usthali and Brahmani rivers have been cleared for settlement and human activities. Thus areas around the settlements show sparse vegetal cover dominated mainly by the domesticated plants.

Usually the Brahmani Plains and the tributary interfluves have fairly dense mixed jungle, mainly sal. The plateau proper formed by the summits of the high hills and ridges like Rambhadebi, Sulla, San Tongri, Dhubuli, Pandrisila, Sinduria, Chhata, Jannu, Katasar, Buriha, Bagdhala and Amgothu is covered with dense mixed jungle, mainly sal. The deep forest of this region mainly belongs to the Gogwa State Forest. Human interference is relatively less as the terrain is physically inaccessible. Thus the plateau proper roughly above 1000 ft shows the highest density of species with relatively pure stands while the heterogenity of species increases downslope in the plateau rim and plains. These hardwood forests could form the most important economic resources of the region if exploitation is done scientifically with proper environmental planning.

Settlements

The region appears to be a very poorly settled one. About half of the total area in the plateau proper with state forests lies entirely uninhabited. The frequency of settlement is relatively higher over the valley plains and the flood plains of the Brahmani river. Rural settlements dot the region mainly in a dispersed pattern. Naikul (L640900) on the right bank of the Brahmani river is the largest settlement. Other important settlements are Suguda (F430080), Ganganan (F500040), Bhatsingh (F598050), Jaragola (L585995), Gogwa (L596963), Nilaposi (L645828), Sasamara (L570808), Khandadhuan (L460799), Kundapitha (F490075), Kulsara (F660102), etc.

Settlements have mainly occupied such sites as river banks (Naikul, Rengala, Bhatsingh, Ganganan, Kodalipal, etc.); flood plains (Kulsara, Daraikela, Dharmapur, Kundapitha, etc.); forest clearings (Nilaposi, Gudamal, Utunia, Phulibari, etc.); forest fringe (Rengalpali, Palsama, Khiloi, Malarbahal, etc.); piedmont areas (Lakshmipur, Kantabahal, Chirgunikudar, Shiarimalia, etc.) river confluence (Gogwa, Ganganan, etc.); meander core (San Khairpali, Kadalipal, etc.) contour pattern (Satkiari, Kamarnali, Mahasindhu, Shiarimalia, etc.); low spur (Karadakham, Phuliharia, etc.); low hill summits (Kalkati); and so on. Settlements have also come up based on situations like the ferry ghat (Kesla, Kandsar, Bankakalo, etc.); transport lines (Suguda, Rengata, etc.); transport node (Ganganan, Suguda, Gogwa, Kantabahal, etc.); administrative functions (Naikul); religious places (Gogwa, Bankakalo, etc.); small check dams (Karadakham, Nuadihi, Bhaluchaba, etc.); fording points (Balroi); and so on (Fig. 8.20).

Settlements are commonly small in size and amorphous in geometric pattern. Spectacular

Fig. 8.20 Site and Situation of Settlements: 73 C/15

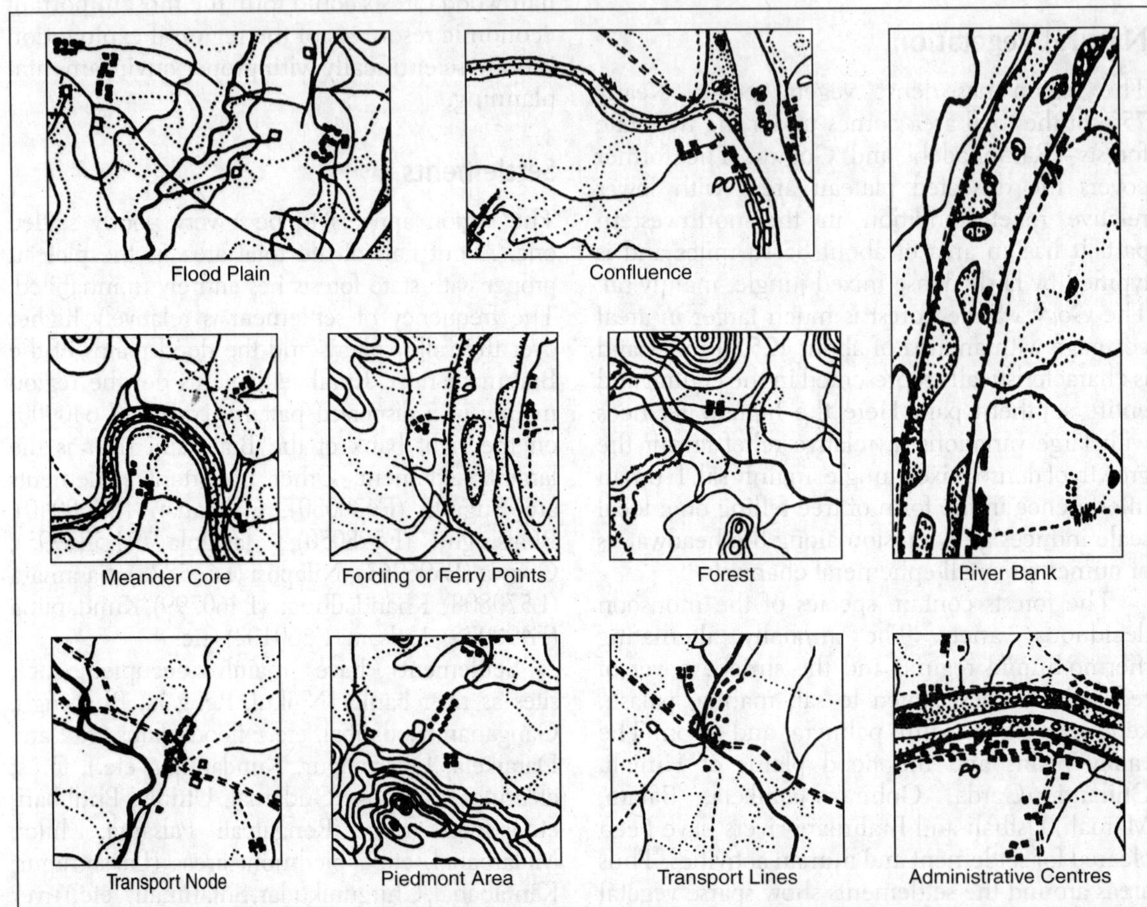

linearity has been formed based on transport lines (Naikul, Kandsar, Baliposi, Suguda, Phulibari, Jaragola, Bhatsingh, etc.) and river bank outlines (Khandam, Kesla, Ganganan, etc.). A really dispersed pattern of settlement is the most common feature. However, dispersed agglomerations have been noted in the Batan-Motuali interfluve, Garda basin and the Usthali-Brahmani interfluve. The forest chowkis in the deep jungle appear as isolated units of settlement (Fig. 8.20 and Fig. 8.21). Settlements are mainly rural, based on local agriculture and other extractive occupations. Naikul has some administrative functions with a post office and a police station. Gogwa is a market place favourably located at the confluence point, on a gravelled motorable road and is connected by ferry service with Kandsar on the left bank of the Brahmani river. It also has a post office. Other important settlements are Bhatsingh, Ganganan, Jaragola and Suguda—all located on the Brahmani-Gohira interfluve, which is physically a rolling plain.

About 50% of the total area lies uninhabited while 24% has a frequency of one settlement/sq mile, 14% two settlements/sq mile, 9% three settlements/sq mile and only 4% four settlement/sq mile. The frequency of settlements increases

Fig. 8.21 Settlement Patterns: 73 C/15

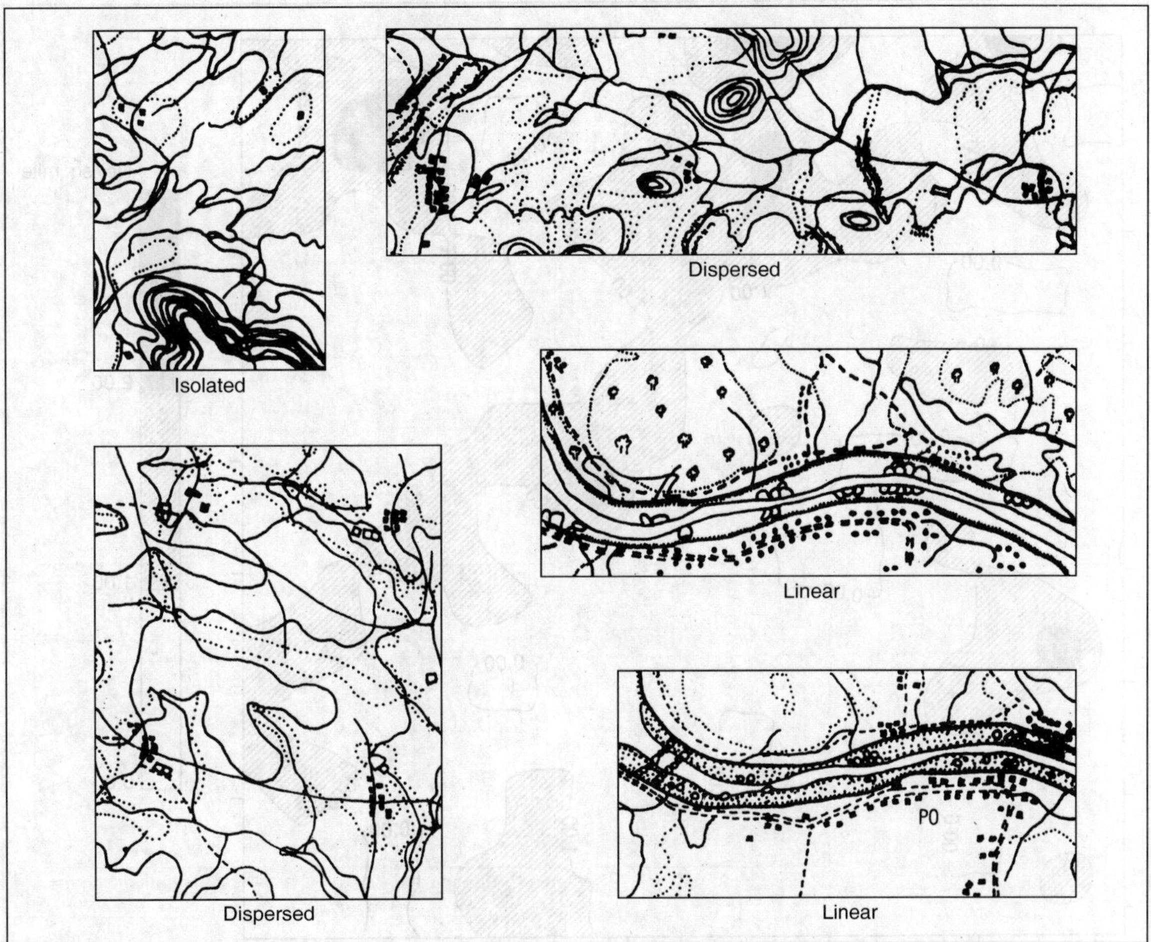

with the decrease in ground slope and relative relief. It is also higher near the perennial rivers and at places, where the frequency of ponds is high, signifying a shallow water table. From the viewpoint of situations, transport and communication have proved to be the most important in the growth of settlements. Excepting the restricted and protected parts, settlements have grown up at convenient locations emphasising once again the capability of man's adjustment to the available local environmental conditions (Fig. 8.22).

Transport and Communication

Typically, the region does not have any well developed transport network of either metalled roads or railway lines. Minor village roads and cart tracks dominate the transport scenario. However, only two segments of gravelled roads, motorable in dry seasons, traverse through the region for a length of about 34 miles. The first one runs from Deogarh to Rengali for a distance of 27 miles through the north-eastern plains and links such major settlements as

Fig. 8.22 Settlement Frequency Map: 73 C/15

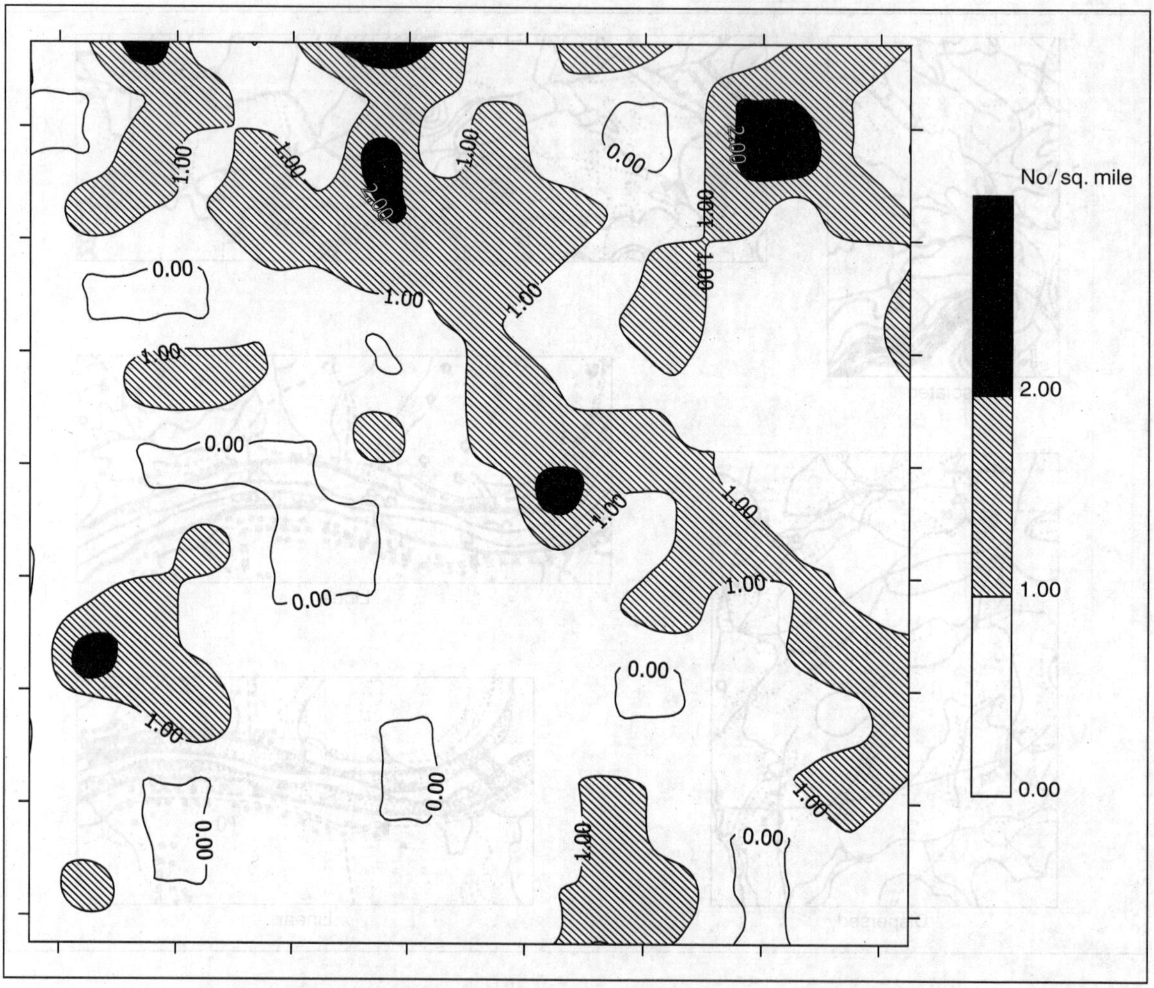

Suguda, Ganganan, Gogwa and Naikul. The second one is a smaller segment of 7 miles running from Basaloi to Sarapal through the southwestern plateau terrain. Besides, ferry services at several points on the Brahmani river reflect the importance of water transportation in the area.

Development of transport network has been non-uniform. About 50% of the total area shows a mean density of less than 1 mile / sq mile, 32% lies within a range of 1–2 miles/sq mile, 17% within 2–3 miles/sq mile while the remaining 1% area within 3–4 miles/sq miles. The plateau sector with moderate-to-high variations in relative relief and dense mixed forests cover is sparsely settled. The north-eastern plains, on the other hand, contain extensive tracts of *reserved* and *protected* forests. Hence, as a whole, transport development has remained at a very poor level except some pockets like the levee and floodplain margins of the Brahmani river and the valley plains of some of the major tributaries (Fig. 8.15). In general, transport density increases with higher settlement frequency.

Fig. 8.23 Transport Density Map: 73 C/15

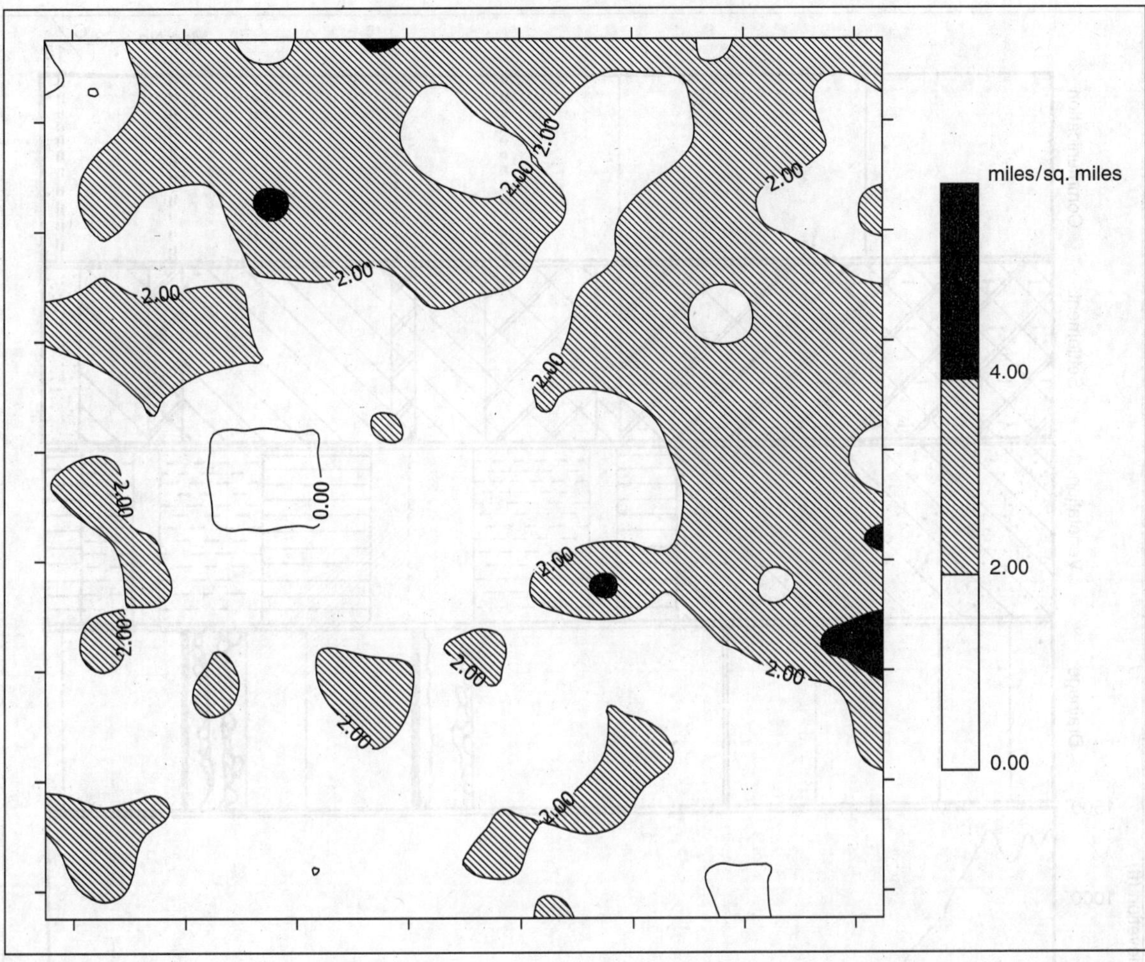

Physical and Cultural Landscapes

The physical landscape of the area mainly covers the districts of Bumra and parts of the districts of Pal Lahara, Rairakhol, and Talcher, Orissa. The NW-SE diagonal roughly divides the area into two macro topographic units—(a) the northeastern part is an erosional plain drained by the south and southeastward flowing Brahmani river and its tributaries and (b) the southwestern part is a dissected plateau with well formed ridges and valleys.

Physiographically, the area contains three distinct morphometric units—plateau proper, plateau rim and erosional plain that respectively cover 10%, 30% and 60% of the total area. The plateau proper occurs roughly along the NW-SE diagonal of the rectangular map, the plateau rim flanks the proper plateau area while the remaining portion especially over the valleys and flood plains in the northeastern half comprise the erosional plain. The relative relief is the highest over the plateau rim and the lowest over the erosional plain, while the dissection index is the highest for the erosional plain and the lowest over the plateau proper area. This is because in the proper

Fig. 8.24 A Transect Chart: 73 C/15

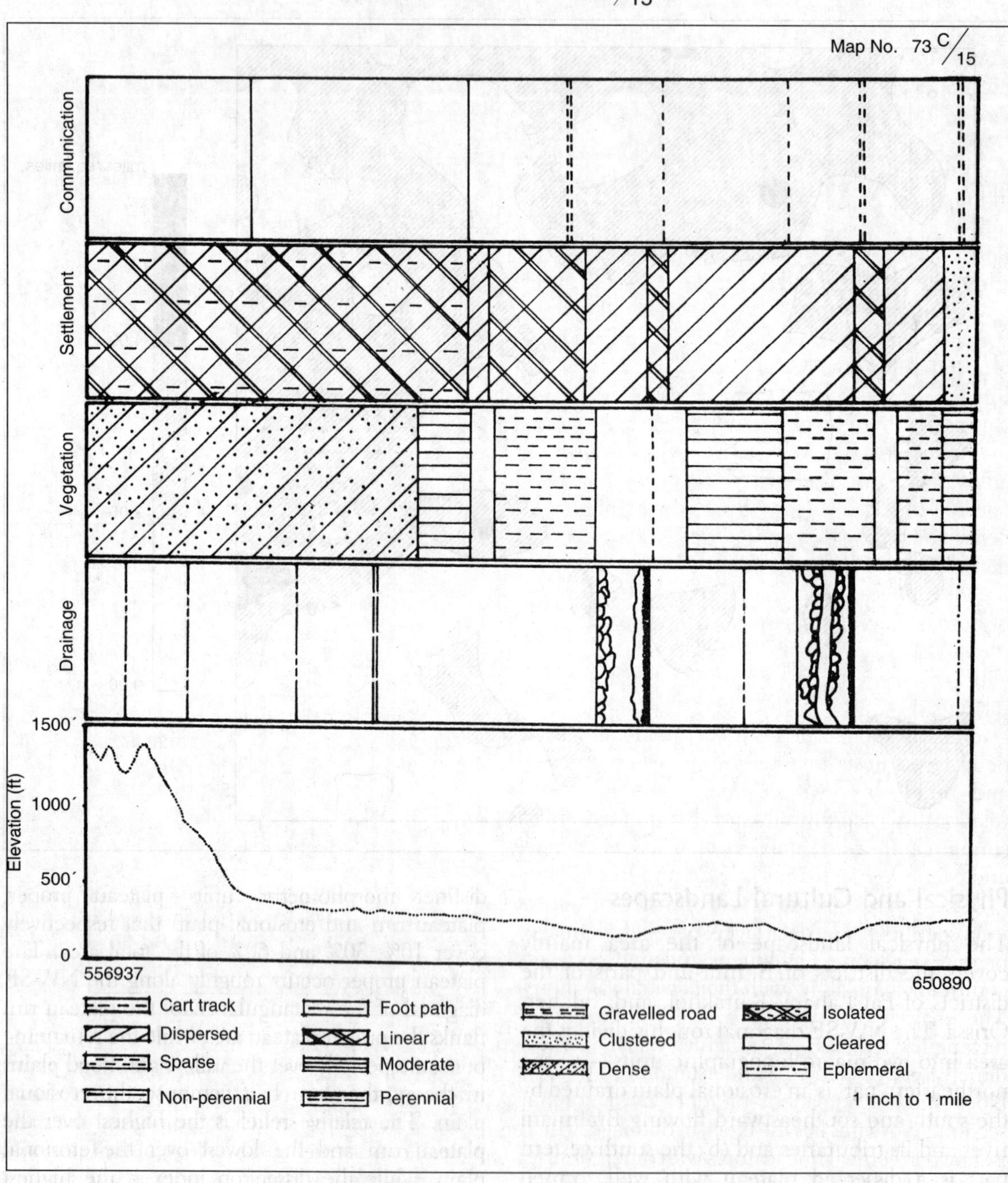

plateau area lower order channels are frequent while large tributaries and subtributaries are more frequent over the plateau rim and large rivers and canals drain the erosional plain. Therefore, the ruggedness of the terrain is very high in the plateau proper and very low over the erosional plain.

Ridges, hills of varying shapes and sizes, flat summits, I-shaped and narrow valleys are found in the plateau proper area, while dissected ridges, hills of varying shapes and sizes, saddles and cols, slopes of varying magnitudes and forms, low spurs and relatively broad valleys are the typical features of the plateau rim areas. The erosional plain contains features like isolated low residual hills, rock outcrops, gullies and badlands, dry beds and silted beds, discontinuous water courses, steep banks, flood plains of varying width, point bars, in-channel deposits, boulder deposits, cross-dams, cultivated areas, etc. The whole of the terrain is covered by forests of varying density, being most dense over the plateau proper and relatively less dense over the erosional plain.

The cultural landscape, being a manifestation of human interactions with the physical environmental factors also presents a pattern that conform with each other because these are intimately interrelated. Within the bounds and constraints of the given surroundings of a physical landscape, man delicately adjusts his life and activities superbly and builds his own cultural landscape. This relationship is commonly shown with the help of simple graphs, e.g., transect charts, scatter plots and overlays (Fig. 8.25, Fig. 8.26 and Fig. 8.27).

Fig. 8.25 Relative Relief (x) and Slope (y)

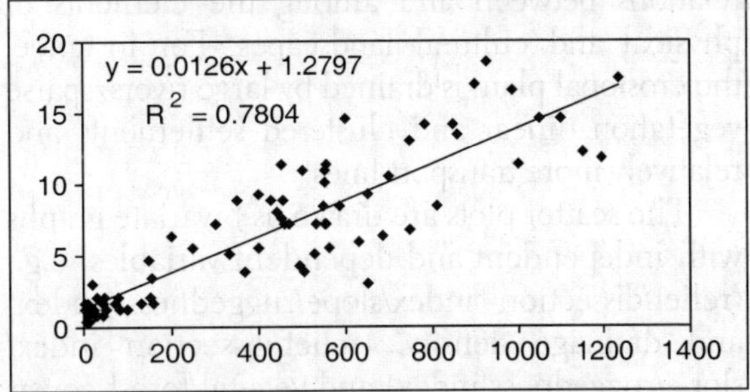

Fig. 8.25 Slope (x) and Transport Density (y)

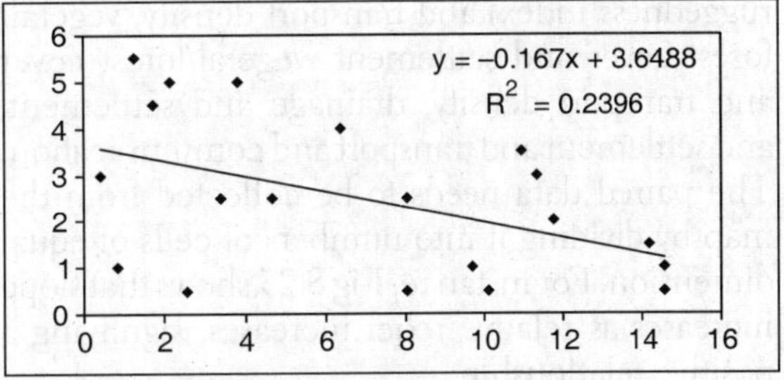

Fig. 8.26 Regions by Operators: Relative Relief and Settlement Frequency

	A	B	C
1	$R_H . S_M$	$R_L . S_H$	$R_L . S_H$
2	$R_H . S_L$	$R_H . S_L$	$R_L . S_H$
3	$R_M . S_L$	$R_M . S_M$	$R_M . S_M$

The transect chart is a construct of superimposed themes or data layers of relief, drainage, natural vegetation, settlement and transport and communication along with their spatial variations. The chart is drawn along a line and comprises rectangular blocks of roughly equal vertical dimensions for clarity of understanding. The base block shows the pattern of relief by a contour section. Above this lies the block of drainage, which is shown as vertical line features with different styles to emphasise the different types of drainage. This is overlaid by a block of natural vegetation which is shown as rectangular blocks with varying degrees of shading to emphasise variations in density and extent. The block of settlement is then drawn again with varying shades to emphasise the areal pattern of settlements. The block of transport and communication is drawn as vertical line features with different styles to emphasise different types of transport and communication (Fig. 8.24). The chart produces a diagrammatic image of the inter-relations between and among the elements of physical and cultural landscapes. For instance, the erosional plain is drained by large rivers, sparse vegetation, linear and clustered settlements and relatively more transport lines.

The scatter plots are drawn as bivariate graphs with independent and dependent variables, e.g., (relief/dissection index/slope/ruggedness index) and drainage density, (relief/dissection index/ slope/ruggedness index) and vegetal/forest cover, (relief/dissection index/slope/ruggedness index) and settlement, (relief/dissection index/slope/ ruggedness index) and transport density, vegetal/ forest cover and settlement, vegetal/forest cover and transport density, drainage and settlement, and settlement and transport and communication. The paired data needs to be collected from the map by dividing it into numbers of cells of equal dimension. For instance, Fig 8.25 shows that slope increases as relative relief increases, signifying a positive relationship:

$$S = 0.0126\ RR + 1.2797$$

goodness of fit = 0.78.

Fig 8.26 shows that density of transport decreases as slope increases, implying a negative relationship:

$$TD = -0.167S + 3.6488$$

goodness of fit = 0.24.

Overlays may also be drawn to cartographically visualise the spatial correspondence between the thematic layers. For this, each grid of sizable dimension needs to be mapped for distributions preferably with 3-class distribution of user's choice describing high (H), medium/moderate (M) and low/less (L). For instance, the map for the line 5' × 5' grids (Fig. 8.27) shows that a more or less perfect relation exists between relief and settlement frequency in A2, B1, B2, B3, C1, C2 and C3 grids, i.e., high relief–less settlement, low relief–high settlement and moderate relief–moderate settlement conditions. In grid A1, although relief is high, moderate number of settlements have come up along the banks and valley floors of Gohra N and Jaraikela N. In grid A3, settlements are relatively less because although relief is moderate, its spatial variation is very high and larger rivers are lacking, thereby restricting the number of settlements.

EXERCISE

1. What is a topographical map?
2. What is meant by map craft?
3. What is the first essential task of topographical map reading?
4. Name the type of information given in the top margin of a topographical map?
5. Name the type of information given in the bottom margin of a topographical map?
6. Name the type of information given in the left margin of a topographical map?
7. Name the type of information given in the right margin of a topographical map?
8. What are 1M sheets? Give their extension and scale. How many of these actually cover India?
9. What are degree sheets? Give their extension and scale.
10. What are quadrant sheets? Give their extension and scale.
11. What are inch sheets? Give their extension and scale.
12. Mention the contour intervals for degree sheets, half-degree sheets and inch sheets.
13. Mention the functional elements of topographical map interpretation.
14. State with illustrations how different relief features are shown in a topographical map.
15. State with illustrations how different drainage features are shown in a topographical map.
16. State with illustrations how different features of natural vegetation are shown in a topographical map.
17. State with illustrations how different features of settlement are shown in a topographical map.

18. State with illustrations how different features of transport and communication are shown in a topographical map.
19. What is a contour section?
20. What is meant by serial profiles?
21. Why are serial profiles important for relief interpretation?
22. Explain the term along with its importance and use: *superimposed profile, projected profile and composite profile.*
23. What is meant by representative section? Why is it drawn and how is its location on the map justified?
24. Draw all necessary profiles for a topographical map and interpret.
25. Find the gradient between the highest and the lowest point on a topographical map.
26. Divide the map into suitable broad physical divisions stating its basis. Give the general morphometric properties of each division in a tabular form.
27. Draw an altimetric frequency histogram and interpret it.
28. Draw a hypsometric curve and interpret it.
29. Explain the terms and write the expressions for measurement: slope, average slope, gradient, relative relief, dissection index, drainage frequency, drainage density, constant of channel maintenance, ruggedness number, drainage texture, length of overland flow, form factor, circularity ratio, elongation ratio, relief ratio, lamniscate ratio, sinuosity index, steam order, bifurcation ratio, length ratio, area ratio, slope ratio, settlement frequency, settlement density, nearest neighbour index, connectivity index, and detour index.
30. Identify the landforms from a topographical map and account for their origin and occurrence.
31. Identify the different forms of slope from a topographical map and explain its origin.
32. Draw a relative relief map and interpret it.
33. Draw a dissection index map and interpret it.
34. Draw a slope map after Wentworth and interpret it.
35. Draw a slope map after Smith and interpret it.
36. Draw a slope map after Raisz and Henry and interpret it.
37. Draw a slope map after Robinson and interpret it.
38. Identify the different drainage patterns from a topographical map and explain its origin.
39. Identify the different channel patterns from a topographical map and explain its origin.
40. Identify drainage features from a topographical map and interpret it.
41. Identify the features formed due to river erosion and deposition from a topographical map.
42. Identify the features of natural vegetation from a topographical map and interpret it.
43. Based on site, identify the different settlement patterns from a topographical map and interpret it.
44. Based on situation, identify different settlement patterns from a topographical map and interpret it.
45. Based on spatial pattern, identify the different settlement patterns and interpret it.
46. Based on the connectivity index, identify the different networks of transport and interpret it.
47. From a topographical map with the help of transect charts establish the relations between: *relief and settlement, relief and drainage, relief and natural vegetation, relief and transport and communication, settlement and drainage, and settlement and transport and communication.*
48. From a topographical map with the help of thematic maps establish the relations between: *relief and settlement, relief and drainage, relief and natural vegetation, relief and transport and communication, settlement and drainage, and settlement and transport and communication.*
49. From a topographical map with the help of scatter diagrams establish the relations between: *relief and settlement, relief and drainage, relief and natural vegetation, relief and transport and communication, settlement and drainage, and settlement and transport and communication.*
50. Interpret a topographical map under the following heads: *relief, drainage, natural vegetation, settlement, and transport and communication.*

9

Remote Sensing and Image Interpretation

HIGHLIGHTS

- Characteristics of Remote Sensing
- Development of Remote Sensing
- Principles of Remote Sensing
- Nature of EM Radiation
- Atmosphere-Radiation Interaction
- Surface-Radiation Interaction
- Types of Remote Sensing
- Aerial Photography
- Types of Aerial Photographs
- Photogrammetry
- Interpretation of Aerial Photographs
- Imaging Systems
- Imaging Sensors
- Image Interpretation
- Visual Image Analysis
- Digital Image Processing
- Image Enhancement
- Image Transformation
- Image Classification

The term remote sensing means obtaining information about an object without touching the object itself. It concerns two essential facets—the *technology of acquiring data* through *a device which is located at a distance from the object,* both being intimately linked with each other. In all practicality, the intervening space between the sensor and the sensed object (i.e., earth) is filled with air (aerial platform) or is even partly vacuum (space platform). Only electromagnetic waves are able to serve as *an efficient link* between the two. Hence, *remote sensing* (also called RS) implies the "data acquisition of electromagnetic radiation from

Major Advantages of Remote Sensing

- **Synoptic Overview:** RS provides a synoptic view, so that environmental data covering a large area of the earth can be captured instantaneously and then be processed to generate map-like products to study the various spatial features in relation to each other, and to delineate its nature of regional trends and orientation.
- **Feasibility Aspect:** Since some areas on the earth's surface are not accessible to ground survey, the only feasible way of obtaining information about such areas may be from RS platforms.
- **Time Saving:** The RS technique saves time and man power, as information about a large area is gathered quickly.
- **Multidisciplinary Applications:** The same RS data can be used by researchers/workers in different disciplines and, therefore, the overall benefit-to-cost ratio is higher.
- **Others:**
  - The photogrammetric systems are marked by analogy to the eye system and have high geometric fidelity.
  - Scanners provide RS data such that the digital information is directly telemetered from space to ground.
  - Imaging radars possess the unique advantages of all-weather and all-time capability.
  - It provides multispectral and multiscale data for GIS database, thereby minimizing the in situ data collection.

sensors flying on aerial or space platforms, and its interpretation for deciphering ground object characteristics." (Gupta 2003). Although remote sensing has been variously defined, the most concise and technical definition is that "it is the practice of deriving information about the earth's land and water surfaces using images acquired from an overhead perspective, using electromagnetic radiation in one or more regions of the electromagnetic spectrum, reflected or emitted from the earth's surface." (Campbell 1997).

CHARACTERISTICS OF REMOTE SENSING

Remote sensing has several unique features that have made it very useful in geographical analysis, particularly *measuring*, *mapping*, *monitoring* and *modelling* the surface features of the earth. It is certainly not just a data-collection process; it also includes data analysis—*the methods and processes of extracting meaningful high-quality spatial (geographic) information from the remotely sensed data for direct input in a modern GIS application*. Traditionally, such environmental data was collected through direct measurements in the field through instrumentation and ground survey. It was very *expensive*, *labour intensive* and *time consuming*. As an alternative, the *science of remote sensing* developed very fast.and has become universal in nature and application. From an orbital platform in space, it provides the *easiest*, *cheapest* and *quickest* means to keep our GIS database up-to-date. As digital *RS data is perfectly compatible with the raster-based GIS data model*, it can easily be *integrated* with other types of raster GIS data for further analysis.

On the earth's surface, *land use and land cover are the two most vital clues* that meaningfully reflect the 'geography of human drama'. Hence, the most common application of RS images and aerial photographs is mapping the land use and land cover along with detecting the changes that have taken place over the year. RS is also used to collect *information on several important bio-physical variables*, which include planimetric location (x, y), topographic/bathymetric elevation (x, y, z), colour and spectral signature, vegetation chlorophyll absorption characteristics, vegetation biomass, vegetation moisture content, soil moisture content, temperature, and texture/surface roughness (Jensen 1983).

DEVELOPMENT OF REMOTE SENSING

Remote sensing is a relatively *young* scientific discipline, and is an area of *emerging technology* that has undergone phenomenal growth during

Remote Sensing: Some Definitions

- ... imagery is acquired by a sensor other than a conventional camera through which a scene is recorded, such as by electronic scanning, using radiations outside the normal visual range of the film and camera—microwave, radar, thermal, infrared, ultraviolet, and multispectral, special techniques are applied to process and interpret remote sensing imagery for the purpose of producing conventional maps, thematic maps, resource surveys, etc. in the fields of agriculture, archaeology, forestry, geography, geology, and others (American Society of Photogrammetry).
- Remote sensing denotes the joint effects of employing modern sensors, data processing equipment, information theory and processing methodology, communication theory and devices, space and airborne vehicles, and large-system theory and practice for the purposes of carrying out aerial or space surveys of the earth's surface (National Academy of Sciences 1970).
- Remote sensing is the acquisition of physical data of an object without touch or contact (Lintz and Simonett 1976).
- Remote sensing includes all methods of obtaining pictures or other forms of electromagnetic records of the earth's surface from a distance, and the treatment and processing of the picture data ... is concerned with detecting and recording electromagnetic radiation from the target areas in the field of view of the sensor instrument. This radiation may have originated directly from separate components of the target area; it may be solar energy reflected from them, or it may be reflections of energy transmitted to the target area from the sensor itself (White 1977).
- Remote sensing is simply the art and science of acquisition, measurement, interpretation and evolution of photographs, imageries and other remotely sensed data (Moffitt and Mikhail 1980).

the last three decades. It has dramatically *enhanced* our capability for resource exploration, mapping and monitoring the earth's environment on *local*, *regional* and *global scales* (Fischer 1975, Williams and Southworth 1984). The locus of its development may be traced as follows.

Period of Initiation (19th Century)

The history of remote sensing dates back to the beginning of the practice of photography in the early 1800s. In 1839 Louis Daguerre publicly reported the result of his experiments with photographic chemicals. The use of photography to record an aerial view of the earth's surface (from a captive balloon) dates back to 1858. In the following years, various improvements were made both in the technology related to photography and in the methods of acquiring photographs of the earth. These aerial images of the earth were the first instances of remote sensing, but it must be noted that these were done mostly out of curiosity and did not form the basis for a systematic field of study. Theoretical knowledge increased with practical research and James Clerk Maxwell developed the theory of Electromagnetic Energy in 1873.

Period of Aerial Photography (1900–1960)

It began with the use of powered airplanes as platforms for aerial photographs. The manoeuvrability of the airplane provided the necessary control of speed, altitude and direction required for a systematic use of the airborne camera. The motion picture of the Italian landscape near Centocelli taken by Wilbur Wright in 1910 is said to be the first aerial photograph taken from an airplane. World War I (1914–1919) marked the beginning of the age of aerial photography. During the war, equipments were designed specifically for it and many people were trained in the acquisition, processing, and interpretation of aerial photographs, many of whom later became pioneers in the application of aerial photography in civilian endeavours. Camera designs were improved and tailored specifically for use in aircrafts. Photogrammetric instruments, specifically designed for the analysis of aerial photographs, were developed for acquiring the accurate measurements from photographs.

Milestones in the History of Remote Sensing

1800	Discovery of infrared by Sir Willam Herschel
1820	Niepce took the first photograph of nature
1839	Practice of photography began
1847	Infrared spectrum shown by Fizeau and Foucault to share properties with visible light
1850–1860	Photography from balloons
1862	Forest mapping from aerial photographs
1873	Theory of electromagnetic energy developed by Maxwell
1909	Photography from airplanes
1910–1920	World War I: Aerial reconnaissance
1920–1930	Development and initial application of aerial photography and photogrammetry
1930–1940	Development of the radar in Germany, United States and United Kingdom
1940–1950	World War II: Application of the non-visible portions of the EMS; training persons in the acquisition and interpretation of aerial photographs
1950–1960	Military research and development
1956	Colwel's research on disease detection with infrared photography
1960–1970	The term remote sensing was first used, TIROS weather satellite, Skylab remote sensing observations from space
1972	Launch of Landsat 1
1978	Launch of Seasat
1970–1980	Rapid advances in digital image processing
1980–1990	Launch of Landsat 4: new generation of Landsat sensors
1986	SPOT (French EO Satellite)
1980s	Development of hyperspectral sensors
1990s	Launch of ERS 1, UARS
2000s	Terra (MODIS), EROS A1, AQUA (MODIS), SPOT5, Quick Bird, IKONOS, Cartosat: Resouresat, GSAT, RiSat, IRNSS, WiniSat, SkySat, etc.

With this another landmark was established—the more or less routine application of aerial photography in government programmes, initially for topographic mapping but later for soil survey, geologic mapping, forest survey and agricultural studies. An increased number of aerial photographs became available as the number of experts trained in the acquisition and analysis of aerial photographs increased. The pace was, however, very slow mainly because of the resistance from traditionalists, imperfections in equipment and training, and some genuine uncertainties regarding the proper role of aerial photography in scientific inquiry and practical application. During World War II (1939–1945), the use of the electromagnetic spectrum was extended from the visible spectrum to the infrared and microwave regions, far beyond the range of human vision. Wartime research and operational experience quickly expanded not only the base of the theoretical and practical knowledge about photogrammetry, but also the pool of experienced persons, such as pilots, camera operators and photo-interpreters. The wartime trend continued during the post-war period and most of the established capabilities found their way into civilian applications. Reconnaissance techniques improved further as the Cold War between the western democracies and the Soviet Union set in. These were often held as defense secrets for many years until they were replaced by more sophisticated techniques. In the civilian sphere, the most significant development was the application of the 'colour infrared film' (popularly known as the camouflage detection film) in 1956 by Robert Colwell for the identification of small grain cereal crops and their diseases. This was definitely a milestone in the development of remote sensing.

Period of Satellite Imagery (1960–1980)

A major landmark in the history of remote sensing was the decision to land man on the moon. It generated the space race between the US and Soviet Union, which finally led to the rapid development of space technology and sensing systems. The US National Aeronautics and Space Administration (NASA), the European Space Agency and the national space agencies of a number of countries, like Canada, Japan, Russia, China, India and Brazil developed many aerial and spaceborne programmes, that made the remotely sensed data available worldwide. All these boosted the remote sensing technology and yielded valuable data and images of the earth from space.

The first space photography of the earth was transmitted by Explorer-6 in 1959. The Mercury Programme provided orbital photography (70 mm colour) from an unmanned automatic camera in 1960. The first meteorological satellite, TIROS-I was launched in April 1960 followed by the ITOS and NOAA series. The newly designed instruments extended the reach of aerial observation outside the visible spectrum into the infrared and microwave regions. It was in this context that the name 'remote sensing' was first used by Evelyn Pruit, a scientist of the US Navy's Office of Naval Research, as the term 'aerial photography' could no longer describe accurately the imaging processes and systems outside the visible region of the spectrum. The Gemini mission (1965) provided a number of good vertical and oblique stereo photographs, that formally demonstrated the potential of RS techniques in the exploration of earth's resources (Lowman 1969). Later, the Apollo Programme experimented on the coverage of the earth by stereo-vertical photography and multispectral 70 mm photography, which paved the way for unmanned space orbital sensors. In 1972, with the launch of the first Earth Resources Technology Satellite ERTS-1, later renamed 'Landsat-1', a new era began. It carried a 4-channel multispectral scanning system (MSS) and a tape recorder on board, which provided extremely valuable data on worldwide distribution.

Table 9.1 Landsat Missions

Satellite	*Period*	*Principal Sensor*
Landsat-1	23.06.1972–06.01.1978	MSS, RBV
Landsat-2	22.01.1975–27.07.1983	MSS, RBV
Landsat-3	05.03.1978–07.09.1983	MSS, RBV
Landsat-4	16.07.1982–	TM, MSS
Landsat-5	01.03.1984–	TM, MSS
Landsat -6 (failed at launch)	05.10.1993	ETM
Landsat-7	1998	ETM+
Landsat-7	follow-on	LATI
MSS	Multispectral Scanning System	
RBV	Return Beam Vidicon	
TM	Thematic Mapper	
TM+	Thematic Mapper plus	
ETM	Enhanced TM	
ETM+	Enhanced TM plus	
LATI	Landsat Advanced Technology Instrument.	

Modern Period of Satellite Imagery (1980–)

The modern period began in the early 1980s, with the introduction of more improved scanning systems. These provided spatial data with finer resolution. The development of electronic technology led to the design of solid state scanners in the 1990s, that became an immediate favourite of the Landsat Series, the French SPOT Series, the Indian IRS Series, the Japanese MOS and Fuyo series, and the China-Brazil series, e.g., TM (Thematic Mapper), TM+ (Thematic Mapper plus), ETM (Enhanced TM), ETM+ (Enhanced TM plus), ATSR-2 (Advanced Along-Track Scanning Radiometer), HRS (High Resolution Stereoscopic), HRG (High Resolution Geometric), LISS (Linear Imaging Scanning System),WiFS (Wide Field Sensor), AVHRR (Advanced Very High Resolution Radiometer), SEAWiFS (Sea-viewing

Wide Field-of-view Sensor), MODIS (Moderate Resolution Imaging Spectrometer), ASTER (Advanced Spaceborne Thermal Emission and Reflection Radiometer), and LATI (Landsat Advanced Technology Instrument). During this period the space transportation system and the reusable space shuttle was developed which provided an easy launching, on-board modular approach and facilitated the trial of sensors from an orbital platform, e.g., experiments with the Metric Camera, the Large Format Camera, the Modular Optoelectronic Multispectral Scanner (MOMS) and the Shuttle Imaging Radar Series (SIR-A, SIR-B, SIR-C). Japan's Fuyo, Canada's Radarsat, and ESA's ERS-1/2 Programmes in the 1990s provided SAR with data of the earth from space, which was another great step in remote sensing application, allowing monitoring of ground terrain from space with accuracy in the range of centimetres.

The development of the *CCD-area array technology* led to the design and fabrication of area-array chips with a large number of detector cells. This *hyperspectral* remote sensing technology is being used by the most modern high-spatial-resolution remote sensing systems, e.g., *IKONOS-2* of Space Imaging-EOSAT with 1m spatial resolution in January 2000. The Shuttle Radar Topography Mission (SRTM) collected 224 hours+ 3D radar images of the earth between 60°N and 54°S using C-band and X-band interferrometric synthetic aperture radar (IFSAR) in February 2000, acquiring topographic data for over 80% of the earth's landmass. In 2001, Earth Watch's *Quickbird-2* was launched to image the earth at 0.61 m resolution for the panchromatic band and 2.5 m for the multispectral bands, with stereoscopic capability, making it the highest spatial resolution civilian imaging satellite system (Lo and Yeung 2002). During the past few years, micro-electronics and computer technology have been revolutionised and image processing facilities are now widely available, which, together, are responsible for greater dissemination of remote sensing image processing knowledge and interest in such studies.

PRINCIPLES OF REMOTE SENSING

The fundamental principle of the remote sensing methods is that *in different wavelength ranges of the electromagnetic spectrum, each type of object reflects or emits a certain intensity of light, which is dependent upon the physical or compositional attributes of the object* (Fig. 9.1). Thus using information from one or more wavelength ranges, it is definitely possible to *differentiate* between different types of objects and to map their spatial distribution on the ground. The curves showing the intensity of light emitted or reflected by the objects at different wavelengths are called *spectral response curves*. These contain all the basic information necessary for the successful planning of a remote sensing mission. The remote sensing data acquired from the aerial/spaceborne sensors are *processed*, *enhanced* and *interpreted* for applications.

Remote sensing typically takes 1 of 3 basic forms depending on the wavelengths of energy detected and the purpose of study. The *simplest* form is one that records the reflection of solar radiation from the earth's surface. It mainly uses energy in the visible and near infrared portions of the spectrum. The *second* form records radiation emitted from (rather than reflected from) the earth's surface. Because the emitted energy is the strongest in the far infrared spectrum, this kind of remote sensing requires special instruments designed to record these wavelengths. It reveals information about the thermal properties of the

Remote Sensing Systems

Passive Systems (energy source is independent of the recording instrument)

- Analogue (AP, VD)
- Digital (MSS, LAS, SR)

Active Systems (send out their own electromagnetic radiation at a specified wavelength to the ground and then sample the portion reflected back to the detecting devices)

- Analogue (SLAR)
- Digital (SAR)

Table 9.2 Simple Taxonomy of the Remote Sensing Systems

	Passive Systems			*Active Systems*	
	Reflected Sunlight	*Thermal Emission*			
		Infrared	*Microwave (Radio)*	*Visible/IR*	*Microwave (Radio)*
Non-imaging		Thermal Infrared Radiometry	Passive Microwave Radiometry	Laser Profiling	Radar Altimetry Microwave Scatterometry
Imaging	Aerial Photography Visible/Near-Infrared Imaging	Thermal Infrared Imaging	Passive Microwave Radiometry		Real Aperture Radar (RAR), Synthetic Aperture Radar (SAR)
Sounding	Ultraviolet Backscatter Sounding	Thermal Infrared Sounding	Passive Microwave Sounding	Lidar	

objects, which can be interpreted to suggest patterns of moisture, vegetation, surface materials and man-made structures. Finally, the *third* form of remote sensing instruments generate their own energy, then record the reflection of that energy from the earth's surface. These are *active* sensors which are best represented by imaging radars that transmit microwave signals towards the earth's surface from an aircraft or satellite, then use the reflected energy to form an image. Because they sense energy provided directly by the sensor itself, such instruments have the capability of operating both at night and during cloudy weather.

Thus, the remote sensing systems are of two types—*passive* and *active*. In the *passive* systems, the energy source is *independent* of the recording instrument. They use the radiation emitted and reflected from ground surfaces, e.g., camera and TIR detectors. These can further be subdivided into two categories—*analogue* (e.g., aerial photographs and videographs) and *digital* (e.g., scanners and spectroradiometers). The *active* systems, on the other hand, send out their own electromagnetic radiation at a specified wavelength to the ground and then sample the portion reflected back to the detecting devices, e.g., an imaging radar. The active systems can also be sub-divided into two categories—*analogue* (e.g., SLAR images of the past) and *digital* (e.g., SAR images).

The electromagnetic radiation travelling through the atmosphere is selectively scattered and absorbed, depending upon the composition of the atmosphere and the wavelength involved. Sensors, such as photographic cameras, scanners or radiometers mounted on suitable platforms, record the radiation intensities in various spectral channels. The remotely sensed data are then digitally processed for rectification and enhancement and then integrated with ground truth and other reference data. The processed products are interpreted for identification/discrimination of ground objects and the thematic data layers are integrated with other multidisciplinary spatial data and ground truth data to be used for decision making by scientists and managers.

NATURE OF EM RADIATION

In remote sensing, electromagnetic (EM) radiation serves as the main communication link between the sensor and the object. The wave nature of EM radiation has been conceptualised by Maxwell as follows: *EM energy propagates in harmonic sinusoidal wave motion, consisting*

Fig. 9.1 Electromagnetic Wave: Electric and Magnetic Components are Perpendicular to Each Other and to the Direction of Propagation

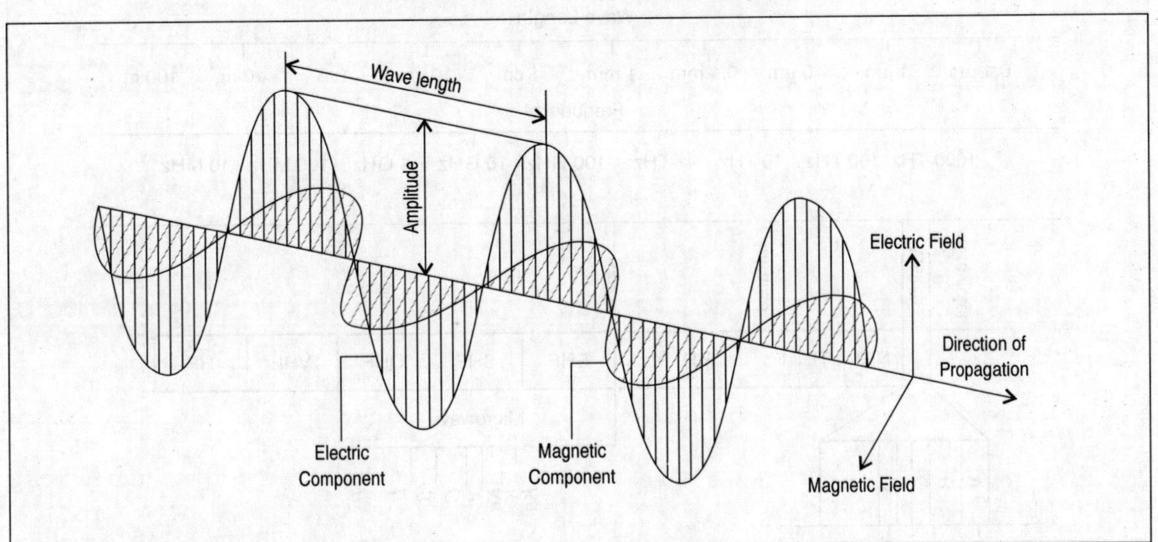

of inseparable oscillating electric and magnetic fields that are always perpendicular to one another and to the direction of propagation. All EM radiation travel with the same speed in a particular medium and is equal to the velocity of light in a vacuum. The fundamental equation is:

$$c = f.\lambda$$

where, c = velocity of light (3×10^8 m/sec), f = frequency of oscillation (hertz or cycles/sec) and l = wave length (μm or 10^{-6} m).

The particle or quantum nature of EM radiation, as postulated by *Max Planck*, states that EM radiation is composed of numerous tiny indivisible discrete packets of energy, called *photons* or *quanta*. The *energy of a photon is given* as:

$$Q = h.f$$

$$\text{or } Q = h.(c/\lambda)$$

where, Q = energy of a photon (Jules), h = Planck's constant (6.626×10^{-34} J sec), and f = frequency.

This means that the *energy of a photon is inversely proportional to its wavelength*. The longer the wavelength (or shorter the frequency) the lower its energy content and vice versa. This explains beautifully the phenomena of *blackbody radiation, selective absorption* and *photoelectric effect*. Hence, naturally emitted long wavelength radiation, such as microwave emission from terrain features, is more difficult to sense than the radiation of shorter wavelengths, e.g., emitted thermal infrared energy. The low energy content of long wavelength radiation means that the systems operating at long wavelengths must view *large areas of the earth* at any given time to obtain a detectable energy signal.

All matter at temperatures above absolute zero (0°K or 273°C) continuously emits EM radiation. The *intensity* and *spectral composition* of the emitted radiation depends upon the *composition* and *temperature* of the body. A blackbody is an ideal body that absorbs all radiation incident on it, without any reflection. It has a continuous spectral emission curve, unlike the natural bodies that emit only at discrete spectral bands, depending upon the composition of the body. The dominant wavelength at which a blackbody radiation curve reaches a maximum is related to its temperature by Wien's displacement law:

$$\lambda_{max} = A/T,$$

where, λ_{max} = wavelength (cm) at which peak radiation occurs, A = a constant (= 0.29 cm K) and T = temperature of the body, K.

Fig. 9.2 Electromagnetic Spectrum

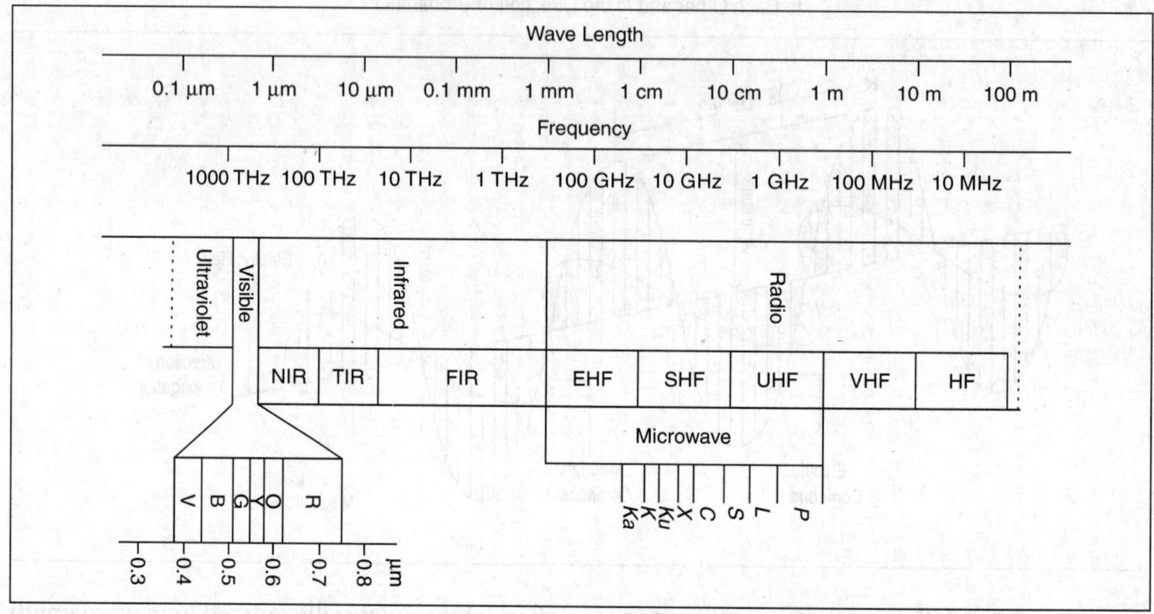

Table 9.3 Principal Divisions of the Electromagnetic Spectrum

Major Division	*Sub-Division*	*Range of Wavelength*	*Remarks*
Gamma rays		below 10^{-4} µm	
X-rays		10^{-2}–10^{-4} µm	
Ultraviolet radiation	UV–C (extreme)	below 0.28 µm	
	UV–B (far)	0.28 – 0.32µm	
	UV–A (near)	0.32 – 0.40 µm	
Visible	Blue	0.4 –0.5 µm	Reflective Range; Optical Range
	Green	0.5 – 0.6 µm	
	Red	0.6 – 0.7 µm	
Infrared radiation	Near infrared (NIR)	0.7 – 1.0 µm	
	Shortwave infrared (SWIR)	1.0 – 3.0 µm	
	Mid infrared (MIR)	3 – 35 µm	Thermal Infrared Range
	Far infrared (FIR)	35 – 1000 µm	
Microwave Radiation		1 mm – 1 m	
Radiowave		above 1 m	

The *Stefan-Boltzman law* gives the total radiation emitted by a blackbody over the entire EM range as:

$$W = \sigma.T^4$$

where, W = the total radiation emitted by a blackbody over the entire EM range (watts/m²), s = Stefan-Boltzman constant, 5.6697×10^{-8} watts/m²/K⁴, T = absolute temperature (K) of the emitting material.

Another important radiation relationship is given by the *Rayleigh–Jeans law*, valid for the microwaves:

$W_\lambda = 2\pi c\ kT / \lambda^4$.

where, W_λ = spectral radiance, i.e., the energy radiated per unit wavelength per second per unit area of the blackbody, and k = Boltzman's constant.

The maximum spectral radiation from earth features occurs at 9.7 μm. As it correlates with terrestrial heat, it is called *thermal infrared energy* which can neither be seen nor photographed but can be sensed with thermal devices like radiometers and scanners. By comparison, the sun has a much higher energy peak that occurs at about 0.5 μm that can both be seen and photographed.

The *electromagnetic spectrum* is the ordering of EM radiation according to its wavelength, in other words, its frequency or energy. The EM spectrum is commonly presented between cosmic rays and radiowaves, the intervening parts being gamma rays, X-rays, ultraviolet, visible, near infrared, thermal infrared, far infrared and microwave. The EM spectrum from 0.02 μm to 1 m wavelength can be divided into two main parts—*optical range* and *microwave range*. The optical range refers to that part of the EM spectrum in which the optical phenomena of reflection and refraction can be used to focus the radiation. The microwave range is from 1 mm to 1 m wavelength. For remote sensing purposes, the most important *spectral regions* are 0.4–14 μm (lying in the optical range) and 2 mm–0.8 m (lying in the microwave range).

ATMOSPHERE-RADIATION INTERACTION

The radiation reflected and emitted by the earth passes through the atmosphere and interacts with its constituents like, gases (CO_2, H_2O vapour, O_3, etc.) and suspended materials (aerosols, dust particles, etc.). In the process, it gets partly *scattered*, *absorbed* and *transmitted*. The degree of such interaction depends on the *path length* (i.e., the distance travelled by the radiation through the atmosphere) and the wavelength. *Path length* depends on the location of the energy source and altitude of the sensor platform. In passive sensing, radiation travels *twice* through the atmosphere—*first* Sun → Earth and *second* Earth → Sensor, while *terrestrial* radiation traverses the atmosphere only *once*.

Sensor platforms may lie at *low altitude* or at *high altitude* or at *space altitude*, affecting thereby the quality of radiation received or sensed. During attenuation some of the wavelengths are transmitted with *higher efficiency*, while others are more susceptible to *scattering* and *absorption*. A remote sensor collects the total radiation composed of two components—*ground radiance* (i.e., emanation from the ground) and *path radiance* (i.e., generated by atmospheric effects). Path radiance tends to mask *ground signal* and acts as a *background noise*, thereby affecting the quality of images and data that the sensors finally generate.

The atmosphere-radiation interactions can be grouped into three physical processes—*scattering, absorption* and *refraction*. Scattering takes place due to diffuse multiple reflections of EM radiation by gas molecules and suspended particles in two basic forms—*nonselective* and *selective*. *Nonselective* scattering occurs when all wavelengths are *equally scattered* by dust, cloud and fog. In this, the *scatterer particles* are much *larger* than the *wavelengths* of radiation.

Fig. 9.3 Types of Scattering

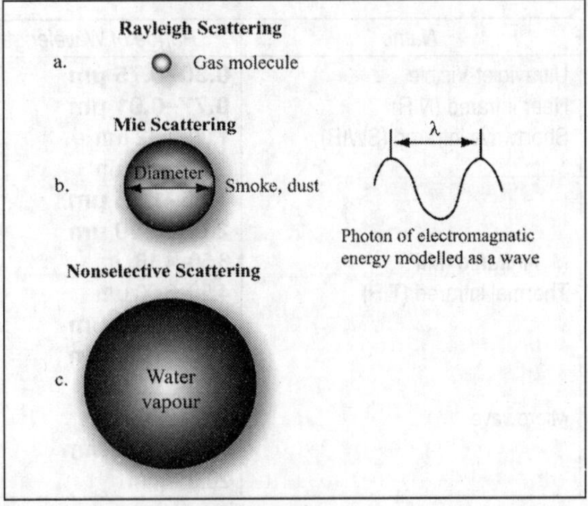

As all visible wavelengths are equally scattered, *clouds and fog appear white. Selective* scattering can take two forms—*Rayleigh scattering* and *Mie scattering*. Rayleigh scattering occurs when atmospheric particles have *diameters* that are *small relative* to the *wavelength* of radiation. It is inversely proportional to the 4th power of the wavelength, which means it causes *severe path radiance* at the ultraviolet and blue end of the spectrum and is *responsible for the blue colour of the sky*. It leads to haze on images and photographs with low contrast and poor sharpness. *Mie* scattering, on the other hand, is caused by large particles, like dust, pollen, smoke and water vapour, which are abundant and significant in lower altitudes, close to the earth's surface. *Diameters* of such particles are *roughly equivalent to the wavelength* of the scattered radiation. It influences the entire spectral region and has a greater effect on larger wavelengths than Rayleigh scattering. Its effect is more manifest in overcast atmospheric conditions.

The atmospheric gases selectively absorb EM radiation. The most effective absorbers are O_2, H_2O vapour, O_3, etc. The *spectral regions of least absorption* are called *atmospheric windows*, as they can be used for looking at the ground from aerial or space platforms through the atmosphere. The visible part of the spectrum is marked by the presence of an excellent atmospheric window. Prominent windows occur throughout the EM spectrum at intervals. The *major* ones are: 0.30–0.75 μm (UV–Visible region), 0.77–0.91 μm (NIR region), 1.55–1.75 μm and 2.05–2.40 μm (SWIR region), 8.00–9.20 μm and 10.2–12.4 μm (TIR region) and 7.5–11.5 mm (Microwave region) (Table 9.4).

SURFACE-RADIATION INTERACTION

The EM energy incident on the earth's surface may be reflected, absorbed and/or transmitted. The interrelationship between these is stated as the following energy balance equation:

$$E_I(\lambda) = E_R(\lambda) + E_A(\lambda) + E_T(\lambda)$$

where, E_I = incident energy, E_R = reflected energy, E_A = absorbed energy and E_T = transmitted energy.

The proportions of energy reflected, absorbed and transmitted certainly vary at different wavelengths, primarily because the different earth features as well as their material type and composition are not identical. The eye utilises the *spectral variations* in the magnitude of reflected energy to *discriminate between various objects* and to *distinguish different features* on an image. As reflected energy predominates in remote sensing, the energy balance equation may be rewritten as:

Table 9.4 Major Atmospheric Windows

Name	*Range of Wavelength*	*Region*	*Remarks*
Ultraviolet-Visible	**0.30–0.75 μm**	Optical Reflective	
Near infrared (NIR)	**0.77–0.91 μm**	Optical Reflective	
Shortwave infrared (SWIR)	1.00–1.12 μm		
	1.19–1.34 μm		
	1.55–1.75 μm	Optical Reflective	*clearer windows in bold*
	2.05–2.40 μm	Optical Reflective	
Mid-infrared (MIR)	3.50–4.16 μm		
Thermal infrared (TIR)	4.50–5.00 μm		
	8.00–9.20 μm	Optical TIR	
	10.2–12.4 μm	Optical TIR	8 – 14μm for *aerial sensing*
	17.0–22.0 μm		
Microwave	2.06–2.22 mm		
	7.50–11.5 mm	Microwave	
	20.0 + mm		

Fig. 9.4 Spectral Reflectance Curves

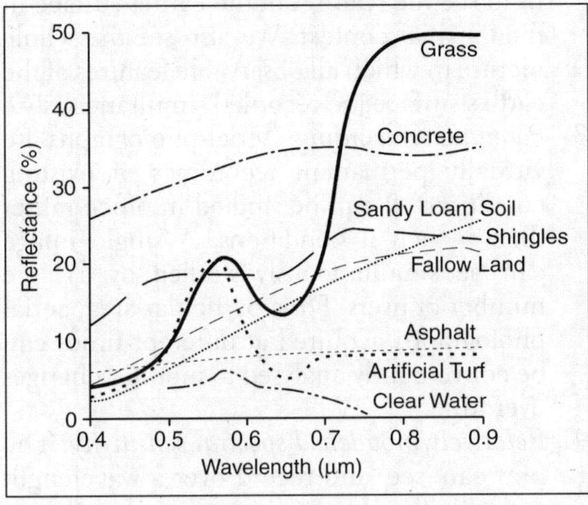

$$E_R(\lambda) = E_I(\lambda) - \{E_A(\lambda) + E_T(\lambda)\},$$

i.e., the reflected energy is *equal to* the energy incident on a given feature reduced by the energy that is either absorbed or transmitted by that feature. The reflectance of the earth's surface features is measured as a function of wavelength and is called *spectral reflectance*,

$$R_\lambda = \frac{E_R(\lambda)}{E_I(\lambda)} \text{ (expressed as a \%).}$$

Fig. 9.5 Spectral Reflectance Curves

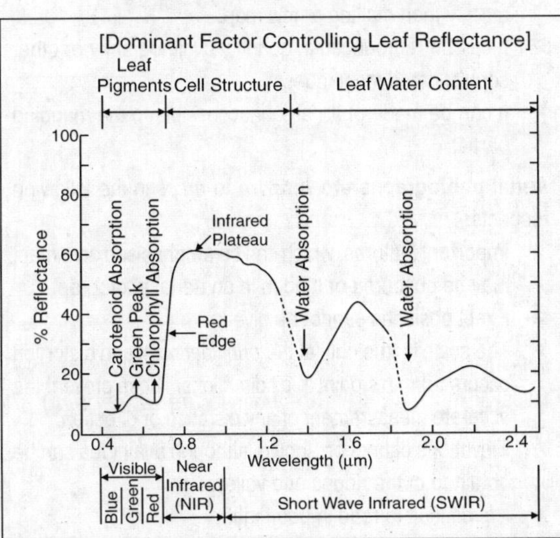

A spectral reflectance curve is the graphical plot of the spectral reflectance of an object as a function of wavelength (Fig. 9.4 and Fig. 9.5). Its geometric configuration describes precisely the spectral characteristics of an object and thus defines the wavelength regions in which the RS data are to be acquired for a particular application.

TYPES OF REMOTE SENSING

Depending on the *location of the camera* or *sensor*, remote sensing may be divided into *four* types: *close-range, terrestrial, aerial* and *space* (Fig. 9.6). The *close-range* type is also known as *bio-stereometrics* in which the camera lies very near to the object. The output is naturally the close-up photographs. In *terrestrial photogrammetry*, the camera lies on a known and fixed point on or near the ground. The ordinary photographs are the resultant products. In *aerial photography* the camera is mounted on an aircraft flying over

Fig. 9.6 Major Remote Sensing Platforms

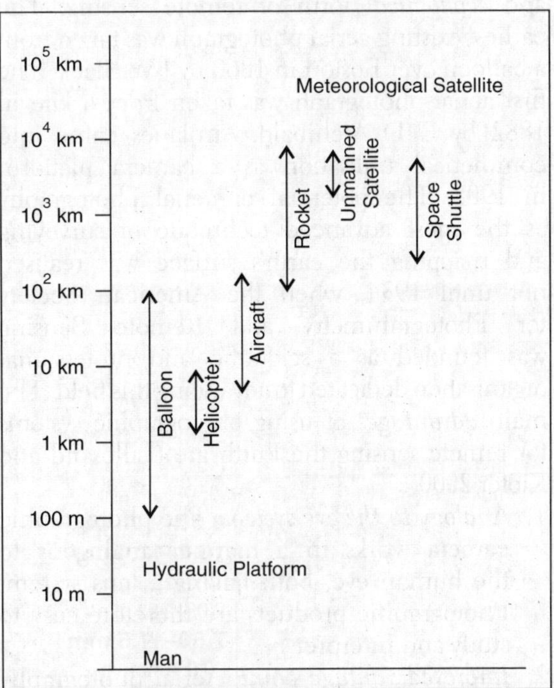

an area and aerial photographs are obviously the products generated. Aerial photography is commonly used to cover a small area for map revision purposes. The photographic scale is usually large, normally not smaller than 1:25,000. *Space shuttles* have also been used to acquire the analogue data of the earth's surface using the Large Format Camera (LFC) (Doyle 1985). In *space imaging*, sensors are mounted on either a space shuttle or an artificial satellite to capture the digital images of the earth's surface. Normally, *an unmanned polar orbiting satellite* (at an altitude range of 700–900 km) is used as a platform so that it can image the whole earth from pole to pole at a fixed interval of time. RS can be *analogue* (e.g., aerial photographs) or *digital* (e.g., soft copies of images) depending on the form of data acquisition. RS can be *small scale* or *large scale* based on the magnitude of coverage.

AERIAL PHOTOGRAPHY

Aerial photography is the most *common*, *versatile*, and *economical* form of remote sensing. The earliest existing aerial photograph was taken from a balloon over Boston in 1860 by J W Black. The first aerial photograph was taken from a kite in 1882 by E D Archibald. Airplanes came into commerical utilisation as a camera platform in 1908. The potential of aerial photography as the most advanced technique of surveying and mapping the earth's surface was realised not until 1934, when the American Society for Photogrammetry and Remote Sensing was founded as a scientific and professional organisation dedicated to advancing this field. The main *advantages* of using photographic systems for remote sensing the Earth are (Lillesand and Kiefer 2000):

1. *Analogy to the eye system*: The photographic camera works in a manner analogous to the human eye, both having a lens system. Photographic products are therefore easy to study and interpret.
2. *Improved vantage point*: Aerial photography gives a bird's eye view of large areas, enabling us to see the features of the earth's surface in their spatial context. We also see the whole picture in which all observable features of the earth's surface are recorded simultaneously.
3. *Permanent recording*: Aerial photographs are virtually permanent recordings of existing conditions. It can be studied in office rather than in actual conditions. A single image can be simultaneously studied by a large number of users. For a particular area, aerial photographs acquired at different times can be conveniently analysed to monitor changes over time.
4. *Relatively broadened spectral sensitivity*: The film can 'see' and record over a wavelength range that is about twice as broad as that of the human eye (0.3–0.9 μm).
5. *Increased spatial resolution and geometric fidelity*: It is a great advantage that interframe

Properties of Aerial Photographs

Aerial photographs are valuable for conveying geographic information for the following reasons:

- It possesses in pictorial effect a wealth of details, which no map can equal.
- It is up-to-date and possesses relative accuracy.
- It can be prepared for use in a short time, much more quickly than making a new map.
- It can be reproduced in quantity by lithography or other sophisticated techniques.
- It can be made of an area inaccessible to the mapping parties.

Aerial photographs are *inferior* to *maps* in the following respects:

- Important features, which can be emphasised on a map, may be obscured or hidden in an aerial photograph.
- Exact positions cannot be given.
- The scale is true only at the *principal point* and distortion occurs from this point in all directions. These prevent the accurate measurement of any distance or direction.
- Elevations cannot be shown, although a fair idea can be obtained of the ridges and valleys.
- It is difficult to read in poor light.

distortions do not occur. With proper selections of camera, film and flight parameters, an aerial photograph records more spatial detail. With proper ground reference data, accurate measurements of positions, distances, directions, areas, heights, volumes and slopes from aerial photographs are possible.

6. *Stereo-capability*: The photographic technique still remains the most widely used for stereoscopic studies. The spatial detail in 3D appearance becomes available to us by viewing aerial photographs under magnification with a stereoscope.

In addition, these systems also have *advantages* associated with all RS techniques, such as, *synoptic overview*, *feasibility aspects*, *time saving capability* and *multidisciplinary applications*. However, the major *limitations* are: a) film quality deteriorates with time, b) quality of photograph deteriorates with duplication, c) photographic films are sensitive only in the 0.3–0.9 μm wavelength range and the rest of the longer wavelengths cannot be photographed, and d) the use of the photographic techniques is still confined to aerial platforms and some selected space missions only.

The working principle of an aerial/space camera is simple: $\frac{1}{u}+\frac{1}{v}=\frac{1}{f}$ where, u = object distance, v = image distance, and f = focal length of the convex lens. If the object lies far away, i.e., u = infinity, the image is formed at the focal plane since, v = f. Cameras for remote sensing utilise this principle and are of fixed focus type. They carry a photosensitive film placed at the focal plane and the objects falling in the field-of-view of the lens are imaged on the film. A photographic system consists of *three* components—*camera*, *filter* and *film*.

Fig. 9.7 A Strip Camera

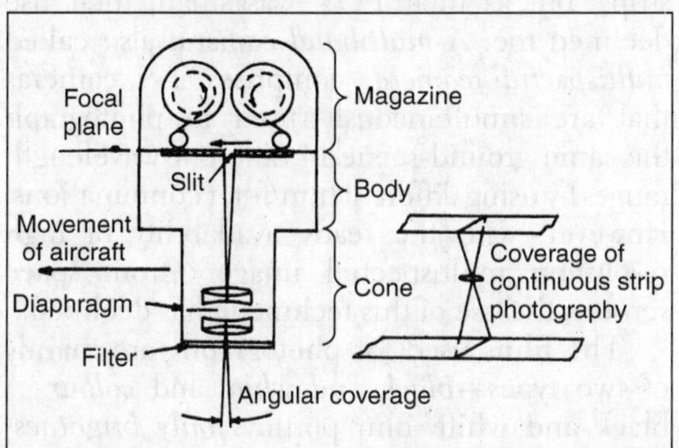

Fig. 9.8 A Frame Camera

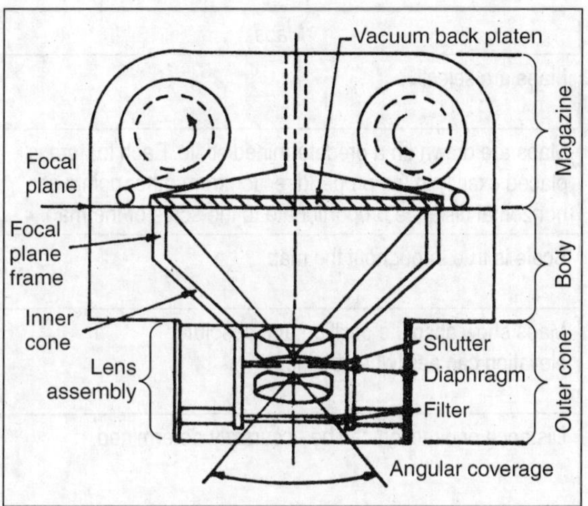

Cameras are precision equipments, normally placed on stable mounts. Depending on the film format, *four* types of cameras are distinguished—*small format* (35 mm), *medium format* (70 mm), *medium to large format* (126/140 mm), and *large format* (240 mm). For low altitude photography with high temporal resolution, *smaller format cameras* are used, while in normal cases, large format cameras are mainly used.

Depending on the construction, objective and working, there are *four* types of aerial cameras—*single lens frame cameras*, *panoramic cameras*, *strip cameras* and *multiband cameras*. Single lens frame cameras are of two types—frame reconnaissance cameras (e.g., Earth Terrain Camera) and mapping cameras. *Reconnaissance cameras* have a long focal length and a narrow FOV. *Mapping cameras*, also called *metric cameras*, *photogrammetric cameras* or *cartographic cameras*, have a very

Table 9.5 Comparison of Properties between Maps and Aerial Photographs

Maps	*Aerial Photographs*
Maps are selective.	Aerial photographs possess a wealth of details in pictorial effect which no map can equal.
Maps are drawn on a predetermined scale. Each feature is placed exactly in the proper direction from other points at a horizontal distance proportionate to the scale of the map.	Aerial photographs lack this quality.
Scale is true throughout the map.	Scale is true only at the principal point, and therefore distortion increases towards the margin.
Maps show absolute position and absolute elevation can easily be obtained.	Neither absolute positions nor absolute elevation can be obtained, although a fair idea can be obtained of the ridges and valleys.
Distance or direction can be accurately determined.	Displacements of positions caused by relief and camera tilt do not permit the accurate determination of either distance or direction.
Relative relief is readily found.	Relative relief is not readily apparent.
Areas inaccessible to surveyors cannot be mapped.	Aerial photographs can easily be taken of areas inaccessible to surveyors.
It takes time to achieve cartographic precision and accuracy.	With freedom of flight, this can be prepared in a short time.
It is easy to read even in poor light conditions.	Because of a lack of contrast in tone, it is difficult to read in poor light.
Maps emphasise certain selected features.	Aerial photographs lack this quality.
Maps are symbolised.	Aerial photographs lack this quality.
Maps are generalised.	Aerial photographs lack this quality.
Maps are usually lettered, titled and labelled.	Aerial photographs lack this quality.
Maps are related to a system of parallels and meridians.	Aerial photographs lack this quality.
Maps often show out-of-date information with no relative accuracy.	Aerial photographs present up-to-date information with relative accuracy.

high geometric accuracy. *Principal point, fiducial marks*, and *reseau marks* are exposed simultaneously with the exposure of the ground scene. Extensive flight and camera data are also shown alongside each frame, making it *very* useful for photogrammetric studies. *Panoramic* cameras were deployed in the past for photo reconnaissance survey, particularly for mapping *dynamic resources*. However, the images are highly geometrically distorted and have been discarded at present. *Strip cameras* were developed for *detailed photography* in selected strips, but as their FOV was small, their use declined too. A *multiband camera*, also called *multispectral camera*, comprises 2–9 cameras that are simultaneously used to photograph the same ground scene in different wavelength ranges by using different film/filter combinations. However, with the ready availability of high resolution multispectral imagery from space sensors, the use of this technique has declined.

The films used for photography are mainly of two types—*black and white* and *colour*. A black and white film portrays *only brightness variations* across a scene. Based on sensitivity, it can be called *panchromatic* (0.3–0.7 μm) or *infrared* (0.3–0.9 μm). Colour films are basically of two types—a) *colour negative films* which result in a colour negative and b) *reversal films* which directly yield a colour positive transparency. *Filters* permit transmission of selected wavelengths of light by absorbing or reflecting the unwanted radiation and are used to enhance image quality. Based on physical property, they are of *three* types—*spectral* filters, *neutral density* filters and *polarisation* filters. *Spectral* filters lead to spectral effects by transmitting some selected wavelengths and blocking the rest. *Neutral density* filters are used to provide uniform illumination intensity over the entire photograph. *Polarisation* filters permit the passage of only those rays that vibrate in a particular plane and block the rest.

The geometric fidelity of aerial photographs is controlled by the orientation of the optical axis of the lens system. If it is *vertical*, fidelity is *high*, and *accurate* geometric measurements such as heights, slopes, and distances (or *x*, *y*, *z*, coordinates of objects in a coordinate system) are possible from the photographs. *Vertical stereoscopic aerial photographs* are the most commonly and widely used RS data products from aerial platforms till now. Aerial photographs are said to be oblique when the optical axis is inclined. *Oblique* photography is done to view a region from a distance for *logistic* or *intelligence* purposes, to study *vertical faces* of escarpments, to read *vertical snow stacks* in snow surveys and to cover large areas in only a limited number of flights. They have *limited use* as geometric distortions are high. *Resolution* of the photographic system is defined by the closest discernible spacing of bright and dark lines of equal width, expressed as *line pairs/mm*. It depends on *lens resolution, film resolution, object contrast ratio, navigational stability, image motion* and *atmospheric condition*. Ground resolution distance (GRD) of an aerial photograph is determined by:

$$GRD = \frac{H.R_s}{f}$$

where, H = flying height, R_s = photographic resolution and f = focal length.

Aerial Photography in India

- In India, the scale of aerial photographs varies between 1:5,000 and 1:50,000.
- The most common scales for studying natural resources from aerial photographs are 1: 15,000 and 1:25,000.
- The most commonly used distortion-free high resolution aerial cameras are WILD RC 20, LMK, KA SERIES, WILD RC 10, WILD RC 10A, RMK 15/23 or ZEISS RMK 30/23.
- Lens specifications are f = 8.8 cm for super wide angle, f = 15 cm for wide angle, f = 21 cm for normal angle and f = 30 cm for narrow angle.
- The most preferred period of aerial photography spans from the end of the rainy season to the beginning of winter.
- Aerial photographs are shot when the elevation of the sun above the horizon is above 30°, i.e., during the period 3 hours before noon to 3 hours after noon, or between 08.00 to 10.00 hrs in the morning and 14.00 to 16.00 hrs in the afternoon.

TYPES OF AERIAL PHOTOGRAPHS

Aerial photographs are normally taken with the help of high precision cartographic aerial cameras like *Reconnaissance Frame Camera, Continuous Strip Camera, Panoramic Camera, Terrestrial Camera* and the *Side-Looking Airborne Radar* (SLAR) (Fig. 9.3, Fig. 9.4 and Fig. 9.5). According to the camera orientation, angular coverage and type of emulsion, aerial photographs may be classified as follows:

A. Camera Orientation (Fig. 9.6b)

i. A *vertical* photograph is that which is taken with the optical axis of the camera held in a vertical position.

Fig. 9.9 Types of Aerial Photographs (a) Vertical (b) Low Oblique (c) High Oblique (d) Convergent

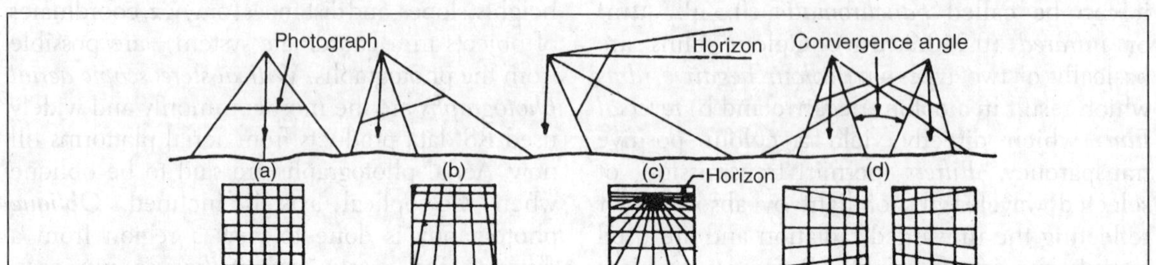

ii. A *high oblique* photograph is that which is taken with the optical axis deliberately tilted to show the earth's horizon.

iii. A *low oblique* photograph is that which is taken in an intermediate situation between i. and ii.

iv. A *convergent* photograph is a sequential pair of low obliques in which the optical axes converge towards one another, each one covering essentially the same portion of the ground.

B. Angular Coverage

(a function of focal length and format size)

i. A *narrow angle* photograph (10°–20°) with f = 610 and 915mm, used for intelligence, general interpretation and mosaics.

ii. A *normal angle* photograph (50°–75°) with f = 210 and 300 mm used, for interpretation, mapping, colour photography, mosaics and orthophotography.

iii. A *wide angle* photograph (85°–95°) with f = 153 mm, used for mapping alone

iv. A *superwide angle* photograph (110°–130°) with f = 88 mm, used for mapping areas with low relief.

C. Emulsion Type

i. A *panchromatic* black and white photograph, used for photogrammetric mapping and interpretation.

ii. A *colour* photograph, used for interpretation and, to a limited extent, mapping.

iii. An *infrared black and white* photograph, used for interpretation and detection of camouflage.

iv. An *infrared colour* photograph, used for interpretation, particularly in the analysis of plant and crop diseases, soil analysis and water pollution.

Identity of the Photograph

This can easily be done from the tilting on the air photograph by noting the Negative Number (printed on the southwest corner of the

Fig. 9.10 Identity of the Aerial Photograph

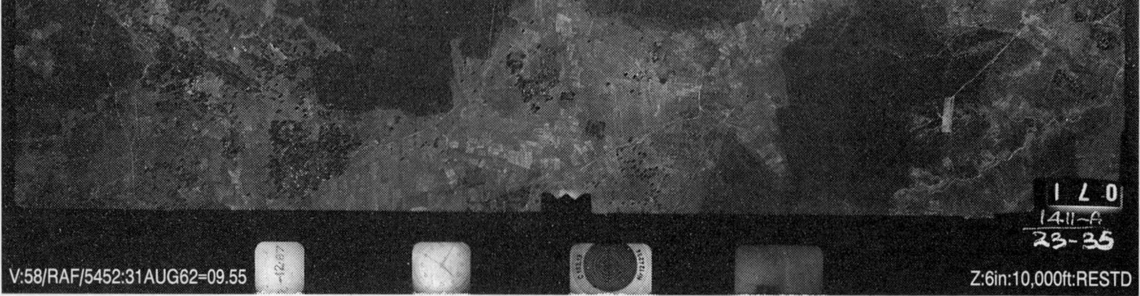

photograph), Camera Position (V–vertical, P–a port facing oblique, S–a starboard facing oblique, N–nose or a forward facing oblique, T–tail or a rear facing oblique, H–handheld), Taking Unit, Service, Sortie Number, Date (followed by =), Time Group and Zone (Z for GMT), Focal Length of lens, Altitude and Security Class. Thus 0700, V:60/RAF/3000:30 JAN 70 = 08.30Z:6 in: 5,000 ft:RESTD means negative number 0070, a vertical air photograph taken by 60 sqn RAF insortie number 3,000 on 30th January, 1970 at 08:30HRS GMT with a 6 inch lens at 15,000 ft and is a restricted one (Fig. 9.10).

PHOTOGRAMMETRY

Photogrammetry is traditionally defined as the *art* and *science* of obtaining reliable measurements by means of photography and has been extended to cover non-photographic and digital images as well (Colwell 1997). Historically, the most common use of photogrammetry has been to produce hardcopy topographical maps. Today, it is widely used to produce a range of GIS data products, e.g., precise raster image backdrops for vector data and digital elevation models (DEM). Thematic data in three dimensions can also be extracted directly from photographs for direct input in a GIS. Traditionally, photogrammetry has been divided into two types—*metrical photogrammetry* and *image interpretation*. The first concerns the quantitative measurements for planimetric and topographic mapping and has been further subdivided into two types—*analogue* and *analytical* photogrammetry. Analogue techniques involve *manual procedures for identifying, measuring and mapping the features from the photographs or images*, while in the analytical procedure, *measurement and data acquisition are performed through rigorous numerical or mathematical modelling*. The image interpretation was previously performed manually using mind-eye-intuition synchronisation. Today, it is performed through digital analysis using a sophisticated image-processing software

Types of Stereoscope

The stereoscope used for interpretation of aerial photographs may be classified as mirror, lens or prism types:

- The **Mirror Stereoscope** has a large field of view and is very adaptable for general military use. The magnifying mirror stereoscope is an improvement. The 4-power binoculars can be used to study a particular object or they may be folded back if a large field of view is desired.
- The **Lens Stereoscope**, which consists simply of spectacles, has the advantage of magnification but has only a limited field of view.
- The **Prism Stereoscope** is of particular use for examining the T-3A photographs as it permits a wide separation of the two overlapping aerial photographs.

Stereovision

- If two overlapping vertical photographs are viewed with a stereoscope, the observer will definitely get the stereovision of the *overlapped* area that helps him to interpret the photographs better.
- A pair of *two overlapping aerial photographs* arranged for stereovision is called **stereogram**.
- Two vertical photographs of an object or group of objects taken from the *same elevation* from *two different* camera positions and with an *overlap of not less than 60% not more than 75%* are known as a **stereopair**.
- *Three* vertical aerial *photographs* such that the *entire area of the central photograph is overlapped by the other two* are called **stereotriplet**.
- An **anaglyph** is a form of stereogram on which a picture is formed by almost superimposing an image in red over one in blue to secure a *stereo* or *perspective* effect when observed through an anaglyphoscope, i.e., spectacles with one blue and one red lens.
- In a **vectograph**, there is a superimposed double image representation of the two images of a stereopair. However, in this case, 'white' light, which has been polarised in two planes at right angles to each other is used. The vectograph picture must be viewed with a special pair of polarised spectacles for stereovision.

Fig. 9.11 Principles of Stereopair

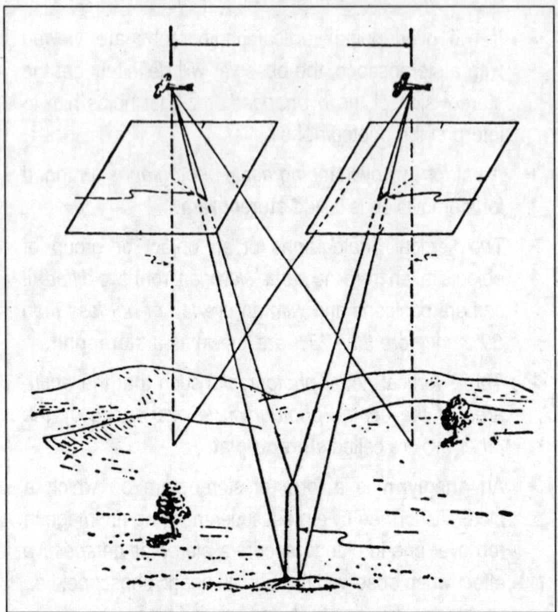

in a computer with the necessary hardware configuration and soft copy images to produce accurate topographical maps or orthomosaic.

Stereoscopic Viewing of Aerial Photographs

In photogrammetry, normally *stereoscopic pairs of vertical* aerial photographs, commonly called *stereopairs* or *stereograms*, are studied to extract maximum spatial information. A pair of aerial photographs, taken from *two* camera stations but covering some common area, constitutes a stereopair which, when viewed through a stereoscope, gives an impression as if a 3D model of the common area is being seen. A stereopair has the following essential characteristics:

- The camera axes of the photographs are co-planer.
- The scales of the two photographs are identical.
- The brightness of the photographs are similar.
- The amount of overlap in the successive photographs is at least 60% (forward) and 25% (lateral).
- The base-height ratio or B/H (where, B = distance between the exposure stations, and H = distance between an object point and the line joining the two stations) is close to 0.25.

For accurate extraction of metric data, the stereomodel needs to be set up correctly. This is done by i) *inner orientation* that sets the correct focal length and ensures that the centre of each photograph is correctly determined, ii) *relative orientation* that creates a tilt-free stereomodel from the pair of aerial photographs and iii) *absolute orientation* that relates the stereomodel to the correct terrestrial coordinate system so that any measurement on the stereomodel gives the correct x, y, and z coordinates (Petrie 1997, Wolf and Dewitt 2000). In a laboratory, the requisites for stereo viewing are a *mirror stereoscope*, a *stereopair* of photographs and *adhesive tape*. The procedure is as follows:

1. Set the mirror stereoscope on the table.
2. Take a stereopair of vertical aerial photographs.
3. By means of their collimating marks, find the principal point on the left photograph and mark it. Find also the conjugate principal point on the right photograph. The line joining these two points produces the line of flight which should be parallel to the eye base while the photographs are viewed through a stereoscope.
4. Orient the two photographs in the direction of flight with the overlapping areas towards each other such that shadows fall towards the viewer.
5. Place the stereopair under the stereoscope so that the left photo (taken by the camera on the left of the overlap area) is observed by the left eye and the right photo by the right eye.
6. Detect the same object on both photographs. Put a finger on this point in the left photograph. Look through the left lens with the left eye and shift the left photograph to position it at the centre of the lens. Fix the corners of the photograph with adhesive.
7. Take the right photograph, put a finger on the same point in it. Look through the right lens with the right eye and shift the photograph to position it at the centre of the lens.

8. Now make fine orientation by shifting and rotating the right photograph if necessary so that the two images coincide or fuse to produce one single 3D image.

Scale of a Vertical Aerial Photograph

A vertical aerial photograph is one in which the optical axis is vertical. The optical axis of a thin lens is defined as the line joining the centres of curvature of the spherical surfaces of the lens. The lens centre acts as the perspective centre through which all rays must pass. Thus the position of points on the image plane can be found by drawing rectilinear rays emanating from the ground points and passing through the lens centre, e.g., ground points 'A' and 'B' are imaged at 'a' and 'b' respectively (Fig. 9.12). The term *principal point* (P) of an aerial photograph is defined by its geometric centre that can be located with the help of marks appearing on its edges or corners, called *fiducial marks*. The point on the ground vertically below the lens centre is called the *nadir point* (N) and the line joining successive nadir points is known as the *nadir line*; it gives the ground track of the flight path. In a remote sensing camera, the negative plate is placed at a distance 'f' (focal length) above the lens centre. For all practical purposes, an equivalent positive plane can be imagined at a distance 'f' below the perspective centre. Thus, the points A and B lying on a flat ground covered in a vertical photograph are imaged at a and b. Therefore, for ground distances, AN and BN, the corresponding map distances are respectively aP and bP on the positive plane.

Fig. 9.12 Scale of an Arial Photograph

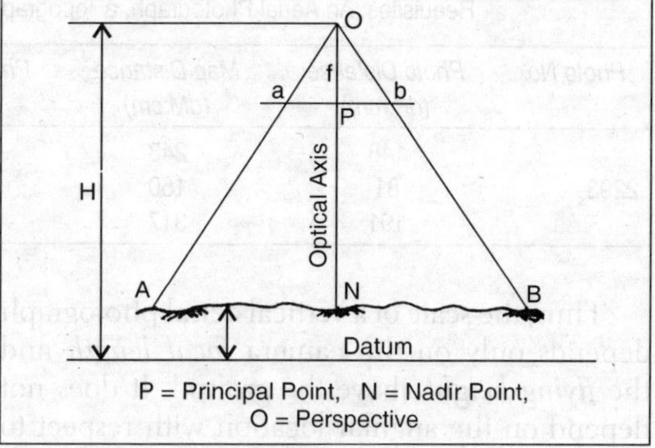

Therefore, aP = bP and AN = BN

From Fig. 9.12

Let OP = Principal Distance
= focal length
= f

The scale of the aerial photograph (Sp) can be computed as (Table 9.6):

Sp (Photo scale) = 1 : Fp

where, Fp (photo scale factor) defined by

$$\frac{AN}{ap} \text{ or } \frac{BN}{bp}$$

From Fig. 9.12, ΔS aOP and AON are similar or also the ΔS bOP and BON

ON = H – h

$$\therefore \quad \frac{AN}{ap} = \frac{ON}{OP} \text{ or } \frac{BN}{bp} = \frac{ON}{OP}$$

$$= \frac{H-h}{f} \qquad = \frac{H-h}{f}$$

Table 9.6 Determination of Photo Scale

Requisites: An Aerial Photograph, a Topographical Map of the Area and a Long Millimetre Ruler

Photo No.	*Focal Length/ Principal Distance (f:mm)*	*Altimeter Reading (hm:m)*	*Height of Terrain (h:m)*	*H = (hm–h:m)*	*Sp = 1:(H/f) × 1000*	*Mean Scale*
	210	6500	1620	4880	1:23238	
2267/1	210	6500	1800	4700	1:22381	1:22730
	210	6500	1760	4740	1:22571	

Table 9.7 Determination of Photo Scale

Requisites: An Aerial Photograph, a Topographical Map of the Area and a Long Millimetre Ruler

Photo No.	*Photo Distance (dP:mm)*	*Map Distance (dM:cm)*	*Photo Scale Factor (Fp = dM/dP)*	*Mean Photo Scale Factor*	*Mean Photo Scale*
	148	243	82095		
2293	81	150	92593	85891	1:85891
	191	317	82984		

Thus, the scale of a vertical aerial photograph depends only on the camera *focal length* and the *flying height* above the ground. It does not depend on the angular location with respect to the principal point in a flat terrain. However, if the terrain has a variable elevation, the scale varies. It is larger for elevated areas and smaller for depressed areas. For all practical purposes, in such a situation, it is convenient to have an *average scale* obtained by using the average elevation of the sensorcraft.

The scale of an aerial photograph can also be found by *comparing distances* on the aerial photograph to corresponding distances on a map with known scale. In this method, control points having nearly identical elevation are first identified and the photo scale is computed as (Table 9.7):

Sp (photo scale) = 1: Fp

where, Fp (photo scale factor) is given by

$$\text{Fp} = \frac{\text{map distance}}{\text{photo distance}} \times \text{denominator of RE}$$

The scale of oblique photographs varies depending on the angle of inclination of the optic axis. Such photographs are first rectified and only then are they used for photogrammetric applications.

Planimetric Measurement on Aerial Photographs

Distance

The *distance* between two points on an aerial photograph is directly measured with a millimetre ruler. The corresponding ground distance is computed as:

Ground distance = dP × Fp

For example, for a photo distance of 182 mm on a photograph with Fp = 48800, ground distance = 48800 × 182 mm ⇒ 8.8816 km.

Area

The *area* on aerial photographs can be measured either graphically using a transparent graph paper or with the help of a planimeter.

Relative height

The *relative height* of an elevated object can be determined in two ways—a) using a single photograph and b) using a stereopair.

Using a single aerial photograph

On a planimetric map, all features are shown in their correct geographical positions because the scale is true all over the map. This is not

Fig. 9.13 Relief Displacement in an Aerial Photograph

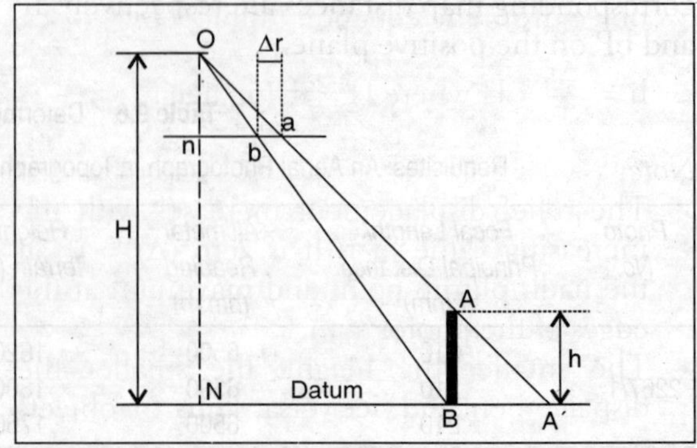

Table 9.8 Determination of the Relative Height of an Object

Aerial Photograph No. 223 Focal Length = 30 cm Photo Scale = 1:25,000 Flying Height = 300 m

Object	Observations			$h = (\Delta r/r) \times H$ (m)	Mean Relative Height (m)
	Distance from Nadir Point		Image Displacement, $\Delta r = (r - r')$ (mm)		
	Image (top) : r (mm)	Image (foot) : r' (mm)			
Cliff	120.50 85.00 102.50	76.50 53.00 64.00	44.00 32.00 38.50	109.54 112.94 112.68	111.72

so in case of aerial photographs due to image displacement or distortion. Relief is the most significant source of image displacement, others being lens distortion and tilt of the camera axis. From an aerial photograph, image displacement due to relief variation (Dr, relief displacement) can easily be identified and measured to find the relative height of a tall object (h). Relief displacement is radial from the nadir/plumb point. The requisites are large scale aerial photographs with images of elevated objects, an mm-ruler, grease pencil and table for recording. The procedure is as follows:

1. Determine the position of the nadir point on the photograph with three or more back rays joining the images of the top and the foot of a vertical object, relief displacement being radial from the nadir point.
2. Measure the distances, r of the image of the top and r of the image of the foot in mm from the nadir point.
3. Find the relief displacement as, Dr = (r – r')
4. Compute the relative height (h) of the tall object from the equation (Table 9.8):

$$h = \frac{\Delta r}{r} \times H \text{ where } H = \text{flying height}$$

Note:

- The relief displacement increases with the increasing value of r. It is zero ($\Delta r = 0$) at the nadir/plumb point and maximum at the edges of the photograph.
- The smaller the height, the smaller the displacement and vice versa. Thus, for objects in the datum plane, the relief displacement is nil (i.e., if $h = 0, \Delta r = 0$).
- As the flying height increases, the relief displacement decreases and this is why the satellite images have very low relief displacement.

Using a stereopair of aerial photographs

A stereoscopic pair of aerial photographs allows the heights of terrain objects to be precisely measured using the theory of stereoscopic parallax. The requisites are a *stereopair of photographs*, a *mirror stereoscope*, a *parallax bar*, a *long ruler* and a table for recording. The apparent image displacement along the x–direction, i.e., flight direction, defines the x–parallax or the *stereoscopic parallax*. It is accurately measured using the 'floating mark' method. When the

Fig. 9.14 Parallax in a Stereopair

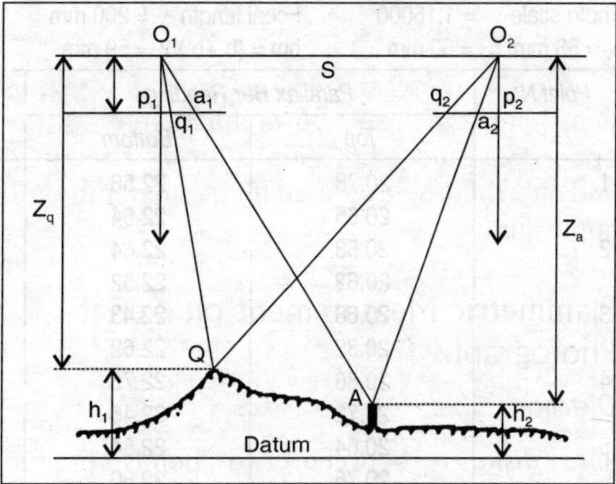

stereopair of aerial photographs is viewed under a stereoscope, the human operator perceives a 3D stereomodel revealing the 3rd dimension of terrain objects. The procedure is as follows:

1. Prepare, orient and fix the photographs correctly.
2. Mark the object for which height measurement is required by encircling it with a wax pencil (about 1 cm diameter) for easy identification.
3. Measure with ruler the distance between the foot of the object A (a') in the left hand photograph and the foot of the object (a") in the right hand photograph.
4. Measure the distance between the two principal points of the photographs (P_1 and P_2).
5. Difference between these two lengths is the stereoscopic parallax of 'a' in the photographs, Pa.
6. Set the index to the centre of the micrometer drum of the parallax bar by rotating it. By loosening the screw holding the left hand glass piece of the parallax bar, make the distance between the floating marks approximately equal to the distance between the left hand principal point and its conjugate image on the right hand photograph. Clamp the screw and take the parallax reading (Pa) at the foot of the object, A.
7. Similarly, take the parallax reading, Pb at the top of the object B, of which the relative height is to be determined.
8. Find $\Delta p = Pb - Pa$.

In a flat terrain, the Pa is pactically the same for all conjugate image points and approximately $= P_1q_2 = q_1P_2$ or the average of the two base values, i.e., $bm = (b_1 + b_2)/2$. The relative height is computed by the formula (Table 9.9):

$$h = \frac{\Delta p.H}{bm + \Delta p}, \text{ where H = flying height}$$

INTERPRETATION OF AERIAL PHOTOGRAPHS

Photo interpretation is the *act* and *art* of identifying the features and surfaces portrayed on stereoscopic aerial photographs through inductive and deductive reasoning and techniques of inference and logical deduction. There are two basic methods of using photographs—the *pattern recognition* method and the *pattern interpretation* method. The first involves the comparison of outstanding visible features with objects already familiar to the user, while the second method concerns not only the identification but also the classification and

Table 9.9 Determination of the Relative Height of an Object (Parallax Bar Method)

Photo scale = 1:15000   Focal length = 200 mm   Flying height, H = (200 mm×15000) = 3000 m

b_1 = 58 mm, b_2 = 60 mm   bm = $(b_1+b_2)/2$ = 59 mm

Point No.	*Parallax Bar Reading*		*Δp = Pb–Pa*	*Mean Δp*	*h = H.Δp/(bm + Δp)*
	Top	*Bottom*			
1.	20.78	22.56	1.78		
	20.65	22.54	1.89		
2.	20.58	22.64	2.06		h
	20.62	22.52	1.90		= (3000 × 1.844)/
3.	20.68	22.43	1.75	1.844	(59 + 1.844)
	20.82	22.62	1.80		
4.	20.86	22.75	1.89		= 5532/60.844
	20.72	22.45	1.73		= 90.92 m
5.	20.64	22.54	1.90		
	20.76	22.50	1.74		

interpretation of the significance of the pattern with all its elements. *Patterns* are fundamentally composed of definite recognisable reflectances that form elements that contrast with surrounding elements and patterns.

Principles of Photo Interpretation

Photo interpretation is based upon factors like *shape, size, shadow, tone and associated features*. While studying the natural terrain patterns, elements of form and reflectance are carefully detected to obtain the pattern which is then interpreted with deductive reasoning. Form includes topographic elements (basing valley, cliffy escarpment, plain, plateau, hill, ridge, mountain, etc.); drainage plans (dendritic, rectangular, parallel, trellis, radial, annular, deranged, braided, meandering, anastomosing, reticulate, etc.); cross-section shape and gully gradients (V-, U-, C-, CU-, CV-, VU- and UVshaped, gentle and steep gradients) and man-made forms. The elements of reflectances are contrasts in vegetative assembly, in landuse, in soil or rock exposures and special reflectance features.

While studying the landuse patterns, factors of shape and reflectance are most important.

Fig. 9.15 Land Cover / Landuse Map Prepared from a Stereopair

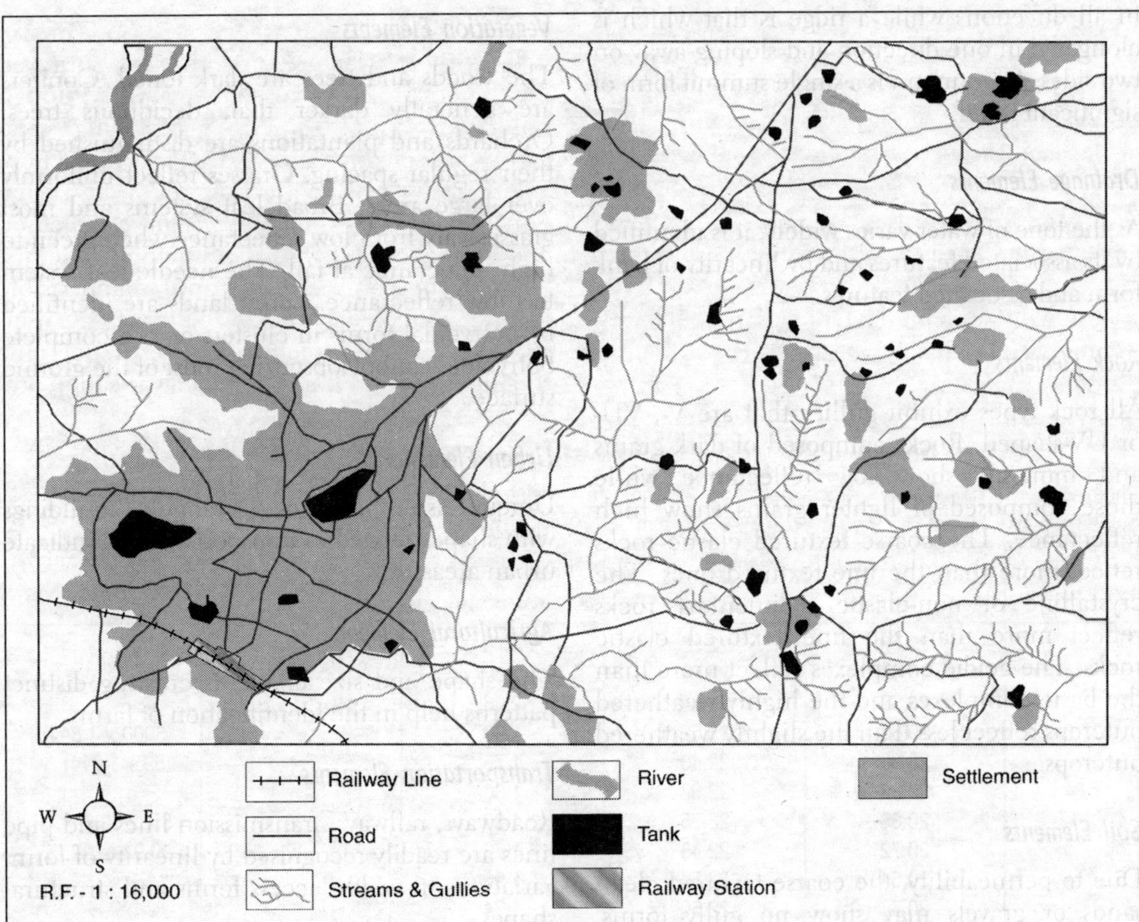

The elements of shape include location, geometric shape, shadows, linearity of form and other associated features. The elements of reflectance include relative values of hue, chromatic saturation and brightness of colour or grey scale simulation.

Relief Element

An enclosed depression without a surface outlet indicates a basin. A linear depression with a surface outlet is a valley. A clifisa relief change of low magnitude while an escarpment is that of high magnitude. An assembly of similar terrains of low to moderate relief indicates a plain while that of moderate to high relief indicates a plateau. A hill is a single summit form sloping in all directions while a ridge is that which is elongated in one direction and sloping away on two sides. A mountain is a single summit form of significant relief.

Drainage Elements

As the tone of water varies widely, it is identified by its associated features and by linearity of bank form and associated features.

Rock Elements

All rock types exhibit gullies that are V-, VU- or UV-shaped. Rocks composed of dark grains and minerals show low reflectance while those composed of lighter grains show high reflectance. The coarse textured clastic rocks reflect more than the fine textured ones. The crystalline or non-elastic sedimentary rocks reflect more than the fine textured clastic rocks. The acidic complexes reflect more than the basic complexes and the highly weathered outcrops reflect less than the slightly weathered outcrops.

Soil Elements

Due to permeability, the coarse textured clean sands or gravels may show no gully forms. However, the finer sands may show V-shaped gullies with steep to very steep gradients. Deep *loess* and deep marine drift sand show U-shaped gullies of flat to gentle gradients. Pyroclastic drifts develop U-shaped gullies with steep to very steep gradients. Cohesive soils of silt, clay and combinations with some sand and gravel develop C-shaped gullies. Gentle gradients are associated with clay and silly clay. Well drained and coarse textured soils reflect more light than poorly drained and fine textured soils. Reflectance decreases with increasing organic matter content. Darker mottlings indicate higher clay content while lighter mottlings indicate higher silt content. Freshly plowed and cloddy soils have lower reflectance.

Vegetation Elements

The woods and trees are dark toned. Conifers are generally darker than deciduous trees. Orchards and plantations are distinguished by their regular spacing. Grasses reflect uniformly over large areas. Broad leaf systems and most grasses vary from low reflectance when green to high reflectance at fall. The needle leaf system has low reflectance. Forest lands are identified by botryoidal forms in clusters or as a complete botryoidal canopy obscuring most of the ground surface.

Urban Elements

Unique assemblage of rectangular buildings with shapes related to transport systems indicate urban areas.

Agricultural Elements

The shape and size elements creating distinct patterns help in the identification of farms.

Transportation Elements

Roadways, railways, transmission lines and pipe lines are readily recognised by linearity of form, variations in width, access forms and structural shapes.

Table 9.10 Differences between an Aerial Photograph and an Image

Photograph	Image
Originally produced in analogue form	Originally produced in digital form
Generated by photographic film systems	Generated from line scanners and digital cameras
Does not have any pixels	Basic element is a pixel
Lacks row and column structure	Possesses row and column structure
Scan lines absent	Scan lines may be observed
Zero indicates no data	Zero is a value that does not indicate absence of data
No numbering at any point	Every point has a certain digital number
Photography is restricted to the photographic range of the EM spectrum	Image can be generated for any part of the EM spectrum or any field
Once a photograph is acquired, colour is specific and cannot be changed	Colour has no specific role and can be changed during processing

Photogrammetric Outputs

The major products of photogrammetry commonly available are the following:

Aerial Photographs

This is used in all sorts of interpretation and general planning.

Photomosaics

It is a collection of overlying aerial photographs assembled to form a composite picture of the terrain. The assembly in which these photographs are laid in such a manner as to allow the photo number and flight number of each one to appear on the finished assembly is called an index mosaic or an index. A mosaic that is assembled from a single strip of photography is called a strip mosaic, while that which is assembled without regard to any plotted control is called an uncontrolled mosaic.

Photomap

A printed photomosaic, on which the background details has been cartographically improved to clarify the interpretation, and to which a grid and map frame have been added, is a photamap. It is very useful and more accurate.

Orthophoto

A picture of the ground prepared from a pair of overlapping photographs in such a manner

Table 9.11 Relative Merits of Imaging Systems over Photographic Sensors

Advantages	Disadvantages
Provides digital information which can be telemetered to the ground from space Problem of film retrieval associated with photographic sensors is absent Remote sensing in extended range of 0.3 µm–1 mm possible in contrast to photographic range of 0.3–0.9 µm Higher spectral resolution Higher radiometric resolution Good spatial resolution from modern sensors can be obtained Information can be stored and is reproducible Amenability of data to digital processing for enhancement and classification Flexibility in the handling of data Repeatability of results	High data rate and technical sophistication required in generating images and data processing

Sensors and Scanners

- Photographic sensors expose the **instantaneous field-of-view** (IFOV) on a light-sensitive emulsion such as a **film**, which serves both as a recording and a storage medium.
- In non-photographic sensors, the IFOV is scanned and the output recorded by microelectronic detectors such as **charge-coupled devices** (CCD).
- The scanner systems build up 2-dimensional images of the terrain for a swath beneath the aircraft using either across-track (**whiskbroom**) or along-track (**pushbroom**) scanners.
- The **whiskbroom** scanner uses a rotating or oscillating mirror to scan the terrain along scan lines that are at right angles to the flight lines. This allows repeated measurement of energy from one side of the aircraft to the other with relatively large IFOV and dwell time producing better resolution.
- The **pushbroom** scanner uses a linear array of CCDs located in the focal plane of the scanner such that each scan line is viewed by all arrays simultaneously. It has higher reliability and longer life expectancy and produces a greater range in signal and stronger signal with large signal-to-noise rate.

Satellite Orbits

- A satellite at 36000 km has the same period as the earth and seems to remain stationary with respect to the earth's surface. Hence, these view the same portion of the earth's surface at all times. Such orbits are called ***geosynchronous orbits*** and the satellites in such orbits are called ***geostationary satellites***, e.g., weather and communications satellites, like NOAA, GOES-e, METEOSAT, INSAT, GMS, GOES-W, etc.
- ***Sun-synchronous*** satellites cover each area of the world at a constant local time of the day with almost no change in solar illumination angle or azimuth, permitting good comparison and mosaicing. It has an inclination that carries the satellite track westward at a rate that compensates for the change in local sun time as the satellite moves from north to south. In such *near-polar orbits*, satellites pass from north to south (*descending node*) on the sunlit side with activated thermal, radar and microwave sensors.
- The path followed by a satellite is called its ***orbit***. An ***orbit cycle*** is completed when the satellite retraces its path, passing over the same point on the earth's surface directly below the satellite, called the ***nadir*** point, for a second time. The time for one orbital cycle defines a satellite's ***period***.
- As a satellite revolves around the Earth, the sensor "sees" a certain portion of the Earth's surface. The area imaged on the surface, is referred to as the *swath*.

that the perspective element has been removed. It is a planimetric map and when contours are superimposed it can be used as a topographic map.

Orthomosaic

An assembly of a series of contiguous orthophotos. It has uniform scale throughout and can be used as a map.

Radarmosaic

A systematically assembled output of a series of radar pictures taken along consecutive flight lines.

Maps

For a given area, as per the user's choice, planimetric, topographic and thematic maps may be prepared with a high degree of accuracy with all kinds of map details.

Digital Terrain Model (DTM)

This is a listing of the XYZ coordinate of a series of points in the form of cards, magnetic tapes or floppy discs for direct input into a data processor or computer.

IMAGING SYSTEMS

The *non-photographic* multispectral imaging systems operate in the optical range of the EM

Fig. 9.16 Along Track Scanner

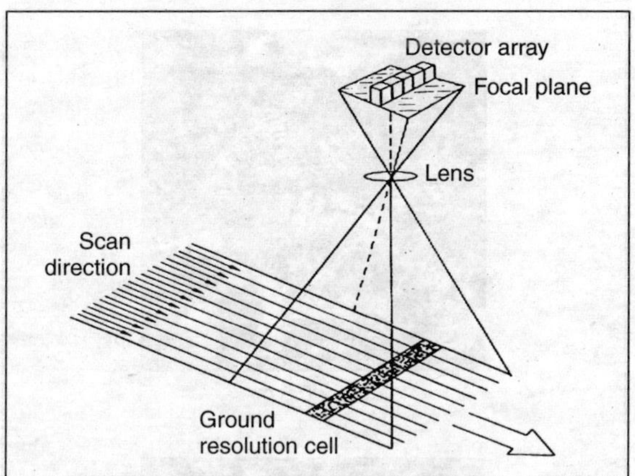

Fig. 9.17 Cross Track Scanner

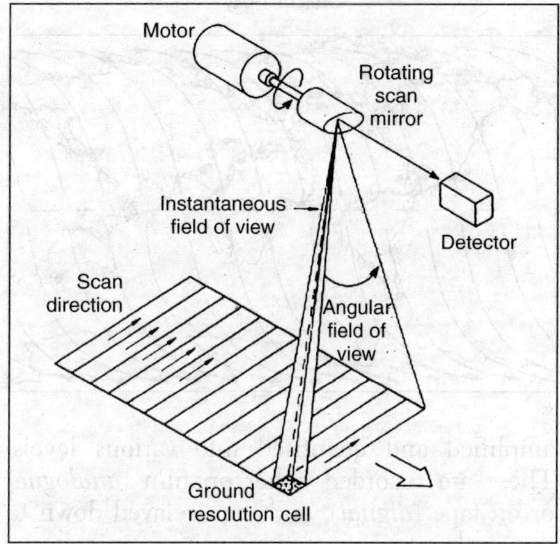

spectrum, particularly between 1.00–1.12 μm and are far *superior* to the photographic sensors. A non-photographic imaging sensor or scanning device consists of two main parts—an *optical part* and a *detector part*. The optical part has two main units—*radiation-collecting* and *radiation-sorting*. The first one comprises *mirrors*, lenses and a *telescopic set-up* to collect the radiation from the ground and focus it onto the radiation-sorting unit that uses various *gratings*, *prisms* and *interferometers* to separate the radiation of different wavelength ranges. After sorting, the radiation of selected wavelength ranges is directed to the detectors. Detectors comprise devices that *transform* this *optical energy* into *electrical signal*, which is subsequently

Fig. 9.18 Satellite Orbits

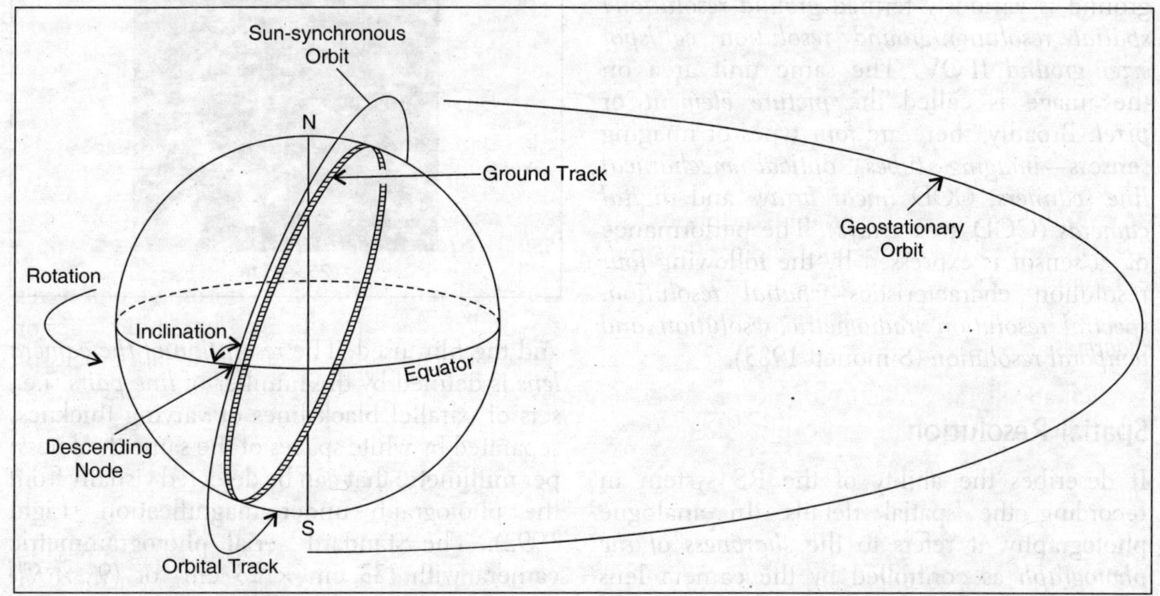

Fig. 9.19 Typical Ground Trace of Sun-synchronous Orbit of 1-day

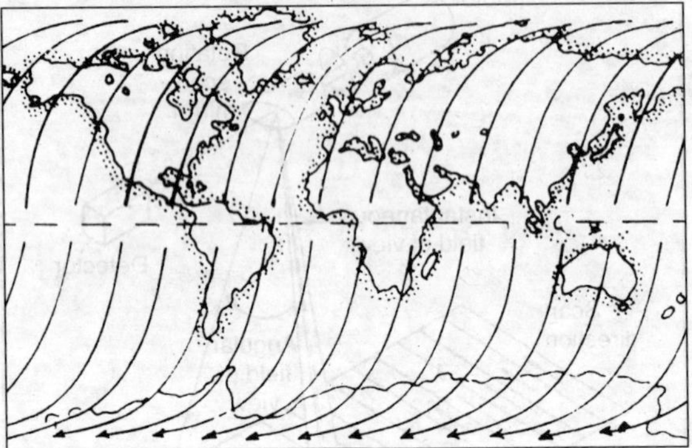

amplified and quantised into various levels. These are recorded either on film (*analogue*) or on tape (*digital*), and then relayed down to the earth receiving stations using a microwave communication link for real-time display and/or subsequent processing and application.

On the ground, the scanner data can be rearranged as a 2-dimensional array of digital numbers (DNs) and be presented as an optical analogue by choosing a suitable grey scale. Therefore, an image is an optical analogue of a 2-dimensional array. The unit area on the ground is variously termed *ground resolution/ spatial resolution/ground resolution cell/spot size/ ground* IFOV. The same unit area on the image is called the *picture element* or *pixel*. Broadly, there are *four* types of imaging sensors—*imaging tubes, optical mechanical line scanners, CCD linear arrays* and *digital cameras* (CCD area arrays). The performance of a sensor is expressed by the following *four* resolution characteristics—*spatial resolution, spectral resolution, radiometric resolution and temporal resolution* (Simonett 1983).

Spatial Resolution

It describes the ability of the RS system in recording the spatial details. In analogue photography, it refers to the *sharpness of the photograph* as controlled by the camera lens

Fig. 9.20 Spatial Resolution

and the film used. The *resolution of the camera lens* is defined by the number of *line pairs* (i.e., sets of parallel black lines of varying thickness separated by white spaces of the same thickness) per millimetre that can be detected visually from the photograph under magnification (Light 1992). The standard aerial photogrammetric camera with (23 cm × 23 cm) or (9" × 9") format produces photographs with a resolution ranging between 20–40 lp/mm. However, modern cameras equipped with forward motion compensation (FMC), gyro-controlled stabilised mounts, and high-resolution films like, WILD RC 30 and Zeiss RMK-TOP can produce photographs with resolutions as high as 60 lp/mm. Ground resolution (GR) depends on both the scale of the photograph as well as the resolution power of lens and is given by (Lillesand and Kiefer 2000):

$$GR = \omega \times \alpha$$

where, ω = width in mm for one line pair of the photographic system and α = scale factor of the photograph.

Thus, for an aerial photograph with scale 1: 20,000 and lens resolution 60 lp/mm, ω = 1/60 mm, α = 20,000. GR = 20,000/60 → 333.4 mm → 0.33 m. For digital images, spatial resolution is described as the instantaneous field of view (IFVO), i.e., the solid angle through which the detector is sensitive to the EM emission and is measured in milliradians. The IFOV depends on the height of the imaging platform, the size of the detector element and the focal length of the optical system. It is given by (Curran 1985) as:

$$D = H \times \beta$$

where, D = ground dimension of the detector element in metres, H = orbital altitude in metres and β = the IFOV in milliradian.

For Landsat MSS detectors, the ground dimension of the detector element, or the spatial resolution is normally 79 m. However, it varies as there occurs changes in the orbital altitude of the satellite (880–940 km). The IKONOS-2 images the earth with 1 m PAN + 4 m MSS spatial resolution, while the Quickbird-2 with 0.61m PAN + 2.5 m MSS spatial resolution (Jensen and Cowen 1999, Lo 1997).

Principal Applications of Bands in Remote Sensing

Spectral Band	*Principal Applications*
Blue	Good water penetration; hence, suitable for coastal mapping and bathymetry. Also useful soil/vegetation/forest type discrimination and identification of cultural features. Sensitive to atmospheric haze.
Green	Has some ability to penetrate water but sensitive to turbidity. High reflectance from vegetation; hence useful for vegetation discrimination and vigour assessment. Also useful for identification of cultural features. Sensitive to atmospheric haze.
Red	High chlorophyll absorption in vegetation and high reflection in soils; hence good for differentiating soil from vegetation. Also useful for delineating snow cover and cultural features.
Near-IR	High reflection from vegetation and absorption in water. Best band for discriminating between vegetation types and vigour. Useful for delineating water bodies and soil moisture.
Mid-IR	In shorter wavelengths, good for vegetation and soil moisture content and for discriminating between snow and clouds. In longer wavelengths, good for discriminating between mineral and rock types and also moisture content.
Thermal	Vegetation stress analysis, soil moisture and thermal mapping.

Spectral Resolution

It refers to the EM radiation wavelengths to which an RS system is sensitive. It has two components—the number of *wavelength bands* (or *channels*) *used* and *the width of each wave band*. A large number of bands and a narrow bandwidth for each band gives higher spectral resolution. The spectral resolution of an aerial photograph is very low as the film sensitised to the visible spectrum has only one wide band. On the other hand, the Landsat TM and ETM+ all detect in seven bands (0.45–0.515 µm, 0.525–0.605 µm, 0.63–0.69 µm, 0.75–0.90 µm, 1.55–1.75 µm, 10.40–12.50 µm, 2.09–2.35 µm) and have a much higher spectral resolution. The use of narrower bandwidths allows more unique spectral signatures of objects to be recorded. Thus it helps to discriminate more subtle differences among these objects. Therefore, it is very important to select the correct spectral resolution for the type of information to be extracted from an image.

Radiometric Resolution

It is the smallest difference in radiant energy that can be detected by a sensor. In aerial photography, it is inversely proportional to the contrast of the film. Therefore, a higher contrast film can resolve smaller differences in exposure by subtle changes in grey tones. For digital images, radiometric resolution refers to the number of discrete levels into which a signal may be divided during the analogue to digital conversion, known as the *quantisation level*. For Landsat MSS detectors, the reflected radiant energy is recorded in 6-bits quantisation levels: (2^6, or 64 quantising levels, giving digital numbers from 0 to 63 to the pixel). With the increased radiometric resolution, Landsat TM detectors are more sensitive to small changes in radiant energy than the Landsat MSS detectors. This is because they use 8-bit *quantisation levels* (from 0 to 255). The AVIRIS data is again superior to all other types in radiometric resolution because of their use of 12-bit *quantisation levels* (from 0 to 4095).

Fig. 9.21 Optical-Mechanical MSS Device

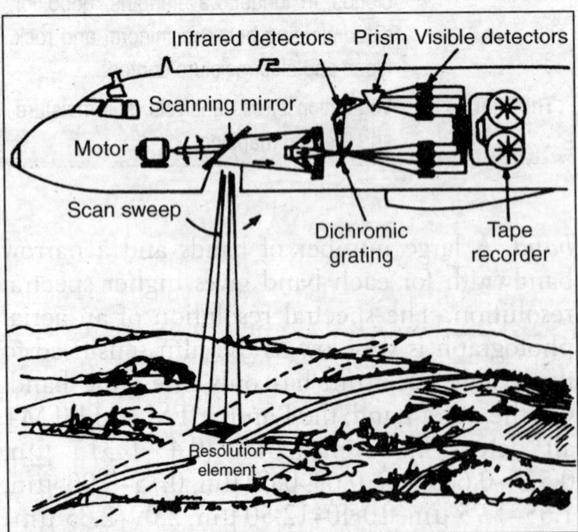

Temporal Resolution

It refers to the frequency of data collection. High temporal resolution facilitates change detection and monitoring of environmental phenomena. For satellite remote sensing, it refers to repeat coverage, i.e., number of days required for an orbiting satellite to return to the same point on earth (Jensen and Cowen 1999).

IMAGING SENSORS

Basically, the space platforms have the advantage that they provide a wider global coverage and allow images of earth to be acquired at regular intervals. Hence the process of the *topographic mapping* of the world has been significantly *enhanced*. The satellite platform is subject to

Fig. 9.22 Design of a Landsat MSS System

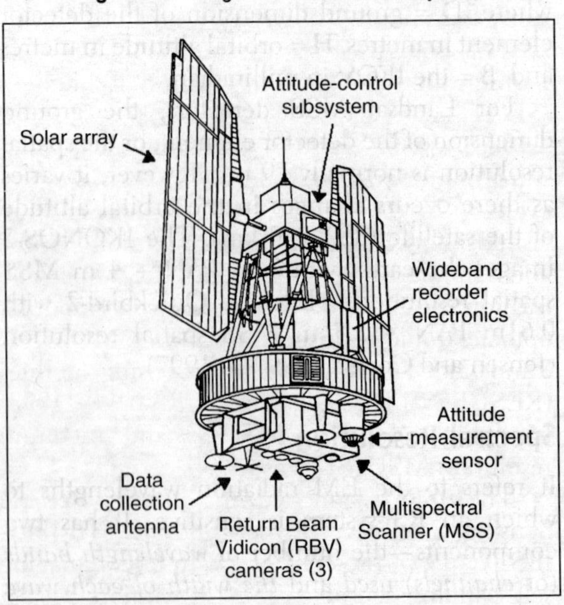

much *less turbulence* because space is *vacuum* and *tilt free*. The very high-orbiting altitude of the satellite means that relief displacement is not as severe as that on an air photo. Imaging sensors with high-resolutions are constantly being improved and developed and a more *automated approach* to large scale (1:50,000) environmental mapping of the earth's surface has become a *reality* of late (Dowman 1994, Konecny and Schiewe 1996).

In the 1970s–80s, the main space sensor on the NASA initiated Landsat Program (orbit = near-polar, near-circular, sun-synchronous, altitude = 900 km:1–3 and 705 km:4–7, repeat cycle = 18 days:1–3 and 16 days:4–7) was MSS (multispectral scanning system: resolution = 79/82 m), RBV (return beam vidicon: resolution = 80 m), TM (thematic mapper: resolution = 30 m), ETM (enhanced thematic mapper: resolution = 30/15 m) and ETM+ (enhanced thematic mapper plus: resolution = 30/15 m). The Earth Observation Satellites (EOS-AM-1) initiated by NASA in 1999 (orbit = near polar, near circular, sun-synchronous, altitude = 705 km, repeat cycle = 16 days) has ASTER (advanced spaceborne thermal emission and reflection) sensor with three radiometer subsystems in the VNIR, SWIR and TIR region.

The French SPOT (Systeme Probatoire de l'Observation de la Terre) series of satellites (SPOT -1:1986, SPOT-2:1990, SPOT-4:1998, SPOT-5:2003, orbit = near-polar, sun-synchronous, altitude = 830 km, repeat cycle = 26 days) used HRV (high resolution visible) and HRVIR (high-resolution visible and infrared) sensors. The ESA-Spacelab mission used MOMS series of sensors (modular optoelectronic multispectral scanner: MOMS-1, 2 and 02P). The Russian remote sensing satellites (RESURS-01-03:1994, RESURS-01–04:1998) have high-resolution MSU–E and medium-resolution MSU–SK sensors.

The ISRO launched the Indian Remote Sensing Satellites (IRS-1A-1988, IRS-1B-1991, IRS-1C- 1995, IRS-ID-1997) with linear imaging self-scanning sensors (LISS-1:resolution = 72.5 m, LISS-II: resolution = 36.25 m and LISS-III: resolution = 23/70 m), panchromatic sensor (PAN: resolution = 5.8 m) and wide image field sensors (WiFS: resolution = 188 m). CARTOSAT (2005) (orbit = polar, sun-synchronous, altitude = 618 km, repeat cycle = 5 days) has two sensors—PAN and MSS, both of 2.5 m resolution. CARTOSAT data is used for large scale mapping, updating topographic maps, disaster assessment, relief planning and management and various other geospatial and mapping applications. Japan's Earth Resources Satellite–1 (renamed Fuyo-1:1992) uses an optical sensor (OPS) and a radar sensor (SAR). The Advanced Earth Observing Satellite (ADEOS:1996) carries two sensors—AVNIR (advanced visible and near infrared radiometer) and OCTS (ocean colour and temperature sensor). The Advanced Land Observation Satellite (ALOS), as a follow up of Fuyo–1 (2001) have three sensors—PRISM (a panchromatic remote-sensing instrument for stereo mapping), AVNIR–2 and PALSAR (phased array-type L band synthetic aperture radar). China and Brazil jointly formed the remote sensing programme, CBRS—1 (1998: sun-synchronous, altitude = 778 km, repeat cycle = 26 days) consists of three main sensors—CCD Camera, IR–MSS (infrared multispectral scanning system) and WFI (wide field sensor). The CBERS–2 (2001) and its follow ups CBERS-3 and 4 have high-resolution sensors (PAN = 5 m, MSS = 10 m, and IR = 40–80 m).

Space Imaging IKONOS–2 (2000), the first commercial high-resolution satellite orbits in a sun-synchronous, near polar orbit (altitude = 681 km, repeat cycle = 11 days, geocode frame = UTM) has two sensors – one PAN with 1 m resolution and another MSS with 4 m resolution. Digital Globe (formerly called Earth Watch) launched Quick Bird –2 (2001), orbit = Sun-synchronous, altitude = 450 km) has two sensors—one PAN with 0.61 m resolution and another MSS with 2.5 m resolution. Orbital Imaging Corporation launched OrbView-1 and 2 (2001) that carry SeaWiFs sensor; Orb View-3 and 4 carry 1 m resolution PAN and 4 m resolution MSS sensor.

Table 9.12 Comparison between Manual and Digital Processes in Image Interpretation

Manual Interpretation	***Digital Processing and Analysis***
It dates back to the early beginnings of RS for air photo interpretation.	It is more recent and has been used since the advent of the digital recording of RS data and the development of computers and image processing software.
It requires, if any, few specialised equipment.	It requires specialised, and often expensive, equipment.
It is often limited to analysing only a single channel of data or a single image at one time.	It is useful for the simultaneous analysis of many spectral bands and can process large datasets much faster than a human interpreter.
It is a subjective process, meaning that the results will vary with different interpreters.	It is more objective, generally resulting in more consistent results.

The ISI-EROS series consists of EROS A1 and A2 satellite (altitude = 480 km) with a CCD area arry sensor with 1.8 m resolution. EROS B1 and B2 use a CCD/TDI sensors of high resolution.

IMAGE INTERPRETATION

Image *processing* and *analysis* is defined as the 'act of examining images for the purpose of identifying objects and judging their significance' (Colwel 1997). An image analyst studies the RS data in detail and through logical processes attempts to *detect*, *identify*, *classify*, *measure* and *evaluate* the significance of physical and man-made features, their patterns and spatial relationship. It is certainly an iterative process to map the spatial pattern of the features on the earth's surface in real time perspective.

An image in analogue format is independent of the type of sensor used to collect the data and the means by which the data was actually acquired. The *hard copy* images are mostly interpreted manually or visually, i.e., by a human interpreter. *Digital images* are represented as arrays of pixels in raster format. In both formats, images can be displayed as black and white or monochrome or as colour images by combining different channels or bands and interpreted accordingly. Digital analysis uses image-processing software and automatically generates data products for further analysis. It may also be used as a prelude to visual interpretation. In fact, visual and digital methods are not mutually exclusive; both have their own merits and in most cases, a mix of both methods is employed for greater accuracy and better precision.

Visual Image Analysis (VIA)

For visual interpretation, a human analyst essentially takes into account the *nine* basic elements of an image, such as, *tone/colour*, *size*, *shape*, *texture*, *pattern*, *height*, *shadow*, *site* and *association* (Estes et. al 1983; Teng et. al 1997). Of these, tone is the most important element in image interpretation and analysis. Size, shape, texture and pattern precisely represent the spatial arrangements of tone and colour. Texture is particularly useful in the analysis of cultural features. For RADAR imagery, it is as important as tone in VIA and DIP. Height and shadow help to measure the parallax element, while site and association provide the basis for the image classifiers to improve the accuracy of classification.

Tone refers to the relative brightness or colour of the objects in an image. It depends on the nature of surface roughness, colour, the atomic structure of the materials, and the range of wavelength used. On a black and white panchromatic film, it varies from black through grey to white in continuous steps, thus making use of the spectral and radiometric resolution of the film. In general,

Fig. 9.23 Land Cover / Landuse Map Prepared from Satellite Image

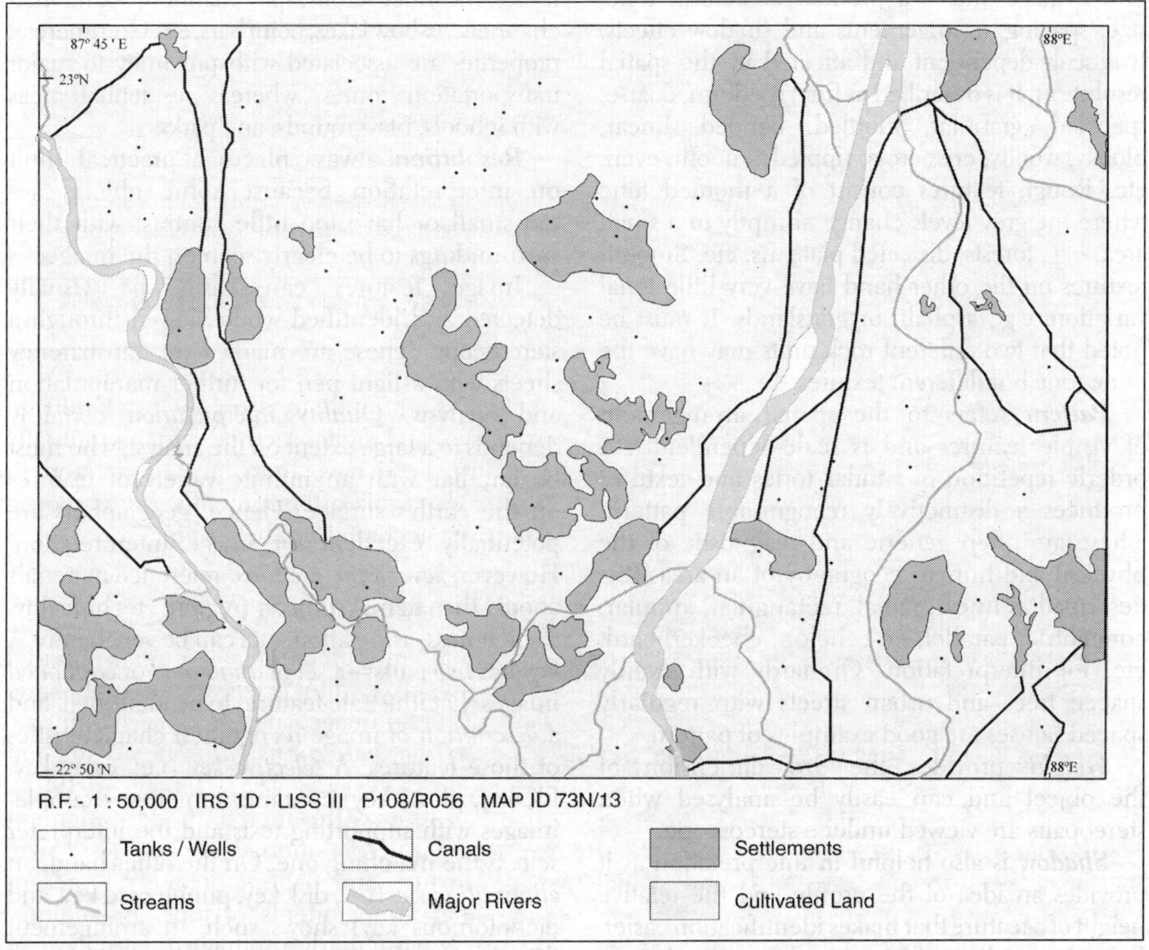

the more light the object reflects, the lighter is its tone on the image. The tonal variation is used to identify the features, particularly land cover and landuse, forest types, or depth of water. It also allows the elements of shape, texture, and pattern of objects to be distinguished.

Size of objects in an image refers to the physical dimensions of the objects and is a function of scale. It is assessed relative to other objects in a scene and in conjunction with shape and association, it becomes a wonderful parameter for quick and better interpretation. For example, if an interpreter had to distinguish zones of landuse, and had identified an area with a number of buildings in it, large buildings, such as factories or warehouses, would suggest commercial property, whereas small buildings would indicate residential use.

Shape refers to the form, structure or outline of individual objects and can be a very distinctive clue for interpretation, e.g., alluvial fans, meander belts, sand dunes, ox-bow lakes, volcanic cones, etc. Straight edge shapes typically represent urban or agricultural landuse, while natural features, such as forest edges, are more irregular in shape, except where man has created a road or clear-cuts.

Texture refers to the arrangement and

frequency of tonal variation in particular areas of an image and is a product of colour, tone, size, spacing, arrangements and shadow effects. It is scale-dependent and affected by the spatial resolution. It is described as fine, medium, coarse, speckled, granular, mottled, banded, linear, blocky, woolly, criss-cross, rippled, smooth, even, etc. Rough textures consist of a mottled tone where the grey levels change abruptly in a small area, e.g., forests, dissected plateaus, etc. Smooth textures on the other hand have very little tonal variation, e.g., asphalt, or grasslands. It must be noted that two different rock units may have the same tone but different textures.

Pattern refers to the spatial arrangement of visible features and is scale-dependent. An orderly repetition of similar tones and textures produces a distinctively recognisable pattern. These are often generic and diagnostic of the physical and human geography of an area. It is described as linear, radial, rectangular, annular, concentric, parallel, en-echelon, checkerboard, etc. for interpretation. Orchards with evenly spaced trees and urban streets with regularly spaced houses are good examples of pattern.

Height provides the 3rd dimension of the object and can easily be analysed when stereopairs are viewed under a stereoscope.

Shadow is also helpful in interpretation as it provides an idea of the profile and the relative height of a feature that makes identification easier. It emphasises linear features and is very important for discriminating man-made features. It works wonders especially when considered together with shape, size, and direction of illumination. It is also useful for enhancing or identifying topography and landforms, particularly in radar imagery. However, an interpreter should be very careful as shadows often mask the smaller targets.

Site refers to the locational characteristics of features in physical perspective.

Association defines the spatial relationship between the 'feature of interest' and 'other recognisable objects' in an area. Certain features are preferentially associated with each other, which offer a good clue for interpretation. For example, alluvial landforms are associated with levees, meander scrolls, spill channels, abandoned channels, ox-bow lakes, point bars, etc. Commercial properties are associated with proximity to major transportation routes, whereas residential areas with schools, playgrounds, and parks.

Resolution always places a practical limit on interpretation because some objects are too small or have too little contrast with their surroundings to be clearly seen on the image.

Image features can easily be visually detected and identified when viewed through a stereoscope. These are marked on transparency sheets with a light pen for further manipulation and analysis. *Quality interpretation* certainly depends to a large extent on the analyst, who must be familiar with an infinite variety of features on the earth's surface. Hence, geographers are potentially excellent for image interpretation. However, *keys* or *pre-prepared* reference materials about the item, subject, region, technicality, discreteness, association, etc. can be very handy. A key has *two* parts—a set of *annotated* or *captioned* images that illustrate features to be identified, and a *description* of image recognition characteristics of those features. A *selective key* (i.e., essay key, file key, photo key) contains numerous example-images with supporting texts and the interpreter selects the matching one. On the other hand, an *elimination key* (i.e., disk key, punch card key, and dichotomous key) shows such an arrangement that the interpreter goes on eliminating features step-by-step until it is discretely identified.

Digital Image Processing (DIP)

Now a days RS data is mostly available in digital format (like Landsat MSS, Landsat TM, Landsat ETM+, SPOT, QuickBird, Cartosat, IKONOS, etc.) These are normally organised in *three* formats, such as, *band interleaved by pixel* (BIP), *band interleaved by line* (BIL), and *band sequential* (BSQ). Thus, an image consisting of *four* spectral channels can be visualised as *four* superimposed images with corresponding pixels in one band registering exactly to those

Fig. 9.24 Flowchart of Digital Image Processing

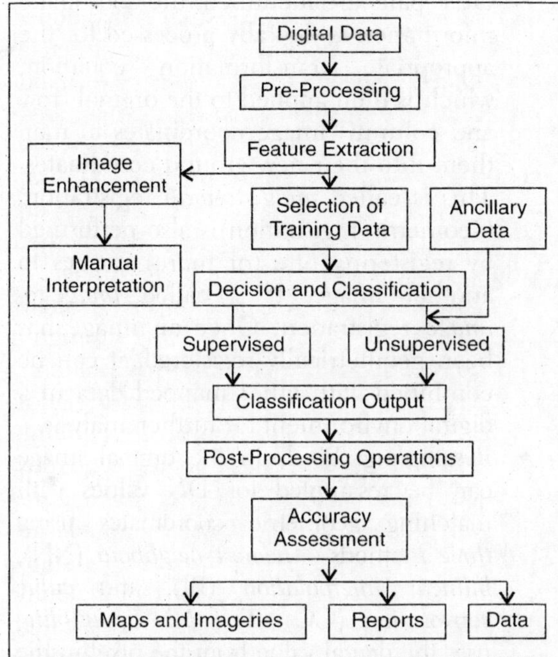

in the other bands. These are digitally processed using a computer system, called 'image analysis system' with appropriate hardware and software (e.g., Gram++, Ilwis, Erdas, JT Maps, Geomatica, ArcGis, etc.) for further analysis and applications.

DIP is a collection of techniques for the manipulation of digital images by computers. It differs from image to image depending on the type of image format, initial condition of the image and the information of interest and the composition of the image scene. The common image processing functions are—

1. *pre-processing*: performed prior to the main DIP generally to make the necessary radiometric, geometric and atmospheric corrections and feature extraction,
2. *image enhancement*: performed solely to improve the appearance of the imagery for VIA,
3. *image transformation*: involves the combined processing of data from multiple spectral bands, and
4. *image classification and analysis*: uses mathematical algorithms to assign individual pixels to specific classes; accuracy of classification is then assessed by ground truthing.

The final result of DIP consists of maps (or images), data and a report, that provides the user with full information about the source data, method of analysis, data products generated and its reliability.

Pre-processing

It is defined as those operations that are preliminary to the main analysis for correcting the inherent sensor-specific and platform-specific distortions of image data. It includes a wide range of operations from the very simple to the very extremes of abstractness and complexity, such as, *radiometric correction, geometric correction, atmospheric correction* and *feature extraction*.

Radiometric Corrections

The malfunctioning of the detectors of the RS systems produce *radiometric errors*, scattering and absorption results in *atmospheric attenuation*, and slopes and terrain aspects cause *topographic attenuation*. Such distortions are of two types:

1. the *relative distribution of brightness* over an image in a given band can be different from that on the ground scene, and
2. the *relative brightness of a single pixel* from band to band can be distorted compared to the spectral reflectance character of the corresponding region on the ground.

Cosmetic operations are performed to remove all such radiometric errors, as follows:

- *Line Dropouts*: It occurs usually when a string of adjacent pixels in a scan line contains *spurious* DN and is corrected *either* by replacing the defective line with a duplicate of preceding or subsequent line, *or* by taking the average of the two.
- *De-Stripping*: Banding or stripping occurs if one or more detectors go *out of adjustment*

in a given band. It is corrected in *two* ways —*histogram matching* and *image regression*. In the *first* method, the histogram of the DN values of one image is correctly matched to that of another image so that the apparent distribution of DN values in the two images are made as close as possible. In the *second* method, the DN value of each pixel of the subject image is related to that of the reference image band by band. It produces a linear regression equation that is solved and the required values regressed.

Elimination of Random Noise: Odd pixels with spurious DN crop up frequently and thereby create *random noise* in an image. It is identified by marked differences in DN values from adjacent pixels in the affected band and is then replaced by an average DN value of the neighbourhood pixels.

Geometric Corrections

Geometric distortions arise from the *earth's curvature*, *platform motion* (altitude, attitude and velocity), *perspective of the sensor optics*, *relief displacement* and *non-linearities in scanning motion*. These are of two types: *systematic* and *non-systematic*.

1. *Systematic Distortions*: These are *caused* by the forward motion of the platform during each mirror sweep, irregular mirror scan and platform velocities. Such distortions are well understood and are corrected by *mathematical modelling* of the sensor and platform motion and the geometric relationship of the platform with the earth.
2. *Non-Systematic Distortions*: These are *caused* by the spacecraft variables, i.e., flying heights (or orbital altitudes), and attitude errors, i.e., roll, pitch and yaw. Such distortions are corrected, as follows:
   - *Locating GCPs*: It employs identification of geographic features on the image called *ground control points* (GCPs) with their known longitudes and latitudes (l, f). For this, an accurate base map or a GPS can be used. Once several well distributed GCP pairs are identified, the coordinate information is digitally processed for the appropriate transformation equation, which is then applied to the orginal (row and column) image coordinates to map them into their new ground coordinates. This is called *image-to-map* registration. Geometric registration is also performed by registering one (or more) images to another image. It is called *image-to-image* registration. Once an image has been geometrically registered, it can be combined with other mapped data in a digital environment for further analysis.
   - *Resampling*: Pixels in the orginal image can be resampled for DN values with matching geometric coordinates using *three* methods— *nearest neighbour* (NN), *bilinear interpolation* (BI), and *cubic convolution* (CC). The *NN resampling* uses the digital value from the pixel in the original image, which is 'nearest to the new pixel location' in the corrected image. It is the simplest method and does not alter the original values, but often results in a disjointed or blocky image appearance. The *BI resampling* takes a 'weighted average' of four pixels in the original image nearest to the new pixel location, thus altering the original pixel values and creating entirely new digital values in the output image. It is undesirable if spectral classification is to be done later on. The CC *resampling* computes a 'distance-weighted average' of a block of 16 (4 × 4) or 25 (5 × 5) pixels in the original image surrounding the new pixel location. It also alters the orginal pixel values, but it produces a much sharper image avoiding a blocky appearance.
   - *Polynomial Mapping*: In this the rectified coordinates are viewed as a *2nd order* polynomial function of the source coordinates. Pixel values are then statistically *regressed* for the rectified locations.

Atmospheric Corrections

Atmospheric particles cause *severe scattering*, especially in short wavelengths and when the atmosphere is *humid*, *smoggy* and *dusty*. It is corrected by using models such as, LOWTRAN, MODTRAN or HITRAN based on the *atmospheric radiative transfer equations*. Haze has an *additive effect* on scene brightness and is corrected by subtracting the intercept from the DN of each band for the entire image.

Feature Extraction

After the above corrections, the analyst now performs *feature extraction*, which is defind as the process of isolating the most useful components of the data for further study while discarding the less useful aspects (errors, noise, etc.). It reduces the *image dimensionality* and thereby *saves time* and *resources* in subsequent analysis.

Image Enhancement

It refers to a set of operations carried out to improve the *quality* and *interpretability* of the image by increasing apparent contrast among the various features in the scene. Techniques for enhancement are:

Spectral Enhancement Techniques

- *Density Slicing*: In this, the DNs of a single band image are *sliced* into distinct classes of grey levels and mapped in the RGB colour cube. Density slicing is particularly useful in displaying thermal maps.
- *Contrast Streching* (CS): As the dynamic range of a sensor is never fully utilised, a dull image is commonly produced. This can be restored by *contrast stretching*, that remaps the DN distribution to the fullest display capabilities of an image processing system. CS is of three types—*linear*, *histogram equalisation* and *Gaussian*. The *linear* CS technique is applied to a single band, greyscale image, where the image data is mapped to the fullest range of the display device. In *histogram* equalisation, the range of pixel values in the input image is spread over the fullest range of the display device so that each level in the displayed image contains an approximately equal number of pixel values and the histogram becomes almost uniform. The *Gaussian* stretch involves the fitting of the observed histogram to a normal or Gaussian histogram.

Multi-Spectral Enhancement Techniques

- *Image Arithmetic Operations*: These are applied to enhance images of separate spectral bands in a single multispectral dataset or to individual bands from multi-dated image sets. *Addition* is done to give a dynamic range of images that equals the input images. Band *subtraction* is performed to coregister scenes of the same area acquired at different times for change detection. *Multiplication* uses a single *real* image and *binary* image made up of ones and zeros. Band *rationing* involves the division of DN values of one spectral band by the corresponding DN value of another band. It eliminates differences in scene illumination due to cloud or topographic shadow and brings out spectral variations in different target materials and is widely applied in geological, ecological and agricultural studies.
- *Principal Component Analysis* (PCA): Spectrally adjacent bands in a multispectral RS image are often highly correlated. PCA is used to *remove* this redundancy and *enhance* the image, *reduce* the dimensionality and *fix* the coefficients that specify the axes pointing to the directions of greatest variability. In this a set of bands is transformed such that the new bands, called *principal components* (PC) are uncorrelated and ordered in terms of the amount of image variation they explain. Thus the PCs are simply the *statistical abstraction* of the variability inherent in the original band set and are more interpretable. Transformation yields coefficients, such as, *eigen values* and *eigen vectors* that are then

applied in a linear fashion to the original pixel values. This transformation is derived from the covariance matrix of the original data set and the resultant coefficients describe the lengths and directions of the principal axes.

- *Decorrelation Stretch*: The principal components can be stretched and transformed back into RGB colours—a process called *decorrelation stretching*. It works best on images of semi-arid areas. In this, the three bands are stretched in such a way that they are at right angles to one another. The amount of decorrelation stretch is measured either from the covariance matrix or the correlation matrix.
- *Canonical Component Analysis* (CCA): It is also called *multiple discriminant analysis* and is appropriate when there is *a priori* information about the particular features of interest. Canonical component axes are transformed to maximise the separability of different user-defined feature types.
- *Hue, Saturation and Intensity* (HIS) *Transform*: In the HIS model, hue is the *dominant wavelength* of the perceived colour represented by angular positions around the top of a hexcone, saturation or purity is given by the *distance* from its central vertical axis and intensity or value is represented by the *distance* above its apex. In HIS transform, the RGB colour cube coordinates are converted to HIS hexcone coordinates and vice versa. It *enhances* an image by yielding a more precise representation of human colour vision than the RGB system and is very useful for geological and geomorphological applications.
- *Fourier Transformation* (FT): It *converts* a single-band image from its *spatial* domain to its equivalent *frequency* domain and vice versa. In this the grey scale values forming a single-band image is viewed as a 3D *intensity surface*, with rows and columns defining the two axes and the grey level value at each pixel giving the third (z) dimension. The FT *provides* the details of the frequency distribution of each of the scale components of the image and also the proportion of information associated with each frequency component.
- *Spatial Processing*: In this, *spatial filters* are used to selectively emphasise or suppress information at different spatial scales over an image. These are of two types—*spatial domain filters* and *frequency domain filters*. Spatial domain filters, also called *convolution filters*, may either be *high-pass* (sharpening) or *low-pass* (smoothing). The *low-pass* filters (e.g., moving-average, median and adaptive) are used to *subtract* the low frequency image from the original image to enhance high spatial frequencies. The *result* is a sharper, de-blurred image with isolated or amplified local detail. *High-pass convolution* filters (e.g., Laplacian filter, 1st Differential, etc.) are often biased in certain directions and are best suited for the enhancement of the *edges of the directional features* in an image. The *frequency domain* filters operate on the amplitude spectrum of an image and remove, attenuate or amplify the amplitudes in specified wavebands. It is a 3-step process: *first*, FT is performed and fourier spectrum is computed from the 2D plots; *second*, an appropriate filter transfer function is selected and the elements of the fourier spectrum are multiplied, and *third*, an inverse FT is performed to return to the spatial domain for display purpose.

Image Transformations

Transformations by *arithmetic operations* of multi-band image data, either from a single-band multispectral image or from two or more multi-dated images of the same area generate *new enhanced* images for further analysis. Image *subtraction* is done to identify changes between geometrically registered multi-dated images. Image *addition* is performed for mapping changes in urban development around cities and for identifying areas of on going deforestation. Image *division* or spectral rationing brings out subtle variations in the spectral responses of various surface covers, e.g., NDVI (*normalised difference vegetation index*). PCA reduces the dimensionality in the data by compressing it into fewer bands and thereby maximises the information (or variance) from the original data into new components (= bands).

Image Classification

Digital image classification is the *process of assigning all pixels in an image* to particular classes or landcover/landuse themes based on the patterns of DN in the data, its spatial relationship with neighbouring pixels, and the relationships between the data acquired on different dates. Classification comprises four steps—

1. pre-processing, i.e., corrections, noise suppression, band rationing, PCA, etc.;
2. selection of the training set for the particular features that best describe the pattern;
3. decision regarding choice of suitable method or classifier, and
4. assessing the accuracy of classification.

There are two generic approaches for this—*supervised* and *unsupervised*, although many user-defined *hybrid* approaches, i.e., a combination of the two, may be devised.

Supervised Classification

It comprises three stages—*training, supervision* and *classification*. In the *training stage*, the analyst identifies the homogeneous representative samples of the different surface cover types (information classes) of interest. These samples are called *training areas*, selected by an analyst based on his familiarity with the geographical features of the area. It is in this way that the analyst is *supervising* the categorisation into specific classes. While *classifying*, the computer uses algorithms to determine the numerical 'signatures' for each training class. Each pixel in the image is then compared to these signatures and labelled as the class it most closely resembles digitally. Thus in supervised classifications, the information classes are first identified, and then they are used to determine the spectral classes that represent them.

The most widely used statistical classifiers are *box* (*parallelopiped*) *classifier, centroid (k-means) classifier* and *maximum likelihood classifier*.

- In *box classifiers*, the maximum and minimum DN for each class in each band is first found and then parallelopipeds/boxes are constructed so as to enclose the scatter in each theme. Each pixel is then tested to see if it falls inside any of the boxes. It has several *limitations*. A pixel may fall outside these boxes and remain unclassified; a pixel vector plotted away from the theme scatter may still fall within a decision box and be classified erroneously, and the boxes may overlap yielding further complications. Hence, it is called a *quick and dirty* method.
- In the *centroid classifier*, the mean vector for each category is determined from the average DN in each band for each class. An unknown pixel is then classified by computing the distance from its spectral position to each of the means and assigning it to the class with the closest mean. The *demerits* are that any pixel, lying further away from the nearest centre than the threshold distance, is left unclassified and, as clusters vary in size, a pixel may easily be misclassified.
- The *Gaussian maximum likelihood classifier* computes the probability that a pixel belongs to a particular class. First, the mean vector and covariance matrix of the spectral cluster of each category of DN are computed. Based on these, the statistical probability of a pixel's brightness value belonging to each category of landuse is computed. The pixel is then assigned to the category with the highest probability. It is regarded as the *best classifier* as it assumes equal probability of occurrence of each category of landuse/landcover.

Unsupervised Classification

It uses *clustering algorithms* to produce natural clusters of pixels of similar brightness from the multispectral data. It is a very *popular* approach as it is easier to learn and the available image-processing software contains menus to run very

Fig. 9.25 Supervised Classification

HIGH RESOLUTION IMAGE OF PORTION OF THE DAMODAR VALLEY (GOOGLE EARTH 2007)

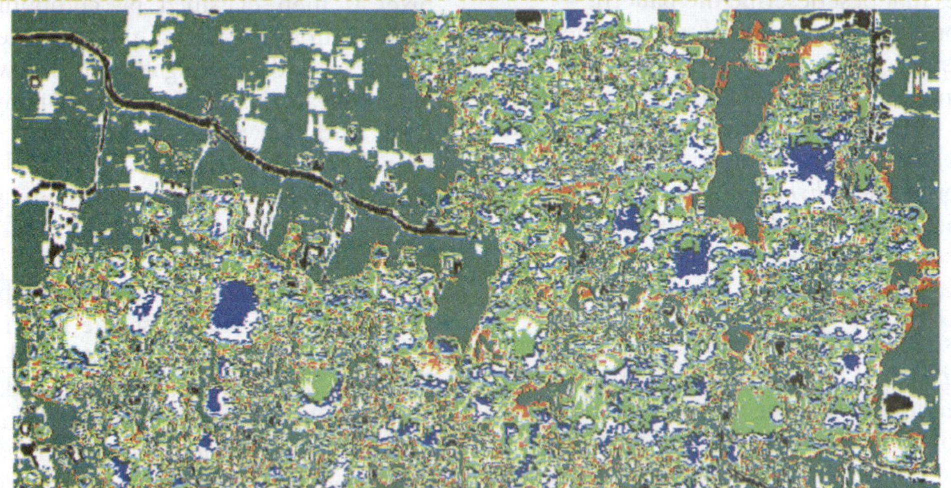

SUPERVISED CLASSIFICATION OUTPUT OF THE ABOVE IMAGE

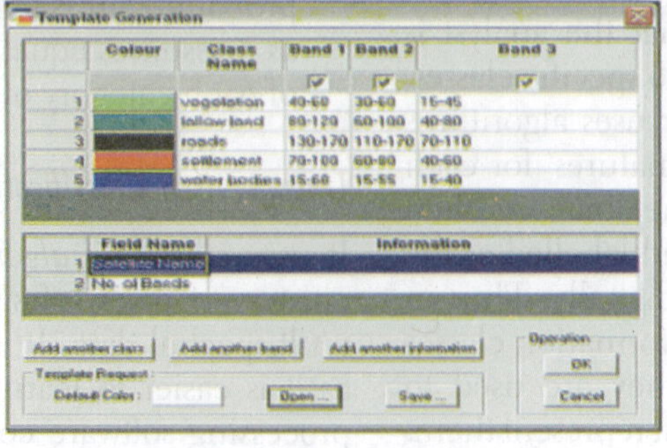

TEMPLATE GENERATED FOR ABOVE CLASSIFICATION

fast clustering programs. These natural clusters are then related to actual landuse/landcover categories after a very careful *ground-truthing*. Spectral classes are grouped first, based solely on the numerical information in the data, and then matched by the analyst to information classes. The iterative process may result in some clusters that the analyst may want to combine subsequently, or clusters that should be broken down further. Both of these require special manipulation to optimise the number of clusters. The most commonly and widely used mathematical strategies are:

Fig. 9.26 Unsupervised Classification

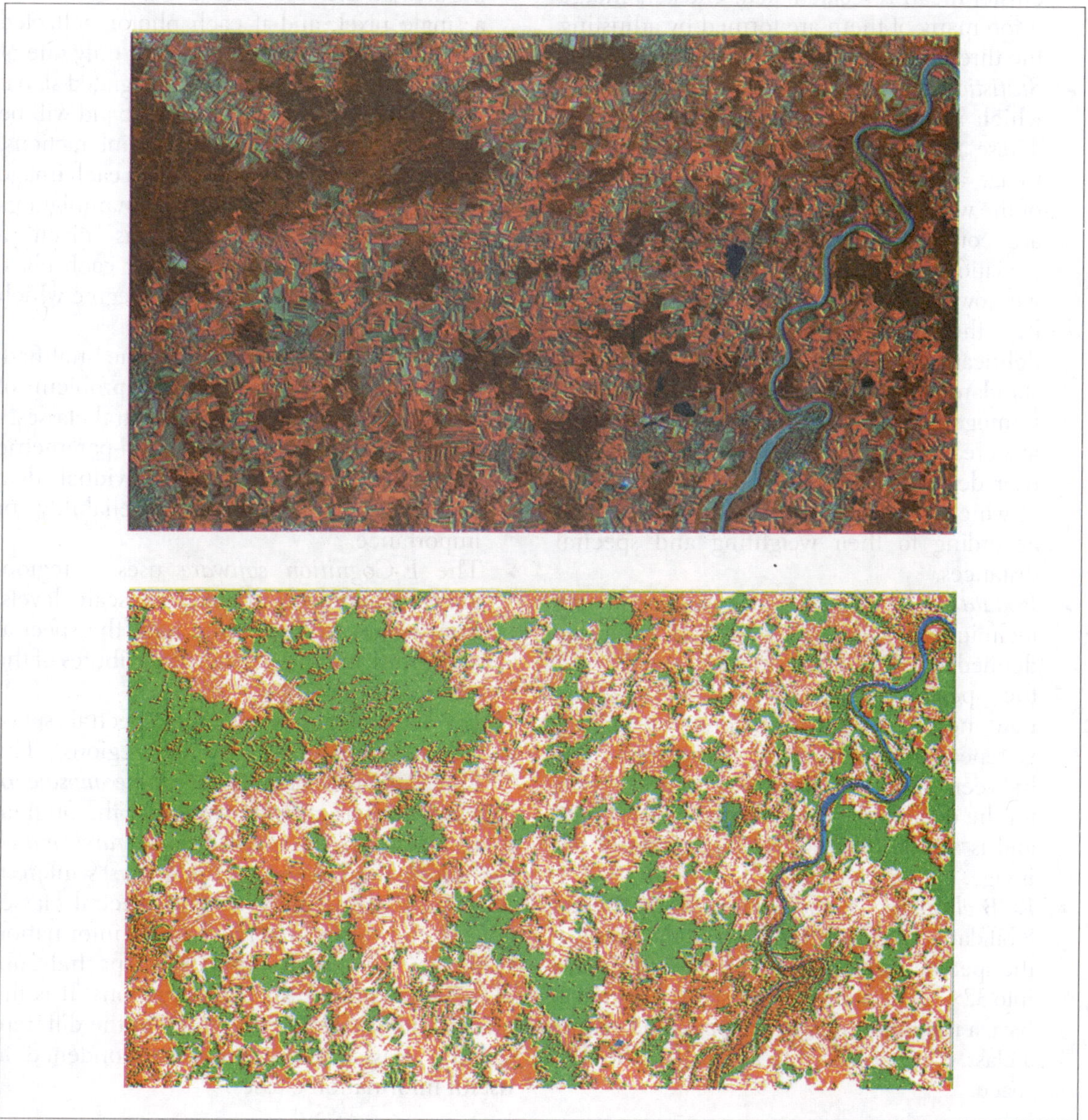

- *Sequential clustering,* where pixels are analysed *one at a time pixel-by-pixel* and *line-by-line*. The spectral distance between each analysed pixel and previously defined cluster means are first computed. If it is greater than a certain threshold value, the pixel begins a new cluster, otherwise it contributes to the nearest existing clusters in which case, the cluster mean is recalculated. Clusters merge if too many of them are formed by adjusting the threshold value of the cluster means.
- *Statistical clustering* uses 3×3 windows in which all pixels have similar vector space. These are moved one at a time through the image avoiding all overlaps. For each band of the window, *mean* and *standard deviations* are computed. The smaller the standard deviations the greater the homogeneity of the window. These values are then compared by the user-specified parameters for delineating the upper and lower limits of the standard deviation. If the window passes the homogeneity test, it forms clusters. Clusters are created *until* their number exceeds the user defined maximum number of clusters at which point some are merged or deleted according to their weighting and spectral distances.
- *Isodata clustering* is an *iterative self-organising* technique. It begins with a set of arbitrarily defined cluster means, located evenly over the spectral space. After each iteration, new means are calculated and the process is repeated until there is some difference between iterations. It produces *good results* for the data that are not normally distributed and is also *not biased* by any section of the image.
- *RGB clustering* is a *quick* method for 3-band 8-bit data. The algorithm plots all the pixels in the spectral space and then divides this space into 32×32×32 clusters. A cluster is required to have a minimum number of pixels to become a class. It is not biased to any part of the data space.

Sophisticated classifiers are continuously being formulated, developed and tried. Some of these are given below:

- *The Neural classifiers* apply feed-forward multi-layer ANN (artificial neural network) using the 'back propagation training methods'.
- *The Linear Mixture Model* (LMM) based on that if there are *n* land cover types present in the area on the land surface that is covered by a single pixel, and if each photon reflected from the pixel area interacts with only one of these *n* cover types, then the integrated signal received at the sensor in a given band will be the linear sum of the *n* individual interactions.
- *The Fuzzy classifiers* do not assign each image pixel to a single class in an unambiguous fashion, rather each pixel is given a 'membership grade' (0 to 1) for each class, that provides a measure of the degree which the pixel belongs to or resembles.
- *The Decision-tree classifiers* use the 'stratified' or 'layered' approach to the problem of discriminating between the spectral classes.
- *Evidential reasoning* is a non-parametric procedure which weights individual data sources according to their reliability or importance.
- The *E-Cognition software* uses a region-growing approach at different scale levels, from coarse to fine, using both the spectral properties and the geometric attributes of the regions.

By classification the large spectral space is reduced into relatively few regions. The classified image becomes *a simple mosaic of pixels* as good as a *thematic map* of the original image. However, the spectral classes *rarely match* the information classes of the analyst's interest. As a result, sometimes, unique spectral classes appear and, alternatively, a broad information class may contain a number of spectral sub-classes with unique spectral variations. It is the analyst's job to decide the utility of the different spectral classes and their correspondence to useful information classes.

EXERCISE

1. What is remote sensing?
2. What is the most common application of aerial photographs and satellite images?
3. State the salient characteristics of remote sensing.
4. State the principles of remote sensing.
5. What is EMR?
6. What is EMS?
7. Which range of EMS is important for aerial photography? What is it called?
8. Discuss how atmosphere affects EMR.
9. What are transmission windows?
10. How are the transmission windows utilised?
11. What are the different classes of remote sensing?
12. What is terrestrial photography?
13. What is bio-stereometrics?
14. What is aerial photography?
15. What is space imagery?
16. What are the different types of remote sensing systems?
17. Discuss the salient features of passive remote sensing systems.
18. Discuss the salient features of active remote sensing systems.
19. What are the imaging characteristics of remote sensing systems?
20. What is meant by spatial resolution? Explain.
21. What is meant by spectral resolution? Explain.
22. What is meant by radiometric resolution? Explain.
23. What is meant by temporal resolution? Explain.
24. What is meant by ground resolution? Explain.
25. What is meant by IFOV?
26. What are quantisation levels?
27. Define photogrammetry. What are its different types?
28. What is digital photogrammetry?
29. Compare vertical aerial photographs with maps.
30. What is meant by relief displacement? What are its uses?
31. Classify aerial photographs and discuss the characteristics of each.
32. Classify aerial photographs based on camera orientation.
33. Classify aerial photographs based on angular coverage.
34. Classify aerial photographs based on emulsion type.
35. What are the common products of photogrammetry? Discuss in detail.
36. Discuss how aerial photographs are interpreted.
37. Discuss the factors influencing the interpretation of aerial photographs.
38. Explain how an aerial photograph is identified.
39. Explain how the scale of an aerial photograph is determined.
40. What is stereoscopy? Explain the minimum conditions for stereoscopic fusion.
41. Write the steps followed for the stereoscopic fusion of a stereo-pair of aerial photographs in a laboratory.
42. What is meant by line of flight?
43. What is meant by principal point?
44. What is meant by conjugate point?
45. What does flying height mean?

46. What is meant by datum?
47. Explain how the tilt displacements are rectified?
48. Define roll, pitch and yaw.
49. What is meant by stereoscopic parallax? How can this be rectified?
50. Explain how relief displacements are geometrically corrected.
51. What is DEM?
52. What are non-photographic remote sensing systems?
53. What are the characteristics of TIR, RADAR, TM, TIMS, ATLAS, SLAR, RAR, and SAR?
54. Define image interpretation.
55. Define image processing. Compare analogue and digital images.
56. Compare the manual and digital processes of image interpretation.
57. What are the basic elements of an image? Explain their role in image analysis.
58. What is meant by interpretation key? What are its different types? How are these used in photo interpretation?
59. Define digital image processing. What are the common image processing functions?
60. What is pre-processing? What is its significance?
61. What are the common forms of radiometric errors? Explain how are these rectified.
62. What are the common forms of geometric distortions? Explain how are these rectified.
63. How are atmospheric corrections performed?
64. What is meant by image enhancement? What are its techniques?
65. Explain in detail the techniques of spectral enhancement.
66. Explain in detail the techniques of multi-spectral enhancement.
67. Why do line dropouts occur in images? How is this rectified?
68. Why does banding appear in images? How is this problem solved?
69. Distinguish between systematic and non-systematic distortions.
70. What is image-to-image registration?
71. What is image-to-map registration?
72. What is meant by GCP? Explain its function in image processing.
73. What is resampling? Explain the various techniques of it.
74. What is meant by image feature?
75. What is density slicing?
76. What is contrast stretching? What are its various techniques?
77. What are the various arithmetic operations performed in an image?
78. Explain how PCA enhances an image.
79. What is decorrelation stretch?
80. What is HIS transform?
81. What is Fourier transform?
82. What is Fourier spectrum?
83. What is spatial processing? Name the different types of spatial filters.
84. What are the differences between a high-pass and a low-pass filter?
85. Name the different high-pass filters.
86. Name the different low-pass filters.
87. What is meant by spatial domain filter?
88. What is meant by frequency domain filter?
89. What is spectral rationing?
90. What is NDVI?
91. What is meant by image classification? What are its different steps?
92. Give the salient features of supervised classification.

93. Distinguish between supervised and unsupervised classification.
94. Explain how supervised classification is performed.
95. Compare the different types of classifiers in supervised approach.
96. Explain why the Gaussian maximum likelihood classifier is one of the best.
97. Discuss the characteristics of unsupervised classification.
98. Explain the application of different clustering algorithms in unsupervised classifications.
99. What is a neural classifier?
100. What is a linear mixture model?
101. What is a fuzzy classifier?
102. What is a decision-tree classifier?
103. What is evidential reasoning?
104. What is ground truthing? Explain its significance in image interpretation.
105. What are the objectives of image processing?

Landuse/Landcover Map using Visual Image Processing

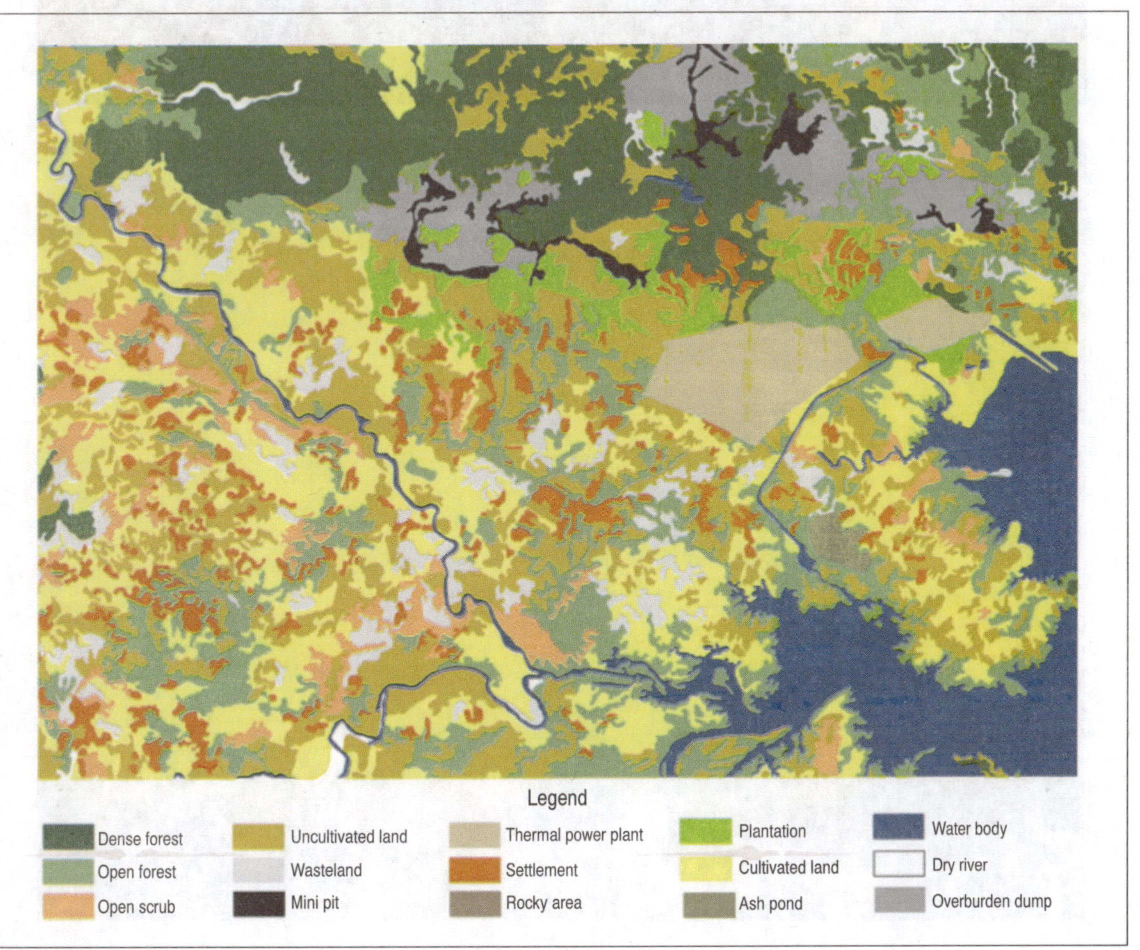

Practice Satellite Image 1 and 2

Practice Stereopair 1A and 1B

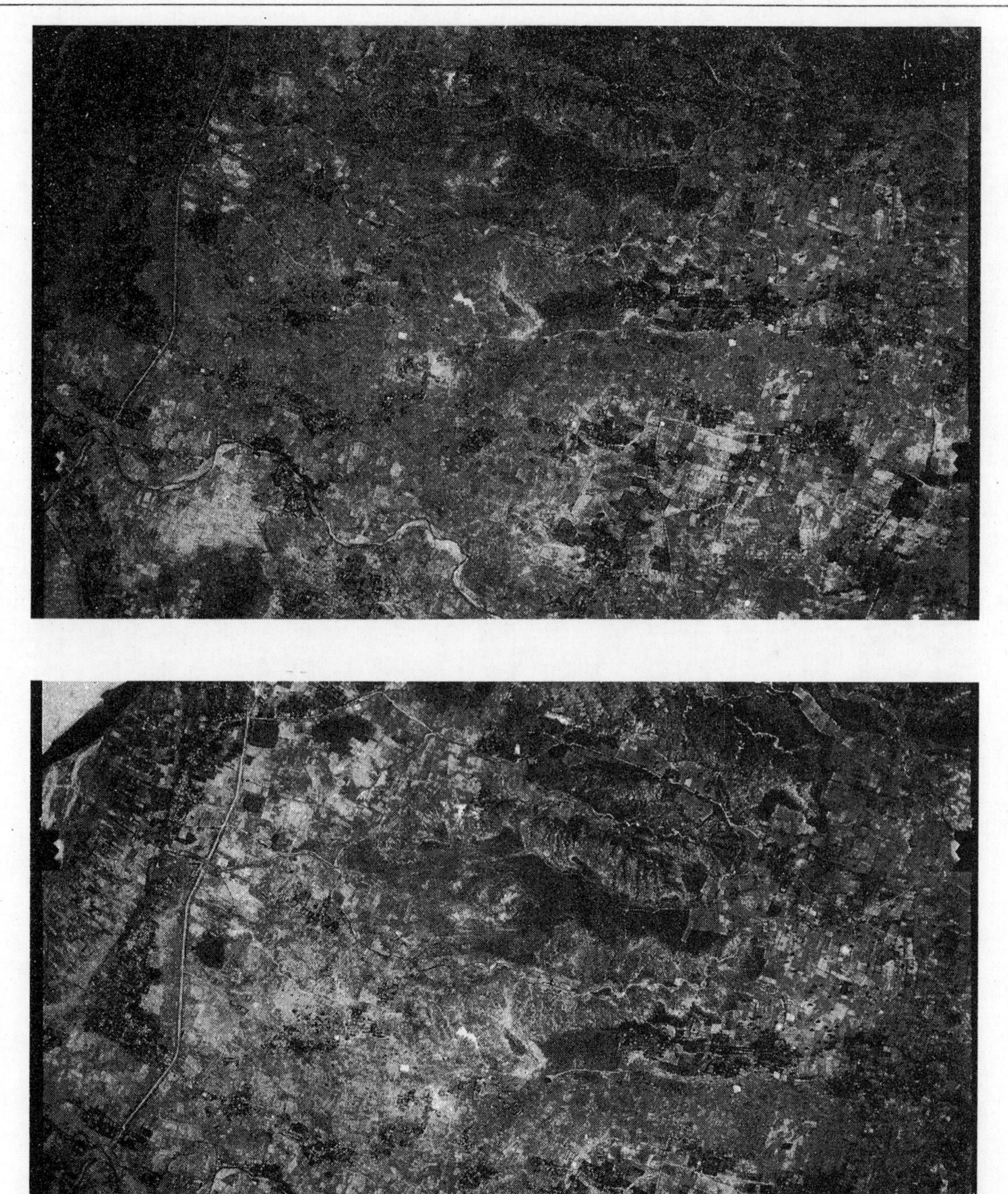

Practice Stereopair 2A and 2B

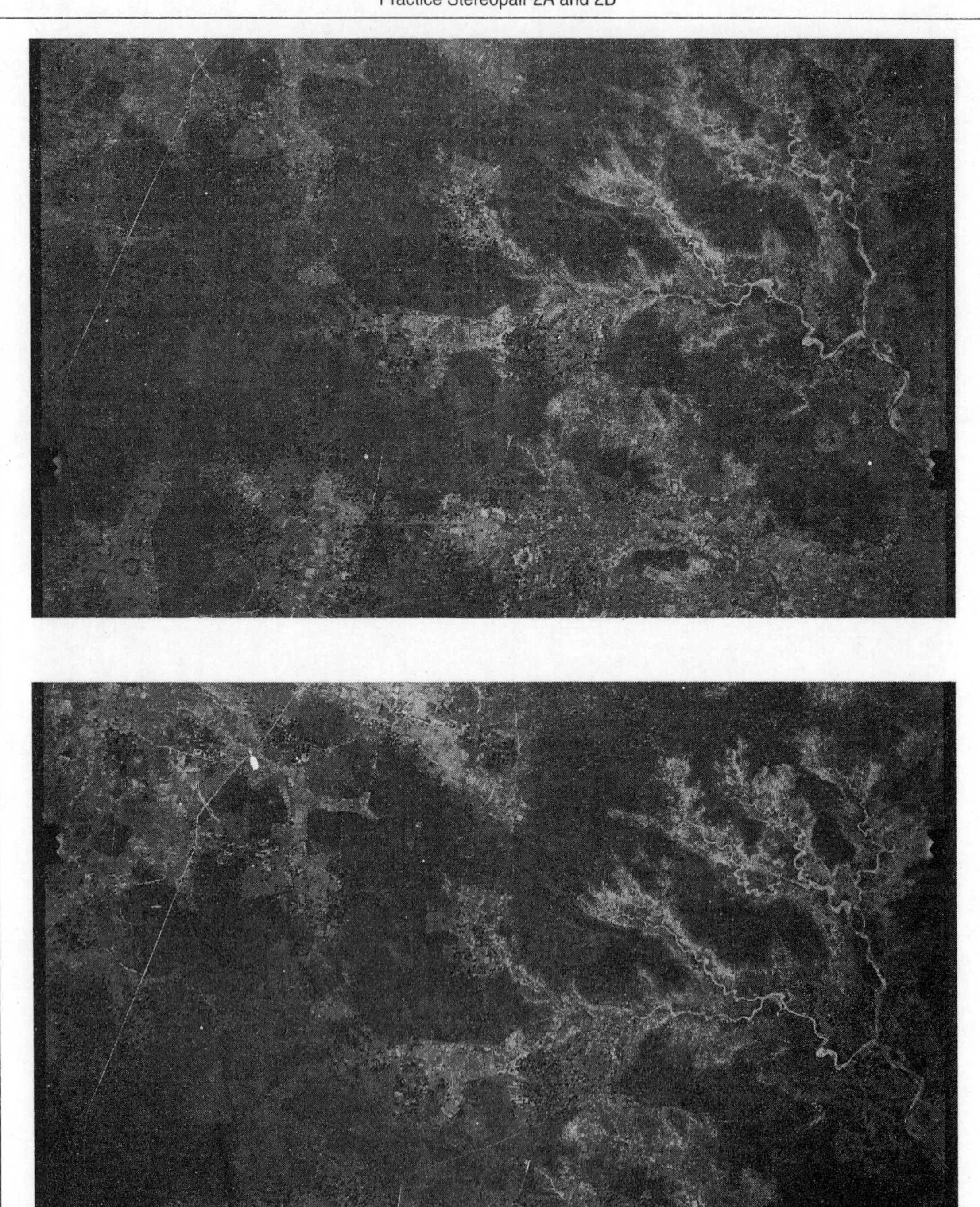

Unit IV

Field Techniques

10

Identification of Minerals and Rocks

HIGHLIGHTS

- 7 Keys to Recognising Minerals
- Selected Specimens of Minerals
- 4 Keys to Recognising Rocks
- Selected Specimens of Rocks

Over 98% of the earth's crust is made up of the eight major elements—oxygen (46.7%), silicon (27.69%), aluminium (8.07%), iron (5.05%), calcium (3.65%), sodium (2.75%), potassium (2.58%), and magnesium (2.08%). Some elements naturally make minerals by themselves (e.g., gold, platinum, copper, sulphur and carbon), but in most cases, two or more elements chemically combine to form rocks and minerals. To a geologist, the term 'mineral' is restricted to a) a substance of inorganic origin, b) a substance having a definite chemical composition, and c) a substance possessing a definite physical property. A mineral may, therefore, be defined

Fig. 10.1 Igneous Activity

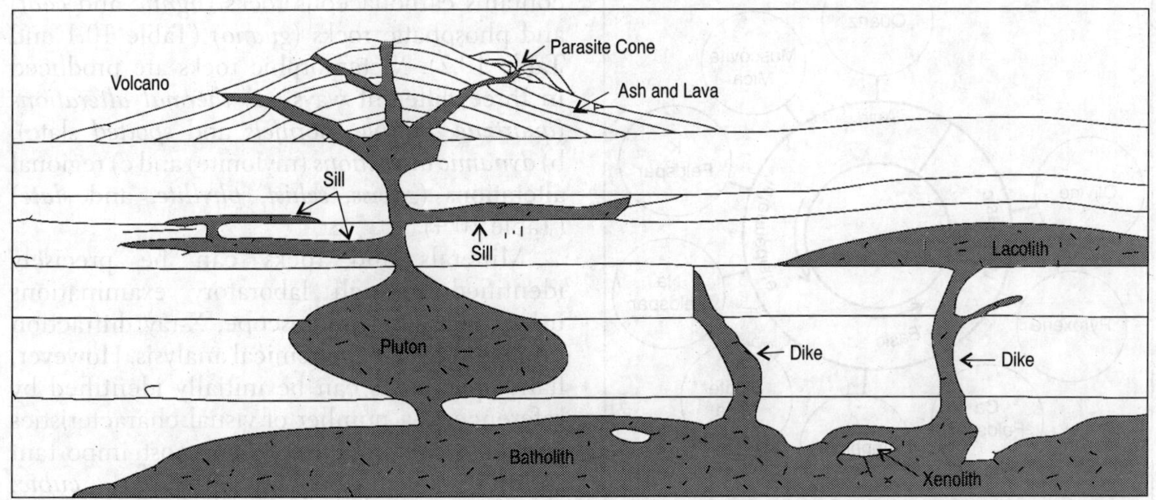

as a naturally occurring isotrophic inorganic substance of definite chemical composition, atomic structure and physical property.

Primarily, rocks are the cohesive aggregates of more than one mineral, which are mechanically bound together. Minerals are certainly compounds of their constituent minerals but rocks are definite mixtures of their component minerals. Thus the mineral quartz is a compound of the elements silicon and oxygen, whereas the rock granite is a mixture of several minerals—quartz, feldspar, mica, etc. Most of the rock forming minerals consist of a) silicates in either their primary forms, such as, feldspar, mica, amphibole, pyroxene and olivine, or in their secondary forms, such as, chlorite, serpentine and various clay minerals (e.g., illite), b) oxides of silicon (quartz) and iron (magnetite and hematite), c) carbonates of calcium (calcite and aragonite), or carbonates of calcium and magnesium (dolomite), d) sulphides of iron (pyrites), e) sulphates of calcium (anhydrite and gypsum) and f) chlorides of sodium (halite).

Genetically rocks are of three types—igneous, sedimentary and metamorphic. Igneous rocks are the primary rocks produced directly from the cooling of magma. They may be plutonic (deep seated), hypabyssal (intermediate), volcanic (extruded) and pyroclastic (products of explosive volcanism) (Fig. 10.1). Slow cooling gives rise to a coarsely crystalline texture (phaneritic), while rapid cooling produces a fine crystalline texture (cryptocrystalline) or even a glassy rock, and the intermediate between the two gives a porphyritic texture (Fig. 10.1). *Granite*, *grano-diorite* and *rhyolites* are acid rocks (> 66% silica content), *syenite*, *diorite*, *trachyte* and *andesite* are intermediate rocks (52–66% silica content), *gabbro*, *dolerite* and *basalt* are basic or mafic rocks (44–52% silica content) and *peridotite* and *serpentinite* are ultra basic or ultramafic (<44% silica content) rocks. Sedimentary rocks primarily fall into three groups—*clastic*, *chemical*, *bio-chemical* and *organic*. In the first group, rudaceous rocks (*conglomerate*, *breccia*, *tillites*) are formed of large rock fragments, arenaceous rocks (*sandstone*) are composed of sand particles and argillaceous rocks (*siltstone*, *mudstone*, *shale*, *claystone*) are formed by fine silts to very fine clay particles. In the second group, the characteristic rocks are—calcareous (limestone, chalk, dolomite), siliceous (*chert*, *flint*), evaporites (*gypsum*, *anhydrite*, *helite*) and ferruginous (*various ironstones*). The third group contains carbonaceous rocks (*lignite* and *coal*) and phosphatic rocks (*guano*) (Table 10.1 and Table 10.2). Metamorphic rocks are produced in three different ways—a) *thermal alterations* (*quartzite*, *marble*, *hornfels* and *spotted* slate), b) *dynamic alterations* (mylonite) and c) regional alterations (*gneiss*, *schist*, *phyllite*, and *slate*) (Table 10.4).

Fig. 10.2 Mineralogy of Igneous Rocks

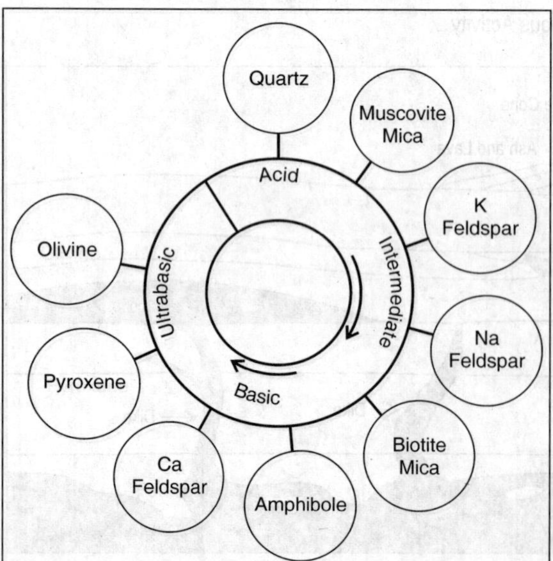

Minerals and rocks can be precisely identified through laboratory examinations using polarizing microscope, X-ray diffraction photography or geochemical analysis. However, hand specimens can be initially identified by reference to a number of visual characteristics and physical properties. The most important of these are the crystal form (e.g., *cubic*, *tetragonal*, *hexagonal*, *trigonal*, *ortorhombic*,

Table 10.1 Grain Size of Sedimentary Rocks

Name Particles	*Diameter of Particles 25.4 mm = 1 inch*	*Sedimentary Rock*
Boulders Cobbles Pebbles Granules	Greater than 256 mm 64–256 mm 4–64 mm 2–4 mm	Conglomerates (rounded) and Breccias (angular)
Very coarse sand Coarse sand Medium sand Fine sand Very fine sand	1–2 mm 1/2–1 mm 1/4–1/2 mm Sandy 1/8–1/4 mm 1/16–1/8 mm	Sandstones
Coarse silt Fine silt	1/64–1/16 mm Gritty 1/256–1/64 mm	Siltstones and Mudstones
Clay	Less than 1/256 mm Smooth	Shales Claystones

Table 10.2 Constituents of Sedimentary Rocks

	Major Constituents	*Accessory Minerals*	
	Abundant	*Less abundant*	*Less than 1%*
Mechanically deposited	Quartz Clay mineral Mica Calcite	K-feldspar Plagioclase Rock fragment	Magnetite Tourmaline Garnet Amphibole Hematite Limonite Others
Chemically deposited	Calcite Dolomite	Quartz (Chert) Gypsum (Anhydrate) Halite Hematite	

Table 10.3 Origin of Metamorphic Rocks

Rock Type	*Increasing Temperature and Pressure*
Shale	Slate → Phyllite → Schist
Sandstone	Quartz
Limestone	Marble
Basalt	Schist → Amphibolite
Granite	Granite → Gneiss
Coal	Peat → Lignite → Bituminous → Anthracite

Table 10.4 Moh's Hardness Scale

Scale	*Mineral*	*Remarks*
1	Talc	
2	Gypsum	← Finger nail
3	Calcite	← Copper coin
4	Fluorite	
5	Apatite	
6	Microcline	← Knife-glass
7	Quartz	← Tool steel
8	Topaz	
9	Corundum	
10	Diamond	

monoclinic and *triclinic*), the nature of the twined crystal (e.g., *contact*, *penetration* and *polysynthetic*), the crystal habit (e.g., *prismatic*, *granular*, *lamellar*, *acicular*, *bladed*, *botryoidal*, *reniform*, *file form*, *nodular*, *dendritic*, *fibrous*, and *amygdaloidal*), cleavage (i.e., *the number of directions along which a mineral splits*), fracture (e.g., *conchoidal*, *sub-conchoidal*, *even*, *uneven*, *hackly*, *earthy*), colour (e.g., *opalescence*, *iridescence*, *schiller*, *tarnish*), streak (i.e., *colour of its powder*) (Fig. 10.4), lustre (e.g., *metallic*, *vitreous*, *resinous*, *pearly*, *silky* and *adamentine*), diaphaneity (e.g., *transparent*, *translucent*, *opaque*), hardness (Fig. 10.3), specific gravity (i.e., *light*, *heavy*, etc.), feel (i.e., *greasy*, *soft*, etc.), magnetism, taste, odour, fusibility, pyro-electricity and piezo-electricity. The general chemical properties tested for identification are—solubility in acid and nature of acid

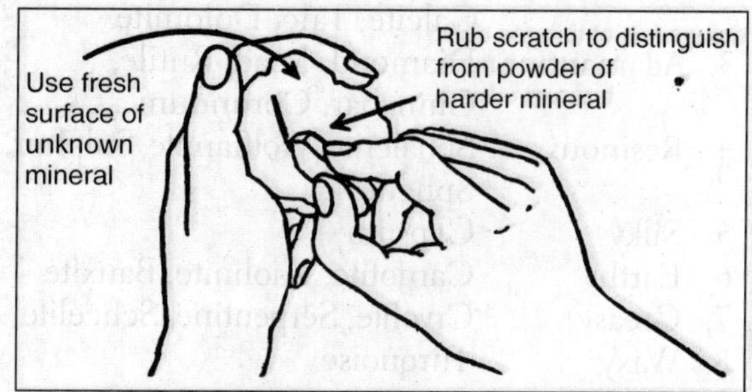

Fig. 10.3 Testing Hardness of a Specimen

Fig. 10.4 Testing Streaks of Minerals

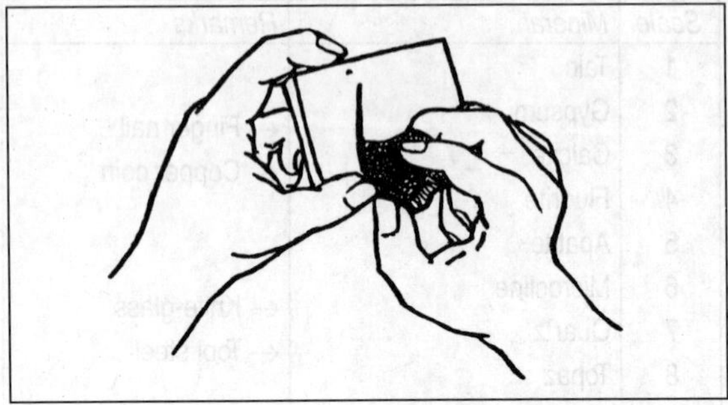

reaction. The diagnostic characteristics of hand specimens of some common minerals and rocks are described in the following paragraphs.

7 KEYS TO RECOGNISING MINERALS

Key No. 1: Lustre

Every mineral has either a metallic or a non-metallic lustre (i.e., reflection of light from the mineral surface). A metallic lustre is typical of a metal, e.g., gold, silver, copper, aluminum, etc. Such minerals are opaque and when crushed they yield a powder which is black or darker in colour than the mineral itself. Minerals having a non-metallic lustre become transparent on a thin edge, and when crushed they yield a powder which is white or lighter than the mineral itself. The common ones are—

1. Vitreous — Quartz, Malachite, Azurite, Barite, Halite, Topaz
2. Pearly — Mica, Chlorite, Gypsum, Calcite, Talc, Dolomite
3. Adamantine — Diamond, Lead, Rutile, Cinnabar, Corundum
4. Resinous — Sphalerite, Wolframite, Sulphur, Sphene
5. Silky — Gypsum
6. Earthy — Carnotite, Kaolinite, Bauxite
7. Greasy — Cryolite, Serpentine, Scheelite
8. Waxy — Turquoise

Key No. 2: Hardness

Hardness is described by Mohs scale. He placed talc, the softest of all minerals as No. 1 in the series and diamond, the hardest of all known minerals as No. 10, as follows—

10	Diamond	5	Apatite
9	Corundum	4	Fluorite
8	Topaz	3	Calcite
7	Quartz	2	Gypsum
6	Feldspar	1	Talc

The scale does not indicate the exact hardness; it only means that any mineral can scratch all those beneath it. Some familiar objects we can use in the field to test are—

6.5 Steel File
5.5 Knife Blade, Window Glass
3.0 Copper Coin
2.5 Fingernail

Minerals under 2.5 will leave a mark on paper; those under 5.5 can be scratched by a knife; and those over 5.5 will scratch glass.

Key No. 3: Colour

There are minerals that are reasonably constant in their colour and are diagnostic of them, as follows—

1. Yellow — Sulphur
2. Pink — Feldspar
3. Blue — Azurite
4. Green — Malachite
5. Brass — Chalcopyrite
6. Auburn — Apatite
7. Bronze — Pyrrhotite
8. Black — Pitchblende
9. White — Kaolinite

Key No. 4: Streak

The colour of a powdered mineral, called streak is obtained by rubbing it against a piece of unglazed porcelain, called streak plate. It is diagnostic of some minerals, as follows—

1. Indian-red — Haematite
2. Black — Magnetite, Graphite

3. Greenish	Galena
4. Greenish-black	Chalcopyrite
5. Gray	Chalcocite
6. Pale Green	Malachite
7. Light Blue	Azurite
8. Scarlet	Cinnabar
9. Orange	Realgar
10. Yellow	Carnotite

Key No. 5: Cleavage

Crystalline minerals are said to cleave or have cleavage when they break in definite directions along smooth surfaces. These are diagnostic of some minerals as follows—

1. Octahedral	Diamond
2. Platy	Covellite, Molybdenite
3. Cubic	Galena, Halite
4. Scaly	Graphite
5. Rhombic	Calcite, Dolomite, Siderite, Magnesite
6. Flaky	Mica, Chlorite
7. Diamond-shaped	Barite
8. Blocky	Anhydrite, Feldspar

Key No. 6: Fracture

Minerals that break in irregular directions are said to fracture. These are also typical of some minerals—

1. Conchoidal	Quartz, Malachite, Azurite, Magnesite, Opal, Tourmaline, Beryl
2. Hackly	Gold, Copper

Key No. 7: Specific Gravity

It implies how heavy it is with respect to equal volume of water and can be diagnostic of some minerals as follows—

1. Very Heavy	Gold, Silver, Cinnabar, Pyromorphite, Wulfenite
2. Heavy	Galena, Copper, Pitchblende, Wolframite, Malachite, Azurite, Siderite, Corundum, Zircon, Garnet
3. Medium	Chalcopyrite, Cuprite, Pyrite, Haematite, Magnesite, Gypsum, Calcite, Talc, Dolomite, Kyanite, Bauxite, Kaolinite, Quartz
4. Light	Graphite, Halite, Stilbite, Natrolite, Sulphur, Borax, Opal

SELECTED SPECIMENS OF MINERALS

Anhydrite [$CaSO_4$]

Its composition is calcium sulphate but does not contain water like gypsum. It is found in veins, cavities (as crystals) and also as evaporate deposits (Fig. 10.5).

Fig. 10.5 Anhydrite

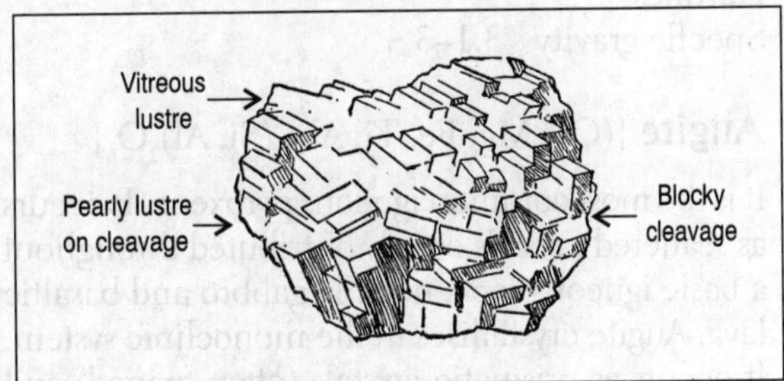

Lustre	Vitreous; Pearly (on cleavage)
Cleavage	Blocky
Colour	White
Streak	White
Hardness	can be scratched by copper coin
Specific	Gravity 2.9 –3.0

Apatite [$Ca_5(PO_4)_3(OH,F,Cl)$]

It is the common phosphate (calcium phosphate with hydroxyl, flourine and chlorine) that occurs as accessory minerals in rocks. It is known as *hydroxyapatite* when rich in hydroxyl, *flourapatite* when rich in flourine and *chlorapatite* when rich in chlorine. Apatite occurs either as transparent, glassy, prismatic or tabular crystals or as dull, fibrous or granular aggregates (Fig 10.6).

Fig. 10.6 Apatite

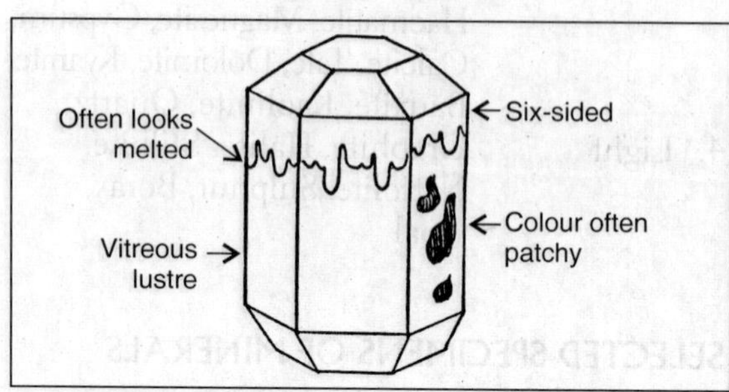

Crystal	hexagonal
Colour	white (if pure) but more often green, brown, yellow or blue
Streak	White
Cleavage	Poor
Fracture	Conchoidal
Hardness	5
Specific gravity	3.1–3.4

Augite [(Ca, Mg, Fe, Ti, Al)$_2$ (Si, Al)$_2$O$_6$]

It is the most common igneous pyroxene. It occurs as scattered crystals evenly distributed throughout a basic igneous rock, notably gabbro and basaltic lava. Augite crystallises in the monoclinic system. It occurs as prismatic crystals (often twined) and as granular masses (Fig. 10.7).

Colour	Brown, green or black
Streak	Colourless
Lustre	Glassy
Opacity	Translucent to opaque
Cleavage	Good (2-dimensional at 87°)
Fracture	Uneven

Fig. 10.7 Augite

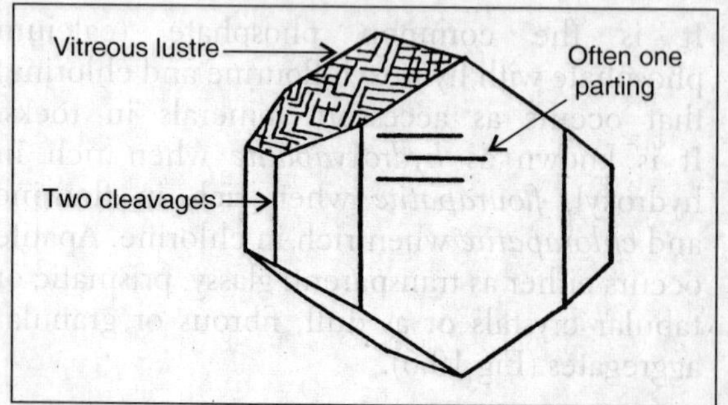

Hardness	5.5–6.0
Specific gravity	3.23–3.52

Azurite [$Cu_3(CO_3)_2(OH)_2$]

It is the basic copper carbonate with distinctive azure blue shades. It occurs only in the oxidized portions of copper ore veins. Azurite commonly occurs as an earthy material. The crystal system is monoclinic with tabular or equidimensional habit (Fig. 10.8).

Fig. 10.8 Azurite

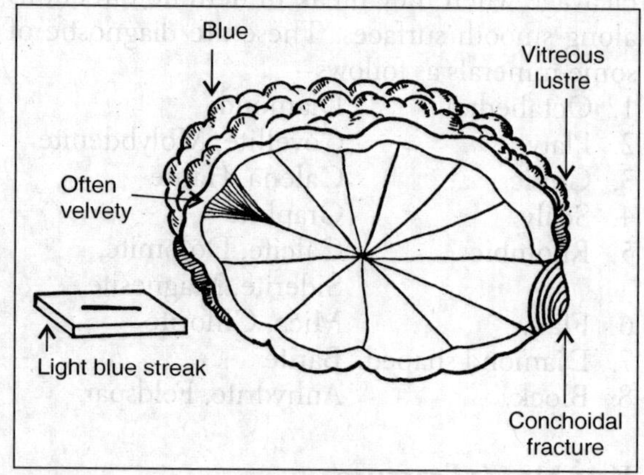

Crystal	Transparent
Colour	Azure blue
Streak	Light blue
Lustre	Brilliant (vitreous to adamantine)
Cleavage	Complex
Fracture	Conchoidal
Hardness	3.5–4.0 (it is very soft and can easily be broken to form gemstones)
Specific gravity	3.77–3.89

Barite [$BaSO_4$]

It is the most common mineral containing barium and one of the most common sulphate minerals. It is heavy and often found as crystals of enchanting blue colour (Fig. 10.9).

Lustre	Vitreous
Cleavage	Diamond-shaped
Colour	Blue, White, also colourless

Fig. 10.9 Barite

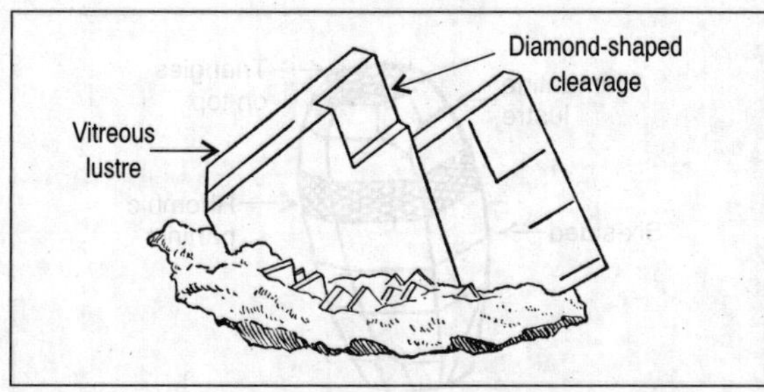

Streak	White
Hardness	can be scratched by copper coin
Specific	Gravity 4.5

Bauxite [Principal Ore of Aluminium]

It is a mixture of diaspores [AlO(OH)], gibbsite, boehmite and other materials. Within this non-crystalline colloidal precipitates, clay, limonite and partly weathered silicates may be present (Fig. 10.10).

Fig. 10.10 Bauxite

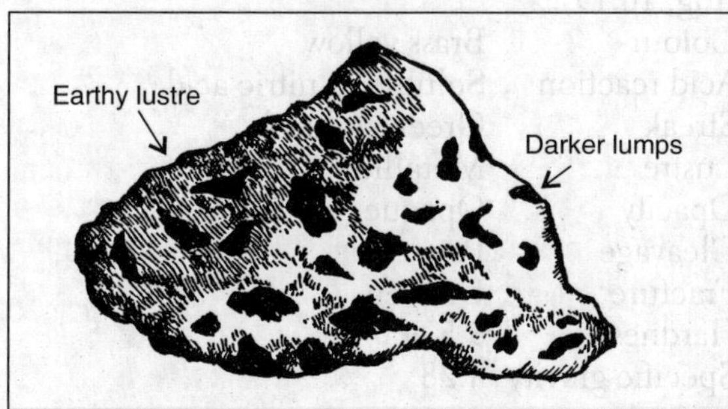

Occurrence	Earthy and massive
Structure	Nobby
Colour	Generally grey and white with reddish brown iron stains
Streak	White

Biotite [$K(Fe,Mg)_2(Si_3Al)O_{10}(OH)_2$]

Biotite (potassium-magnesium-iron-aluminium-silicate) is a type of mica and occurs as tabular prismatic crystals with scattered grains, scales and scaly masses (Fig. 10.11).

Fig. 10.11 Biotite

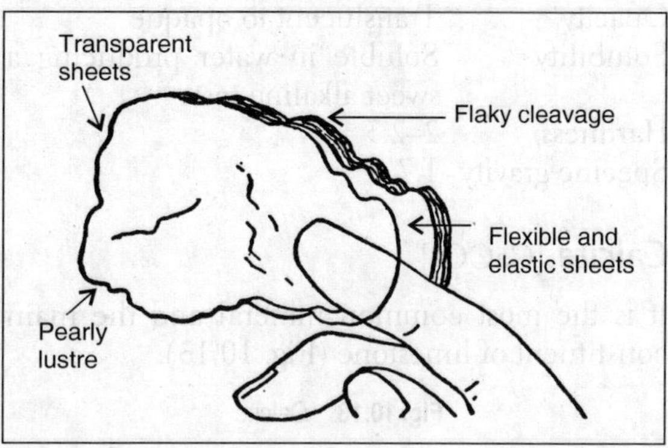

Crystal	Monoclinic
Form	Scaly
Colour	Black, dark green or brown
Lustre	Generally glossy with white streaks and platy structures
Opacity	Opaque to transparent
Cleavage	Perfectly 1-dimensional (splits into thin elastic sheets)
Hardness	2.5–3.0

Borax [$Na_2B_4O_7,10H_2O$]

It is the most widespread borate (hydrous sodium borate). It occurs in large deposits in the dry beds of salt lakes in arid and semi arid regions. Borax occurs as prismatic crystals, crusts and porous masses (Fig. 10.12).

Crystal	Monoclinic
Colour	Usually colourless, white or tinted
Streak	White

Fig. 10.12 Borax

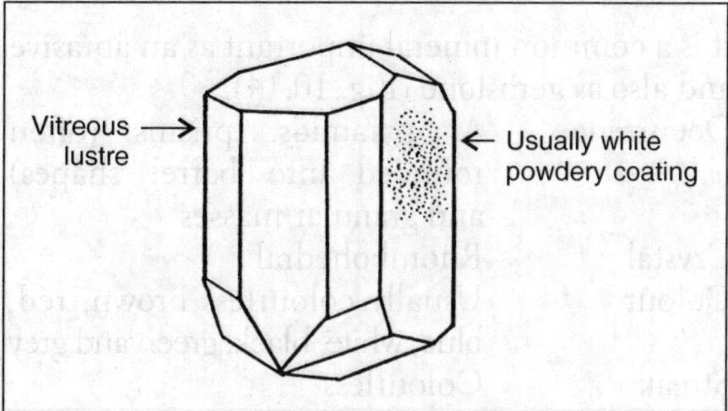

Lustre	Glassy or resinous
Opacity	Translucent to opaque
Solubility	Soluble in water producing a sweet alkaline taste
Hardness	2–2.5
Specific gravity	1.7

Calcite [$CaCO_3$]

It is the most common mineral and the main constituent of limestone (Fig. 10.13).

Fig. 10.13 Calcite

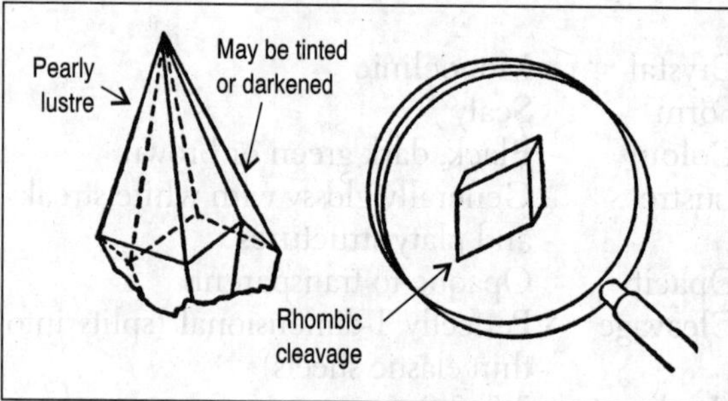

Occurrence	Dog-tooth spar crystals
Crystal	Rhombohedral
Colour	White or grey but sometimes colourless or tinted yellow, blue, green and pink
Acid reaction	Quick effervescence with hydrochloric acid
Streak	White
Lustre	Vitreous to earthy
Cleavage	Perfectly 3-dimensional
Hardness	3

Corundum [Al_2O_3]

It is a common mineral important as an abrasive and also as gemstone (Fig. 10.14).

Occurrence	As pyramids, prisms (often rounded into barrel shapes) and granular masses
Crystal	Rhombohedral
Colour	Usually colourless brown, red, blue, white, black, green and grey
Streak	Colourless

Fig. 10.14 Corundum

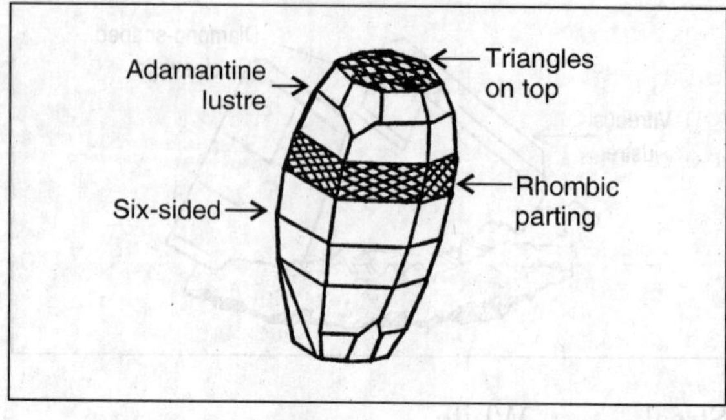

Lustre	Brilliant to glossy
Fracture	Conchoidal
Hardness	9
Specific gravity	4.0

Chalcopyrite [$CuFeS_2$]

It is the most important copper mineral and ore of copper. Chalcopyrite (copper iron sulphide) occurs as tetrahedral or spheroidal crystals (often twined) and as granular and compact masses (Fig. 10.15).

Colour	Brass yellow
Acid reaction	Soluble in nitric acid
Streak	Greenish-black
Lustre	Metallic
Opacity	Opaque
Cleavage	Poor
Fracture	Uneven
Hardness	3.5–4.0
Specific gravity	4.28

Fig. 10.15 Chalcopyrite

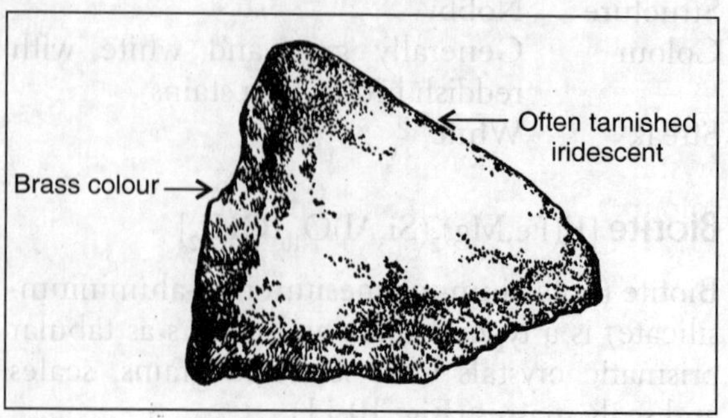

Cuprite [Cu_2O]

It is known as ruby copper and occurs in veins as cubes, octahedra, dodecahedra and combinations of both and also as needles, fibrous and earthy masses (Fig. 10.16).

Fig. 10.16 Cuprite

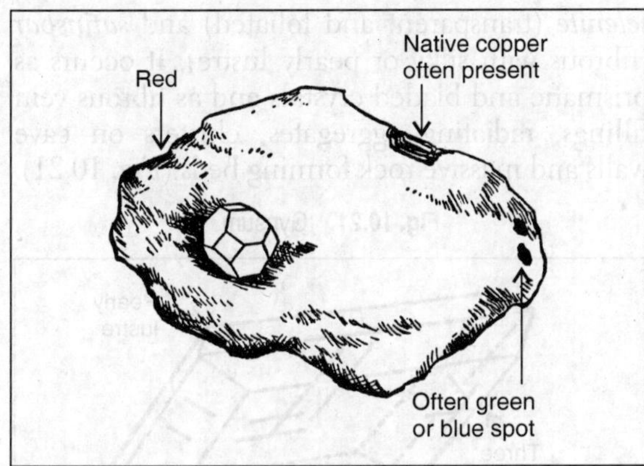

Crystal	Cubic
Colour	Red to black
Acid reaction	Produces white precipitates of cuprous chloride when dissolved in hydrochloric acid
Streak	Brownish red
Lustre	Brillant to earthy
Hardness	3.5–4.0
Specific gravity	6.0

Diamond [C]

It is the hardest known mineral and the most precious gemstone. Diamond is the natural crystalline allotropic form of carbon with a very high refractive index and dispersive power. It occurs as octahedral crystals, some with curved faces and striations (Fig. 10.17).

Crystal	Cubic
Colour	Normally colourless but may be yellow, brown, black, blue, green or red depending on the impurities
Lustre	Brilliant or greasy
Opacity	Transparent to translucent

Fig. 10.17 Diamond

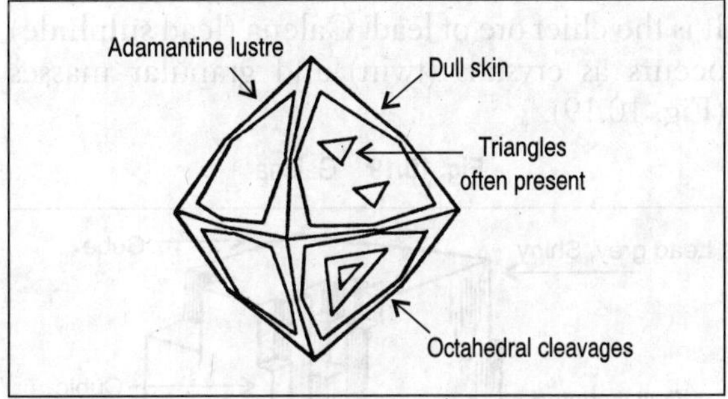

Hardness	10
Specific gravity	3.5

Dolomite [$CaMg(CO_3)_2$]

Dolomite (calcium magnesium carbonate) is structurally similar to calcite but with Mg^{++} and Ca^{++} in alternating metal positions. The form of occurrence is commonly massive, sometimes granular and may also be crystals (Fig. 10.18).

Fig. 10.18 Dolomite

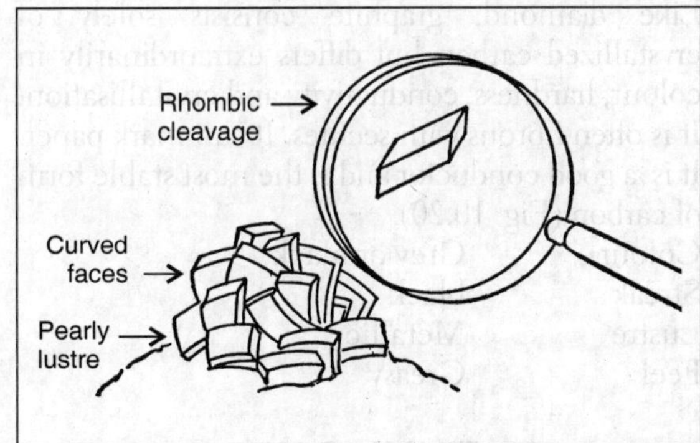

Colour	White, yellow, brown, red, green, grey or black
Acid reaction	It reacts only with warm hydro chloric acid
Lustre	Vitreous or pearly
Cleavage	Perfectly 3-dimensional
Fracture	Conchoidal
Hardness	3.3–4.0
Specific gravity	2.8–2.9

Galena [PbS]

It is the chief ore of lead. Galena (lead sulphide) occurs as crystals (twin) and granular masses (Fig. 10.19).

Fig. 10.19 Galena

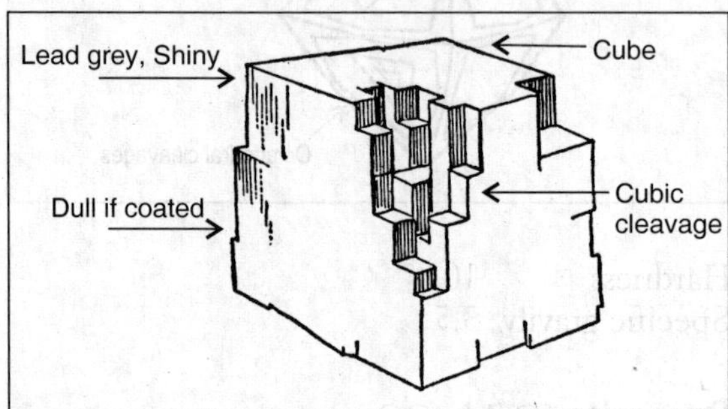

Colour	Lead grey
Lustre	Metallic
Cleavage	Perfectly cubic and
Streak	Grey
Specific gravity	7.57 (very heavy)

Graphite [C]

Like diamond, graphite consists solely of crystallized carbon but differs extraordinarily in colour, hardness, conductivity and crystallisation. It is often fibrous with sectiles. It can mark paper. It is a good conductor and is the most stable form of carbon (Fig. 10.20).

Colour	Grey or black
Streak	black
Lustre	Metallic
Feel	Greasy

Fig. 10.20 Graphite

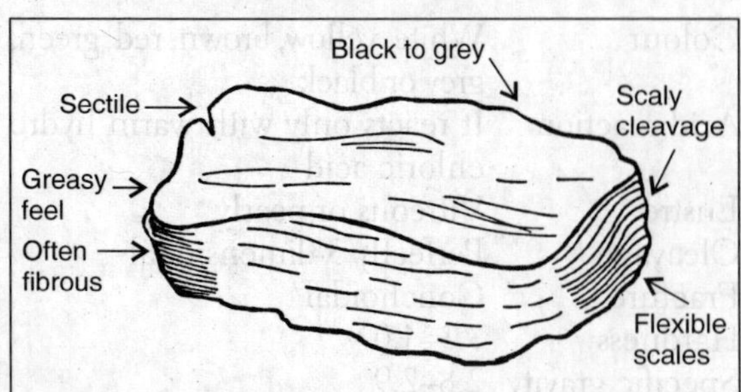

Cleavage	1-directional scaly (flexible scales)
Specific gravity	2.1–2.2

Gypsum [$CaSO_4$, $2H_2O$]

It is the most common sulphate. Gypsum (hydrous calcium sulphate) is of three varieties with distinctive habits—*alabaster* (massive), *selenite* (transparent and foliated) and *satinspar* (fibrous with silky or pearly lustre). It occurs as prismatic and bladed crystals and as fibrous vein fillings, radiating aggregates, clusters on cave walls and massive rock forming beds (Fig. 10.21).

Fig. 10.21 Gypsum

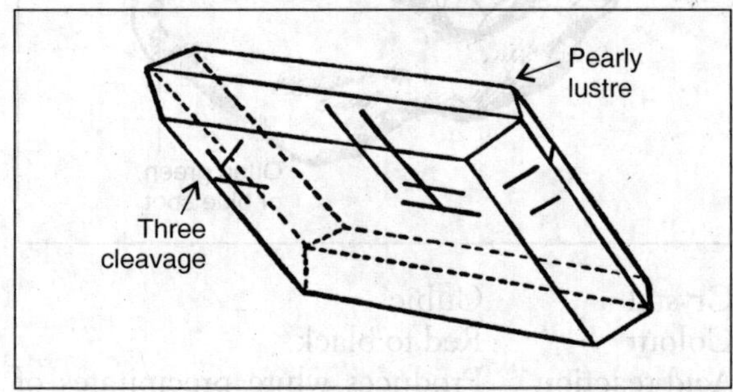

Colour	White or grey but may be colourless (in large crystals), pink (in alabaster) or brown and yellow (in massive beds)
Lustre	Silky or earthy
Cleavage	Distinct
Hardness	2
Specific gravity	2.3

Haematite [Fe_2O_3]

It is the major ore of iron. Spectacular haematites occur as brilliant black tabular crystals, commonly foliated. Generally three varieties are recognized—*red haematite* (as columnar or radiating masses and fibrous clusters), *kidney ore* (as kidney shaped masses) and *earthy* or *ocherous haematite* (as dull yellowish, or in oolitic, earthy form) (Fig.10.22).

Crystal	Rhombohedral
Acid reaction	Soluble in concentrated hydro chloric acid

Fig. 10.22 Haematite

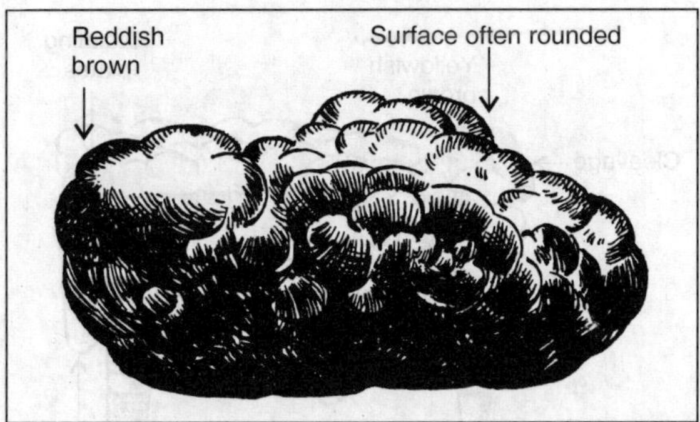

Streak	Blood red to brownish red
Magnetism	Attracted by magnets
Hardness	5.26
Specific gravity	5.26

Hornblende

$[(Ca,Na,K)_{2\text{-}3}(Mg,Fe,Al)_5(Si\ Al)_8O_{22}(OH,F)_2]$

It occurs as fibrous, stubby prismatic crystals (Fig. 10.23).

Crystal	Monoclinic
Colour	Green-black
Streak	Colourless
Lustre	Glassy
Cleavage	2-dimensional at 56°
Fractures	Uneven
Hardness	5–6
Specific gravity	3–3.5

Fig. 10.23 Hornblende

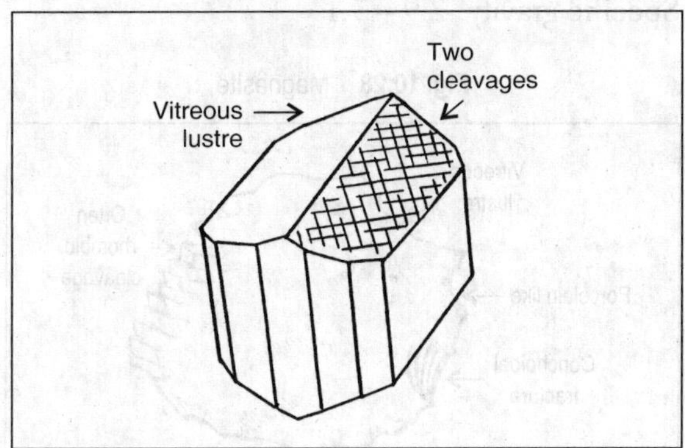

Ilmentite $[FeTiO_3]$

It is the major source of titanium and occurs as plate-like crystals in veins or disseminated grains, compact masses and placer deposits (black beach sands) (Fig. 10.24).

Fig. 10.24 Ilmentite

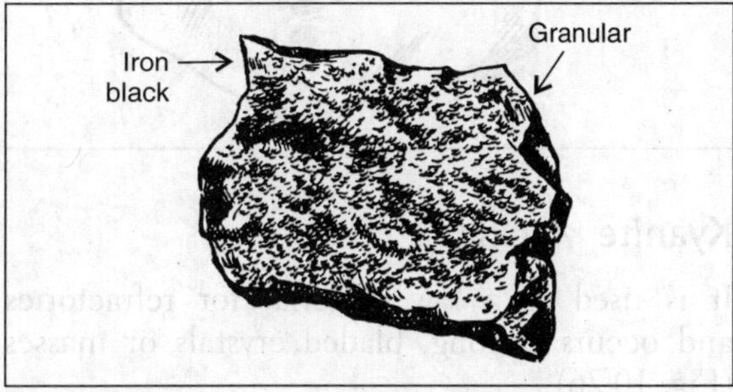

Crystal	Rhombohedral
Colour	Black
Acid reaction	Soluble in hot hydrochloric acid
Streak	Black
Lustre	Metallic
Opacity	Opaque
Magnetism	Attracted by magnet
Hardness	5.6
Specific gravity	4.79

Kaolinite $[Al_2Si_2O_5(OH)_4]$

It is an important raw material used in pharmaceuticals, china and whitewares and whitening agents. Kaolinite (hydrous aluminium silicate) occurs as tabular crystals, worm-like aggregates, clay-like masses and scattered particles. It is commonly plastic when wet (Fig. 10.25).

Crystal	Triclinic
Feel	Greasy
Small	Earthy
Colour	White (if pure), yellowish or grey
Streak	White
Lustre	Dull
Cleavage	Perfectly in 1-direction
Hardness	2.0–2.5
Specific gravity	2.6

Fig. 10.25 Kaolinite

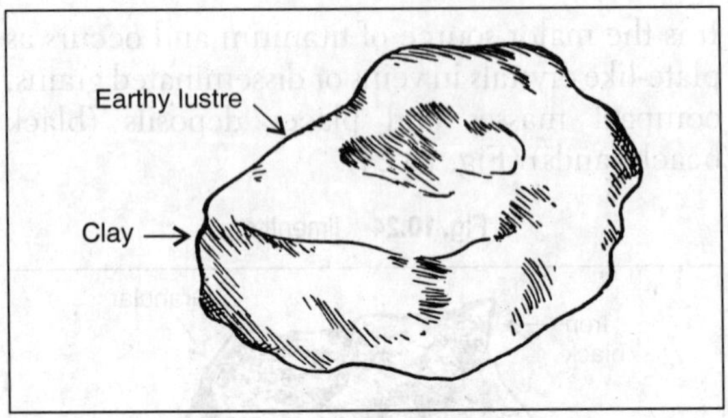

Kyanite [Al_2OSiO_4]

It is used as a raw material for refractories and occurs as long, bladed crystals or masses (Fig. 10.26).

Fig. 10.26 Kyanite

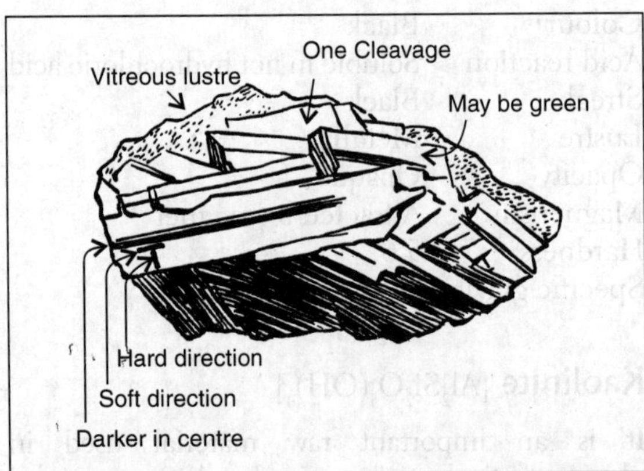

Crystal	Triclinic
Colour	Distinctive blue, but may also be white, grey, green, brown or black
Lustre	Glassy
Opacity	Translucent to transparent
Cleavage	2-directional, non-orthogonal
Hardness	5–7
Specific gravity	3.53–3.65

Limonite [$2Fe_2O_3$, $2H_2O$]

It is the non-crystalline iron hydroxide occurring as massive, crusty, stalactitic cavity fillings and as varnish-like coatings on rocks (Fig. 10.27).

Fig. 10.27 Limonite

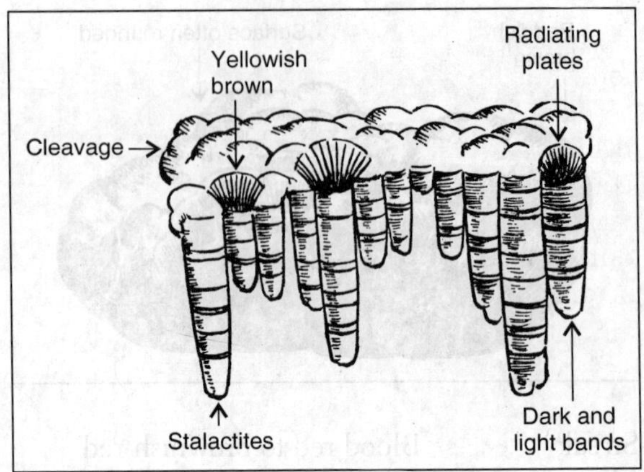

Form	Amorphous
Streak	Yellowish brown
Lustre	Glassy to dull
Hardness	5.5
Specific gravity	2.7–4.3

Magnesite [$MgCO_3$]

It is simply magnesium carbonate and fizzes in acid (preferably warm). It occurs both as crystals and also in the form of cleavable masses (Fig. 10.28).

Lustre	Vitreous
Cleavage	Rhombic
Colour	Green
Fracture	Conchoidal
Streak	White
Hardness	can be scratched by knife blade
Specific gravity	2.9 – 3.1

Fig. 10.28 Magnesite

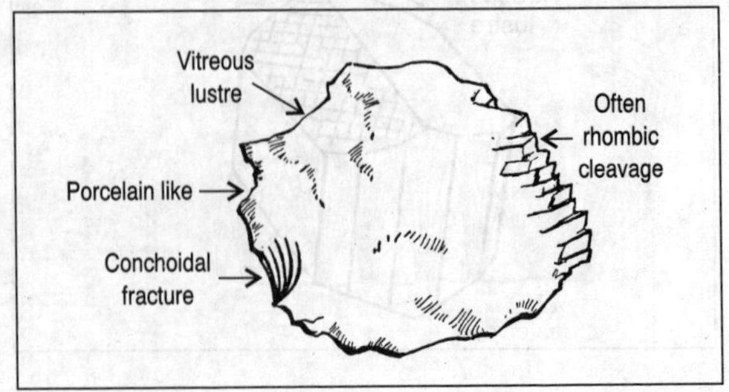

Magnetite [Fe_3O_4]

It contains both Fe^{++} (ferrous) and Fe^{+++} (ferric) ions. It is the most valuable iron ore and occurs as octahedral and dodecahedral crystals (often twined) and granular masses. Its magnetism, black colour, black streak and hardness (5.5 to 6.5) are diagnostic. It is brittle, opaque and metallic to dull in lustre. It dissolves slowly in hydrochloric acid. Specific gravity is 5.20 (Fig. 10.29).

Fig. 10.29 Magnetite

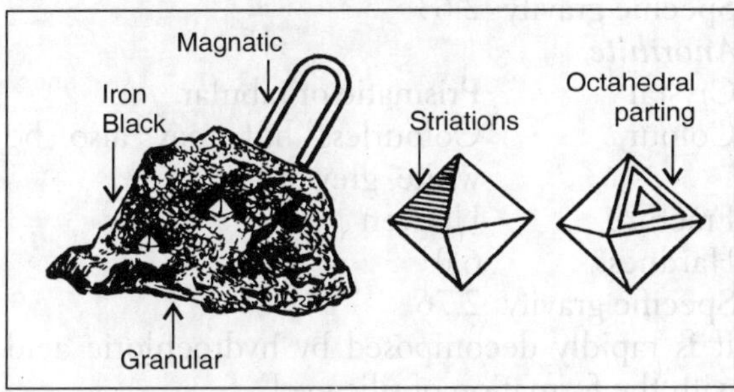

Malachite [$Cu_2CO_3(OH)_2$]

It is the basic copper carbonate that occurs as massive deposits or as botryoidal, fibrous or stalactitic masses or as slender needle-like crystal aggregates (Fig. 10.30).

Crystal	Monoclinic
Colour	Green
Streak	Light green
Lustre	Silky
Hardness	3.5–4.0
Specific gravity	3.9–4.0

Fig. 10.30 Malachite

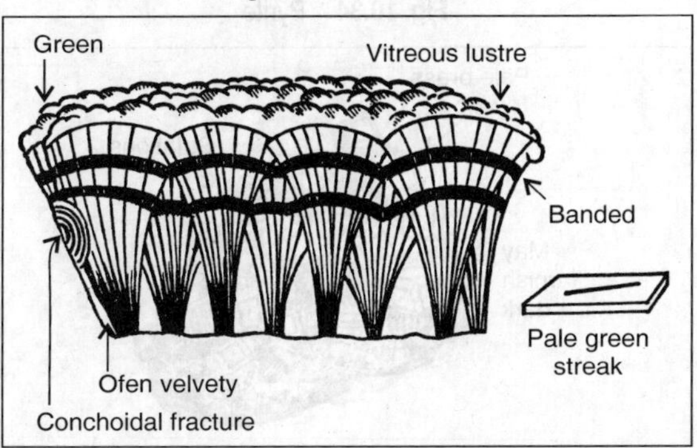

Monazite [$(Ce,La,Th)PO_4$]

Monazite (cereum-lanthanum-thorium phosphate) occurs as equi-dimensional or elongated crystals or as placer deposits (river and beach sands) (Fig. 10.31).

Fig. 10.31 Monazite

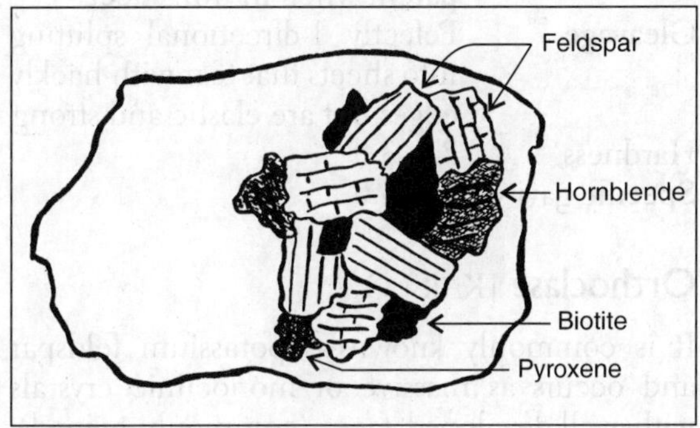

Crystal	Monoclinic with well developed faces
Colour	Red or brown
Lustre	Resinous
Opacity	translucent
Cleavage	1-directional
Fracture	Conchoidal or uneven
Hardness	5.0–5.5
Specific gravity	5.0–5.3

Muscovite [$KAl_2(Si_3Al)O_{10}(OH)_2$]

It is the commonest mica. Muscovite (hydrous potassium aluminium silicate) occurs as tabular crystals often pseudo-hexagonal (Fig. 10.32).

Fig. 10.32 Muscovite

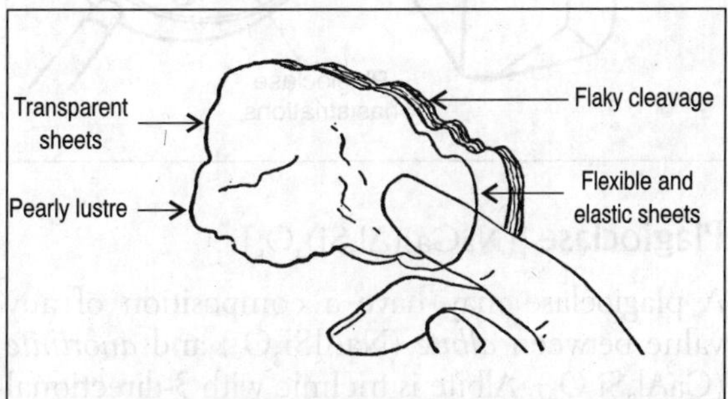

Crystal	Monoclinic
Colour	Usually colourless, grey, pale green, brown, yellow, pink or violet
Streak	White
Lustre	Glassy or pearly
Opacity	Opaque to translucent, trans parent when in thin sheets
Cleavage	Pefectly 1-directional splitting into sheets that tear with hackly edges but are elastic and strong
Hardness	2.5–3.0
Specific gavity	2.7–3.0

Orthoclase [$KAlSi_3O_8$]

It is commonly known as potassium feldspar and occurs as massive or monoclinic crystals with well-developed faces (commonly twined), having a distinctive cross section (Fig. 10.33).

Colour	Flesh-red or grey
Streak	Colourless
Lustre	Vitreous and pearly (on cleavage faces)
Cleavage	Perfect
Hardness	6.0
Specific gravity	2.57

Fig. 10.33 Orthoclase

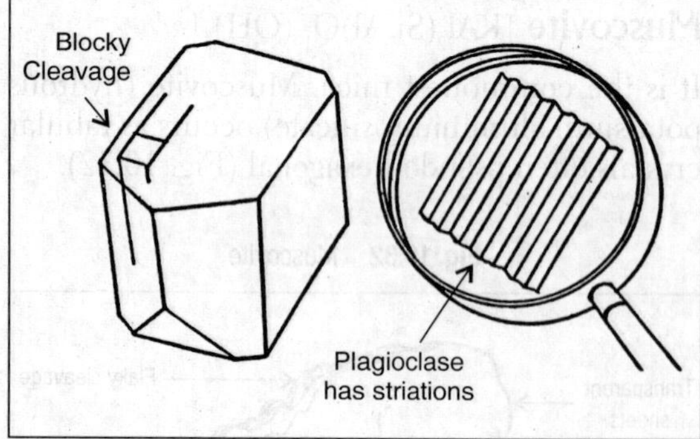

Plagioclase [$(NaCa)(Al,Si)_4O_8$]

A plagioclase may have a composition of any value between *albite* ($NaAlSi_3O_8$) and *anorthite* ($CaAl_2Si_2O_8$). Albite is triclinic with 3-directional cleavage—two at about 86° and one poor in a third direction. It occurs as tabular or elongated crystals intergrown with quartz and mica. Anorthite is structurally similar to albite. Lustre, cleavage and opacity are same as that of albite.

Albite

Colour	Flame yellow
Lustre	Vitreous
Opacity	Transparent to translucent
Cleavage	Striated
Hardness	6.5
Specific gravity	2.63

Anorthite

Crystal	Prismatic or tabular
Colour	Colourless but may also be white, grey or blue
Fracture	Uneven
Hardness	6.0
Specific gravity	2.76

It is rapidly decomposed by hydrochloric acid with the formation of silica gel.

Pyrite [FeS_2]

Pyrite (iron sulphide) means *fire mineral* and is also known as *fool's gold*. It is harder and more brittle than gold. It occurs as cubes, pyritohedra and octahedra (often twined) and as granular, radiating, globular and stalactitic masses (Fig. 10.34).

Colour	Brass yellow which tarnishes to brown
Streak	Brown-black
Cleavage	Indistinct
Solubility	Soluble in concentrated HNO_3

Fig. 10.34 Pyrite

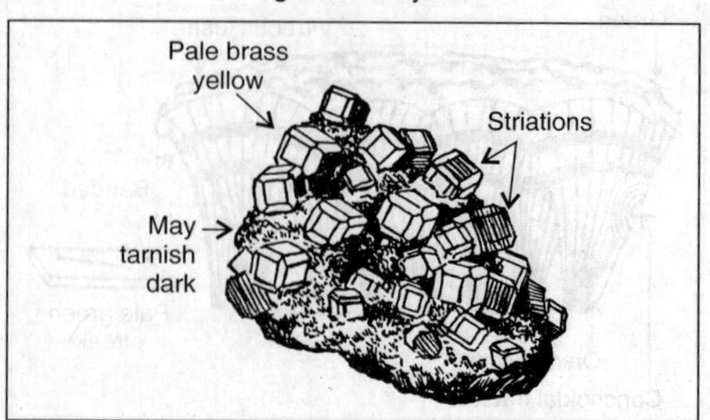

Hardness	6–6.5
Specific gravity	5.01

Quartz [SiO_2]

It is the principal constituent of glass and occurs as irregular grains intergrown with other minerals or as rounded grains or as microscopically grained specimens or as prismatic crystals (often twined) (Fig. 10.35).

Crystal	Hexagonal
Colour	Colourless, if pure, and any colour, if not pure
Opacity	Normally transparent
Fracture	Conchoidal
Hardness	7
Specific gravity	2.65

Fig. 10.35 Quartz

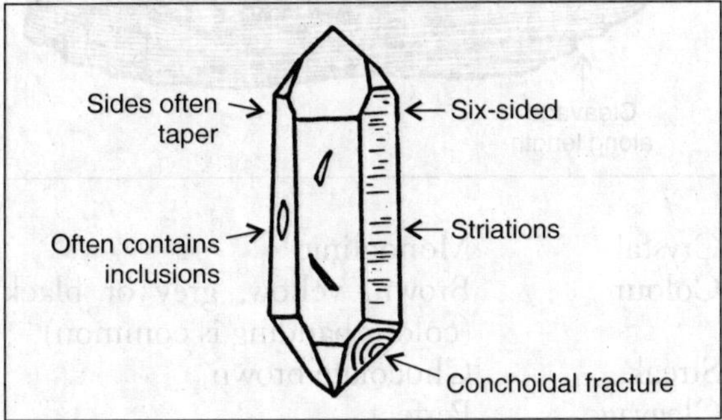

Talc [$Mg_3Si_4O_{10}(OH)_2$]

Talc (hydrous magnesium silicate) occurs as foliated, radiating and compact masses. It has a 3-layer sheet structure. Two silica tetrahedral layers enclose an octohedral layer in which all octahedral positions are filled with magnesium ions. Because of weaker sheet-bondage, talc is a better lubricant and feels greasier (Fig. 10.36).

Colour	White, green, blue or brown
Streak	White
Lustre	pearly
Opacity	Translucent to transparent
Cleavage	Perfectly 1-directional
Fracture	Irregular
Hardness	1
Specific gravity	2.58–2.83

Fig. 10.36 Talc

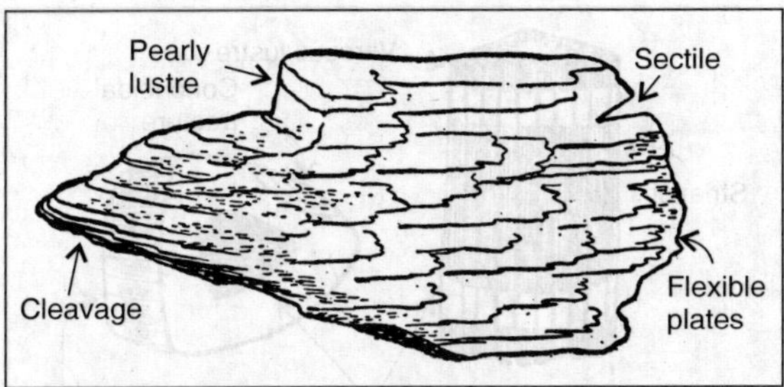

Topaz [$Al_2(SiO_4)(OH,F)_2$]

Topaz (aluminium fluro silicate) is often used as gemstones. It is highly durable and has a high index of refraction. It occurs as crystals (column prisms with striated faces) and as granular masses. It is colourless and brittle (Fig. 10.37).

Crystal	Orthorhombic
Streak	Pale
Lustre	Glossy
Opacity	Transparent
Cleavage	Perfectly 1-directional
Hardness	8
Specific gravity	3.49

Fig. 10.37 Topaz

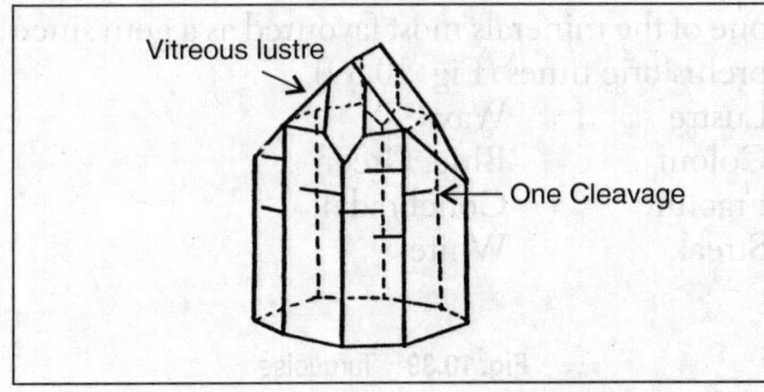

Tourmaline

Tourmaline is an aluminium silicate with a very complex chemical formula. It occurs in a wide variety of colours—rubellite, indicolite, achroite, dravite, etc. The colours are often zones either along the length or across the width of the unique rounded- triangular crystals. Watermelon tourmaline has a green exterior, surrounding first a white zone and then a red core. The jet black

Fig. 10.38 Tourmaline

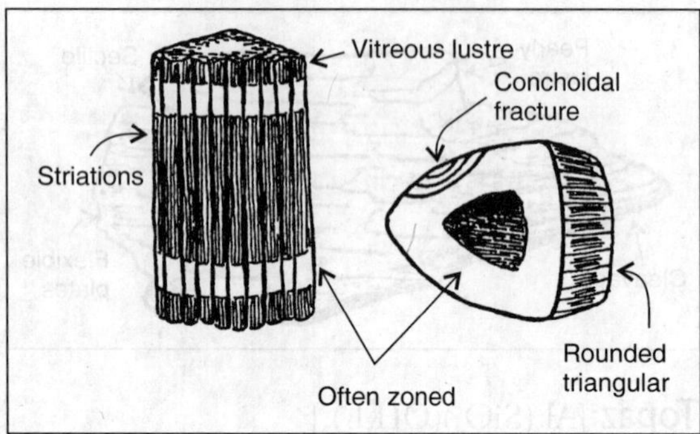

specimens, which resemble hornblende, tend to fracture like coal (Fig. 10.38).

Streak	Leaves white marks or scratches on the streak plate
Colour	Black, pink, green
Lustre	Non-metallic; vitreous
Cleavage	Poor
Fracture	Conchoidal
Striations	Well developed
Specific gravity	2.9–3.2

Turquoise [$CuAl_6(PO_4)_4(OH)_8$, $4H_2O$]

It occurs in stringers and small nodules and is one of the minerals most favoured as a gem since prehistoric times (Fig. 10.39).

Lustre	Waxy
Colour	Blue, Green
Fracture	Conchoidal
Streak	White

Fig. 10.39 Turquoise

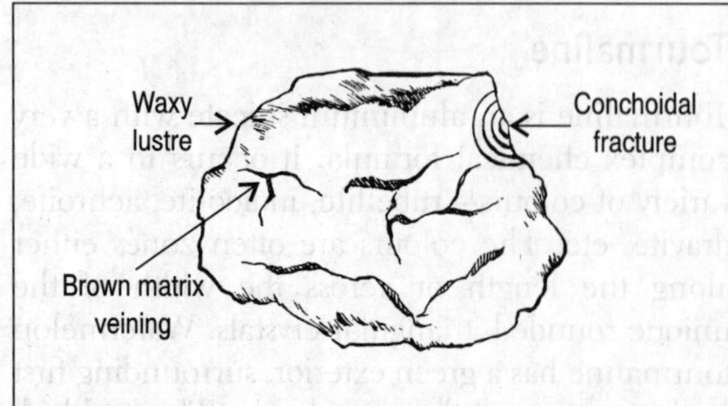

Hardness	can scratch glass and be scratched by quartz
Specific gravity	2.6–2.8

Wolframite [Fe,Mn)WO_4]

It occurs as prismatic or tabular crystals, groups of bladed crystals, massive granular groups or intergrowths of needle-like crystals (Fig. 10.40).

Fig. 10.40 Wolframite

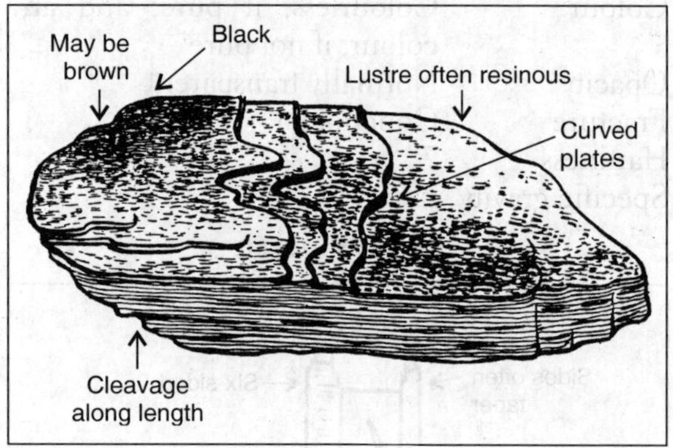

Crystal	Monoclinic
Colour	Brown, yellow, grey or black (colour banding is common)
Streak	Chocolate-brown
Cleavage	Perfect
Fracture	Uneven
Hardness	4.45
Specific gravity	7.1–7.9

Zircon [$ZrSiO_4$]

Zircon (zirconium silicate) occurs as small crystals in the form of prism-pyramid combination and also as rounded grains. It is brittle and very resistant to weathering (Fig. 10.41).

Crystals	Tetragonal
Colour	Colourless, yellow, grey, brown, blue or green
Lustre	Brilliant
Fracture	Conchoidal
Hardness	7.5
Specific gravity	4.2–4.9

Fig. 10.41 Zircon

4 KEYS TO RECOGNIZING ROCKS

Key No. 1: Texture and Structure

These can be conveniently considered to identify specimens of rocks, as follows—

Texture	*Rocks*
Cryptocrystalline or Glassy	Basalt
Medium-grained	Dolerite
Phaneritic	Granite, Diorite, Gabbro, Peridotite
Argillaceous	Shale, Mudstone
Arenaceous	Sandstone
Rudaceous	Conglomerate
Cataclastic	Mylonite
Hornfelsic	Hornfels
Granular	Marble, Quartzite
Slaty	Slate
Phyllitic	Phyllite
Schistose	Schist
Gneissose	Gneiss

Key No. 2: Colour

Although less reliable a parameter, sometimes it can well be taken as one of the diagnostic parameters to identify a specimen in the field, as follows—

Colour	*Rocks*
Light	Granite, Pegmatite, Syenite
Dark	Gabbro, Peridotite
Black	Basalt, Coal, Obsidian
Gray	Pumice, Slate
Gray/Black	Shale
White	Limestone, Quartzite
White, Gray	Marble, Felsite
White, Brown, Red	Sandstone

Key No. 3: Acid Test

The carbonate rocks, such as limestone and marble will fizz or effervesce with acid due to formation of CO_2.

Key No. 4: Mineral Content

Rocks can be identified by examining their composition with naked eye, as follows—

Mineral Composition	*Rock*
Calcite	Marble
Clay	Shale, Slate
Feldspar	Felsite
Glass	Obsidian
Quartz	Sandstone, Quartzite
Calcite, Dolomite	Limestone
Pyroxene, Olivine	Basalt
Mineral/ Rock Fragments	Conglomerate
Olivine, Pyroxene, Hornblende	Peridotite
Feldspar, Quartz, Muscovite	Pegmatite
Feldspar, Biotite, Hornblende	Syenite
Feldspar, Hornblende, Biotite, Pyroxene	Monzonite
Feldspar, Pyroxene, Olivine, Hornblende	Gabbro
Feldspar, Quartz, Biotite, Hornblende	Gneiss, Granite

SELECTED SPECIMENS OF ROCKS

Basalt

Basalts make up over 98% of all volcanic rocks. It is a very fine grained (cryptocrystalline) and dark coloured (melanocratic) mafic (44–52% silica content) igneous rock with typical volcanic texture, very high specific gravity (2.9–3.1) and high hardness (cannot be scratched by knife). It is very compact and is characterised by conchoidal fractures. Individual minerals cannot be identified by the naked eye. The characteristic minerals are calcic-plagioclase, pyroxene and iron ore with or without olivine. They may be vescicular or the vescicules may be filled with secondary minerals forming amygdales and thereby giving rise to *amygdaloidal structure*. The basaltic magma which erupted under water forms pillow-like masses giving rise to a basaltic pillow lava. The individual pillows have a glassy crust and a more or less crystalline interior which may show radiating contraction cracks. Two main types of basalts may be recognized—*tholeiitic basalt* (rich in clinopyroxenes and poor in calcium) and *alkali basalt* (rich in lime and olivine) (Fig. 10.42).

Coal

It is regarded as a sedimentary rock because it is found in layers or beds. It is lighter in weight, black in colour and often friable. It has banded structure (Fig. 10.43).

Fig. 10.42 Basalt

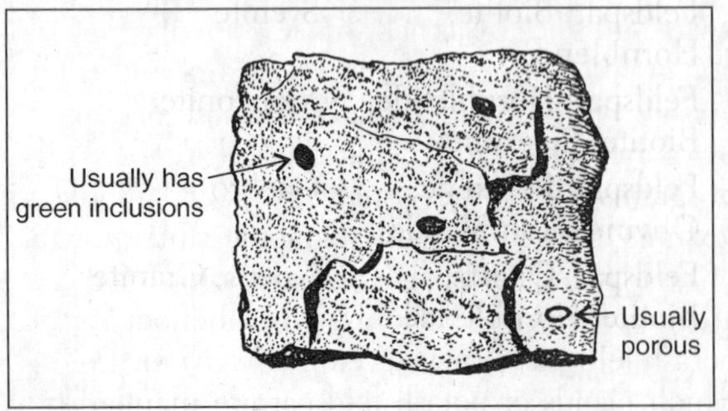

Fig. 10.43 Coal

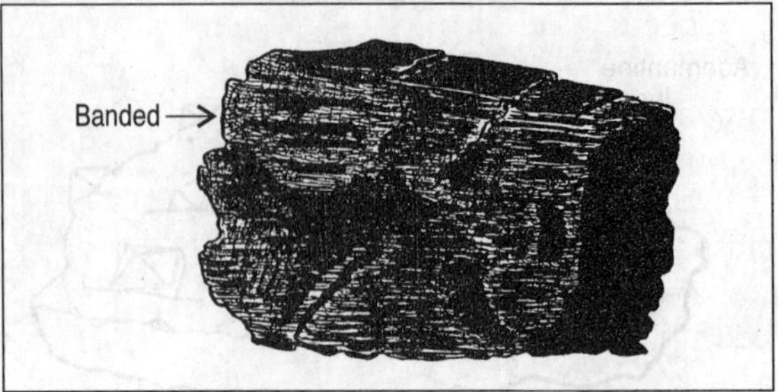

Conglomerate

Conglomerates are clastic sedimentary rocks of rudaceous type. These are pebbly rocks of medium to fine in size (over 2 mm in diameter). These are variegated in colour with moderate to high specific gravity and moderate compactness. Being mostly of fluvial, glacial and shoreline origin, these consist of typical mud-free gravels or their lithified equivalents (i.e., grains of quartz, jasper, feldspar and rock fragments of various sizes, shapes and colours). The texture is typically clastic as lithic clasts are the essential constituents. The pebbles, cobbles, gravels, rock fragments and granules are cemented together in a *fine-grained matrix* or ground mass of *ferruginous* (reddish brown), *siliceous* (light colour) *argillaceous* (muddy odour) and *calcareous* (reaction with dilute hydrochloric acid) sediments (Fig. 10.44).

Fig. 10.44 Conglomerate

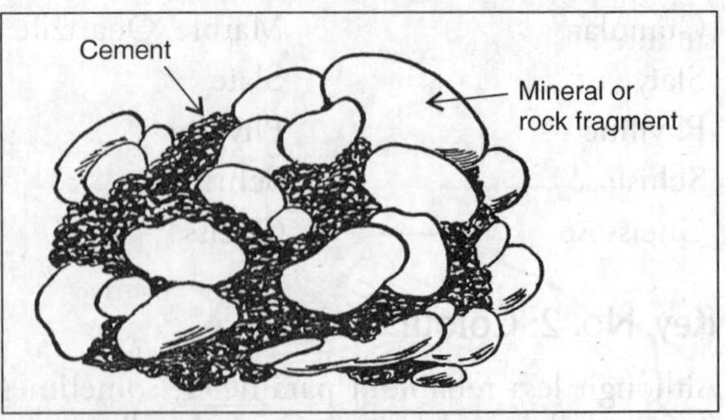

Dolerite

Dolerites are the hypabyssal representatives of the gabbros and basalts, finer grained than the former but coarser grained than the latter. They occur mostly as dykes and sills and their texture is frequently ophitic or subophitic. They are dark coloured (melanocratic) mafic igneous rocks with high specific gravity (2.64–3.12), high hardness and high compactness. Feldspar laths are randomly oriented forming a criss-cross arrangement—the interspaces are filled up by ferro-magnesian minerals in which traces of cleavage can be seen. As it is prone to weathering, local patches of reddish brown tint can often be distinguished. Generally two types of dolerites are recognised—*tholeiitic* or *quartz dolerite* (rich in clino and ortho-pyroxene and calcicplagioclase) and *alkali dolerite* (rich in lime).

Gabbro

Gabbros are coarse or medium grained, melanocratic, mafic and plutonic igneous rocks with high specific gravity (2.9–3.2), high compactness and high hardness. The essential constituents are calcic plagioclase feldspar and one or more ferromagnesians like orthopyroxene, clinopyroxene, olivine, hornblende, etc. Some gabbros have a subhedral-granular texture while others have a poikilitic or a subophitic texture in which plates of pyroxene or amphibole partially enclose the crystals of plagioclase. Many gabbroic instrusions have well marked banded or layered structures while a few have orbicular structures (Fig. 10.45).

Fig. 10.45 Gabbro

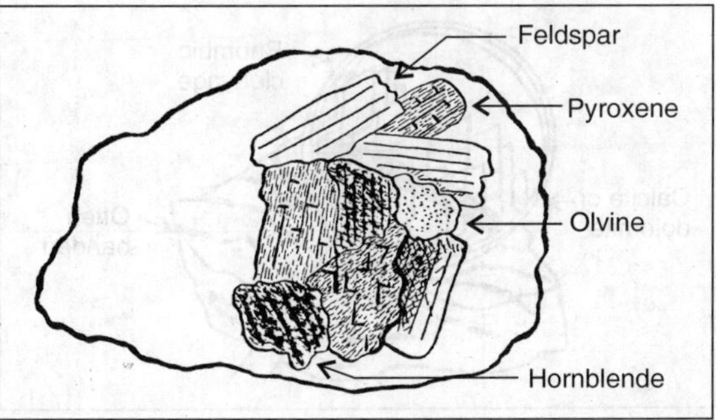

Gneiss

Gneisses are medium coloured, moderately coarse grained, hard and compact metamorphic rocks of the highest grade with medium specific gravity. It is a coarsely crystalline rock of a granular texture. It is typically characterised by alternate bands of *light* and *dark* coloured minerals. The lighter bands are composed of quartz and feldspar with an equant granular habit while the *darker bands* are defined by the preferred orientation of flaky and platy ferro-magnesian minerals like mica, amphibole, hornblende, augite, etc. (Fig. 10.46).

Fig. 10.46 Gneiss

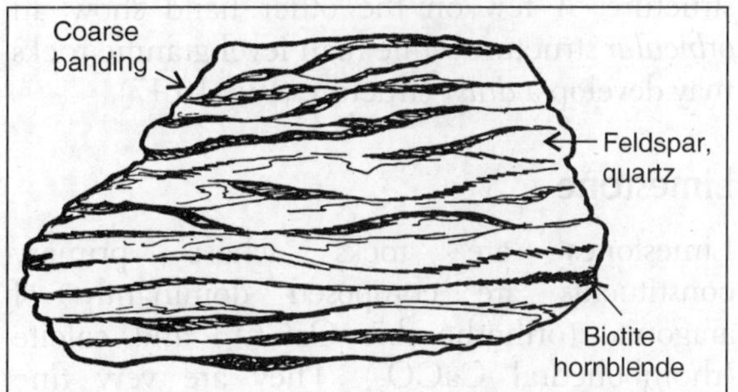

Granite

Granites are coarse grained (*phaneritic*), light coloured (*leucocratic*), acidic (>66% silica content) and *holocrystalline plutonic* igneous rocks with medium specific gravity (2.63–2.75). They are hard and compact rocks composed essentially of quartz, feldspar and mica, usually with some ferro-magnesian minerals. Feldspar is identified by its lath shape in which cleavage may be present, while quartz is identified by its vitreous lustre and mica by its flaky habit. Both plagioclase and mafic minerals tend to form subhedral crystals while intergranular spaces are occupied by anhedral quartz. This typical interlocking of grains form the usual (subhedral-angular) texture. Sometimes micrographic intergrowth of quartz and feldspar is found. Some of these are *porphyritic*, commonly with phenocrysts of potash feldspar. A few develop *rapakivi* texture in which ovoids of potash feldspar are mantled by

Fig. 10.47 Granite

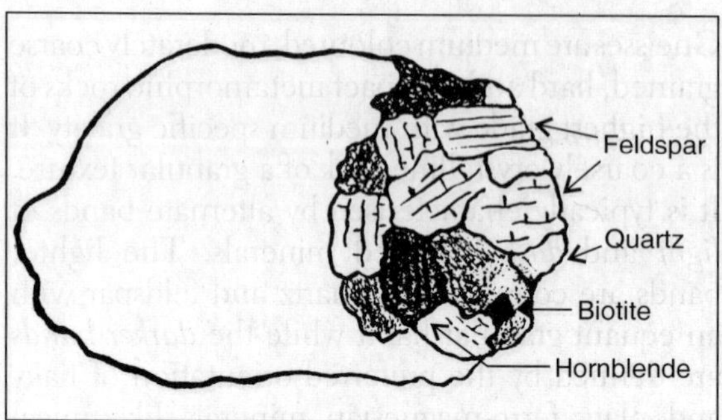

lime-bearing plagioclase. Regarding the larger scaled structures, some have a *banded* or layered structure. A few on the other hand show an *orbicular* structure while high level granitic rocks may develop a *drusy* structure (Fig. 10.47).

Limestone

Limestones are rocks whose primary constituents are composed dominantly of aragonite (orthorhombic $CaCO_3$) and calcite (rhombohedral $CaCO_3$). They are very fine grained, light coloured with moderate specific gravity (2.5–2.8) very compact but of low hardness (easily scartched by a knife or even fingernails). Texture is typically non-clastic—calcirudite (pebble-sized grain), *calcarenite* (sand-sized grain), *calcisiltite* (silt-sized grain) and *calcilutite* (clay-sized grain). It strongly reacts with cold and dilute hydrochloric acid. A salient feature is that most of the limestones are particulate and bimodal, i.e., they consist initially of relatively large carbonate particles mixed in various proportions with carbonate mud. Compositional layering characterised by different colour bands are often present (Fig. 10.48).

Fig. 10.48 Limestone

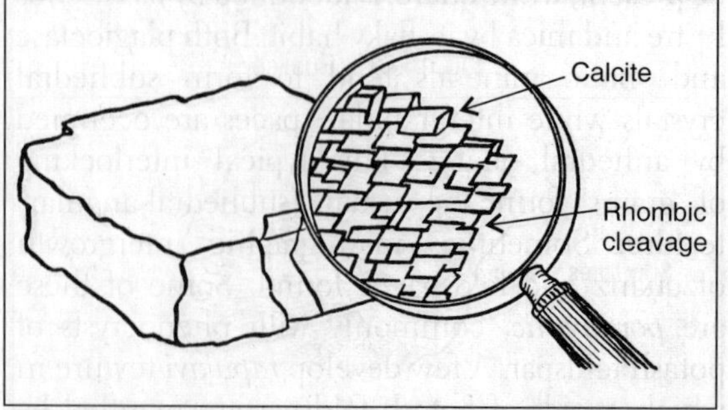

Marble

Marbles are coarse grained, light coloured rocks with high compactness, low hardness and medium specific gravity. They are produced by the metamorphism of limestones and hence are dominantly composed of calcite. They strongly effervesce with dilute hydrochloric acid. Generally, they show an even grained texture. Some show a crude foliation due to segregation of minor silicate constituents or in some cases a parallel alignment of lustrous flakes of graphite while some bear the imprint of stain in the form of lensoid outlines (Fig. 10.49).

Pegmatite

Pegmatites are very coarse grained, hard, moderately compact leucocratic igneous rocks with medium specific gravity. They occur as dykes or veins in the plutonic bodies or as marginal segments to such plutonic intrusions into the adjoining country rock. They are dominantly composed of quartz and feldspar and also contain minerals like tourmaline, fluorite, topaz, lepidolite, etc. They show a typical interlocking texture. Pegmatite minerals are oriented perpendicular to the walls of the dykes and some

Fig. 10.49 Marble

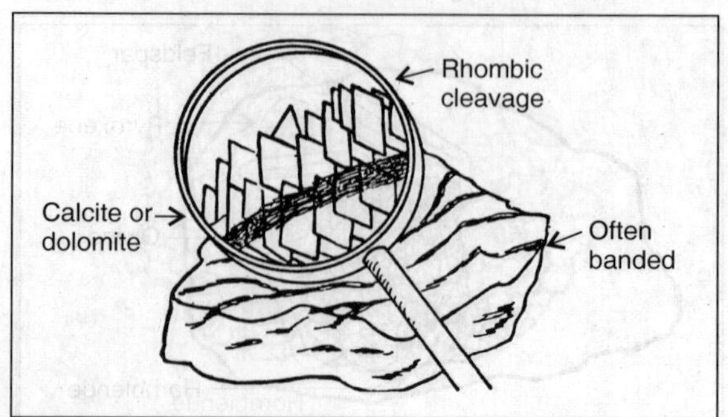

Fig. 10.50 Pegmatite

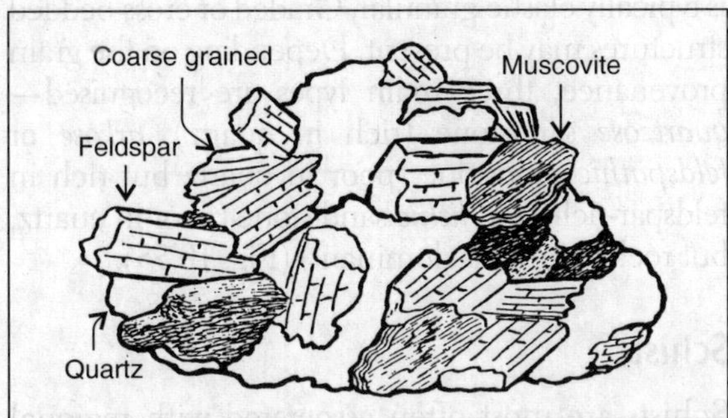

of the rocks are banded. These have the dark minerals concentrated at the margins, leaving an almost purely feldspathic centre (Fig. 10.50).

Peridotite

Peridotites are dark rocks resembling gabbrs but heavier and contain little or no plagioclase feldspar. They have a comparatively high content of iron, and the colour is dark green or black. The dominant mineral is olivine. Pyroxene and hornblende are less in quantities. Minerals are mutually intergrown but can easily be seen with an unaided eye. It does not fizz in acid (Fig. 10.51).

Phyllite

Phyllites are fine grained, light coloured metamorphic rocks with moderate compaction,

Fig. 10.51 Peridotite

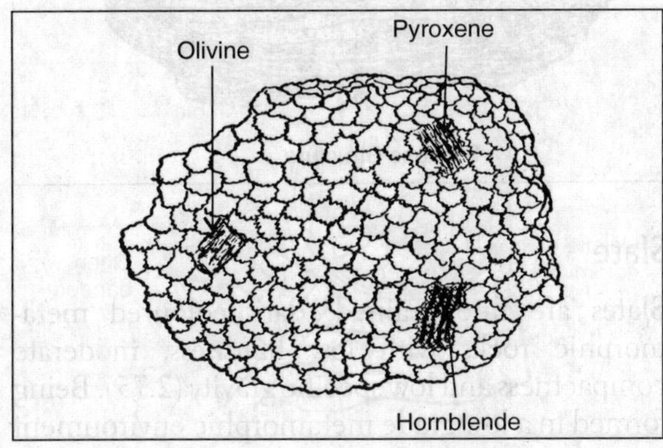

Fig. 10.52 Phyllite

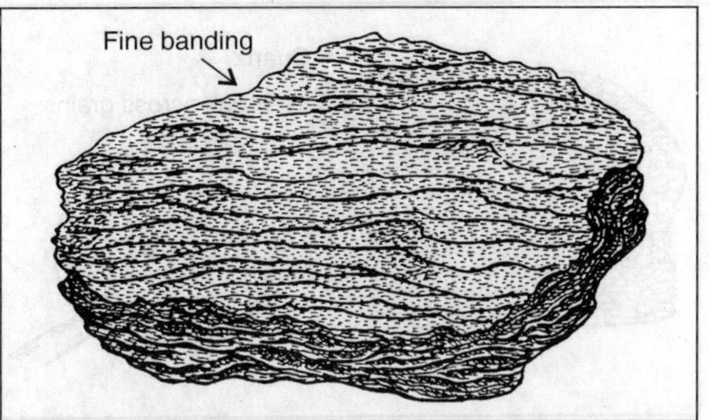

low hardness and medium specific gravity. The grain size is perceptibly coarser than that of slate and broken surfaces show a lustrous sheen (due to coarsening of grain) imparted by the flakes of mica and chlorite aligned along the foliation plane. The foliations may show complications by transverse secondary S-surfaces, lineations and incipient segregated laminations. The texture is, hence, typically *phyllitic*. The presence of mica gives a silvery lustre while a bluish tint is added if chlorite is present (Fig. 10.52).

Pumice

Expanded by the explosion of steam as it escapes from a volcano and puffed up into a froth of glass, pumice is a foamy mass of silky glass shards. These shards may be intermixed or else drawn out in parallel strands. Its porous nature and the many isolated cells of air enable pumice to float on water. Although most pumice is light-coloured, some of it is brownish or occasionally red. The sharp cutting edges of the bits of glass make it very abraisive (Fig. 10.53).

Fig. 10.53 Pumice

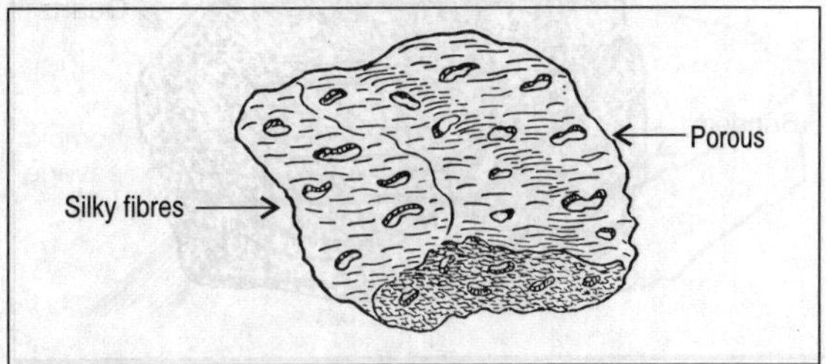

Fig. 10.54 Quartzite

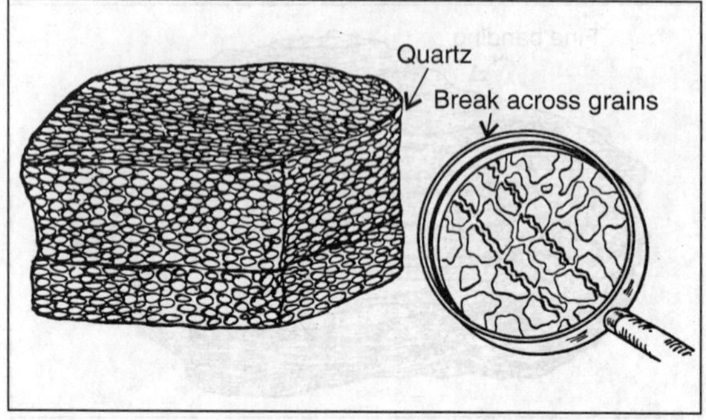

Quartzite

Quartzites are very hard granular rocks produced by the recrystallisation of sandstone. They are light coloured, fine to medium grained rocks with high compactness, low to medium specific gravity, a *granoblastic* texture and an overall vitreous lustre without any cleavage. They are non-foliated and composed largely of quartz (Fig. 10.54).

Sandstone

Sandstones are very fine to very coarse grained (1/16 mm to 2 mm), light coloured, moderately compact and moderately hard sedimentary rocks with a rough feel and medium specific gravity. It is dominantly composed of detrital sand grains that pack together *grain-against-grain* to form a framework that provides the primary solid support in the deposit. The framework is cemented together in a matrix of ferruginous, siliceous,

Fig. 10.55 Sandstone

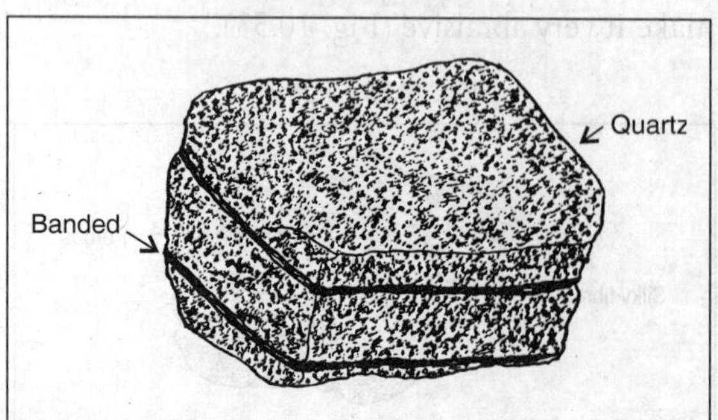

argillaceous and calcareous sediments. The texture is typically clastic-granular. Graded or cross-bedded structures may be present. Depending on the grain provenance, three main types are recognised—*quartzose* sandstone (rich in quartz), *arkose* or *feldspathic* sandstone (poor in quartz but rich in feldspar-rich) and *lithic* sand-stone (poor in quartz, but rock fragments dominant) (Fig. 10.55).

Schist

Schists are most often associated with regional metamorphism in medium grade metamorphic environments. They are medium grained, light coloured rocks with low hardness, low compactness and medium specific gravity. They are dominantly composed of flaky minerals (mica) which are oriented in a preferred direction imparting foliations in the rock. The equant grains of quartz and feldspar are interspaced with the foliation. The texture is typically schistose formed due to the parallel arrangement of flaky minerals. They are markedly fissile rocks breaking readily into parallel-sided slabs. This is due to the presence of *phyllosilicates* (muscovite, biotite and chlorite) and the parallel alignment of amphibole prisms and fibres (Fig. 10.56).

Fig. 10.56 Schist

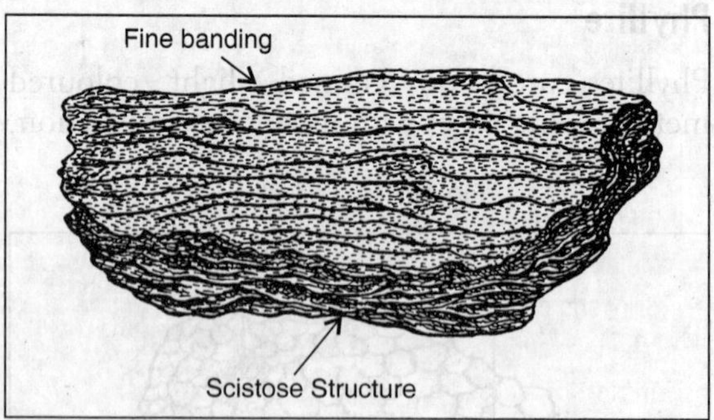

Slate

Slates are fine grained, dark coloured metamorphic rocks with low hardness, moderate compactness and low specific gravity (2.75). Being formed in a low grade metamorphic environment they show perfect planar schistosity or slaty cleavage (closely spaced foliation planes) which is commonly inclined at high angles to relict bedding. The principal minerals are colourless phyllosilicates and chlorites. It produces metallic sound and muddy odour (Fig. 10.57).

Fig. 10.57 Slate

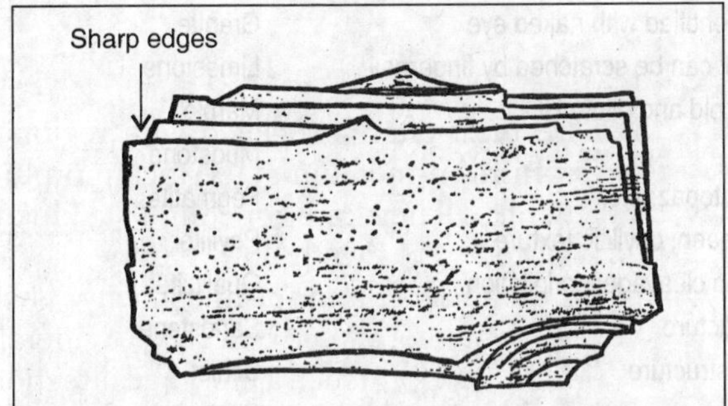

Shale

Shales are fine grained argillaceous sedimentary rocks with low compactness, low hardness (can be scratched by fingernails) and low specific gravity.

Fig. 10.58 Shale

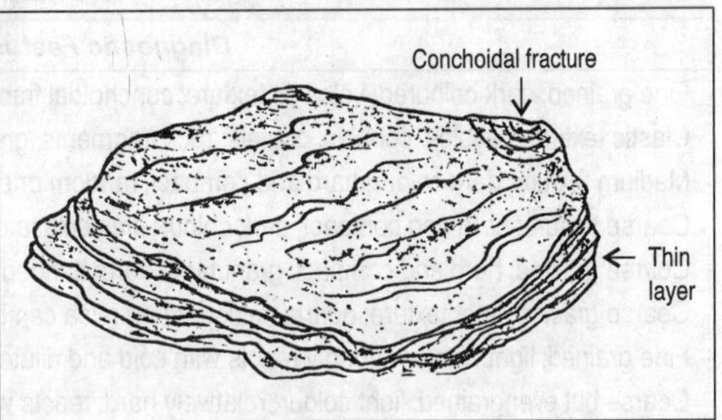

They are characterised by a non-clastic texture and are dominantly composed of a mixture of clay minerals (illite, montmorillonite, kaolinite and chlorite) and silt-sized particles of quartz, feldspar, mica, fossil remains, pyroclastic grains, etc. Shales usually display little bedding plane orientation of the clay minerals but they part easily along planes marked by enrichment in clay, organic matter or mica. These planes actually represent intermittent phases of sedimentation at a reduced rates. Banding of different colours may also be present (Fig. 10.58).

Common Tests for Identification of Mineral Specimens

Diagnostic Features	*Mineral*
Earthy lustre; massive occurrence; nobby structure; generally grey and white with reddish brown iron stain	Bauxite
Scaly form; 1D cleavage; platy structure; glossy lustre; black, dark or brown	Biotite
Quick effervescence with dilute HCl; perfect 3D cleavage; vitreous lustre	Calcite
Brass yellow; green-black streak; metallic lustre; soft	Chalcopyrite
Cubic crystal; brilliant lustre; hardest; colourless	Diamond
Effervescence with warm dilute HCl; perfect 3D cleavage; vitreous lustre; conchoidal fracture	Dolomite
Metallic lustre; cubic cleavage; grey streak; lead grey colour; soft	Galena
Silky and pearly lustre; soft; distinct cleavage; prismatic and bladed crystals; fibrous vein fillings	Gypsum
Magnetic attraction; brownish streak	Hematite
Triclinic crystal; glassy lustre; 2D non-orthogonal cleavage; azure blue	Kyanite
Magnetic attraction; black streak	Magnetite
Scaly form; 1D cleavage; platy structure; usually colourless	Muscovite
Massive with well developed faces; flesh red or grey; vitreous lustre; perfect cleavage	Orthoclase feldspar
Tabular or blocky; flame yellow; vitreous lustre; striated cleavage	Plagioclase feldspar
Colourless; hexagonal crystal; conchoidal fracture; very hard	Quartz
Greasy feel; pearly lustre; very soft; white streak; 1D cleavage	Talc

Common Tests for Identification of Rock Specimens

Diagnostic Features	*Rock*
Fine grained; dark coloured; volcanic texture; conchoidal fracture; amygdaloidal structure	Basalt
Clastic texture; pebbles, cobbles, gravels, rock fragments, granules are cemented together	Conglomerate
Medium texture; dark colour; hard and compact; random orientation of feldspar laths	Dolerite
Coarse texture; hard and compact; dark colour; plagioclase constitution	Gabbro
Coarse grained; hard and compact; granular texture; banded structure	Gneiss
Coarse grained; light texture; quartz, feldspar and mica can be identified with naked eye	Granite
Fine grained; light colour; strongly reacts with cold and dilute HCl; can be scratched by fingernail	Limestone
Coarse but evengrained; light colour; relatively hard; reacts with cold and dilute HCl	Marble
Fine grained; non-clastic texture; very soft; muddy odour	Mudstone
Coarse grained; interlocking texture; quartz, feldspar, tourmaline, topaz, etc.	Pegmatite
Fine grained; light colour; silvery lustre with bluish tint; lustrous sheen; phyllitic texture	Phyllite
Light colour; very compact; granoblastic texture; vitreous lustre; no cleavage; no foliation	Quartzite
Sand grains; clastic granular texture; graded or cross-bedded structure	Sandstone
Light colour; flaky minerals; foliations; markedly fissile; schistose structure	Schist
Fine grained; soft; clayey; muddy odour; thin layers	Shale
Fine grained; dark colour; planar schistocity; slaty cleavage; metallic sound; muddy odour	Slate

EXERCISE

1. Define rocks. What are the different types of rocks?
2. What are the characteristics of igneous rocks?
3. What are the characteristics of sedimentary rocks?
4. What are the characteristics of metarmorphic rocks?
5. Define minerals. How are these distinguished from rocks?
6. What are the common tests for identifying a rock?
7. What are the common tests for identifying a mineral?
8. Write two diagnostic characteristics to identify: basalt, conglomerate, dolerite, gabbro, gneiss, granite, limestone, marble, phyllite, quartzite, sandstone, schist, slate and shale.
9. What are minerals? What are the characteristics of minerals?
10. Write two diagnostic characteristics to identify: bauxite, biotite, calcite, chalcopyrite, dolomite, galena, gypsum, hematite, kyanite, magnetite, muscovite, orthoclase, plagioclase, quartz, and talc.

11

Field Study

HIGHLIGHTS

- Basic Principles of Fieldwork
- Choosing a Topic
- Data Collection
- Data Processing and Presentation
- Ideas for Field Study
- Methods of Setting a Map in the Field

Geography is described as a *natural science* or an *earth science* or a *regional science* and/or a *field science*. Its domain is the earth and any segment of the earth's surface provides an *open natural laboratory* to geographers who seek to identify and explain the *spatial organisation* of geographical features on the earth's surface through a careful study of their *patterns* and *processes*.

Approches to the Field Study

Field study may be oriented in two ways:

- **Area-specific study:** Usually a balanced geographical account with all elements is prepared to present an integrated form of the problems in their totality. Hence, necessary regional development plans can be formulated/suggested in a realistic and feasible manner.
- **Problem-specific study:** It is better known as a dissertation thesis that explores a geographically relevant problem of man related to his habitat, economy and society. Naturally, it is explored in a problem-solving environment using suitable techniques of data collection, analysis and presentation.

Books certainly teach us the basic theoretical principles of geography. Along with journals, they enrich students with discussions of the investigations carried out and the conclusions arrived at by researchers in geography. Field studies, on the other hand, allow students to share their experiences of the geographical research with others. It also gives them a scope to unearth the geographical facts hitherto undiscovered. To a geographer, therefore, fieldwork is of utmost importance. It involves the observation of the landscape in minute detail. The art of observation is based purely on geographical perspectives.

BASIC PRINCIPLES OF FIELDWORK

Fieldwork is an essential element in the learning process of geography and regional studies. These demand an understanding of the places—their spatial relationships, their physical structures and human contexts. Field visits enable students to interact with a specific type of environment and learn pro-actively by doing so. It is a form of active learning within the ambit of the environment. It helps students to find, by direct investigation, the answers to a number of fundamental questions: *what*, *where*, *when*, *why* and *how*. These are questions that can be applied to any of our fieldwork destinations.

Fieldwork is collecting information and a good fieldwork is collecting information for a purpose (Greasley 1984). Geographers need to collect information for two basic reasons—to *test* whether an idea they have had is correct and to *find* the answer to a question they have asked as a result of reading *or* something they have observed.

The whole course of fieldwork may be divided into a sequence of eight steps that together build up the universal flow chart of a field study project (Fig. 11.1). In the *first step*, an idea is conceived to be tested in the field and is then written as a statement. In the *second step*, following geographical logic, this statement is broken up into a set of questions in such a way that they together compose the basic theme of the statement. In the *third step*, hypotheses are built up based on the postulations for each of the above questions. In the *fourth step*, it is decided what data and information are actually required for the above purpose. In the *fifth step*, the relevant data and information are acquired from all kinds of sources in the field. In the *sixth step*, the data and information is statistically processed and manipulated and then cartographically presented by means of charts,

Fig. 11.1 Flowchart of a Field Study

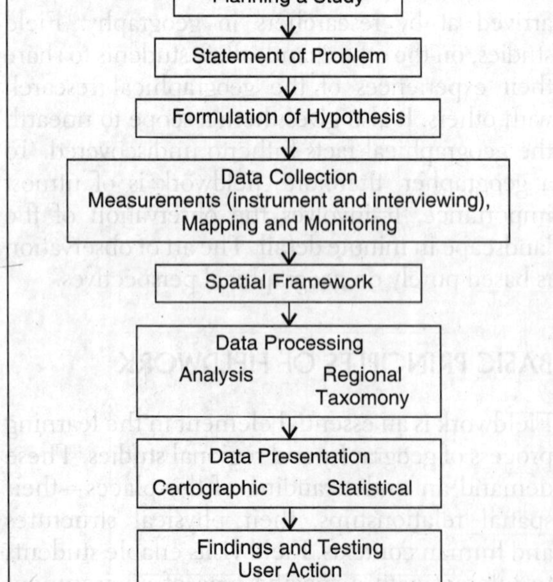

Sequence of Steps in a Fieldwork

Steps	*Work-theme*
I	Choose an idea to be tested and write it in the form of a statement.
II	Break the statement into forms of several questions following the suit of geographical logic.
III	For each question, postulate a geographically relevant hypothesis.
IV	It is now to be decided what data and information need to be collected, how and from where.
V	Collect the required data and information.
VI	Process and present the information (cartographically and statistically).
VII	Analyse the results of fieldwork to answer the individual questions and to test the main idea.
VIII	If it is so required, modify the original statement and write a new one, then follow the same steps to reach your goal.

Checklist of Equipment for Fieldwork

- Comfortable and proper clothing, comfortable and strong shoes, water bottle, first-aid kit, clipboard and pencils, umbrella or cap, wrist watch, etc.
- Maps: administrative map, topographical map, cadastral map, Google map, etc.
- Camera, drawing sheets, accessories and items, etc.
- Instruments for land survey, e.g., abney level, clinometer, compass, dumpy level, tape, ranging rod, GPS, total station, etc.
- Kits: soil testing kit, water testing kit, etc.
- Blank field books, field data sheets, questionnaires and schedules for land, household, market, traffic survey, etc.
- All necessary equipments for any special kind of survey.

Fig. 11.2 Flowchart for Choosing a Topic

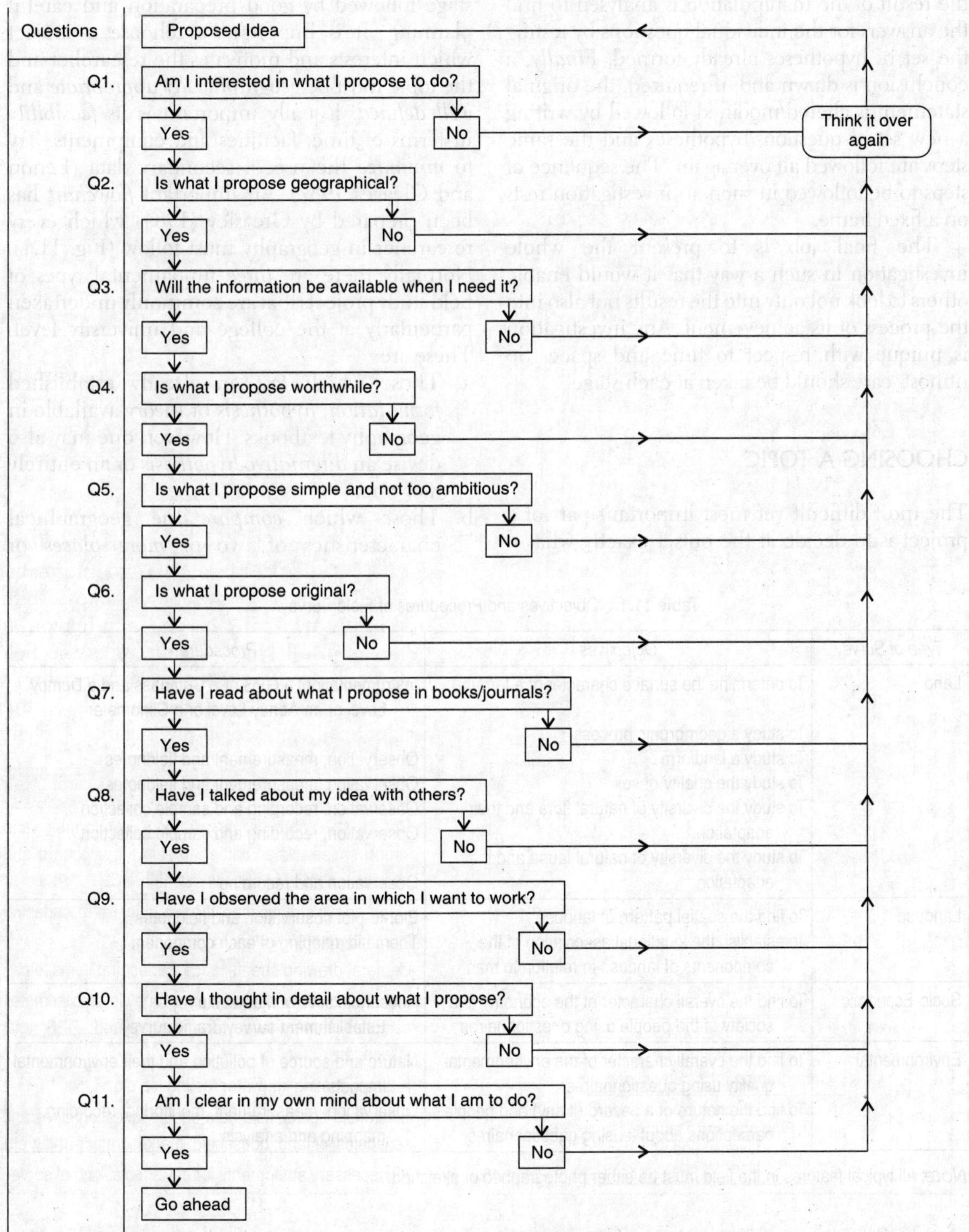

graphs, diagrams and maps. In the *seventh step*, the result of the manipulation is analysed to find the answers for the individual questions by testing the set of hypotheses already formed. *Finally*, a conclusion is drawn and, if required, the original statement is altered/modified followed by writing a new set of questions/hypotheses and the same steps are followed all over again. The sequence of steps to be followed in such an investigation rests on a fixed frame.

The final job is to present the whole investigation in such a way that it would enable others to look not only into the results but also into the process of its achievement. Any investigation is unique with respect to time and space. So utmost care should be taken at each stage.

CHOOSING A TOPIC

The most difficult yet most important part of a project is to decide at the outset exactly what is to be studied. A lot of thought should go into this stage followed by good preparation and careful planning. It is important to choose a subject which interests and motivates the researcher and the topic must be *geographically appropriate* and *well defined*. Equally important is its *feasibility* in terms of time, facilities and equipments. Try to *minimize* the use of secondary data (Lenon and Cleaves 1982). An important *flowchart* has been prepared by Greasley (1984) which every researcher in geography must follow (Fig. 11.1). Normally there are *three* fundamental types of field study projects that are commonly undertaken particularly at the college and university level. These are:

a. Those which *test* an already established *formulation*, *hypothesis* or *theory* available in geography textbooks. However, one may also devise an *alternative hypothesis* or an entirely *new* one.
b. Those which *compare* the geographical characteristics of two *or more places* or

Table 11.1 Objectives and Procedures of Field Survey

Type of Survey	*Objectives*	*Procedures*
Land	To determine the surface character of a terrain To study a geomorphic process To study a landform To study the quality of soil To study the diversity of natural flora and their adaptation To study the diversity of natural fauna and their adaptation	Instrumentation: a GPS or a Compass and a Dumpy Level or an Abney Level or a Clinometer Observation, measurement and fieldnotes Observation, measurement and fieldnotes Observation, recording and sample collection Observation, recording and sample collection Observation and recording
Landuse	To find the spatial pattern of landuse To establish the locational association of the components of landuse in relation to man	Plot-to-plot observation and fieldnotes Thematic mapping of each component
Socio-Economic	To find the overall character of the economy and society of the people using questionnaires	Household survey, Market survey, Establishment survey, Traffic survey
Environmental	To find the overall character of the environmental quality using questionnaires To find the nature of a hazard (if any) and people's perceptions about it using questionnaires	Nature and source of pollution and their environmental impacts Observation, measurement, monitoring, recording, mapping and analysis

Note: All typical features in the field must be either photographed or sketched.

phenomena. A common variant may be a *comparison* of the geographical characteristics of *one* place or phenomena at *two* or more stages of time.

c. Those which study a *problem* or *landuse conflict* such as natural hazard, environmental impact analysis, regional and social inequality, etc.

DATA COLLECTION

The data required to test ideas or questions would depend very much upon the specific idea or question being tested. It is most important that the body of information collected is relevant and capable of providing the evidence necessary for the study. Normally *two types* of data are collected—*primary* and *secondary*. The *first* one concerns the first hand information collected in the field while the *second* one relates to the second hand information collected from the extra-field sources.

The primary data is obtained from the *actual fieldwork* by a series of thorough *surveying* or by a set of *sampling* (line or random point or systematic grid) sufficient to draw statistical inferences at *desired* significance levels. This includes

Types of Field Survey

Land Survey: Abney Level, Clinometer, Prismatic Compass, Dumpy Level, Tape, Ranging Rod, GPS, Camera, etc.

- Measurement of ground slope by profile levelling
- Micro-level study of a landform by observation, field sketches, and measurements of its geometric and morphological properties

Soil Survey: Soil sampler, soil testing kit, field book, camera, etc.

Weather Observation: Mini Weather Station

Land Cover Survey: Topographical map, large scale land cover map, camera, field observation, etc.

- Mapping the locations of the boundaries of the natural features like, plateau, plain, forest, water body, rock outcrop, sheet rock, sand deposit, etc.

Landuse Survey: Cadastral map, camera, etc.

- Plot-wise observation and recording of information on landuse using a cadastral map
- Plot-wise information on landuse obtained from the office of land records

Household Survey: Questionnaires, ID cards, cameras, etc.

- Door-to-door enumeration for data and information regarding economy and society using a well-laid questionnaire

Market Survey: Large scale map, cadastral map, sketch map, camera, clip board, etc.

- Locating the market area on a large scale map by observation
- Locating different types of shops and establishments by observation

Traffic Survey: Field book, watch, camera, etc.

- Counting the number of vehicles (type-specific, direction-specific, date-specific, and time-specific) plying along a road (a road junction at vantage location is a preferred point of study)

Special Survey: Large scale map, cadastral map, sketch map, camera, drawing board, clip board, instruments, etc.

- Geomorphic study: fluvial, marine, aeolian, karst, glacial, periglacial processes
- Environmental study: pollution of land, water, air, etc.
- Social study: population groups, tribes, etc.
- Environmental impact assessment: thermal power plant, deforestation, surface mining, tube well irrigation, bank erosion, flood, landslide, manufacturing, etc.

surveying (land, landuse, market, household, etc.) *questionnaires*, *photographs*, *field sketches*, *transects*, *measurements* and *interviews*. This is an essential part of field study. The checklist of equipments for the collection of primary data consists of instruments (such as, Dumpy Level, Clinometer, Geological Hammer, Shovel, Prismatic Compass, Tape, Pole, Staff, Sextant, Soil Sampler, Sediment Sampler, Graduated Ruler, Thermometer, Camera, Soil Testing Kit, Geological Map, Topographical Map, GPS, Total Station, etc.), data recording articles (such as, board, clip, field book, graph paper, tracing paper, chart paper, pen, pencil, crayon box, semiautomatic and/or automatic data loggers, etc.) and wellplanned questionnaries. The data sheets should be well constructed in tabular form, each with a title. The date and time of survey should be clearly mentioned.

The secondary data, on the other hand, refers to those *collected from* secondary sources like books, journals, census handbooks, district gazetteers, village directories, statistical abstracts, government statistics, company publications, newspapers, television, radio, maps, aerial photographs, satellite imageries, internet, etc. Such information may not only be used as data for testing but also for comparison.

DATA PROCESSING AND PRESENTATION

There are a large variety of techniques and procedures of data processing and presentation. A lot of these have been discussed in this book. However, any other standard book on practical geography, cartography, spatial analysis and statistical methods may be consulted. It is important that the techniques of statistical analysis and cartographic representation be chosen based on the *type of data* and the *type of tests* desired. While doing this, a researcher should be organised and disciplined in mind and thought. As a geographer is concerned with the processes and patterns, it is important to try and explain the patterns *in terms of* the processes in operation as observed in the field. For this, one must not forget to consult the local history section of a library, to interview the old people of the local area and to guess intelligently.

Writing a Field Report

This is an important part of the whole project and may well be divided into five sections or chapters. The first section should be devoted to the *conceptual frame* or *background information* which helped provide the setting for the work. *Prefield investigations* and *review* of similar research work can be discussed here. This would lead to a *statement of the ideas* to be tested and should be written clearly and broken down into testable questions. The second section should contain the *geographical identity* and a brief account of the *natural setting* of the area concerned. The third section should contain a description of *how* the investigation was carried out and should include information about *how* each idea was tested and the equipments used and the procedures followed. Reasons for the *choice* of sampling techniques and *methods* of data analysis must also be incorporated here. The fourth section usually contains the *results* of each of the surveys—the processing, presentation and analysis of data. An *interpretation* of each must be clearly given as accurately as possible. In the final section all the evidences collected to test each question should be *compiled* to test the *main hypothesis*.

Organising a Field Report

The whole write-up should be presented in a systematic way. At the beginning there should be a *title page* containing the title of the work, object of submission, name of the researcher, name of the institution and the data of survey. It the next page *acknowledgements* due to all, i.e., the supervisor, library staff, the people in the field, the people interviewed and others should be mentioned. Next comes the *preface* explaining the significance of the study and a

Organising a Field Report

	TITLE OF THE FIELD PROJECT
	ACKNOWLEDGEMENT
	PREFACE
	CONTENT
	LIST OF TABLES
	LIST OF FIGURES
	LIST OF PHOTOGRAPHS
Section I	INTRODUCTION (Conceptual background, Reasons for the investigation, Objectives, Hypothesis)
Section II	NATURAL SETTING OF THE STUDY AREA (Regional identity, Location, Environmental setting)
Sectio]n III	METHODOLOGY (Methods used to investigate each hypothesis, equipments used, choice of techniques of sampling and field measurements)
Section IV	RESULTS (Interpretation of cartographic and statistical analysis of data)
Section V	CONCLUSION (Testing each hypothesis for significance using the evidences collected; difficulties faced in the present investigation; ideas for further investigation)
	BIBLIOGRAPHY
	APPENDIX

chapter-wise or section-wise organisation of the field report. The following page should deal with the *content* showing the main constituents of each chapter and the page number should be included alongside. A *list of tables* with table numbers, title and page numbers should come immediately afterwards. This should be followed by a *list of diagrams*, *sketches* and *maps* with figure numbers, title and page numbers. Similarly, another page should be devoted to a detailed *list of photographs*. This should be followed by the *body* of the field report. At the end, a *bibliography* is essential, containing a list of all the books, journals, pamphlets and printed material used in the report generation. This should be laid out with the name of the author, followed by the title, date of publication, volume and number, and page number of the journal (in case of articles) and the publishers (in case of books) respectively. The authors' names should be in alphabetical order. Long tables and original survey or field books should be *appended* after the bibliography. The whole work may either be *neatly handwritten* or *typed* and may be *presented* in a *folder* or as a *bound volume* (soft, hard or spring).

IDEAS FOR FIELD STUDY

A short *list* of selected topics, which students may consider, is given in the following paragraphs. However, many more can well be added to this:

Physical Geography

- Influence of rock type and structure on landform
- Influence of slope on landuse
- Influence of slope on drainage and channel pattern
- Influence of rock type on soil
- Relationship between rainfall intensity and run off
- Study of climatic trends and fluctuations
- Study of micro climates
- Problem of gully erosion
- Problem of bank erosion by channel migration
- Changes in soil characteristics along a slope
- Influence of soil characteristics on crop yields
- A physical comparison of two valleys
- Morphometric classification of a region
- A comparison of structural characteristics of two drainage networks
- A comparison of TDCNs (Topologically Distinct Channel Networks) in two drainage nets

- Relationship between landform and accessibility
- Altitudinal variation in temperature, rainfall and vegetation
- Deforestation and soil erosion
- Beach profile, wave character and sediment texture
- Influence of relief on settlement pattern
- Track of a cyclone
- Study of river regimes
- Storm hydrograph at a point
- Changes in slope and cross-section along a river
- Rock stability and slope failure
- Environmental impacts of landslides, thermal power plants, surface mining or dams
- Changes in cross-section and hydraulics on the two sides of a bridge across a river
- Changes in slope, soil texture and landuse away from an alluvial channel
- Variations in soil profiles along a slope
- Geomorphic mapping of a terrain in field
- Measurement, mapping and monitoring of a beach profile, a bank profile, a low residual hill, a channel bar, bedforms on a tidal channel bed or a gully network in a bedland
- Rainfall variability and irrigation
- Seasonal weather and cropping practices
- Environmental perception of people
- Study of the quality of rural and urban environment
- Mapping and analysis of waterlogged areas/ floodprone areas with RS and GIS
- Mapping and analysis of degraded lands/ waste lands with RS and GIS
- Change detection and analysis of landuse/ land cover pattern with RS and GIS
- Change detection and analysis of forest cover with RS and GIS techniques
- Identification and geomorphic analysis of palaeochannels using RS data
- Study of alluvial fans: geometry, origin and significance
- River corridor studies using modern tools
- Watershed management/floodplain management studies using modern tools
- Changes in stream courses using open source images and topographical maps
- Hydrogeomorphological investigations
- Geomorphic signatures of underlying structures using Images and topomaps
- Characterisation and mapping of soil resources with particular reference to ersosion and conservation
- Terrain analysis of a drainage basin
- Studies in thematic cartography
- Geomorphic analysis of longitudinal profiles of rivers
- Fluvial analysis of cross profiles of channels
- Channel pattern studies using satellite images
- Erosion studies using remote sensing techniques
- Hypsometric analysis of a drainage basin
- Analysis of flow characteristics of channels
- Landscape analysis

Human Geography

- Industrial profile of a region
- Town size and quality of life
- Functional hierarchy of settlements
- Rank-size hierarchy of towns
- Size and spacing of towns
- City-size distribution and economic development
- Classification of towns
- Verification of Burgess' model of urban structure or Christaller's central place model
- Economic regionalisation of a country
- Urbanisation and social change
- Changes in the entropy of a settlement pattern
- Pattern of urban growth
- Levels of urbanisation
- Demographic transition
- Trends of urbanisation
- Inter-class mobility of towns
- Comparison of nearest neighbour parameters of settlement patterns in two or more regions
- Identification of the Central Business District
- Variation of urban density from city core
- Suburban housing boom
- Population redistribution in a metropolis

- Diffusion of agricultural technology
- A study of urban slums
- Crop productivity regions
- Population growth and productivity
- A study of rural-urban fringe or rural-urban migration
- Delineation of umland or sphere of influence
- A study of social segregation
- Changing pattern of age-structure or habitational density or sex ratio or literacy of population
- Land capability classification
- Crop diversification in a region
- Settlement growth and landuse
- Diurnal population mobility in a city
- Identification of randomness in settlement patterns
- Transport network structure and economic development
- Distribution of per capita income and population pressure
- Family size and quality of living
- Growth poles and regional developments
- Changes in occupational patterns of small and medium towns
- Consumer behaviour in the periodic markets
- A study of the diurnal traffic flow along a highway
- Factors of location of an economic activity
- Study of social affinity
- Comparision of cultural landscapes of two areas
- Patterns of transhumance of nomads
- Tourism potential of a region
- Agricultural or industrial landscape of a region
- Application of RS and GIS in infrastructure development
- Landuse planning using RS and GIS
- Ecological planning of landscapes
- Optimum location of an economic activity/service using GIS
- Geographical analysis of urban growth and development
- Analysis of urban corridors and expansion
- Analysis of industrial expansion
- Industrialisation and urbanisation
- Economics of floriculture
- Changing agricultural productivity
- Gender development and disparity
- Analysis of migration pattern and growth of border towns
- Perception analysis: environment, landscape, tourism and voting
- Analysis of sex ratio and literacy
- Analysis of house types: patterns and relations
- Changes in cropping patterns
- Analysis of landuse/land cover dynamics
- Industrial landscape: patterns, growth and relations
- Studies on urban modelling

METHODS OF SETTING A MAP IN THE FIELD

Compass Observation Method

Accessories needed—a Compass and a Map, in which Magnetic meridians are drawn.

It is the most satisfactory method to be used when a compass is available.

1. Lay the map flat on the ground
2. Place the Compass on it close to the magnetic meridian drawn on it
3. Turn the map gently until the magnetic meridian is parallel to the Compass needle

Sun Observation Method

Accessories needed—a Watch and a Map in which Geographical meridians are drawn.

It is the simplest method that uses a Watch.

1. Place the Watch horizontally on the map with the small (hour) hand pointing at the Sun.
2. The line from the centre of the watch point to a midway between the hour hand and twelve o' clock will point approximately South and the North in the opposite direction along the true meridian
3. Turn the map gently until the true meridian (or geographical N-S line) is parallel to the N-S line derived from the Watch and the Sun

Fig. 11.3 Sun Observation Method

S
N
SUN
North
South Line
N
S

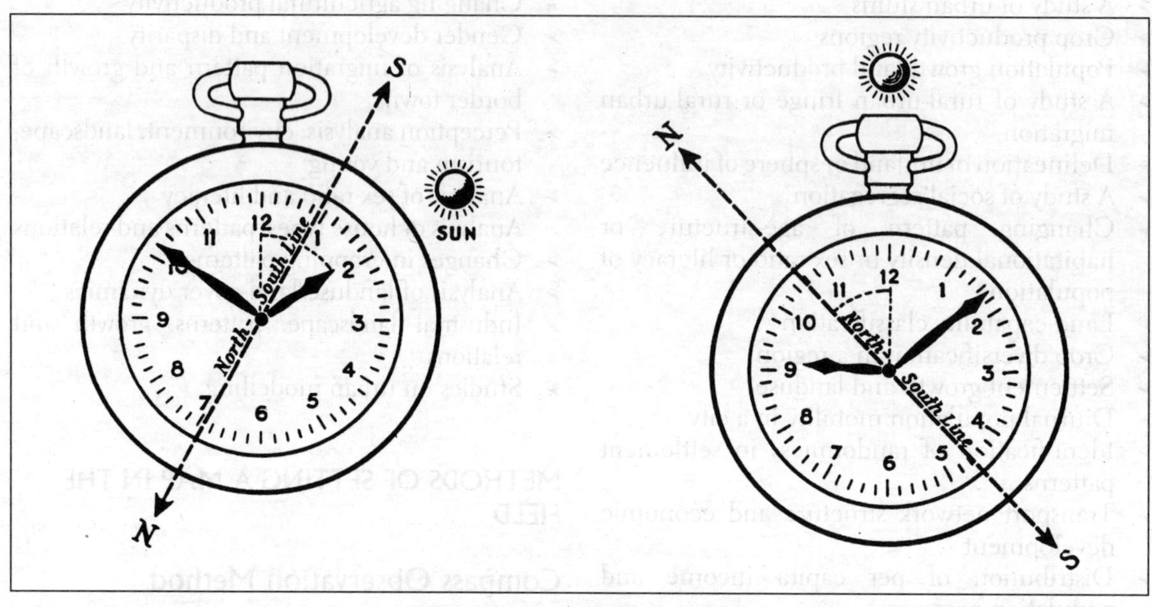

Fig. 11.4 Field Observation Method

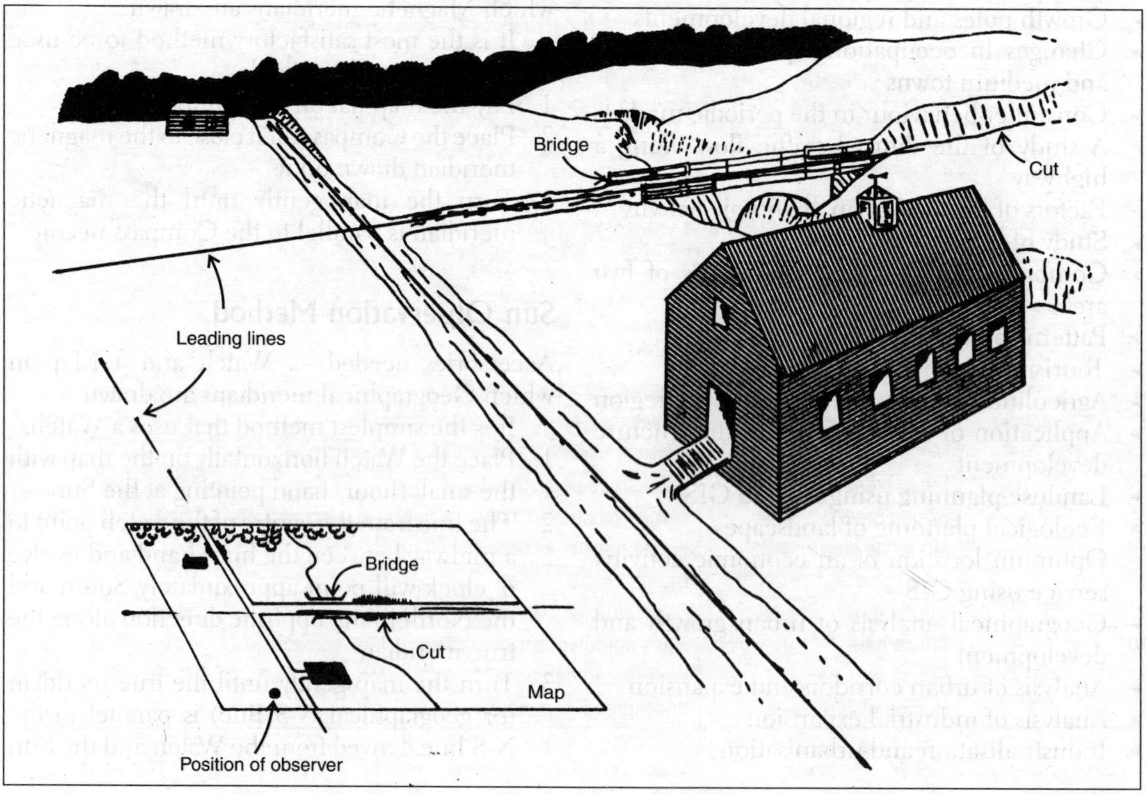

Bridge
Cut
Leading lines
Bridge
Cut
Map
Position of observer

Field Observation Method

Although it is the simplest of all methods, a great skill is required to properly orient the map.

1. Carefully note some distinct features, at least four in number, nearby the point of observation on the map
2. Draw two criss-cross lines through them, called Leading lines
3. Observe and identify the same features on the field
4. Turn the map gently until the Leading lines are oriented

Table 11.2 Slope/Gradient Table (for Field Observation)

Slope	*Gradient*	*Steepness*	*Remarks*
35°–45°	1 in 1 or 100%	Very Steep	Looks very steep; needs hands to climb
25°–35°	1 in 2 or 50%	Steep	Ascends using zigzag course; used by foresters
18°–25°	1 in 3 or 33%	Steep	Permanent grass; walking requires great effort
11°–18°	1 in 5 or 20%	Fairly Steep	Limit of ploughed land; problem for commercial road traffic
6°–11°	1 in 10 or 10%	Moderate	Walkers notice uphill climb
3°–6°	1 in 20 or 5%	Gentle	Acceptable to farmers and builders; drainage good
1°–3°	1 in 60 or 1.5%	Flat	Rather too steep for some railway operations
0.5°		Very flat	Below this drainage ditches are essential

Table 11.3 Transect Table

(from the River Bank to the Metalled Road)

Recorder: Oyntai Jhankar, Priti Sen, and others					
Segment	Distance	Slope	Aspect	Soil	Landuse
1	10 m	2°	south	fine, silty loam	cultivated land (rice)
2	22 m	4°	south	medium, loam	cultivated land (rice & vegetables)

Table 11.4 Beaufort Scale of Wind Force

Beaufort No.	*Description*	*Effect on Land Features*
0	Calm	Smoke rises vertically
1	Light air	Direction shown by smoke but not by wind vanes
2	Light breeze	Wind felt on face; leaves rustle; vane moves
3	Gentle breeze	Leaves and twigs in constant motion
4	Moderate breeze	Raises dust and paper, small branches move
5	Fresh breeze	Small trees begin to sway
6	Strong breeze	Large branches in motion; whistling in telephone wires
7	Moderate gale	Whole trees in motion
8	Fresh gale	Breaks twigs off trees
9	Strong gale	Slight structural damage to roofs, etc.
10	Whole gale	Trees uprooted; considerable structural damage
11	Storm	Widespread damage
12	Hurricane	Widespread devastation

Table 11.5 Landuse Survey Schedule

Surveyor: Kamalika Chatterjee, Sreya Chatterjee, and others
Date: 23.12.2005

Position: Northern Part of the Village
Time: 9-00 a.m.

Plot ID (Field ID with Location and No)	*Existing Use*	*Plot ID (Field ID with Location and No)*	*Existing Use*	*Plot ID (Field ID with Location and No)*	*Existing Use*
123	Vegetables	123/1	House	124	Orchard
-	-	-	-	-	-

Table 11.6 Traffic Survey Schedule

Surveyor: Wriju Sarkar, Sraman De, and others
Date: 23.12.2005
Direction of Traffic: Towards BBD Bag (City Centre)

Position: CR Avenue– Harrison Road Crossing
Time: 9-00 a.m. to 12-00 noon

Bicycle	*Motorcycle/ Scooter*	*Car*	*Light Commercial Vehicles—vans, minibuses, etc.*	*Heavy Commercial Vehicles–lorries, buses, etc.*	*Remarks, if any*
(Tally Marks) Total =	(Tally Marks) Total =	(Tally Marks) Total =	(Tally Marks) Total =	(Tally Marks) Total =	

Table 11.7 Market Survey Schedule

Surveyor: Supati, Khurshida, and others
Date: 23.12.2005

Time: 9-00 a.m. to 12-00 noon

Position: Nehru Market along Lalbagan Rasta (Left Side)
Direction of Survey: Centre to Fringe

Shope or Establishment ID	*Description*	*Shop, or Establishment ID*	*Description*	*Shop or Establishment ID*	*Description*
1	Grocery	2	Hardware	3	Clothing
-	-			-	-

Table 11.8 Household Survey Schedule (part only)

1. HOUSEHOLD ID NO. _____
2. Length of Stay: _________ (years) Tenancy Status : Own / Tenant Caste: G / SC/ST/OBC
3. Religion: ______________________________
4. No of Members: Male _______, Female _______, Children _______ Total: _______
   0–14: _____, 15–30: _____, 31–45: _____, 46–59: _____, 60+: _____
5. No. of Earning Members: ____________ Approx. Monthly Income: ____________
6. Amount of Land Possessed: ____________
7. Name of Occupations: ____________, ____________, ____________
8. Crops grown (with annual production): ____________
9. Crop Rotation (with season): Y / N (______________________________)
10. Fertiliser: Y / N HYV Seeds: Y / N Pesticides / Insecticides: Y / N
11. Irrigation: Y / N Source of Irrigation (with methods of lift): ____________
12. Place of Work : ______ (kms) Mode of Journey : a) to Work _______ b) School: _______
13. No. of Literate Males : _________, No. of Literate Females : _________
14. Madhyamik Level: ____, H.S. Level: ____, College Level: ____, University Level: ____
15. No. of Rooms (with types): ______________________________
16. Wall Material: _________, Floor Material: _________, Roof Material: _________
17. Domestic Fuel used: ______________________________
18. Recreation: ______________________________
19. Type of Ailments (with seasonal incidence): ______________________________
20. Social Practices/Rituals, if any: ______________________________
21. Nearest (a) Health Centre: ___ (km) (b) Market: ___ (km) (c) Post Office / P.C.O. : ___ (km)
22. Amenity/Facility:
    (a) Water Supply: Y / N (b) Electricity: Y / N (c) Fridge : Y / N
    (d) TV with Cable: Y / N (e) Sanitation: Y / N (f) Kitchen Garden: Y / N
    (g) Bi-cycle: Y / N (h) 2 - wheeler: Y / N (i) Waste Disposal: Y / N

EXERCISE

1. Why is field study important in geography?
2. What are the principles of field study?
3. How is the theme of a field study chosen?
4. What are the types of projects usually undertaken in a field study?
5. How is the primary data about land collected in a field study?
6. How is the primary data about landuse collected in a field study?
7. How is the primary data about society collected in a field study?
8. How is the primary data about economy collected in a field study?
9. What are the sources of secondary data in a field study?
10. What are the objectives of land survey in a field study?
11. What are the objectives of landuse survey in a field study?
12. What are the objectives of socio-economic survey in a field study?
13. What are the objectives of environmental survey in a field study?
14. State how a land survey is done in field.
15. State how a landuse survey is done in field.
16. State how a socio-economic survey is done in field.
17. State how an environmental survey is done in field.

Unit V

Advanced Tools of Analysis

12

Geographical Information System

HIGHLIGHTS

- Definitions of GIS
- Nature of GIS
- Evalution of GIS
- Components of GIS
- Functionality of GIS
- Application of GIS
- GIS in India

Data refers to a collection of facts or figures that pertains to places, people, things, objects, events and concepts. These can generally be represented in the following basic forms—*numerical values*, *alphanumeric characters*, *symbols* and *signals*. When data is transformed through processes such as *structuring*, *formatting*, *conversion* and *modelling* to a form that is meaningful to a user, it is referred to as *information* (Fig. 12.1). Therefore, information is the *processed* or *value-added* data that has certain perceived values to a user or a community of users (Lo and Yeung 2002).

The transformation of data into useful information is the core function of an *information system* (Fig. 12.2). From the functional perspective, an information system is built up to achieve the specific objectives of collecting, storing, analysing and presenting information in a systematic manner to be used in some decision making processes (Calkins and Tomlinson 1977).

Fig. 12.1 The Data-Information Relation

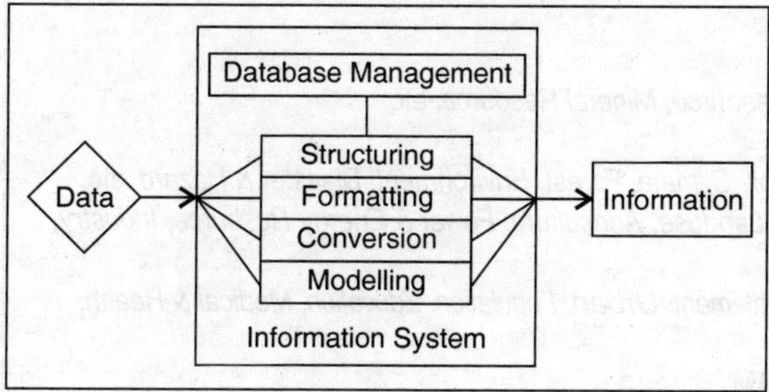

Fig. 12.2 A Simplified Information System

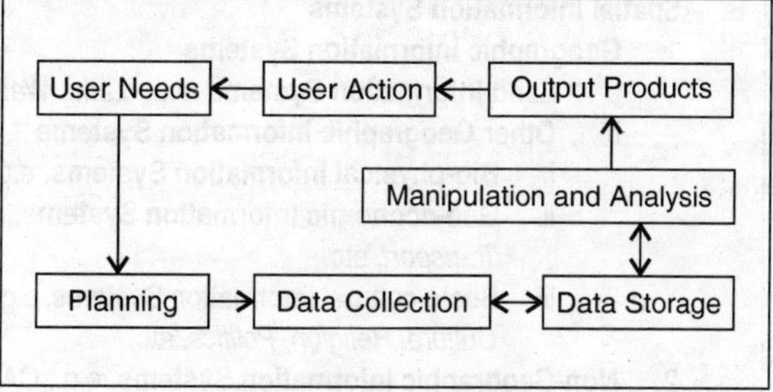

Structurally, an information system is made up of inter-related components that include a combination of data, technical resources and human resources. It can also be perceived as being made up of input, processing, and output sub-systems, all working according to a well defined set of operational procedures (Fig. 12.2). Individual information systems can be operated independently, and at the same time linked with other information systems through standard communications protocols to form an *information system network* (ISN).

The geographical information system (or GIS) is a special class of information systems (Table 12.1). The term 'geographical' conveys two meanings—*earth* and *geographic space*. The term 'earth' implies that all data in the system are pertinent to the earth's features and resources, including human activities. By *geographic space* it means that the commonality of both the data and the problems that the GIS is designed and developed to solve is *geography*, i.e., *location*, *distribution*, *pattern* and *relationship* within a specific geographical reference framework.

Geographic data is certainly a special form of spatial data with two important properties—*first*, the reference to geographic space, which means that the data is registered to an accepted geographical coordinate system of the earth's surface, so that data from different sources can be spatially cross referenced and integrated, and *second*, the representation at geographic scale, which means that the data is normally recorded at relatively small scales and, as a result, must be generalised and symbolised. Basically, it is this specific focus on geographic data and their applications for solving the spatial problem, that makes the GIS unique among the information systems.

DEFINITIONS OF GIS

It is really very difficult to build an all-comprehensive definition of GIS. GIS can be viewed as *a toolbox*, as *a database* and as *an organisation*. Based on this, the most common definitions are:

1. *Toolbox-based definitions*

   "… a powerful set of tools for collecting, storing, retrieving at will, transforming and displaying spatial data from the real world." (Burrough 1986)

   "… a system for capturing, storing, checking, manipulating, analysing and displaying data which are spatially referenced to the earth." (DOE 1987)

   "… an information technology which stores, analyses, and displays both spatial and non-spatial data." (Parker 1988)

   "… a system of hardware, software and procedures designed to support capture, management, manipulation, analysis, modelling and display of spatially referenced

Table 12.1 Types of Information Systems

A. **Non-spatial Information Systems**, e.g., *accounting, banking*, etc.

B. **Spatial Information Systems**

   1. **Geographic Information Systems**

      a. **Land Information Systems**, e.g., *Land, Water Resource, Mineral Resource*, etc.

      b. **Other Geographic Information Systems**

         i. **Bio-physical Information Systems**, e.g., *Soil, Climate, Forest, Environment, Disaster & Hazard*, etc.

         ii. **Geo-economic Information Systems**, e.g., *Landuse, Agriculture, Power & Energy Resource, Industry, Transport*, etc.

         iii. **Socio-cultural Information Systems**, e.g., *Settlement, Urban, Population, Education, Medical & Health, Culture, Religion, Politics*, etc.

   2. **Non-Geographic Information Systems**, e.g., CAD/CAM

data for solving complex planning and management problems." (Rhind 1989)

"... a computer system capable of assembling, storing, manipulating and displaying geographically referenced information." (USGS 1997)

2. *Database definitions*

"... an internally referenced, automated, spatial information system." (Berry 1986)

"... a database system in which most of the data are spatially indexed, and upon which a set of procedures operates in order to answer queries about spatial entities in the database." (Smith et al 1987)

"... any manual or computer based set of procedures used to store and manipulate geographically referenced data." (Aronoff 1989)

3. *Organisation-based definitions*

"... an automated set of functions that provides professionals with advanced capabilities for the storage, retrieval, manipulation and display of geographically located data." (Ozemoy, Smith and Scherman 1981)

"... an institutional entity, reflecting an organisational structure that integrates technology with a database, expertise and continuing financial support over time." (Carter 1989)

"... a decision support system involving the integration of spatially referenced data in a problem solving environment." (Cowen 1988)

Thus, it is not easy to select the best perspective, as it is still a growing concept. However, the ideas expressed by Rhind (1989) and USGS (1997) have been adopted by the authors of textbooks on GIS, notably, Burrough (1986), Aronoff (1989), DeMers (2000) and Clarke (2001). Simply stated, GIS is a *set of computer based systems for managing geographic data* and using these data *to solve spatial problems* (Lo and Yeung 2002). It may well be viewed as *a field of academic study,* as *a branch of information technology* and also as *a data institution* (Table 12.2). GIS is an important component of 'geomatics', or 'geomatic engineering', or 'geo-informatics', that represents the general approach that geographic information is collected, managed and applied. GIScience (i.e., geographic information science) is defined as the set of basic research issues raised by the handling of geographic information, e.g., the unique characteristics of geographic data, the distinct nature of research that requires geographic problem solving, the

Table 12.2 Perceptions of GIS: Concepts and Techniques

GIS as a field of academic study	*GIS as a branch of information technology*	*GIS as a data institution*
• Cartography	• Cartography	• Law
• Remote Sensing	• Remote Sensing	• Information Technology
• Mathematics	• Computer Programming	• Sociology
• Statistics	• Software-specific Training	• Anthropology
• Computer Science	• Workshops	• Cognitive Science
• Information Technology	• Laboratories	• Economics
• Geography		• Political Science
• Urban Planning		• Public Administration
• Resource Management		
GIS Concepts and Techniques		
Data Acquisition/Modelling ⇔ Geo-referencing ⇔ Data Standards ⇔ Data Processing/Analysis ⇔ Computer Hardware/ Software ⇔ Applications ⇔ Systems Development ⇔ Issues and Problems		

interaction between geographic information, research and related academic disciplines, as well as the impact of using geographic information on society (Goodchild 1992). GIScience aims to provide the theoretical and organisational coherence for the scienctific study of geographic information and is never intended as a substitute for GIS both in terminology and in practice.

NATURE OF GIS

A GIS may either be *manual* (or *analogue*) or *automated* (or *digital /computer-based*). Normally, in a manual GIS, several maps (on the thematic attributes of topography, soil, vegetation, drainage, temperature, rainfall, lithology, etc. as per the user's choice and needs), aerial photographs and hard copies of a satellite imagery are used as data sets. These are called *data layers* or *data planes*. These reveal several kinds of information, each of which is called a *theme*. All such maps come from different sources and, hence, they differ in scale and the time they were last updated. Therefore, for a meaningful GIS operation they need to be *carefully*, *skilfully* and *cartographically mainpulated*. The usual manual techniques for this involve the use of a plastic film or a tracing paper. This process is called *registration*. The set of these spatially registered data layers thus becomes the most useful geographic database in which further manipulation is easier. With the manual GIS, the analyst can now draw some new features as per the user's need and choice on another set of transparencies/films, overlaying the set of other data planes. The analyst-planner now lays a coarse grid over the database and marks on it the conjugation of different features to draw/derive the *draft-plan* for a particular parameter.

A typical automated GIS today has the following technical and operational characteristics that are self-explanatory for its present day fields of application and growing popularity:

- It is a part of a computer network that uses advanced processors and memory systems to share input/output connectivity and data storage facilities with different types of mainstream information technology applications.
- It is made up of multi-tier software components that allow geographic information processing across multiple processors, computer clusters and data storage systems.
- It is connected to the internet for access to data and technical resources available on the www and for distributing the geographic information to others across the world.
- It is developed using industry standards for computer systems development methodology, software development tools, database connectivity and communications protocols as well as user interfaces.
- It contains software applications to perform common data processing functions and also a scripting language for the users to develop task-specific applications quickly.
- It uses data stored locally in its own hard drives and data drawn from a local server through LAN, as well as data obtained from national, regional and corporate data warehouses through the internet.
- It uses both raster and vector data and integrates geographic data with other forms of business data.
- It is capable of presenting the results of information analyses using multimedia technologies.
- It is coupled with other software applications, e.g., statistical analysis and visualisation, which enhance its spatial analysis functionality, and spreadsheet and word processing, which are capable of using its output in other desktop applications.
- It supports multi-user information needs ranging from sophisticated 'spatial problem solving and complex business decision making' to relatively simple 'information query and browsing'.

A modern GIS, thus, is a high-tech mapping and decision making tool. It is a complex but organised collection of computer hardware, software, liveware and geographical data,

Why Geographic Information is Special

- It is multi-dimensional because two coordinates must be specified to define a location, whether they be x and y or longitude (λ) and latitude (ϕ).
- It is voluminous, since a geographic database can easily reach a terabyte in size.
- It may be represented at different levels of spatial resolution.
- It may be represented in different ways inside a computer and the way it is done can strongly influence the ease of analysis and the end results.
- It must be projected onto a flat surface applying principles of projection because (1) a flat paper is used as a medium for inputting data to GIS and also for outputting data in a map or image from and (2) rasters are inherently flat.
- It can be time consuming to analyse.
- The process of updating geographic information is complex and expensive.
- Display of geographic information in the form of maps requires the retrieval of large amounts of data.

designed by experts to efficiently capture, store, relationally and hierarchically manage, manipulate, analyse the spatial data and display the results into sets of various forms—graphics and multi-coloured thematic maps to be used for solving spatial geographical problems related to humanity. These may also be associated with multimedia techniques to make the presentations even more lively.

EVOLUTION OF GIS

Cartography forms the base of GIS. The first maps of spatial distribution of rocks or soil, or plants, or communities of people were *qualitative*. The first developments in appropriate mathematics for spatial problems came in the 1930s and 1940s parallel to the developments in statistical methods and time series analysis. The American statistician, Hollerith (1860–1929), the father of automated geoprocessing, used the electro-mechanical data processing technology for *quick*, *accurate* and *cost-effective collection*, *analysis* and *distribution of spatially disposed information* (Streich 1986). Practical progress, however, was completely blocked by the lack of suitable computing tools. It is only since the 1960s, with the availability of digital computers, that both the conceptual methods for spatial analysis and the actual potential for quantitative thematic mapping and spatial analysis were able to flourish (e.g., Norbeck and Rystedt 1972, Joumel and Huijbregts 1978, Cliff and Ord 1981, Davis 1986, Isaaks and Srivastava 1989, Cressie 1991). The development of GIS may be traced as a sequence of evolution through the phases—the *formative period*, the *period of maturing technology* and the *age of geographical information infrastructure* (Table 12.3).

The Formative Period

The 1960s and 1970s represented the important formative years of GIS. Hundreds of software packages for handling and analysing geographic information were developed (Marble 1980). For example, Computer Assisted Geographic Data Processing at MIT (1955), Database Management System by General Electric (1965), CGIS in Canada (1964), GIRAS by USGS (1973), SLDB in Sweden (1973), LAMIS and JIS in Great Britain (1973), ODESSEY at the Harvard Laboratory, MLMIS at the University of Minnesota, High-quality Digital Map Generation by ECU at the Royal College of Art, London, GIMMS at the University of Edinburgh (Tomlinson et al 1977; Mitchell et al 1977; Rhind 1981; Andersson 1987).

By 1977, high-end digital techniques were developed and computer cartography took a big leap forward—to make existing maps more *quickly* and more *cheaply*, to make maps for *specific* user *needs*, to make map production possible in situations where *skilled staff are unavailable*, to *allow experimentation* with different *graphical representations* of the same data, to *facilitate map making* and *updating* when the data is already

Table 12.3 Evolution of GIS

Stage of Development	*Formative Period*	*Period of Maturing Technology*	*Age of GI Infrastructure*
Time Frame	*1960–1980*	*1980–Mid 1990s*	*Mid 1990s–till date*
Technical environment	Mainframe computers, proprietary software, proprietary data structure, mainly raster-based	Mainframe and mini computers,geo-relational data structure, graphical user interface (GUI), new data acquisition technologies (GAP, redefinition of datum, remote sensing)	Workstations and PCs, network/ internet, open systems design, multimedia, data integration, enterprise computing, object-relational data model
Major users	Government, universities and military	Government, universities, utilities, business, and military	Government, universities, schools, utilities, business, military, and the general public
Major application areas	Land and resource management, census, and surveying and mapping	Land and resource management, census, surveying and mapping, facilities management and market analysis	Land and resource management, census, surveying and mapping facilities management, market analysis, utilities, and geographic data browsing

in digital form, to *facilitate analyses of data* that demand interaction between statistical analyses and mapping, to *minimise* the *use* of the *printed map* as a data store and thereby to minimise the *effects of classification and generalisation* on the quality of the data, to create maps that are *difficult to make by hand* (e.g., 3D maps or stereoscopic maps), and to create maps in which *selection and generalisation* procedures are explicitly defined and consistently executed. The introduction of automation led to the review of the whole map-making process, which finally led to a good deal of savings and improvements. The early generations of the systems were developed and used mainly by government, agencies and universities, for very specific data management and research objectives. Application driven systems were used to run in mainframe computers with proprietary software using raster-based proprietary data structure.

The Period of Maturing Technology

The major milestone in GIS development during this period was the application of the *topological concepts* in representing geographic data by means of simple geo-relational data models and *vector-based* systems used to run on minicomputers using geo-relational data structure and graphical user interface. Examples are ArcInfo by ESRI, INFOMAP by Synercom (USA) and CARIS by Universal Systems Ltd (Canada). The ability to port GIS applications to a microcomputer platform in the late 1980s led to the development of MapInfo, SPANS, PC-ArcInfo, etc. Focus now shifted to the methods of data collection maintaining its quality and standards, data analysis and data organisation. GIS technology has now been further developed to integrate raster with vector data and also to integrate geographic data with other business data. This has led to the development of the concept of *open GIS* that is capable of sharing technical and data resources with one another.

There was a phenomenal growth of computer technology during the 1990s and GIS became multi-platform application that ran on different classes of computers as stand alone applications and also as time sharing systems. The increasing access to computers and the urgent need for effective geographic data management together pushed GIS to a new height. Applications quickly and increasingly extended to new and diversified fields, e.g., facility management, vehicle navigation, market research and decision support in business management. GIS applications were designed to meet the demands of corporate business goals and information requirements. Hence, the GIS industry proliferated as a specialised sector in the traditional computer industry. By the end of the 20th century, GIS became relatively matured both in terms of technology and application. Today all image processing software, contain modules of GIS for further geoprocessing, manipulation and analysis. Internet GIS and web cartography have become the order of the day.

The Age of GI Infrastructure

The concept of geographic information infrastructure has brought about a dramatic philosophical and technological revolution in the development of GIS since the mid-1990s. GIS is now a gateway for accessing and integrating geographic data from different sources, located both locally and globally, being increasingly used for interactive visualisation of scenarios resulting from different business decisions, as well as for the communication of spatial knowledge and intelligence among people all over the globe.

More recently GIS has popularised the use of geographic information among the ordinary individuals, corporate houses and government organisations as well. People use GIS to check the weather and traffic conditions, to locate the nearest ATM counter, to find information about a place whenever they intend to visit and so on. Business people rely on GIS to identify locations to set up new shops, new project sites, housing complexes, to determine the best route to deliver their goods and services. Government officials use it to manage land and natural resources, monitor the environment, formulate economic and community development strategies, enforce law and order, and deliver social services. Thus, GIS plays a very critical and important role in the daily life of an individual, business personnel and government officials. Hence, the term 'infrastructure' is used to describe what these systems really are today.

COMPONENTS OF GIS

Geographical information system is made up of *four* important components—*computer hardware*, *computer software*, *data* and *liveware*—that need to be in balance as one single perfectly synchronised unit, if the system is to function satisfactorily (Fig. 12.3).

Computer Hardware

It refers to the computer *components* that form the *physical framework* on which the GIS runs and on which manipulations and analyses are performed. It also includes the devices for storing data, displaying analyses, and creating output. A GIS can either be operated in a *stand alone* environment or in a *distributed* one consisting of a series of PCs or workstations connected by a network. In both the cases, the GIS consists of

Fig. 12.3 The Components of GIS

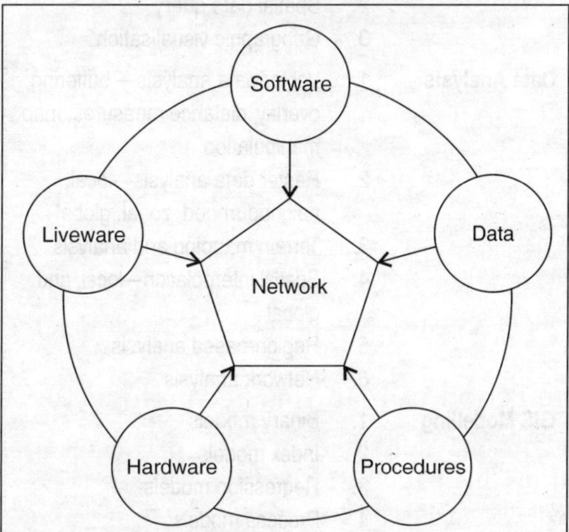

a central processing unit (CPU), a visual display unit (VDU or Monitor) and a Keyboard. The peripheral hardware components are digitiser, scanner, external data storage device (e.g., hard disk, floppy disk, CD-ROM, optical disk, flash drive and tape), hard copy output device, i.e., printer (e.g., dot-matrix, ink-jet, bubble-jet, laser, xerographic or optical) or plotter (e.g., analogue or incremental).

The CPU is linked to a disk drive storage unit which provides space for storing data and programs. The digitiser or the scanner converts data from maps and documents into digital form and sends it to the CPU. The plotter or printer presents the result of the data processing and a CD-ROM drive is used for storing data and programs as well as for communicating with other systems. An inter-computer communication can also take place via a networking system over special data lines or over telephone lines using a device called a modem. The user controls the computer and the peripherals via the VDU, and the keyboard, aided by a mouse, together known as the *terminal*.

GIS Activity Stream

Spatial Data Input	1.	Data entry: use of existing data and creating new data
	2.	Data editing
	3.	Projection and re-projection
	4.	Geometric transformation
Attribute Data Management	1.	Data entry
	2.	Database management
Data Display		Use of tables, graphs, diagrams, and maps
Data Exploration	1.	Attribute data query
	2.	Spatial data query
	3.	Geographic visualisation
Data Analysis	1.	Vector data analysis— buffering, overlay, distance measures, map manipulation
	2.	Raster data analysis—local, neighbourhood, zonal, global
	3.	Terrain mapping and analysis
	4.	Spatial interpolation—local, and global
	5.	Region-based analysis
	6.	Network analysis
GIS Modelling	1.	Binary models
	2.	Index models
	3.	Regression models
	4.	Process models

Computer Software

The software package for a GIS consists of five basic technical modules, which are subsystems for—*data input* and *verification*, *data storage* and *database management*, *data output* and *presentation*, *data transformation* and *interaction with the user* (Fig. 12.4). Some of the readily available softwares are IDRISI, GRASS, GRAM++, MapInfo, ArcView, ARC/INFO, ArcGIS, etc. Data input covers all aspects of transforming data captured in the form of existing maps, field observations, and sensors (including aerial photography, satellites and recording instrument) into a compatible digital format. A wide range of tools are available for this purpose, e.g., the interactive computer screen and mouse, the digitiser, the word processor and spreadsheet programmes, scanners (in satellites or aeroplanes for direct recording of data or for converting maps and photographic images) and devices necessary for reading data already written on magnetic

Fig. 12.4 The Main Software Components of GIS

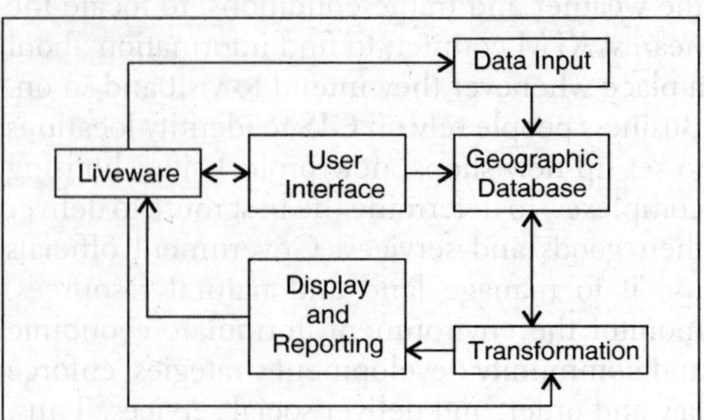

media such as tapes, disks or CD-ROMs. Data storage and database management concern the way in which data about the location, linkages (topology), and attributes of geographical elements (points, lines, areas, and ore complex entities representing objects on the earth's surface) are structured and organised, both with respect to the way they must be handled in the computer and how they are perceived by the users of the system. The computer program used to organise the database is called Database Management System (DBMS or RDBMS).

Data output and presentation concern the ways the data is displayed and how the results of the analysis are reported to the users. Data may be presented as maps, tables and figures (graphs, diagrams and charts) in a variety of ways ranging from the ephemeral image on the computer screen, through hard copy output drawn on printer or plotter to information recorded on magnetic media in digital form. Data transformation embraces two classes of operation, i.e., transformations needed to remove errors from the datasets or to bring them up-to-date or to match them to other datasets and the large array of analysis methods that may be applied to the data in order to achieve answers to the questions asked of the GIS. Transformations can operate on the spatial and non-spatial data either separately or in combination. Many of these transformations are so general that almost all GIS softwares contain this module. There are manipulations that are extremely application-specific and its inclusion in a GIS may only be to satisfy the particular user of that system. The user interacts with the GIS usually through the simplest menu driven commands with a mouse. There are occasions when users write their own programs (e.g., using PCRaster or MMS or any other language) to meet their specific needs. The user-system interaction is most important, as the user must ensure proper formulation according to an agreed set of rules, otherwise the results will be spurious.

Data

The universe of potential data for GIS is, in fact, infinite (Fig. 12.5). The creation and gaining access to reliable and accurately collected and maintained spatial data is very important. The most expensive part of a GIS is associated with data capture, indexing, processing and storage of digital spatial data. Hence, it is mandatory that the data must be multipurpose in nature, usable by many unrelated independent potential users. The data-providing community must maintain a standard for quality data so that the benefits of multiple use can be maximised. Metadata should be standardised and accompanied by data quality and data lineage information. Data exchange standards must be enhanced so that data can be transferred quickly and without loss between systems. As the whole process contains a fixed and measurable amount of potential error, it is crucial that a statement of data quality accompany any file of digital data. It would give

Fig. 12.5 The GIS Data Stream

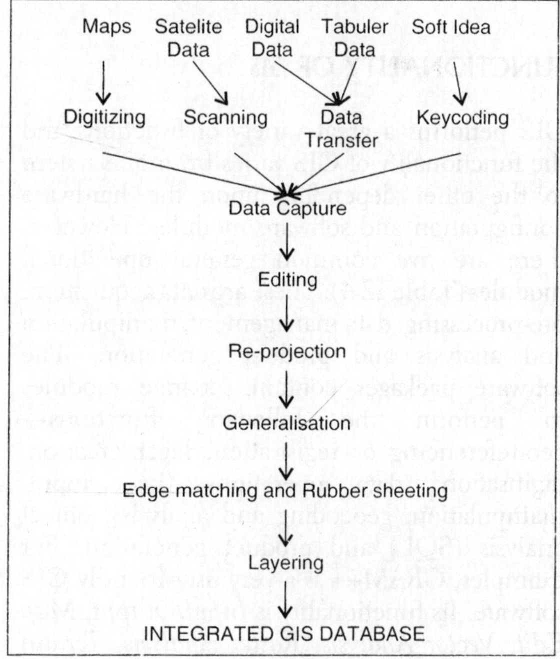

reliability, precision, accuracy, completeness and consistency of the data in the digital file.

Liveware

It refers to the operators or peoples who actually run the system. Liveware is the most expensive component as it constitutes specialised people. These people must be computer literate, data analysts (DA), GIS experts, and above all, a complete geographer with sufficient expertise in theories, models and methods of geography, cartography and remote sensing. Operators must understand the functions available in the GIS software. They must be well trained as digitiser operators, data collectors, data editors and taggers, internet data miners and relational spatial data managers who are able to understand the overall system to make efficient use of the modern GIS technology to solve the problems of humanity on a geographical frame. Thus will be born the community of GIS professionals, who are most wanted in the development and planning of a nation.

FUNCTIONALITY OF GIS

GIS performs a great variety of functions and the functionality of GIS varies from one system to the other depending upon the hardware configuration and software modules. However, there are *five* common general operational modules (Table 12.4). These are data acquisition, pre-processing, data management, manipulation and analysis and product generation. The software packages contain separate modules to perform the following functions—geo-referencing or registration, layer creation, digitisation, data operations (i.e., input, mainpulation, geocoding and analysis), object analysis (SQL) and product generation. For examples, *GRAM++* is a very user-friendly GIS software. Its functionality is *Input/Output, Map Edit, Vector Analysis, Raster Analysis, Terrain Analysis, Image Processing, Statistics, Geo-SQL,*

Table 12.4 Functionality of GIS

Functionality of GIS
Data Input
Data acquisition (both geographical and statistical) from cadastral maps, topomaps, aerial photographs, satellite imageries and field survey (ground, landuse and household)
Data transfer
Data verification
Data editing
Pre-processing
Format conversion, data reduction and generalisation, error detection and editing, merging of points into lines, and points and lines into polygons, edge matching, registration, interpolation and photo-interpretation and
Data Storage
Device: disk, magnetic tape, CD-ROM, and optical disk
Structure: raster and vector format
Data Management
DBMS and RDBMS
Manipulation and Analysis
Cartographic functions (scale changes, projection changes, and map embellishment with scale, title, north point and legend)
Data integration (map overlay, reclassification, spatial aggregation, and spatial transformation)
Feature measurement (number of features, and calculation of distance, area, volume and shape indices)
Spatial analysis—measurement of spatial parameters, spatial operations and modelling
Statistical analysis (descriptive statistics, cross-tabulation, correlation, and inference)
Product Generation
Data presentation
Hard copy (maps, graphs, tables, and text)
Virtual copy (maps, graphs, tables, and text)
Data transfer in various formats with easy interoperability

GramNet, *Raster Layout*, and *Vector Layout*. The operational steps for thematic mapping in GRAM++ are scanning the base map, registering the base map, creating layers, digitising layers, adding labels, permanently attaching labels, attaching data to a layer, performing analysis (raster, vector, terrain, tin, image, spatial statistics), and finally thematic mapping. *MapInfo* is also a very popular GIS software with the following functionalities—*registration of raster image, digitisation of raster image, editing of objects, adding data to the map, geocoding, table operations, SQL, object analysis, graphics,* and *thematic mapping*. *ArcView* is another very popular GIS software with a large hierarchy of operational modules and add-on extensions. It has the following functionalities: *view, creation of themes, digitation, editing of the layers, table creation and operations, SQL, overlay analysis, buffer analysis, network analysis, map layout,* and *thematic mapping* (see Appendix).

Data Acquisition

Data acquisition is simply the *process* of identifying and gathering data from various sources (like topographical maps, large scale maps, field survey, statistical abstracts, aerial photographs, satellite imageries, etc.) in a wide range of methods and formats (as raster and vector datasets of nominal, ordinal, interval and the ratio data in both analog and digital format). Apart from the existing datasets, there are times and circumstances when the user needs to develop his/her own dataset accurately through a precise sampling design. A GIS is of no use until the relevant data is identified and located. This phase is very important in terms of budgeting with respect to time and money.

Data Structure

The contents of a spatial database concerns the characteristic and meaningful entities of the earth's surface. Spatial objects are the delimited geographic areas with different kinds of associated attributes (e.g., point, line and polygon). A *point* is a spatial object with no area while a *line* is a spatial object constituted by the connected sequence of points. *Nodes* are special kinds of points, usually indicated by the junctions of lines or the ends of line segments. A *polygon* is a closed area (divided or undivided) made of different characteristics, while *chains* are special kinds of line segments corresponding to a portion of the bounding edges of a polygon. Actually, in each data plane encoding takes place as per the three geometric entities of points, lines or polygons. While designing a GIS, the choice of a particular spatial data structure is very important. It can affect both the data storage volume and the processing efficiency of the system. There are a variety of ways available to organise data in the GIS (Table 12.5 and Fig. 12.6).

Raster Data Structure (RDS)

In a raster (or cellular) structure, a value of the chosen parameter is developed *for every cell* in a *regular array over space* (Fig. 12.7 and Fig. 12.8). In simple raster data structure, the horizontal dimension along the rows is often oriented in an east-west direction. Along this, the raster elements, called *samples*, are numbered usually from the left and the columns are numbered from the top. Thus, the origin of the raster is the upper left corner. Although a simple rectangular RDS is a very popular approach, it

Table 12.5 GIS Data Structures

Raster (RDS)
Simple (SRDS)
Hierarchical (HRDS)
Vector (VDS)
Whole Polygon (WPVDS)
DIME (DVDS)
Arc-Node (ANVDS)
Relational (RVDS)
Digital Line Graph (DLGVDS)

Fig. 12.6 Approaches to Modelling the Real World

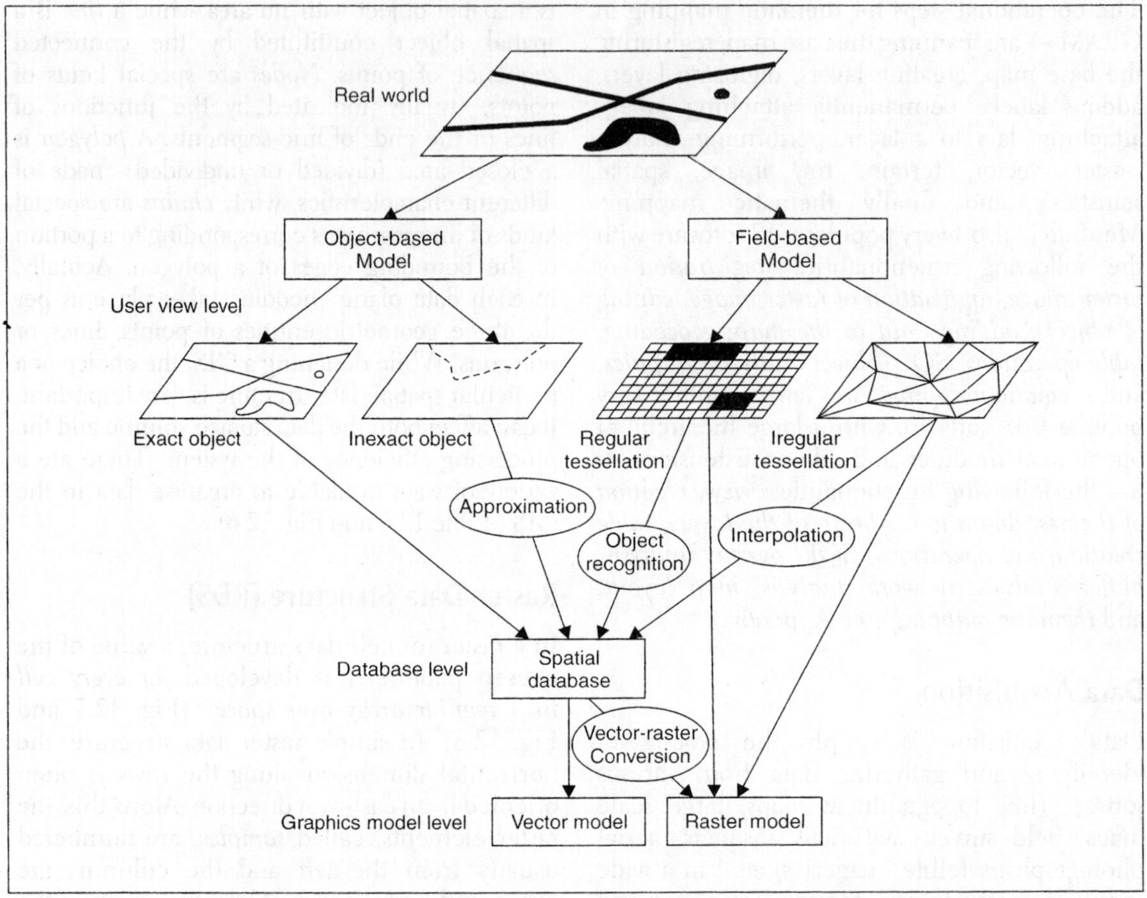

Real world
Object-based Model
Field-based Model
User view level
Exact object
Inexact object
Regular tessellation
Iregular tessellation
Approximation
Object recognition
Interpolation
Database level
Spatial database
Vector-raster Conversion
Graphics model level
Vector model
Raster model

Fig. 12.7 A Simple Raster Data Structure: (a) entity model (b) cell values (c) file structure

0 0 0 0 0 0 0 0 0 0
0 0 0 1 1 1 1 1 0 0
0 0 0 1 1 1 1 1 1 0
0 0 0 0 1 1 1 1 1 0
0 0 0 0 0 1 1 1 0 0
0 0 0 0 0 1 1 1 0 0
0 0 0 0 0 1 1 0 0 0
0 0 0 0 0 0 0 0 0 0
0 0 0 0 0 0 0 0 0 0
0 0 0 0 0 0 0 0 0 0
10,10,1
0,0,0,0,0,0,0,0,0,0
0,0,0,1,1,1,1,1,0,0
0,0,0,1,1,1,1,1,1,0
0,0,0,0,1,1,1,1,1,0
0,0,0,0,0,1,1,1,0,0
0,0,0,0,0,1,1,1,0,0
0,0,0,0,0,1,1,0,0,0
0,0,0,0,0,0,0,0,0,0
0,0,0,0,0,0,0,0,0,0
0,0,0,0,0,0,0,0,0,0
(a)
(b)
(c)

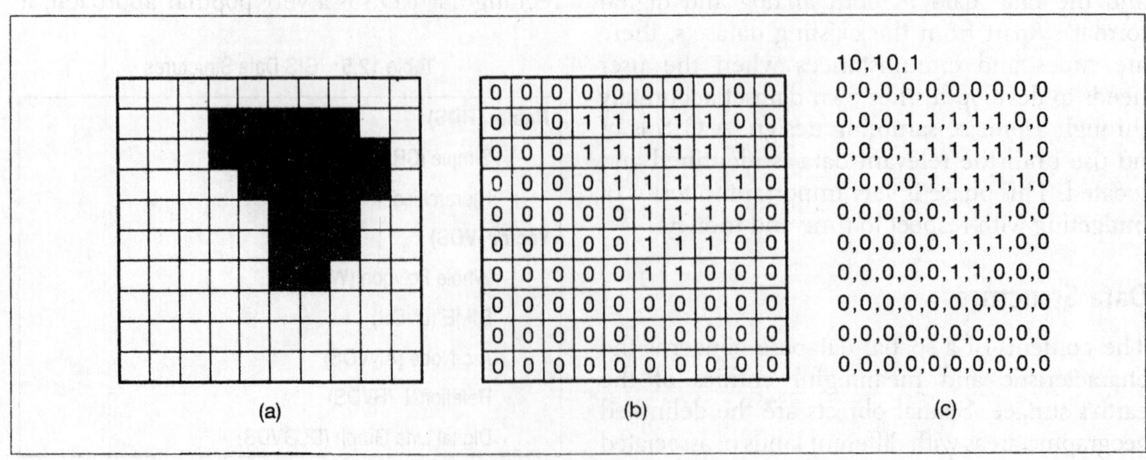

Fig. 12.8 Characteristics of Raster Data

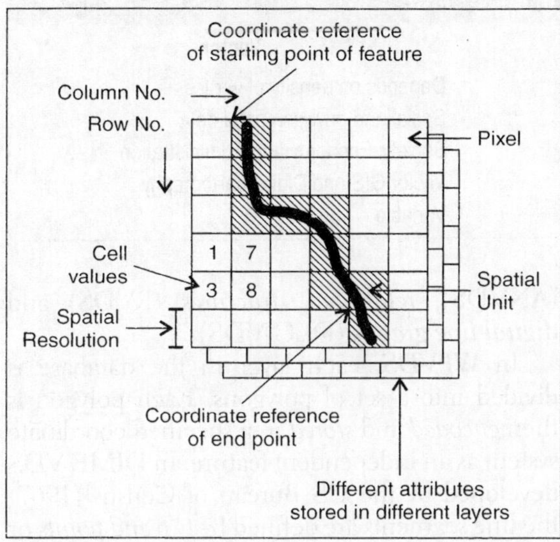

has two important limitations. *First,* there is a finite limit to specify locations. Samples are normally within one cell or another—there is nothing in between. *Second,* adjoining cells may not be evenly spaced. The size of the raster cell must be less than the size of the *resolution element* (or resel) in a dataset to obtain better performance. Geometrical figures shaping the raster cell and completely covering a flat surface are called *tessellations*. These may be *squares* or *triangles* or *hexagons*. The large datasets often need to be compressed. The two widely used algorithms for such purposes are *run-length encoding* and *chain codes*.

In hierarchical RDS, information is stored in several interrelated layers. In this, spatial resolution decreases at each higher layer due to spatial averaging. This continues till the highest level where the single *pixel* (i.e., picture element) represents the numerical average of all the data in the original first layer. It is also called the *pyramidal data structure*. If each higher level layer has pixels that are exactly twice as wide as the previous (and four times the area), a *quadtree data structure* (QHRDS) is formed. A common modification of QHRDS is the *maximum block representation* where all the redundant information in the tree is eliminated by leaving the nodes out.

Vector Data Structure (VDS)

Unlike raster data structures (which are based on the decomposition of the plane), vector data structures are based on *elemental points* whose locations on the plane are known. VDS is most suitable as many computer graphics and computer-aided-designs (CAD) use vector-like-models using combinations of points, lines and polygons (Males 1977, Peuquet 1977). In GIS

Fig. 12.9 The Quadtree Data Structure

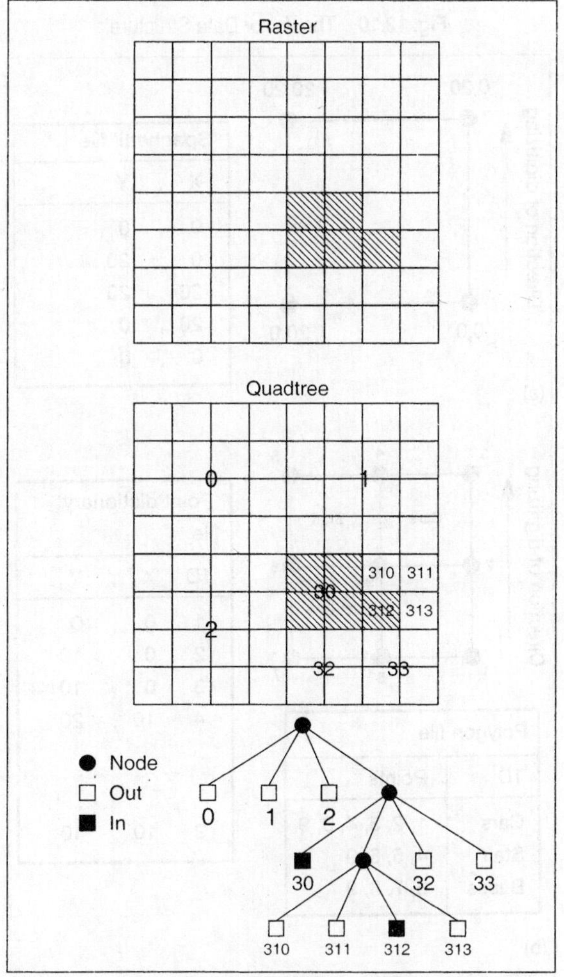

Comparison of Raster and Vector Data

Issue	Raster	Vector
Volume of data	Depends on cell size	Depends on density of vertices
Sources of data	Remote sensing and Imagery	Social and environmental data
Applications	Resources and Environment	Society, Economy and Administration
Software	Raster GIS and Image processing	Vector GIS and Digital cartography
Resolution	Fixed	Variable

there are five commonly used VDS. These are *whole polygon structures* (WPVDS) (Fig.12.9 and Fig. 12.10), *dual independent map encoding file structures* (DIME VDS), *arc-node structures* (ANVDS), *relational structures* (RVDS) and *digital line graphs* (DLG-VDS).

In WPVDS, each layer in the database is divided into a set of polygons. Each polygon is then *encoded* and *stored* in a specified coordinate system as an independent feature. In DIME VDS developed by the US Bureau of Census (1967) the line segments are defined by *two end points* or *nodes shared by adjacent polygon units* and stored with components like segment names, node identifiers and polygon identifiers. In *ANVDS*, the objects in the database are hierarchically structured. In this, the points are the *elemental basic components*, the arcs are *individual line segments* and the nodes are the *intersections between the arcs or the ends of the arcs*. Essentially polygons are areas completely bound by a set of

Fig. 12.10 The Vector Data Structure

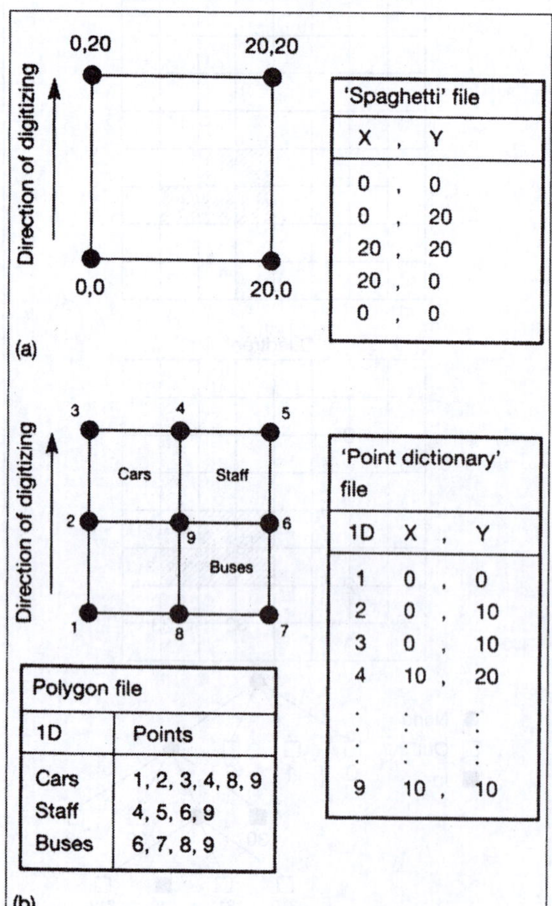

Merits and Limitations of Raster and Vector Data

Raster Model

Merits: Simple data model, Use of cheap technology, Ease of data collection and Ease of data processing

Limitations: No topological processing, Limited attribute data handling, Less compact data structure and Low cartographic output quality

Vector Model

Merits: More compact data structure, Topological processing, High cartographic quality and Sophisticated attribute data handling

Limitations: Complex data model, Difficult overlay processing, Difficult presentation of spatial variability, Expensive data collection and Use of expensive technology

arcs. Unlike WPVDS, the points are stored only once. RVDS is a form of ANVDS in that the attribute values are stored separately in relational tables. The data contents in a *DLG VDS* (developed by the USGS) are subdivided into different thematic layers, i.e., *1st layer*—boundary information, *2nd layer*—hydrographic features, *3rd layer*—transport network, *4th layer*—land survey (Allder and Elasal 1984). The point, line and polygon elements provide the information regarding topology and location. Besides, there is an elaborate system of coding attribute information for the above elements.

Pre-processing

Pre-processing involves *manipulation* of data for permanent storage in the GIS database. The ultimate product is a coordinated set of thematic data layers. The essential tasks are—*format conversion, data reduction and generalisation, error detection and editing, merging of points into lines and points and lines into polygons, edge matching, registration, interpolation* and *photo interpretation*.

Format conversion is done in terms of *data structure* (DS) and *data media*. The former involves modification of one DS into another, which is done by employing BSQ (band sequential), BIP (band interleaved by pixel) or BIL (band interleaved by line) techniques (Fig. 12.5, Fig. 12.6 and Fig. 12.7). In BSQ, the data of each variable is kept in a separate file while in BIP all measurements from a single pixel are kept together. In BIL, the adjacent ground locations for a single theme lie adjacent in the datafile and the subsequent themes are sequentially recorded for the same line. As most of the spatial data is not available in computer compatible formats, a medium of conversion is needed. It is very expensive and time consuming and is done with a digitising tablet either in point mode or in stream mode.

Data reduction involves *changes of scale* in spatial data and data generalisation means *developing a less precise representation* from the original source data. It is achieved by employing a numerical algorithm or by thinning, averaging and resampling. *Error detection* and *editing* are generally done with a workable computer software. *Merging* involves the process of building more complex objects from the elemental points in a systematic way. When creating digital database from maps, *edge matching* is

Table 12.6 Raster-based Geographic Data-Processing Techniques

	Local Operations	*Neighbourhood Operations*	*Extended Neighbourhood Operations*	*Regional Operations*
Logical Operations	Reclassification			
Arithmetic Operations	Reclassification	Aggregation Filtering	Statistical analysis	
Overlay Operations	Logical Arithmetic			Category-wise overlay
Geometric Property Operations		Slope and aspects	Distance, proximity and connectivity	Area Perimeter Shape
Geometric Transformation Operations			Rotation Translation Scaling	
Geometric Derivation Operation			Buffering Viewshed analysis	Identification and reclassification

Table 12.7 Vector-based Geographic Data-Processing Techniques

	Non-topological	Topological	
		Feature-based	Layer-based
Logical Operations	Attribute database query		Reclassification and aggregation
Arithmetic Operations	Change mapping Summary statistics		
Overlay Operations Geometric Property Operations	Address geocoding Calculation of area, perimeter and distances	Overlay analysis	Overlay analysis Network analysis
Geometric Transformation Operations	Coordinate and geometric transformation Surface interpolation		
Geometric Derivation Operation		Buffering	

commonly done by digitising or scanning each sheet separately. The analyst, using cartographic license, manually adjusts vectors that cross the boundary. *Rectification* involves manipulating the raw dataset by a specific coordinate system to delineate the spatial arrangement of objects in the data. Registration, on the other hand, is simply the changing of one of the views of surface spatial relationships to agree with the other. These are usually done by *methods* like ground control point rectification (GCPR), rubber sheeting operation (RSO), nearest-neighbour interpolation (NNI), bilinear interpolation (BI), cubic interpolation (CI) and the rotation, translation and scaling of coordinates (RTS).

Geographic data is often taken at irregular intervals in space. Therefore, values for locations where measurements are not available need to be estimated. This is done by *interpolating values* employing models explaining best-fit statistical relationships between data points (weighing functions or trend surfaces) (Nagy and Wagle 1979). *Photo-interpretation* is basically the process of extracting useful information from aerial photographs or satellite imageries which is done either manually by applying the principle of visual image processing or digitally by using sophisticated image processing software.

Data Management

A database may be thought of as a collection of logically linked datafiles. A database management system (or DBMS) is the *software* (developed by ASHTON-TATE) that allows the user to work efficiently with the data. It provides the means for creating datafiles, specifying field name(s), field type and field width, sorting them, setting relations between them, deleting old data, editing the data, adding new data and virtually everything the user would want to do with ease and confidence. There are *two* basic approaches in DBM technology, namely, conventional DBMS and spatial DBMS. The first one is again divided into *two* systems—relational DBMS and navigational DBMS. Users of GIS must have a clear understanding of the system's internal structure and capabilites to work with the dBase files.

Manipulation and Analysis

Manipulation and analysis is often called *geoprocessing*. It is *concerned* with the analytical operators that process the database contents with different algorithms to derive information as per the user's need and choice. *Manipulation* involves reclassification and aggregation (Fig. 12.11

Fig. 12.11 Attribute Generation: (a) Species Map (b) Species Class (c) Elimination of Redundant Boundary

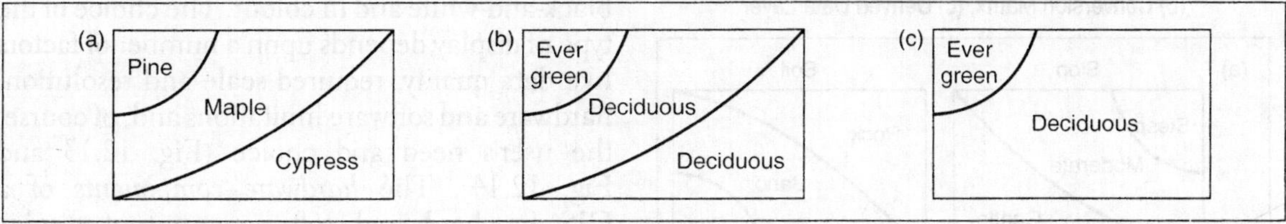

Table 12.8 Vector-based Output Functions in GIS

Method	*Features*	*Advantages*	*Disadvantages*
Graduated symbols	Locations Lines Areas	Association between size of symbols and magnitude of spatial features Easy to show spatial distribution	Difficult to read if too many features are present
Graduated colours	Continuous phenomena Areas	Easy to recognise spatial patterns	Difficult to associate colours with values they represent
Charts	Locations Areas	Easy to show magnitude and categories of spatial features	Difficult to identify spatial patterns and distribution of spatial features
Contours	Continuous phenomena	Easy to show rate of change and trend across an area	Difficult to identify spatial patterns and distribution of spatial features
3D perspective views	Continuous phenomena Locations Areas	Easy to show rate of change Association between 3D values to quantities that they represent	Difficult to read values of individual features Difficult to construct

Table 12.9 Symbols for Mapping Spatial Objects

Spatial object type	Attribute		
	Nominal	Ordinal	Interval / Ratio
Point (0-D)	Symbol map (each category a different class of symbol—colour, shape, orientation), and / or use of lettering: e.g., presence/absence of settlement	Hierarchy of symbols or lettering (colour and size): e.g., small/medium/large settlements	Graduated symbols (colour and size): e.g., pollution concentrations
Line (1-D)	Network connectivity map (colour, shape, orientation): e.g., presence/ absence of connection	Graduated line symbology (colour and size): e.g., classes of roads	Flow map with width or colour lines proportional to flows (colour and size): e.g., traffic flows
Area (2-D)	Unique category map (colour, shape, orientation, pattern): e.g., soil/forest types	Graduated colour or shading map: e.g., rice yield-low/medium/high	Continuous hue/shading: e.g., dot-density or choropleth map
Surface (2.5-D)	One colour per category (colour, shape, orientation, pattern): e.g., relief classes-mountain/valley etc.	Ordered colour map: e.g., areas of gentle/ steep/very steep slopes	Contour map: e.g., isobars / isohyets

Fig. 12.12 Overlay Operation: (a) Input Map Data, (b) Conversion Matrix, (c) Derived Data Layer

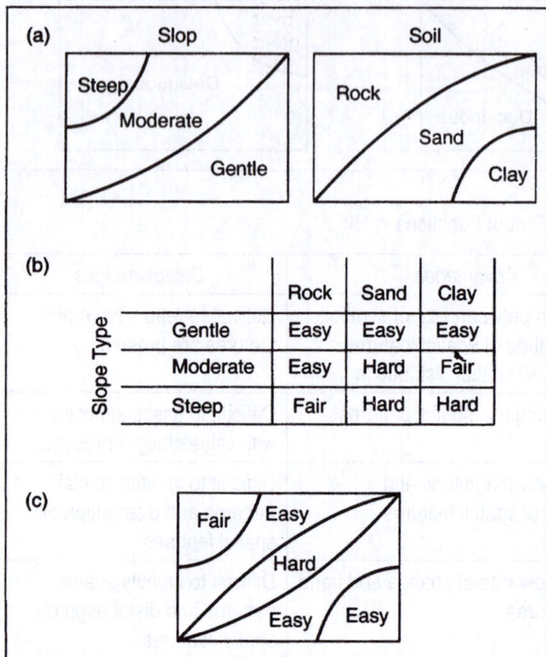

and Fig. 12.12), geometric operations, centroid determination and data structure conversion. The *first one* is done by attribute operation and spatial aggregation while the *second one* by rotation, translation and scaling (RTS), rectification and registration. The average location of a line or polygon is described by the centroid that can be determined by employing several mathematical techniques. The conversion of data structure may then be done, if necessary. The analysis *involves* spatial operations, measurements, statistical analysis and modelling. The *first one* is done by connectivity and neighbourhood operations while the *second one* by measuring distance and direction. The statistical techniques (descriptive, correlational and inferential) of various types are then applied to extract and test ideas for modelling.

Product Generation

A GIS programme provides facilities to display the derived information in the form of maps, graphs and tables on a variety of output media both in black-and-white and in colour. The choice of the type of display depends upon a number of factors like data quality, required scale and resolution, hardware and software limitations and, of course, the user's need and choice (Fig. 12.13 and Fig. 12.14). The *hardware components* of a GIS for hard and soft copy output display are character and graphic printers, pen and electrostatic plotters, colour and grey scale CRT, film recorders and computer output microfilm devices.

With the help of computer cartography, high quality maps of various kinds may be created to show spatial objects and the spatial relationships between and among objects. These are *thematic maps* (showing the spatial variation of a single phenomenon or the relationship between several phenomena), *choropleth maps* (showing the relative magnitude of continuous variables as they occur within the boundary of an unit area), *proxymal* or *dasymetric maps* (showing the location and magnitude of areas

Fig. 12.13 DEM Generated by GIS

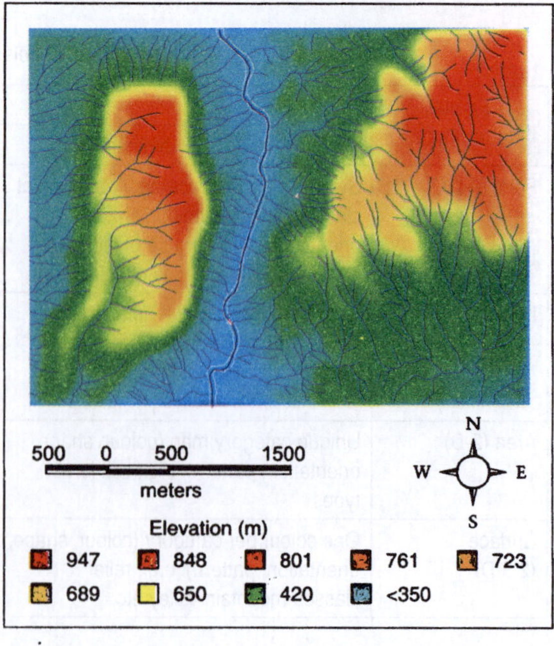

Fig. 12.14 (a) Conventional Model of Cartographic Communication (b) New Model of GIS Information Communication

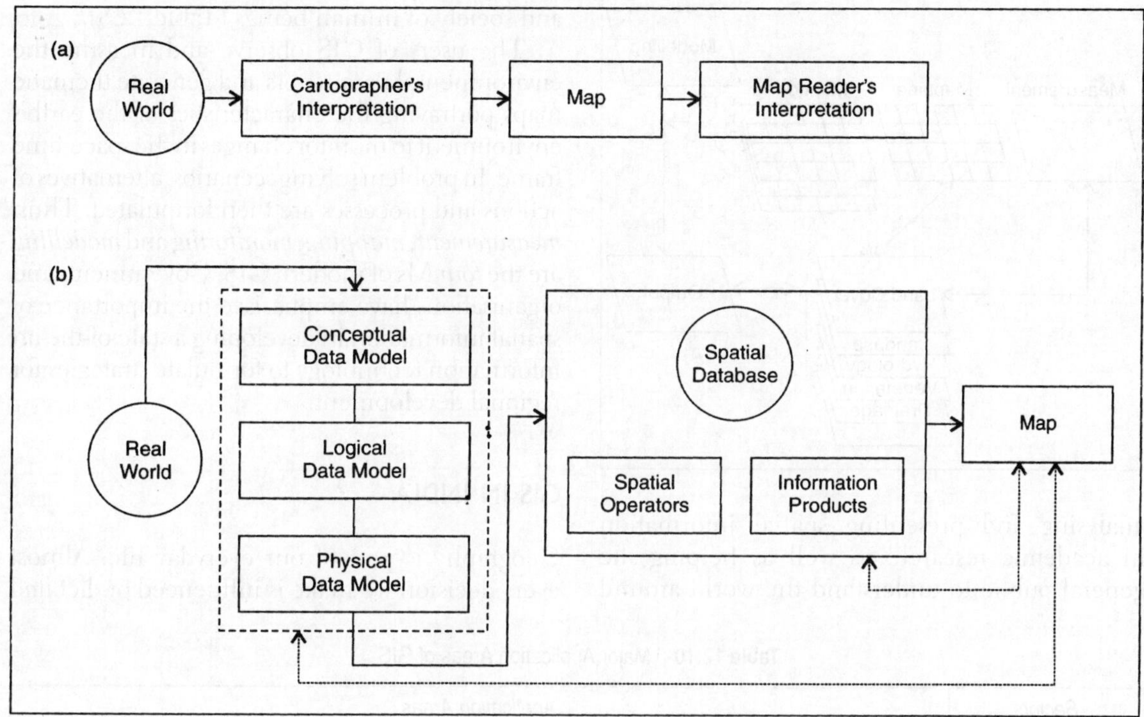

with relative uniformity), *contours or isarithmic maps* (showing quantities by lines of equal value and thus reflecting gradients), *dot and symbol maps* (showing spatial distribution by varying numbers of uniform dots and varying sizes of geometric figures), *line maps* (showing the direction and magnitude of flow or potential), *landform maps* (showing the earth's surface from an oblique aerial perspective), *cartograms* (showing the distribution by the topological distortion of size of the unit area) and *animated maps* (showing the sequences through time by movie loops). GIS also has the facility to create a wide range of computer graphics to represent the information, such as, bar diagrams, pie diagrams, scatter graphs and almost all types of graphs and diagrams. Tabular information of processed data can also be extracted from the DBMS.

APPLICATION OF GIS

A mountain of research regarding the technical development of GIS during the last few decades has made it a worldwide phenomenon. Since 1986, hundreds of books have been published on various aspects of GIS, hundreds of conferences have been held across the globe, and there appeared quite a significant number of important scientific and trade journals devoted entirely to the design, technology, use and management of GIS. Till date, it is a rough estimation that these technical solutions have been installed at about a lakh sites across the globe. Although, the USA, Canada, the European Countries and Australia dominate the user scenario, it is attracting quite a sizeable number of countries each year. GIS is now an essential software tool for delivering government services, making business decisions,

Fig. 12.15 The Four Ms of GIS

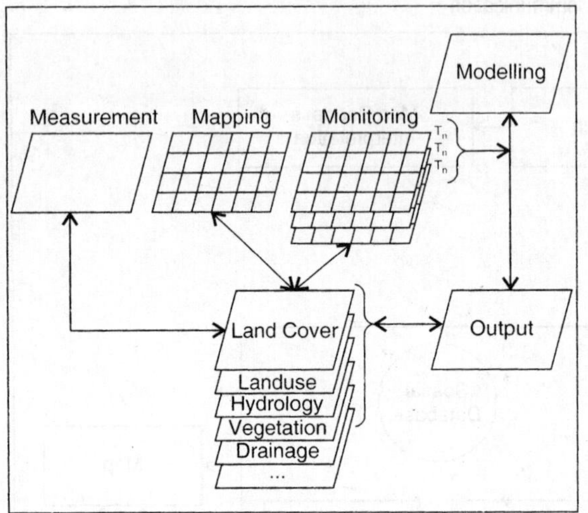

analysing and presenting spatial information in academic research, as well as helping the general public to understand the world around them. It is increasingly used in many different fields directly related to the habitat, economy and society of human beings (Table 12.8).

The users of GIS observe and measure the environmental parameters and generate thematic maps portraying the characteristics of the earth's environment to monitor changes in the space-time frame. In problem solving scenarios, alternatives of actions and processes are then formulated. Thus, *measurement*, *mapping*, *monitoring* and *modelling* are the *four* Ms of modern GIS. Governments and organisations have emphasised the importance of spatial information in developing a state-of-the-art information technology to formulate strategies for regional development.

GIS IN INDIA

Geography is part of our everyday life. Almost every decision we make is influenced or dictated

Table 12.10 Major Application Areas of GIS

Sectors	*Application Areas*
Academic	Research in humanities, science and engineering; primary and secondary schools (school district delineation, facilities management, bus routing, spatial digital libraries).
Business	Banking and insurance; real estate (development project planning and management, sale and renting services, building management); real and market analysis; delivery of goods and services; market share analysis; insurance; retail site selection; direct marketing.
Government	Union government (national topographic mapping, resource and environmental management, weather services, public land management, population census, election and voting); state government (surveying and mapping, land and resource management, highway planning and management); local government (social and community development, land registration and property assessment, water and waste water services); public safety and law enforcement (crime analysis, deployment of human resources, community policing, emergency planning and management, health care); international development and humanitarian relief
Industry	Engineering (surveying and mapping, site and landscape development, pavement management); transportation (route selection for goods delivery, public transit, vehicle tracking); utilities and communications (electricity and gas distribution, pipelines, telecommunication networks); forestry (forest resource inventory, harvest planning, wildlife management and conservation); mining and mineral exploration; systems consulting and integration
Military	Training; command and control; intelligence gathering; target site identification; Tactical support planning; intelligence data integration; mobile command modelling
Environmental management	Landfill site selection; waste disposal system; pollution monitoring; resource management; natural hazard management; environmental impact assessment

Future of GIS

Maps have traditionally been used to explore the Earth. GIS technology has enhanced the efficiency and analytical power of traditional cartography. As the scientific community recognises the environmental consequences of human activity, GIS technology is becoming an essential tool in the effort to understand the *processes of global change*. Maps and satellite information sources can be combined in models to simulate the interactions of the complex natural systems. Through a process known as *visualisation*, a GIS can be used to produce images—not just maps, but drawings, animations, and other cartographic products. These images allow researchers to view their subjects in ways that they never could before. A GIS can display the Earth in realistic 3D perspective views and animations in real time environment that convey information more effectively and to wider audiences than the traditional static 2D maps. Today, through internet map server technology, spatial data can be accessed and analysed over the internet anytime, anywhere. At the present moment the whole world is amazingly applying the smart Mobile GIS with more intelligent components.

Together with cartography, remote sensing, global positioning systems, photogrammetry and geography, GIS has evolved into a discipline with its own research base, known as *geographic information sciences*. An active GIS market has resulted in lower costs and continual improvements in GIS hardware, software, and data. These developments have led to a much wider application of the technology throughout government, business, and industry. GIS and related technology help analyse large datasets, allowing a better understanding of terrestrial processes and human activities to improve economic vitality and environmental quality.

Often, GIS takes the centre stage and geography loses. Actually, many GIS users have no knowledge of geography. With a few clicks of a button, a GIS jockey can easily describe a 'data distribution' but cannot explain why things are the way they are. If things go on like this, it would be a disaster for both geography and GIS. A split may seem imminent to many and definitely it would hurt geography. Many departments receive a large share of their funding due to the work they do with GIS. Geographers need GIS to survive in the existing global socio-econo-political scenario and GIS needs geography to give it a special meaning that no other subject can.

by some fact of geography. Hence GIS is rapidly growing in popularity. Over the past couple of years, India has also entered the GIS scene. Till date, thousands of copies of GIS packages like MapInfo, ArcView, ARC/INFO, ArcGis have been sold in India. The Space Application Centre has developed ISRO-GIS, the Indian Institute of Technology has developed GRAM, the Indian Space Research Organisation (ISRO) has developed GEOSPACE and the USA has developed GRASS. Training programmes both in government and private institutions or organisations have been introduced. The community of Indian geographers is also showing keen interest in GIS as an essential tool of research.

In spite of all these, the development of GIS in India is not very bright. The main reason behind this is that GIS packages are prohibitively expensive, there is a lack of GIS infrastructure (computers, printers, plotters, satellite imageries, etc.) at the village or block level, a general lack of trained personnel, an unavailability of umbrella organisations supervising GIS projects in India, a lack of coordination between different departments or organisations which produce thematic and other maps at entirely different scales, a non-availability of digitised data, maps or imageries of border regions, an obsession about the secrecy of data in both government and semi-government institutions or departments or organisations, and a lack of adequate funds in research. However, there is absolutely no doubt that GIS is a crucial decision making tool in this age of technological civilisation and its application and use are growing steadily.

EXERCISE

1. Define the terms data and information.
2. Distinguish between data and information.
3. Classify data with examples.
4. Discuss the characteristics of geographical data.
5. What is an information system?
6. What are the components of an information system?
7. Classify information systems with examples.
8. Explain the characteristics of GIS as an information system.
9. Critically discuss the toolbox-based definition of GIS.
10. Define GIS as a database system.
11. Discuss the organisation-based definitions of GIS.
12. What do geomatics, geomatic engineering and geo-informatics mean?
13. Explain the scope and content of GIScience after Goodchild (1992).
14. Explain GIS as a field of academic study.
15. Explain GIS as a branch of information technology.
16. Explain GIS as a data institution.
17. Discuss the nature of GIS.
18. Analyse the technical and operational characteristics of a modern GIS.
19. Trace the evolution of the concept of GIS in detail.
20. Analyse the sector-wise major application areas of GIS.
21. Explain: GIS expanded at a tremendous pace as digital cartography developed rapidly.
22. What are the components of GIS?
23. Discuss the role of computer software as a component of GIS.
24. Discuss the role of liveware as a component of GIS.
25. Discuss the significance of data as a component of GIS.
26. Discuss the role of computer hardware as a component of GIS.
27. Explain the functionality of GIS.
28. Explain how data acquisition takes place in a GIS.
29. What is raster data? Discuss its various types with illustrations.
30. What is vector data? Explain how vector data can be filed in various structures.
31. Distinguish between raster data and vector data.
32. Compare the characteristics of raster data and vector data.
33. What is pre-processing? Explain how it is performed.
34. Explain in detail what database management means.
35. What are the essential tasks of spatial data manipulation in a GIS?
36. Explain how spatial data analysis is performed in a GIS?
37. Discuss in detail the various forms of product generation in a GIS.
38. What are the active domains of GIS?
39. Give the main producers and sources of geographical data.
40. What are the main types of geographical data?
41. Specify some important current applications of GIS.
42. Give the names of the currently available popular GIS softwares with versions and makers.
43. Project: With suitable geographical data, map, photograph and imagery, go to all the modules and perform all the operations given in the menu bar and taskbar options of any available GIS software, e.g., Gram++, IDRISI, MapInfo, ArcView, ARC/INFO, ArcGis, etc.
44. Explain: India is certainly lagging behind in GIS applications.

45. Explain the terms: Cadastre, Coordinate System, API, ASCII, Address, Data Transformation, Azimuth, Buffer, Database design, CORBA, DDL, DML, Data Mining, DDE, DEM, DTM, DSS, Data Model, Data Warehouse, Data Exchange Standard, QHRDS, ANVDS, RVDS, DLGVDS, DIGEST, DCW, DTM, DIME, DLG, DRG, 4GL, FTP, GPS, GUI, Image Map, ISDN, Krigging, Metadata, TIN, OLE, Open GIS, Reclassification, Resampling, Scanning, Digitisation, TIGER, Tessellation and WYSIWYG.
46. Explain the four Ms of a GIS.

Appendix

1.0 The Seven Base SI Units

Quantity	Symbol	Unit	Quantity	Symbol	Unit
Mass	kg	Kilogram	Temperature	K	Kelvin
Length	m	Metre	Luminous Intensity	Cd	Candela
Time	s	Second	Amount of Substance	mol	Mole
Electric Current	A	Ampere			

2.0 Number Terms—Great and Small

Prefix	Symbol	Meaning	Prefix	Symbol	Meaning
Tera	T	1 million million ($=10^{12}$)	Deci	d	1 – tenth ($=10^{-1}$)
Giga	G	1 thousand million ($=10^{9}$)	Centi	c	1–hundredth ($=10^{-2}$)
Mega	M	1 million ($=10^{6}$)	Milli	m	1–thousandth ($=10^{-3}$)
kilo	K	1 thousand ($=10^{3}$)	Micro	μ	1–millionth ($=10^{-6}$)
Hecto	H	1 hundred ($=10^{2}$)	Nano	n	1–thousand millionth ($=10^{-9}$)

3.0 Units Used in Remote Sensing

Length

Unit	Distance	
Kilometre	= 1000 m	$= 10^{3}$ m
Metre (m)	= 1.0 m	
Centimetre (cm)	= 0.01 m	$= 10^{-2}$ m
Millimetre (mm)	= 0.001m	$= 10^{-3}$ m
Micrometer (ìm)#	= 0.000001 m	$= 10^{-6}$ m (micrometer was formerly called 'micron')
Nanometer (nm)		$= 10^{-9}$ m
Angstrom Unit (A)		$= 10^{-10}$ m

Frequency (cycles/second)

Hertz (Hz)	1	Kilohertz (kHz)	10^{3} (=1000)
Megahertz (MHz)	10^{6} (=1000000)	Gigahertz (GHz)	10^{9} (=1000000000)

Computer Storage

Bit	= A binary digit (0 or 1)	Byte	= 8 bits, 1 character
Kilobyte (KB)	$= 2^{10}$ (=1024) bytes	Megabyte (MB)	$= 2^{20}$ bytes
Gigabyte (GB)	$= 2^{30}$ bytes	Terabyte (TB)	$= 2^{40}$ bytes

4.0 Official Shapes of Ellipsoids

Name	Date	Equatorial Radius (m)	Polar Radius (m)	Polar Flattening
WGS 84	1984	6,378,137	6,356,752.3	1/298.257
WGS 72	1972	6,378,135	6,356,750.5	1/298.26
AUSTRALIAN	1965	6,378,160	6,356,774.7	1/298.25
KRASOVSKY	1940	6,378,245	6,356,863	1/298.3
INTERNATIONAL	1924	6,378,388	6,356,911.9	1/297
CLARKE	1880	6,378,249.1	6,356,514.9	1/293.46
CLARKE	1866	6,378,206.4	6,356,583.8	1/294.98
BESSEL	1841	6,377,397.2	6,356,079.0	1/299.15
AIRY	1830	6,377,563.4	6,356,256.9	1/299.32
EVEREST	1830	6,377,276.3	6,356,075.4	1/300.8

5.0 Measurements on Ellipsoids

Length of a Degree of Geodetic Latitude on the WGS 84 Ellipsoid			Length of a Degree of Longitude on the WGS 84 Ellipsoid		
Latitude	Kilometres	Miles	Latitude	Kilometres	Miles
0°	110.57	68.71	0°	111.32	69.17
10°	110.61	68.73	10°	109.64	68.13
20°	110.70	68.79	20°	104.65	65.03
30°	110.85	68.88	30°	96.49	59.95
40°	111.04	68.99	40°	85.39	53.06
50°	111.23	69.12	50°	71.70	44.55
60°	111.41	69.23	60°	55.80	34.67
70°	111.56	69.32	70°	38.19	23.73
80°	111.66	69.38	80°	19.39	12.05
90°	111.69	69.40	90°	0	0

6.0 Conversion Factors

	Inch	Feet	Yards	Chains	Furlongs	Miles	Metres
1 Inch (inch)	1	0.0833	0.02778				0.0254
1 Foot (ft)	12	1	0.33				0.3048
1 Yard (yd)	36	3	1				0.9144
1 Chain (ch)	792	66	22	1	0.1	0.0125	20.1168
1 Furlong	7920	660	220	10	1	0.125	201.168
1 Mile	63360	5280	1760	80	8	1	1609.34
1 Fathom	72	6	2				1.8288
1 Cable	8640	720	240				219.456
1 Nautical Mile (NM)	72913.32	6076.11	2025.37				1853.18
1 Metre (m)	39.370078	3.28084	1.09361	0.0497097	0.0049709	0.0006213	1
1 Kilometre (km)	39370.078	3280.84	1093.6130	49.70968	4.970968	0.621371	1000

7.0 Metric Conversion Factors

Length

1 mile	= 1.60934 km	1 km	= 0.621371 mile	1 km	= 10 hectometre (hm)
1 furlong	= 0.201168 km	1 hm	= 10 decametre (dm)	1 dm	= 10 m
1 chain	= 20.1168 m	1 chain	= 4 poles	1 chain	= 66 ft
1 chain	= 100 links	1 link	= 0.201168 m	1 m	= 10 decimetre (dcm)
1 yd	= 0.9144 m	1 yd	= 3 f	1 m	= 1.09361 yd
1 ft	= 0.3048 m	1 ft	= 12 inch	1 cm	= 10 mm
1 inch	= 2.54 cm	1 cm	= 0.393701 inch	1 dcm	= 10 cm
1 fathom	= 1.8288 m	1 fathom	= 6 ft		

Area

1 sq mile	= 2.58999 sq km	1 sq mile	= 640 acres (ac)
1 sq km	= 0.386 sq mile	1 sq km	= 247.105 acres (ac)
1 acre	= 4840 sq yd	1 acre	= 4046.86 sq m
1 acre	= 4 roods	1 rood	= 40 sq rod
1 rood	= 1011.71 sq m	1 acre	= 10 sq chain
1 sq m	= 1.19599 sq yd	1 sq rod	= 30.25 sq yd
1 sq yd	= 0.836127 sq m	1 sq yd	= 9 sq ft
1 sq ft	= 144 sq inch	1 sq ft	= 0.092903 sq m
1 sq chain	= 484 sq yd	1 ha	= 10000 sq m = 11960 sq yd
1 sq inch	= 6.4516 sq cm	1 sq cm	= 0.15500 sq in
1 sq cm	= 100 sq mm	1 sq dcm	= 100 sq cm
1 sq m	= 100 sq dcm	1 are (a)	= 100 sq m
1 hectare (ha)	= 100 ares (a)	1 sq km	= 100 ha

Volume

1 cubic yard (cu yd.)	= 0.764555 cu m	1 cu m	= 1.30795 cu yd
1 cubic feet (cft)	= 0.0283168 cu m	1 cu m	= 35.3147 cft
1 cubic inch (cu inch)	= 16.3871 cc (cubic cm)	1 cc	= 0.0610237 cu inch
1 gallon (gal)	= 0.00454609 cu m	1 cc	= 1 ml (millilitre)
1 gal	= 4.54609 litre	1 litre	= 1000 ml
1 litre	= 0.2200 gal		

Velocity

1 mile/hr (mph)	= 1.60934 km/hr (kmph)	1 kmph	= 0.621371 mph
1 ft/sec (ft/s)	= 0.3048 m/sec (m/s)	1 m/s	= 3.28084 ft/s

Mass

1 ton / 1 metric ton	= 10 quintals	1 lb	= 0.45359237 kg
1 kg	= 2.20462 lb	1 ton / 1 metric ton kg	= 1000 kg (kilogram)
1gm	= 1000 milligram (mg)	1kg	= 1000 gm (gram)
1 ton	= 20 cwt (hundred weight)	1 cwt	= 8 stones
1 stone	= 14 lb (pound)	1 lb	= 16 oz (ounce)

Force

1 lbf	= 4.44822 N (Newton)	1 N	= 0.224809 lbf
1 kgf	= 9.80665 N	1 kgf	= 2.20462 lbf

Pressure

1 lbf/ft^2	= 47.8803 N/m^2	1 N/m^2	= 0.000145038 lbf/inch^2
1 lbf/inch^2	= 6894.76 N/m^2	1 kgf/cm^2	= 98.0665 kN/m^2
1 kgf/m^2	= 9.80665 N/m^2		

1 inch of Mercury column	= 33.86395 millibar of pressure (mb)
1 cm of Mercury column	= 13.33171 millibar of pressure (mb)
1 bar	= 1000 bar
1 millibar (mb)	= 1000 dynes / sq cm
Normal Sea Level Atmospheric Pressure	= 29.92 inches of Mercury column
	= 760 millimetres of Mercury column
	= 1013.2 millibar
	= 1.0132 bar

Temperature

Temperature in °C (Celsius)	= (°F – 32) / 1.8
Temperature in °F (Fahrenheit)	= (°C × 1.8) + 32
Temperature in °K (Kelvin)	= °C + 273.16

Some Indian Measures

1 sq yd	= 4 sq cubit	1 bigha	= 1600 sq yd	1 sq. mile	= 640 acre
1 chhatak	= 20 sq cubit	1 bigha	= 14400 sq ft	1 sq. mile	= 1936 bigha
1 chhatak	= 5 sq yd	1 bigha	= 33 shatak	1 katha	= 1.65 shatak
1 chhatak	= 418 sq m	1 bigha	= 20 katha	1 katha	= 16 chhatak
40 acre	= 121 bigha	1 bigha	= 1337.808 sq m	1 katha	= 80 sq yd
1 acre	= 3 bigha 8 chhatak	1 chain	= 484 sq yd	1 katha	= 66.89 sq m
1 acre	= 4046.86 sq m	1 shatak	= 436 sq ft	1 katha	= 720 sq ft
1 acre	= 10 chain × 1 chain	1 ha	= 7.47 bigha	1 ha	= 100 are
1 acre	= 60.5 katha	1 ha	= 2.47 acre		

8.0 Greek Alphabet

Small letters	*Cap*	*Name*	*Small letters*	*Cap*	*Name*
α	Α	Alpha	ν	Ν	Nu
β	Β	Beta	ξ	Ξ	Xi
γ	Γ	Gamma	ο	Ο	Omicron
δ	Δ	Delta	π	Π	Pi
ε	Ε	Epsilon	ρ	Ρ	Rho
ζ	Ζ	Zeta	σ	Σ	Sigma
η	Η	Eta	τ	Τ	Tau
θ	Θ	Theta	υ	Υ	Upsilon
ι	Ι	Iota	φ	Φ	Phi
κ	Κ	Kappa	χ	Χ	Chi
λ	Λ	Lambda	ψ	Ψ	Psi
μ	Μ	Mu	ω	Ω	Omega

9.0 Critical Values of Student's t

Degrees of freedom	Significance level (one-tailed) 0.05	0.025	0.01	0.005	0.00005
	Significance level (two-tailed) 0.1	0.05	0.02	0.01	0.001
001	6.31	12.71	31.82	63.66	636.62
002	2.92	4.30	6.97	9.93	31.60
003	2.35	3.18	4.54	5.84	12.92
004	2.13	2.78	3.75	4.60	8.61
005	2.01	2.57	3.37	4.03	6.86
006	1.94	2.45	3.14	3.71	5.96
007	1.89	2.37	3.00	3.50	5.41
008	1.86	2.31	2.90	3.35	5.04
009	1.83	2.26	2.82	3.25	4.78
010	1.81	2.23	2.76	3.17	4.59
011	1.80	2.20	2.72	3.11	4.44
012	1.78	2.18	2.68	3.05	4.32
013	1.77	2.16	2.65	3.01	4.22
014	1.76	2.15	2.62	2.98	4.14
015	1.75	2.13	2.60	2.95	4.07
016	1.75	2.12	2.58	2.92	4.01
017	1.74	2.11	2.57	2.90	3.97
018	1.73	2.10	2.55	2.88	3.92
019	1.73	2.09	2.54	2.86	3.88
020	1.73	2.09	2.53	2.85	3.85
021	1.72	2.08	2.52	2.83	3.82
022	1.72	2.07	2.51	2.82	3.79
023	1.71	2.07	2.50	2.81	3.77
024	1.71	2.06	2.49	2.80	3.75
025	1.71	2.06	2.49	2.79	3.73
026	1.71	2.06	2.48	2.78	3.71
027	1.70	2.05	2.47	2.77	3.69
028	1.70	2.05	2.47	2.76	3.67
028	1.70	2.05	2.46	2.76	3.66
030	1.70	2.04	2.46	2.75	3.65
040	1.68	2.02	2.42	2.70	3.55
060	1.67	2.00	2.39	2.66	3.46
120	1.66	1.98	2.36	2.62	3.37
μ	1.65	1.96	2.33	2.58	3.29

Reject H_0 if calculated value of t is greater than critical value at chosen significance level.

10.0 Critical Values of Pearson's Product—Moment Correlation Coefficient

Degrees of freedom	Significance level (one-tailed) 0.05	0.025	0.01	0.005
	Significance level (two-tailed) 0.1	0.05	0.02	0.01
001	0.9877	0.9969	0.9995	0.9999
002	0.900	0.950	0.980	0.990
003	0.805	0.878	0.934	0.959
004	0.729	0.811	0.882	0.917
005	0.669	0.755	0.833	0.875
006	0.622	0.707	0.789	0.834
007	0.582	0.666	0.750	0.798
008	0.549	0.632	0.716	0.765
009	0.521	0.602	0.685	0.735
010	0.497	0.576	0.658	0.708
011	0.476	0.553	0.634	0.684
012	0.458	0.532	0.612	0.661
013	0.441	0.514	0.592	0.641
014	0.426	0.497	0.574	0.623
015	0.412	0.482	0.558	0.606
016	0.400	0.468	0.543	0.590
017	0.389	0.456	0.529	0.575
018	0.378	0.444	0.516	0.561
019	0.369	0.433	0.503	0.549
020	0.360	0.423	0.492	0.537
025	0.323	0.381	0.445	0.487
030	0.296	0.349	0.409	0.449
035	0.275	0.325	0.381	0.418
040	0.257	0.304	0.358	0.393
045	0.243	0.288	0.338	0.372
050	0.231	0.273	0.322	0.354
060	0.211	0.250	0.295	0.325
070	0.195	0.232	0.274	0.302
080	0.183	0.217	0.257	0.283
090	0.173	0.205	0.242	0.267
100	0.164	0.195	0.230	0.254

Reject H_0 if calculated value of r is greater than critical value at chosen significance level (in absolute terms).

11.0 Critical Values of Spearman's Rank Correlation Coefficients r_s

Degrees of freedom	Significance level (one-tailed)			
	0.05	0.025	0.01	0.005
	Significance level (two-tailed)			
	0.1	0.05	0.02	0.01
004	1.000			
005	0.900	1.000	1.000	
006	0.829	0.886	0.943	1.000
007	0.714	0.786	0.893	0.929
008	0.643	0.738	0.833	0.881
009	0.600	0.683	0.783	0.833
010	0.564	0.648	0.745	0.794
011	0.523	0.623	0.736	0.818
012	0.497	0.591	0.703	0.780
013	0.475	0.566	0.673	0.745
014	0.457	0.545	0.646	0.716
015	0.441	0.525	0.623	0.689
016	0.425	0.507	0.601	0.666
017	0.412	0.490	0.582	0.645
018	0.399	0.476	0.564	0.625
019	0.388	0.462	0.549	0.608
020	0.377	0.450	0.534	0.591
021	0.368	0.438	0.521	0.576
022	0.359	0.428	0.508	0.562
023	0.351	0.418	0.496	0.549
024	0.343	0.409	0.485	0.537
025	0.336	0.400	0.475	0.526
026	0.329	0.392	0.465	0.515
027	0.323	0.385	0.456	0.505
028	0.317	0.377	0.448	0.496
029	0.311	0.370	0.440	0.487
030	0.305	0.364	0.432	0.478
035	0.282	0.336	0.399	0.442
040	0.263	0.314	0.373	0.413
045	0.248	0.296	0.351	0.388
050	0.235	0.280	0.332	0.368
055	0.224	0.267	0.317	0.351
060	0.214	0.225	0.303	0.335
065	0.206	0.245	0.291	0.322
070	0.198	0.236	0.280	0.310
075	0.191	0.228	0.271	0.300
080	0.185	0.221	0.262	0.290
085	0.180	0.214	0.254	0.281
090	0.174	0.208	0.247	0.273
095	0.170	0.202	0.240	0.266
100	0.165	0.197	0.234	0.259

Reject H_0 if calculated value of r_s is greater than the critical value at the chosen significance level (in absolute terms).

For degrees of freedom greater than 30 other critical values can be found from the following relationship:

$$rs = z\sqrt{1/(n-1)}$$

where r_s is the critical value, n is the number of individuals in the data set (the degrees of freedom), and z is the appropriate critical value of a standard normal deviate (from Appendix 4). For a two-tailed test at the 0.01 level the appropriate value of z is 2.576, so the critical value of r_s, with 72 degrees of freedom is:

$$2.576\ 1/(72-1) = 2.576\sqrt{0.014}$$
$$= 2.576 \times 0.119$$
$$= 0.306$$

12.0 Critical Values of a Standard Normal Deviate z

	Significance level (one-tailed)				
	0.1	0.05	0.01	0.005	0.001
z	1.282	1.645	2.326	2.576	3.090
–z	–1.282	–1.645	–2.326	–2.5746	–3.090
	Significance level (two-tailed)				
	0.1	0.05	0.01	0.005	0.001
z	1.645	1.960	2.576	2.813	3.291
–z	–1.645	–1.960	–2.576	–2.813	–3.291

13.0 Critical Values of D for the Kolmogorov-Smirnov Goodness of Fit Test

Degrees of freedom	Significance level (one-tailed)				
	0.2	0.15	0.10	0.05	0.01
1	0.900	0.925	0.950	0.975	0.995
2	0.684	0.726	0.776	0.842	0.929
3	0.565	0.597	0.642	0.708	0.829
4	0.494	0.525	0.564	0.624	0.734
5	0.446	0.474	0.510	0.563	0.669
6	0.410	0.436	0.470	0.521	0.618
7	0.381	0.405	0.438	0.486	0.577
8	0.358	0.381	0.411	0.457	0.543
9	0.339	0.360	0.388	0.432	0.514
10	0.322	0.342	0.368	0.409	0.486
11	0.307	0.326	0.352	0.391	0.468
12	0.295	0.313	0.338	0.375	0.450
13	0.284	0.302	0.325	0.361	0.433
14	0.274	0.292	0.314	0.349	0.418
15	0.266	0.283	0.304	0.388	0.404
16	0.258	0.274	0.295	0.328	0.391
17	0.250	0.266	0.286	0.318	0.380
18	0.244	0.259	0.278	0.309	0.370
19	0.237	0.252	0.272	0.301	0.361
20	0.231	0.246	0.264	0.294	0.352
25	0.21	0.22	0.24	0.264	0.32
30	0.19	0.20	0.22	0.242	0.29
35	0.18	0.19	0.21	0.23	0.27
40	0.17	0.18	0.19	0.21	0.25
50	0.15	0.16	0.17	0.19	0.23
60	0.14	0.15	0.16	0.17	0.21
70	0.13	0.14	0.15	0.16	0.19
80	0.12	0.13	0.14	0.15	0.18
90	0.11	0.12	0.13	0.14	0.17
100	0.11	0.11	0.12	0.14	0.16
For degrees of freedom greater than 100, D_{crit} =	$\frac{1.07}{\sqrt{n}}$	$\frac{1.14}{\sqrt{n}}$	$\frac{1.22}{\sqrt{n}}$	$\frac{1.36}{\sqrt{n}}$	$\frac{1.63}{\sqrt{n}}$

Reject H_0 if calculated value of D is greater than the critical value at the chosen significance level.

14.0 Critical Values of chi Square

Degrees of freedom	Significance level (one-tailed)				
	0.1	0.05	0.01	0.005	0.001
1	2.71	3.84	6.64	7.88	10.83
2	4.60	5.99	9.21	10.60	13.82
3	6.25	7.82	11.34	12.84	16.27
4	7.78	9.49	13.28	14.86	18.46
5	9.24	11.07	15.09	16.75	20.52
6	10.64	12.59	16.81	18.55	22.46
7	12.02	14.07	18.48	20.28	24.32
8	13.36	15.51	20.09	21.96	26.12
9	14.68	16.92	21.67	23.59	27.88
10	15.99	18.31	23.21	25.19	29.59
11	17.28	19.68	24.72	26.76	31.26
12	18.55	21.03	26.22	28.30	32.91
13	19.81	22.36	27.69	30.82	34.53
14	21.06	23.68	29.14	31.32	36.12
15	22.31	25.00	30.58	32.80	37.70
16	23.54	26.30	32.00	34.27	39.29
17	24.77	27.59	33.41	35.72	40.75
18	25.99	28.87	34.80	37.16	42.31
19	27.20	30.14	36.19	38.58	43.82
20	28.41	31.41	37.57	40.00	45.32
21	29.62	32.67	38.93	41.40	46.80
22	30.81	33.92	40.29	42.80	48.27
23	32.01	35.17	41.64	44.18	49.73
24	33.20	36.42	42.98	45.56	51.18
25	34.38	37.65	44.31	46.93	52.62
26	35.56	35.88	45.64	48.29	54.05
27	36.74	40.11	46.96	49.65	55.48
28	37.92	41.34	48.28	50.99	56.89
29	39.09	42.56	49.59	52.34	58.30
30	40.26	43.77	50.89	53.67	59.70
40	51.81	55.76	63.69	66.77	73.40
50	63.17	67.51	76.15	79.49	86.66
60	74.40	79.08	88.38	91.95	99.61
70	85.53	90.53	100.43	104.22	112.32
80	96.58	101.88	112.33	116.32	124.84
90	107.57	113.15	124.12	128.30	137.21
100	118.50	124.34	135.81	140.17	149.45

Reject H_0 if calculated value of chi square is greater than the critical value at the chosen significance level.

15.0 Critical Values of the Nearest-Neighbour Index, R (One-Tailed)

	Clustered pattern					Dispersed pattern				
n	0.1	0.05	0.01	0.005	0.001	0.1	0.05	0.01	0.005	0.001
2	0.527	0.392	0.140	0.048		1.473	1.096	1.860	1.952	2.139
3	0.614	0.504	0.298	0.223	0.071	1.386	1.497	1.702	1.777	1.930
4	0.666	0.570	0.392	0.327	0.195	1.335	1.430	1.608	1.673	1.805
5	0.701	0.616	0.456	0.398	0.280	1.299	1.385	1.544	1.602	1.720
6	0.727	0.649	0.504	0.451	0.343	1.273	1.351	1.497	1.550	1.657
7	0.747	0.675	0.540	0.491	0.392	1.253	1.325	1.460	1.509	1.609
8	0.764	0.696	0.570	0.524	0.431	1.237	1.304	1.430	1.476	1.569
9	0.777	0.713	0.595	0.551	0.463	1.223	1.287	1.406	1.449	1.537
10	0.789	0.728	0.615	0.574	0.491	1.212	1.272	1.385	1.426	1.509
11	0.798	0.741	0.633	0.594	0.515	1.202	1.259	1.367	1.406	1.486
12	0.807	0.752	0.649	0.612	0.535	1.193	1.248	1.351	1.389	1.465
13	0.815	0.762	0.663	0.627	0.554	1.186	1.239	1.337	1.373	1.447
14	0.821	0.770	0.675	0.640	0.570	1.179	1.230	1.325	1.360	1.430
15	0.827	0.778	0.686	0.653	0.584	1.173	1.222	1.314	1.348	1.416
16	0.833	0.785	0.696	0.664	0.598	1.167	1.215	1.304	1.337	1.403
17	0.838	0.792	0.705	0.674	0.610	1.162	1.209	1.295	1.327	1.391
18	0.842	0.797	0.713	0.683	0.621	1.158	1.203	1.287	1.317	1.380
19	0.847	0.803	0.721	0.691	0.631	1.154	1.197	1.279	1.390	1.369
20	0.850	0.808	0.728	0.699	0.640	1.150	1.192	1.272	1.301	1.360
21	0.854	0.812	0.735	0.706	0.649	1.146	1.188	1.266	1.294	1.351
22	0.857	0.817	0.741	0.713	0.657	1.143	1.183	1.259	1.287	1.343
23	0.861	0.821	0.746	0.719	0.664	1.140	1.179	1.254	1.281	1.336
24	0.864	0.825	0.752	0.725	0.671	1.137	1.176	1.248	1.275	1.329
25	0.866	0.828	0.757	0.731	0.678	1.134	1.172	1.243	1.269	1.322
26	0.869	0.831	0.762	0.736	0.684	1.131	1.169	1.239	1.264	1.316
27	0.871	0.835	0.766	0.741	0.690	1.129	1.166	1.234	1.259	1.310
28	0.874	0.838	0.770	0.746	0.696	1.127	1.163	1.230	1.254	1.304
29	0.876	0.840	0.774	0.750	0.701	1.124	1.160	1.226	1.250	1.299
30	0.878	0.843	0.778	0.754	0.706	1.122	1.157	1.222	1.246	1.294
31	0.880	0.846	0.782	0.758	0.711	1.120	1.155	1.219	1.242	1.289
32	0.882	0.848	0.785	0.762	0.715	1.118	1.152	1.215	1.238	1.285
33	0.884	0.850	0.788	0.766	0.720	1.117	1.150	1.212	1.234	1.280
34	0.885	0.853	0.791	0.769	0.724	1.115	1.148	1.209	1.231	1.276
35	0.887	0.855	0.794	0.773	0.728	1.113	1.145	1.206	1.228	1.272
36	0.889	0.857	0.797	0.776	0.732	1.112	1.143	1.203	1.224	1.268
37	0.890	0.859	0.800	0.779	0.735	1.110	1.141	1.200	1.221	1.265
38	0.892	0.861	0.803	0.782	0.739	1.109	1.140	1.197	1.218	1.261
39	0.893	0.862	0.805	0.785	0.742	1.107	1.138	1.195	1.216	1.258
40	0.894	0.864	0.808	0.787	0.746	1.106	1.138	1.192	1.213	1.255
41	0.896	0.866	0.810	0.790	0.749	1.105	1.134	1.190	1.210	1.252
42	0.897	0.867	0.812	0.792	0.752	1.103	1.133	1.188	1.208	1.249
43	0.898	0.869	0.815	0.795	0.755	1.102	1.131	1.186	1.205	1.246

44	0.899	0.870	0.817	0.797	0.757	1.101	1.130	1.183	1.203	1.243
45	0.900	0.872	0.819	0.799	0.760	1.100	1.128	1.181	1.201	1.240
46	0.901	0.873	0.821	0.802	0.763	1.099	1.127	1.179	1.199	1.237
47	0.903	0.875	0.823	0.804	0.765	1.098	1.126	1.178	1.196	1.235
48	0.904	0.876	0.825	0.806	0.768	1.098	1.124	1.176	1.194	1.232
49	0.905	0.877	0.826	0.808	0.770	1.097	1.123	1.174	1.192	1.230
50	0.905	0.878	0.828	0.810	0.772	1.095	1.122	1.172	1.190	1.228
55	0.910	0.884	0.836	0.819	0.783	1.090	1.116	1.164	1.182	1.217
60	0.914	0.889	0.843	0.826	0.792	1.086	1.111	1.157	1.174	1.208
65	0.917	0.893	0.849	0.833	0.800	1.083	1.107	1.151	1.167	1.200
70	0.920	0.897	0.855	0.839	0.808	1.080	1.103	1.145	1.161	1.193
75	0.923	0.901	0.860	0.845	0.814	1.077	1.099	1.141	1.156	1.186
80	0.925	0.904	0.864	0.850	0.820	1.075	1.096	1.136	1.151	1.180
85	0.928	0.907	0.868	0.854	0.825	1.073	1.093	1.132	1.146	1.175
90	0.930	0.909	0.872	0.858	0.830	1.071	1.091	1.128	1.142	1.170
95	0.931	0.912	0.875	0.862	0.835	1.069	1.088	1.125	1.138	1.165
100	0.933	0.914	0.878	0.865	0.839	1.067	1.086	1.122	1.135	1.161
200	0.953	0.939	0.914	0.905	0.886	1.047	1.061	1.086	1.095	1.114
300	0.961	0.950	0.930	0.922	0.907	1.039	1.050	1.070	1.078	1.093
400	0.967	0.957	0.939	0.933	0.920	1.034	1.043	1.061	1.067	1.081
500	0.970	0.962	0.946	0.940	0.928	1.030	1.039	1.054	1.060	1.072

To test for clustering: reject H_0 if calculated value of R is less than critical value at chosen significance level.

To test for dispersion: reject H_0 if calculated value of R is greater than critical value at chosen significance level.

16.0 Critical Values of the Nearest-Neighbour Index, R (Two-Tailed)

	Significance Level									
n	0.1		0.05		0.01		0.005		0.001	
2	0.392	1.608	0.276	1.725	0.048	1.952	2.040	2.217		
3	0.504	1.497	0.409	1.592	0.223	1.778	0.151	1.849	0.007	1.993
4	0.570	1.430	0.488	1.512	0.327	1.673	0.265	1.735	0.140	1.860
5	0.616	1.385	0.542	1.458	0.398	1.602	0.343	1.658	0.231	1.769
6	0.649	1.351	0.582	1.418	0.450	1.550	0.400	1.600	0.298	1.702
7	0.675	1.325	0.613	1.387	0.491	1.509	0.444	1.556	0.350	1.650
8	0.696	1.304	0.638	1.362	0.524	1.476	0.480	1.520	0.392	1.608
9	0.713	1.287	0.659	1.342	0.551	1.449	0.510	1.490	0.427	1.574
10	0.728	1.272	0.676	1.324	0.574	1.426	0.535	1.465	0.456	1.544
11	0.741	1.259	0.691	1.309	0.594	1.406	0.557	1.443	0.481	1.519
12	0.752	1.248	0.704	1.296	0.611	1.389	0.576	1.425	0.504	1.497
13	0.762	1.239	0.716	1.284	0.627	1.374	0.592	1.408	0.523	1.477
14	0.770	1.230	0.726	1.274	0.640	1.360	0.607	1.393	0.540	1.460
15	0.778	1.222	0.736	1.265	0.652	1.348	0.620	1.380	0.556	1.444
16	0.785	1.215	0.744	1.256	0.663	1.337	0.632	1.368	0.570	1.430
17	0.792	1.209	0.752	1.249	0.674	1.327	0.643	1.357	0.583	1.417
18	0.797	1.203	0.759	1.242	0.683	1.317	0.654	1.347	0.595	1.406

n	Significance Level									
	0.1		0.05		0.01		0.005		0.001	
19	0.803	1.197	0.765	1.235	0.691	1.309	0.663	1.337	0.605	1.395
20	0.808	1.192	0.771	1.229	0.699	1.301	0.671	1.329	0.615	1.385
21	0.812	1.188	0.777	1.224	0.706	1.294	0.679	1.321	0.625	1.375
22	0.817	1.183	0.782	1.219	0.713	1.287	0.687	1.314	0.633	1.367
23	0.821	1.179	0.786	1.214	0.719	1.281	0.693	1.307	0.641	1.359
24	0.825	1.176	0.791	1.209	0.725	1.275	0.700	1.300	0.649	1.351
25	0.828	1.172	0.795	1.205	0.731	1.269	0.706	1.294	0.656	1.344
26	0.831	1.169	0.799	1.201	0.736	1.264	0.712	1.288	0.663	1.337
27	0.835	1.166	0.803	1.197	0.741	1.259	0.717	1.283	0.669	1.331
28	0.838	1.163	0.806	1.194	0.746	1.255	0.722	1.278	0.675	1.325
29	0.840	1.160	0.810	1.190	0.750	1.250	0.727	1.273	0.681	1.320
30	0.843	1.157	0.813	1.187	0.754	1.246	0.732	1.269	0.686	1.314
31	0.846	1.155	0.816	1.184	0.758	1.242	0.736	1.264	0.691	1.309
32	0.848	1.152	0.819	1.181	0.762	1.238	0.740	1.260	0.696	1.304
33	0.850	1.150	0.822	1.178	0.766	1.235	0.744	1.256	0.701	1.300
34	0.853	1.148	0.824	1.176	0.769	1.231	0.748	1.252	0.705	1.295
35	0.855	1.145	0.827	1.173	0.772	1.228	0.752	1.249	0.709	1.291
36	0.857	1.143	0.829	1.171	0.776	1.225	0.755	1.245	0.713	1.287
37	0.859	1.141	0.832	1.169	0.779	1.221	0.758	1.242	0.717	1.283
38	0.861	1.140	0.834	1.166	0.782	1.219	0.762	1.239	0.721	1.279
39	0.862	1.138	0.836	1.164	0.784	1.216	0.765	1.236	0.725	1.276
40	0.864	1.136	0.838	1.162	0.787	1.213	0.768	1.233	0.728	1.272
41	0.866	1.134	0.840	1.160	0.790	1.210	0.770	1.230	0.731	1.269
42	0.867	1.133	0.842	1.158	0.792	1.208	0.773	1.227	0.735	1.266
43	0.869	1.131	0.844	1.156	0.795	1.205	0.776	1.224	0.738	1.262
44	0.870	1.130	0.846	1.155	0.797	1.203	0.778	1.222	0.741	1.259
45	0.872	1.128	0.847	1.153	0.799	1.201	0.781	1.219	0.744	1.257
46	0.873	1.127	0.849	1.151	0.802	1.199	0.783	1.217	0.746	1.254
47	0.875	1.126	0.851	1.150	0.804	1.197	0.786	1.215	0.749	1.251
48	0.876	1.124	0.852	1.148	0.806	1.194	0.788	1.212	0.752	1.248
49	0.877	1.123	0.854	1.146	0.808	1.192	0.790	1.210	0.754	1.246
50	0.878	1.122	0.855	1.145	0.810	1.191	0.792	1.208	0.757	1.243
55	0.884	1.116	0.862	1.138	0.819	1.182	0.802	1.198	0.768	1.232
60	0.889	1.111	0.868	1.132	0.826	1.174	0.810	1.190	0.778	1.222
65	0.893	1.107	0.873	1.127	0.833	1.167	0.818	1.182	0.787	1.213
70	0.897	1.103	0.878	1.123	0.839	1.161	0.824	1.176	0.794	1.206
75	0.901	1.099	0.882	1.118	0.845	1.156	0.830	1.170	0.801	1.199
80	0.904	1.096	0.886	1.115	0.850	1.151	0.836	1.164	0.808	1.192
85	0.907	1.093	0.889	1.111	0.854	1.146	0.841	1.160	0.814	1.187
90	0.909	1.091	0.892	1.108	0.858	1.142	0.845	1.155	0.819	1.181
95	0.912	1.088	0.895	1.105	0.862	1.138	0.849	1.151	0.824	1.177
100	0.914	1.086	0.898	1.103	0.865	1.135	0.853	1.147	0.828	1.172
200	0.939	1.061	0.928	1.073	0.905	1.095	0.896	1.104	0.878	1.122
300	0.950	1.050	0.941	1.059	0.922	1.078	0.915	1.085	0.901	1.099
400	0.957	1.043	0.949	1.051	0.933	1.067	0.927	1.074	0.914	1.086
500	0.962	1.039	0.954	1.046	0.940	1.060	0.934	1.066	0.923	1.077

17.0 Worksheet for Crude and Urban Density of Population, West Bengal 2001

Districts	Area (sq km : A)	Urban Area (sq km : UA)	Population (P)	Urban Population (U)	Population Density (persons / sq km : P/A)	Urban Density (persons / sq km : U/UA)
Darjeeling	3149	75.23	1609172	520432	511	6918
Jalpaiguri	6227	121.07	3401173	606882	546	5013
Coochbehar	3387	41.34	2479155	225618	732	5458
U Dinajpur	3140	44.36	2441794	294443	778	6638
D Dinajpur	2219	22.7	1503178	196854	677	8672
Malda	3733	25.37	3290468	240940	881	9497
Murshidabad	5324	128.89	5866569	732734	1102	5685
Birbhum	4545	49.98	3015422	258420	663	5170
Burdwan	7024	800.18	6895514	2547048	982	3183
Nadia	3927	210.17	4604827	979519	1173	4661
24 Parganas (N)	4094	499.56	8934286	4850947	2182	9710
24 Parganas (S)	9960	176.76	6906689	1086220	693	6145
Howrah	1467	219.24	4273099	2151990	2913	9816
Hooghly	3149	198.81	5041976	1687749	1601	8489
Kolkata	185	185	4572876	4572876	24718	24718
Bankura	6882	61.49	3192695	235248	464	3826
Puruliya	6259	79.37	2536516	255426	405	3218
Midnapore	14081	385.22	9610788	983905	683	2554
	88752	3324.74	80176197	22427251	903	6746

Density Class (persons / sq km)	Shading Scheme	Districts	Remarks
1 - 500		Bankura, Puruliya	Very Low Density
501 - 1000		Darjeeling, Jalpaiguri, Coochbehar, U Dinajpur, D Dinajpur, Malda, Birbhum, Burdwan, 24 Parganas (S), Midnapur	Low Density
1001 - 1500		Murshidabad, Nadia	Moderate Density
1501 - 2000		Hooghly	High Density
Above 2000		24 Parganas (N), Howrah, Kolkata	Very High Density
2001 - 4000		Bankura, Puruliya, Midnapur	Very Low Density
4001 - 6000		Jalpaiguri, Coochbehar, Murshidabad, Birbhum, Nadia	Low Density
6001 - 8000		Darjeeling, U Dinajpur, 24 Parganas (S)	Moderate Density
8001 - 10000		D Dinajpur, Malda, 24 Parganas (N), Howrah, Hooghly, Kolkata	High Density
Above 10000			Very High Density

18.0 Worksheet for the Mean Centres of Population, West Bengal

District	Coordinates (cm)		Population		Products			
	x	y	2001 : (P_{01})	1991 : (P_{91})	$x.P_{01}$	$y.P_{01}$	$x.P_{91}$	$y.P_{91}$
Darjeeling	6.5	15.3	1609172	1299919	10459618.0	24620331.6	8449473.5	19888760.7
Jalpaiguri	8.4	14.5	3401173	2800543	28569853.2	49317008.5	23524561.2	40607873.5
Coochbehar	9.0	13.5	2479155	2171145	22312395.0	33468592.5	19540305.0	29310457.5
U Dinajpur	5.6	12.5	2441794	1926729	13674046.4	30522425.0	10789682.4	24084112.5
D Dinajpur	7.0	10.8	1503178	1200924	10522246.0	16234322.4	8406468.0	12969979.2
Malda	6.0	10.0	3290468	2637032	19742808.0	32904680.0	15822192.0	26370320.0
Murshidabad	6.3	7.0	5866569	4740149	36959384.7	41065983.0	29862938.7	33181043.0
Birbhum	4.8	6.7	3015422	2555664	14474025.6	20203327.4	12267187.2	17122948.8
Burdwan	5.2	5.2	6895514	6050605	35856672.8	35856672.8	31463146.0	31463146.0
Nadia	7.0	5.5	4604827	3852097	32233789.0	25326548.5	26964679.0	21186533.5
24 Pgs (N)	7.5	3.6	8934286	7281881	67007145.0	32163429.6	54614107.5	26214771.6
24 Pgs (S)	7.0	2.0	6906689	5715030	48346823.0	13813378.0	40005210.0	11430060.0
Howrah	5.8	2.8	4273099	3729644	24783974.2	11964677.2	21631935.2	10443003.2
Hooghly	5.8	3.8	5041976	4355230	29243460.8	19159508.8	25260334.0	16549874.0
Kolkata	6.7	3.0	4572876	4399819	30638269.2	13718628.0	29478787.3	13199457.0
Bankura	3.5	4.5	3192695	2805065	11174432.5	14367127.5	9817727.5	12622792.5
Purulia	1.6	5.0	2536516	2224577	4058425.6	12682580.0	3559323.2	11122885.0
Midnapore	4.2	2.3	9610788	8331912	40365309.6	22104812.4	34994030.4	19163397.6
Sum Total			80176197	68077965	480422678.6	449494033.2	406452088.0	376931416.0

In 1991, the location of the Mean Centre of Population is given by:

$$\bar{x} = (\Sigma(x.P_{91}) / \Sigma P_{91}) = 406452088/68077965 = 5.97 \text{ cm}$$

$$\bar{y} = (\Sigma(y.P_{91}) / \Sigma P_{91}) = 376931416/68077965 = 5.53 \text{ cm}$$

In 2001, the location of the Mean Centre of Population is given by:

$$\bar{x} = (\Sigma(x.P_{01}) / \Sigma P_{01}) = 480422678.6/ 80176197 = 5.99 \text{ cm}$$

$$\bar{y} = (\Sigma(y.P_{01}) / \Sigma P_{01}) = 449494033.2/80176197 = 5.60 \text{ cm}$$

19.0 Worksheet for the Nearest-Neighbour Analysis of Settlement Pattern (Grid A, Toposheet No. 73 E/12)

Settlement with Nearest-Neighbour	Nearest-Neighbour Distance (NND: cm)	Settlement with Nearest-Neighbour	Nearest-Neighbour Distance (NND: cm)
1–2	1.25	23–27	2.60
2–1	1.25	24–25	1.60
3–2	2.10	25–24	1.60
4–5	2.00	26–21	1.40
5–6	1.70	27–31	1.70
6–5	1.70	28–25	1.80
7–8	2.70	29–30	1.20
8–7	2.70	30–35	1.00
9–11	2.20	31–36	1.20
10–14	1.00	32–33	1.10
11–17	1.70	33–32	1.10
12–16	0.90	34–29	1.30
13–16	2.50	35–30	1.00
14–15	1.00	36–31	1.20
15–14	1.00	37–31	1.20
16–12	0.90	38–35	1.70
17–11	1.70	39–40	1.60
18–19	1.40	40–39	1.60
19–18	1.40	41–42	3.30
20–44	2.10	42–41	3.30
21–26	1.40	43–38	2.20
22–27	2.50	44–15	1.70

Computation

$\Sigma NND = 73.5$ cm

Mean NND $= \bar{r}_o$

$= (\Sigma NND / 44)$

$= 1.67$ cm

$= 0.835$ Km

Density of Settlement:

P $= (N / A)$

$= (44/72.285)$ / sq km

$= 0.6087$ / sq km

Mean Expected:

NND $= \bar{r}_e$

$= 1 / (2\sqrt{P})$

$= 1 / (2\sqrt{0.608070})$

$= 0.641$ km

R_n $= (\bar{r}_o / \bar{r}_e)$

$= (0.835 / 0.641)$

$= 1.30$

Therefore, the settlements are more random than dispersed.

20.0 Worksheet for plotting the Rank-Size Curve (Arithmetic Graph)

Rank (r)	Population (Pr)	1/(r)	Computation of Estimated Population of the 1st ranked Town	Computation of Estimated Population of Towns, $Pr = P_1 / r$
1	4580544	1		3292543
2	1008704	0.5	Estimated Population of the	1646271
3	492996	0.333333333	1st Ranked town/city	1097514
4	486304	0.25	is given by:	823135
5	441956	0.2		658508
6	392150	0.166666667	$P_1 = \Sigma(Pr) / \Sigma(1/r)$	548757
7	389214	0.142857143	= 13362768 / 4.0585	470363
8	348379	0.125	= 3292543	411567
9	336390	0.111111111		365838
10	314334	0.1		329254
11	290067	0.090909091		299322
12	285871	0.083333333		274378
13	284615	0.076923077		253272
14	271781	0.071428571		235182
15	261575	0.066666667		219503
16	250615	0.0625		205784
17	231515	0.058823529		193679
18	220032	0.055555556		182919
19	215432	0.052631579		173292
20	207984	0.05		164627
21	202095	0.047619048		156787
22	197955	0.045454545		149661
23	185660	0.043478261		143154
24	170695	0.041666667		137189
25	170201	0.04		131702
26	167848	0.038461538		126636
27	165222	0.037037037		121946
28	162166	0.035714286		117591
29	161448	0.034482759		113535
30	160168	0.033333333		109751
31	155503	0.032258065		106211
32	153349	0.03125		102892
	13362768	4.058495195		

21.0 Worksheet for plotting the Rank-Size Curve

(Log-Log Graph: Power Regression)

Rank (r)	Population (Pr)	log (r)	log (Pr)	sq of log (r)	log (r) × log (Pr)	Solution of the Rank-Size Equation: $P_r = P_1 (r)^b$
1	4580544	0	6.660917059	0	0	
2	1008704	0.301	6.003763743	0.090619058	1.807312974	The normal equations are:
3	492996	0.4771	5.692843396	0.227644692	2.716176584	$\Sigma \log P = N.a + b\Sigma \log r$
4	486304	0.6021	5.686907842	0.362476233	3.423859686	
5	441956	0.699	5.645379034	0.488559067	3.945950608	$\Sigma(\log P. \text{Log } r) = a.\Sigma \log r + b\Sigma(\log r)^2$
6	392150	0.7782	5.593452219	0.605519368	4.352551838	
7	389214	0.8451	5.590188453	0.714190697	4.724257305	From Table:
8	348379	0.9031	5.542051968	0.815571525	5.00497164	$\Sigma \log P = 174.1988872$
9	336390	0.9542	5.526843077	0.910578767	5.273948607	$\Sigma \log r = 35.42$
10	314334	1	5.497391359	1	5.497391359	
11	290067	1.0414	5.462498323	1.084498725	5.688605797	$\Sigma(\log P. \text{Log } r) = 189.6336701$
12	285871	1.0792	5.456170101	1.164632162	5.888196448	$\Sigma(\log r)^2 = 43.48870256$
13	284615	1.1139	5.454257785	1.240869792	6.075734201	
14	271781	1.1461	5.434219092	1.313609474	6.228310854	Thus,
15	261575	1.1761	5.417596234	1.38319065	6.371587576	$174.1988872 = 32a + 35.42b$
16	250615	1.2041	5.399007061	1.449904933	6.501052289	$189.6336701 = 35.42b$
17	231515	1.2304	5.364579134	1.514004548	6.60084061	$+ 43.48870256b$
18	220032	1.2553	5.342485846	1.575709062	6.706275592	
19	215432	1.2788	5.333310213	1.635210772	6.81998964	Eliminating coefficients of 'a',
20	207984	1.301	5.318029926	1.69267905	6.918916452	$5.443715225 = a + 1.106875b$
21	202095	1.3222	5.305555569	1.748263863	7.015107942	$5.353858557 = a + 1.122780075b$
22	197955	1.3424	5.296566476	1.802098654	7.110230968	
23	185660	1.3617	5.268718346	1.854302699	7.174560432	Subtracting and transposing,
24	170695	1.3802	5.2322208	1.904983072	7.221569967	$b = -0.743073$
25	170201	1.3979	5.230962107	1.954236268	7.312571214	$a = 6.26620417$
26	167848	1.415	5.224916171	2.002149575	7.393117127	
27	165222	1.4314	5.218067875	2.048802225	7.468953275	$P_1 = \text{antilog } (6.26620417)$
28	162166	1.4472	5.209959804	2.094266368	7.539635174	$= 1845883$
29	161448	1.4624	5.208032669	2.138607904	7.616216549	
30	160168	1.4771	5.204575753	2.181887201	7.687789466	Therefore, $P_r = 1845883(r)^{-0.743073}$
31	155503	1.4914	5.191738772	2.224159702	7.742760329	
32	153349	1.5051	5.185680948	2.265476457	7.805227567	
	Total	35.42	174.1988872	43.48870256	189.6336701	

22.0 Worksheet for Functional Classification of Towns/Cities

Towns	Proportion of Workers employed (%)							Nomenclature of Towns
	Mining (M)	Household Industry (HI)	Manu-facturing Industry (Mf)	Trade & Commerce (Tc)	Transport & Communi-cation (Tr)	Professional Services (P)	Others (O)	
T1	8.8	16.3	22.2	10.8	12.4	**25.2 (1)**	4.3	P_1 : Service Town
T2	4.5	18.3	16.7	22.3	20.8	12.3	5.1	*Diversified*
T3	7.8	9.8	10.9	**31.6 (1)**	27.4	7.7	4.8	Tc_1 : Trade & Commerce Town
T4	3.9	12.4	8.6	22.9	**34.2 (1)**	9.8	**8.2 (1)**	Tr_1 O_1: Transport Town and Others
T5	2.1	9.6	15.8	**31.8 (1)**	18.7	11.5	**10.5 (1)**	Tc_1 Tr_1: Trade & Commerce and Transport Town
T6	0	**34.2 (2)**	8.2	7.2	**31.2 (1)**	18.2	1.0	HI_1 : Household Industry
T7	0	**32.6 (2)**	6.4	8.5	**30.7 (1)**	11.9	**9.9 (1)**	HI_1 Tr_1 O_1: Household Industry, Transport Town and Others
T8	5.6	9.9	18.4	22.1	18.2	19.6	6.2	*Diversified*
T9	**12.3 (1)**	11.3	24.5	9.5	**34.2 (1)**	6.1	2.1	M_1 Tr_1: Mining and Transport Town
T10	**21.1 (2)**	17.2	**26.9 (1)**	7.3	12.7	11.1	3.7	M_1 Mf_1: Mining and Manufacturing Town
T11	9.7	10.6	**28.2 (1)**	6.9	18.9	18.4	7.3	Mf_1: Manufacturing Town
T12	3.4	8.6	14.4	18.4	20.5	**30.5 (1)**	4.2	P_1: Service Town
T13	2.1	12.7	**30.8 (1)**	12.8	17.2	23.1	1.3	Mf_1: Manufacturing Town
T14	2.5	5.8	23.8	24.4	20.2	22.4	0.9	*Diversified*
T15	3.9	21.5	14.6	16.5	21.1	18.6	3.8	*Diversified*
T16	8.3	15.2	15.4	21.2	12.3	20.6	7	*Diversified*
Mean	6	15.4	17.9	17.1	21.9	16.7	5.0	
Standard Deviation	5.4	8.1	7.5	8.5	7.4	6.9	3.0	

23.0 Worksheet for Dominant and Distinctive Analysis of Cropping Pattern, West Bengal 2001

Districts	Production of Crops ('000 tonnes)					Proportion of Production of Crops (% of Total)				
	Rice	Pulses	Oilseeds	Jute	Potato	Rice	Pulses	Oilseeds	Jute	Potato
24 Pgs (N): J	856.4	4.8	37.7	886.6	138.1	**44.5**	0.2	2.0	*46.1*	7.2
24 Pgs (S): R	1003.7	6.6	4.6	16.6	83.5	*90.0*	0.6	0.4	1.5	7.5
Bankura	1222.4	0.4	18	2.4	522.9	**69.2**	0.0	1.0	0.1	29.6
Birbhum: R, P	1157.4	17.9	38.8	7.7	273.2	*77.4*	*1.2*	2.6	0.5	18.3
Burdwan	1930.6	3.5	44.5	400.9	1217.3	**53.7**	0.1	1.2	11.1	33.8
Coochbehar: J	466.7	5.7	7.2	904.3	275.1	28.1	0.3	0.4	*54.5*	16.6
D Dinajpur	425.3	2.2	17.1	182.5	66.5	**61.3**	0.3	2.5	26.3	9.6
Darjeeling: Po	59.2	1.3	0.3	30.9	98.6	31.1	0.7	0.2	16.2	*51.8*
Hooghly: Po_3	847.7	0.3	27.6	623.5	2356.4	22.0	0.0	0.7	16.2	***61.1***
Howrah	287.6	0.2	3.7	80.9	198.4	**50.4**	0.0	0.6	14.2	34.8
Jalpaiguri	410.1	1.9	6.3	510.3	314.9	33.0	0.2	0.5	**41.0**	25.3
Malda: P_3	536.2	25.6	37	370.8	44.5	**52.9**	***2.5***	3.6	36.6	4.4
Midnapore	2710.8	14.4	79.5	167.1	1823.4	**56.5**	0.3	1.7	3.5	38.0
Murshidabad: J	1085.4	34.7	58.6	2010	174.4	32.3	1.0	1.7	*59.8*	5.2
Nadia: J, O_2	958.7	38.4	1116.4	2330.6	106.2	21.1	0.8	***24.5***	*51.2*	2.3
Purulia: R_2	745.1	6.5	2.3	0	12.7	**97.2**	0.8	0.3	0.0	1.7
U Dinajpur: Po	554.4	10.7	29.7	611.1	1161.3	23.4	0.5	1.3	25.8	*49.1*
Note: Dominant Crops = underlined & bold				Mean		49.7	0.6	2.7	23.8	23.3
Distinctive Crops = in the following order: *italics* 1st order, **bold** 2nd order, ***bold & italic*** 3rd order				Stand. Dev		23.6	0.6	5.7	20.8	18.9

Dominant Crops:

(1) Rice: North 24 Parganas, South 24 Parganas, Bankura, Birbhum, Burdwan, D Dinajpur, Howrah, Malda, Midnapur and Purulia
(2) Jute: Coochbehar, Jalpaiguri, Murshidabad and Nadia
(3) Potato: Darjeeling, Hooghly and U Dinajpur

Distinctive Crops :

(1) Rice: South 24 Parganas and Birbhum (1st order); Purulia (2nd order)
(2) Pulses: Birbhum (1st order); Malda (3rd order)
(3) Oilseeds: Nadia (3rd order)
(4) Jute: North 24 Parganas, Coochbehar, Murshidabad and Nadia (1st order)
(5) Potato: Darjeeling, U Dinajpur (1st order); Hooghly (3rd order)

24.0(a) Worksheet for Location Quotients of Urban Population, West Bengal 2001

Districts	Urban pulation, 2001 : U	Population, 2001: P	U / TU	P / TP	LQ = (U / TU) / (P / TP)
Purulia	255239	2535233	0.01135	0.03160	0.36
Bankura	235264	3191822	0.01046	0.03978	0.26
Midnapore	1010954	9638473	0.04495	0.12014	0.37
Birbhum	258479	3012546	0.01149	0.03755	0.31
Burdwan	2572423	6919698	0.11439	0.08625	1.33
Nadia	979047	4603756	0.04353	0.05738	0.75
U Dinajpur	294471	2441824	0.01309	0.03043	0.43
Malda	240915	3290160	0.01071	0.04101	0.26
Hooghly	1687410	5040047	0.07504	0.06282	1.19
Howrah	2153571	4274010	0.09575	0.05327	1.79
Murshidabad	732343	5863717	0.03256	0.07309	0.44
Darjeeling	520877	1605900	0.02316	0.02001	1.16
Kolkata	4580544	4580544	0.20307	0.05709	3.56
D Dinajpur	196643	1502647	0.00874	0.01873	0.47
24 Parganas (S)	1089730	6909015	0.04846	0.08612	0.56
24 Parganas (N)	4849218	8930295	0.21565	0.11132	1.94
Jalpaiguri	603847	3403204	0.02685	0.04242	0.63
Coochbehar	225506	2478280	0.01002	0.03089	0.32
	TU = 22486481	TP = 80221171			

24.0(b) Distribution of Location Quotients

Range of LQs	Districts	Remarks
below 1	Purulia, Bankura, Midnapur, Birbhum, Nadia, Malda, U Dinajpur, D Dinajpur, Murshidabad, South 24 Parganas, Jalpaiguri, Coochbehar	Low
1 to 2	Burdwan, Hooghly, Howrah, Darjeeling, North 24 Parganas	Moderate
Above 2	Kolkata	High

25.0 Specifications for Estimating Wind Speed over Land or Sea

Beaufort No.	Descriptive term	Specification		Speed of wind at 10 m above ground	
		Land	Sea	Knots	km/hr
0.	Calm	Calm; smoke rises vertically.	Sea like a mirror.	Less than1	Less than 1
1.	Light air	Smoke bends from the vertical and drifts slowly with the winds; wind vane not affected.	Ripples with the appearance of scales are formed but without foam crests.	1–3	1–5
2.	Light breeze	Wind felt on face; leaves rustle, ordinary vanes moved by wind.	Small wavelets, still short, but more pronounced crests have a glassy appearance and do not break.	4–6	6–11
3.	Gentle breeze	Leaves and small twigs in constant motion; wind extends light flag.	Large wavelets (crests begin to break); foam of glassy appearance; perhaps scattered white horses.	7–10	12–19
4.	Moderate breeze	Raises dust and loose paper; small branches are moved.	Small waves becoming longer; fairly frequent white horses.	11–16	20–28
5.	Fresh breeze	Small trees begin to sway; crested wavelets form on inland waters.	Moderate waves taking a more pronounced form; many white horses are formed (chances of some spray).	17–21	29–38
6.	Strong breeze	Large branches in motion; whistling heard in telegraph wires; umbrellas used with difficulty.	Large waves begin to form; the white foam crests are more extensive everywhere (probably some spray).	22–27	39–49
7.	Near gale	Whole trees in motion; inconvenience felt when walking against the wind.	Sea heaps up and the white foam of the breaking waves begins to be blown in streaks along the direction of the wind (spin drift begins to be seen).	28–33	50–61
8.	Gale	Breaks twigs of trees; generally impedes progress due to difficulty experienced in walking against wind.	Moderately high waves of greater length; edges of crest break; foam is blown in well marked streaks along the direction of the wind.	34–40	62–74
9.	Strong gale	Slight structural damage occurs (chimney pots and slates on roofs removed).	High waves; dense streaks of foam along the direction of the wind; crests of waves begin to topple, tumble and roll over; spray may affect visibility.	41–47	75–88
10.	Storm	Trees uprooted and considerable structural damage occurs, for instance, kuccha houses blown down (seldom experienced inland).	Very high waves with long overhanging crests. The resulting foam, in great patches, is blown in dense white streak along the direction of the wind. On the whole, the surface of the sea takes a white appearance. The tumbling of the sea becomes heavy and shock-like. Visibility is affected.	48–55	89–102

Beaufort No.	Descriptive term	Specification		Speed of wind at 10 m above ground	
		Land	Sea	Knots	km/hr
11.	Violent storm	Widespread damage (very rarely experienced).	Exceptionally, high waves (small and medium-sized ships might for a time get lost behind the waves). The sea is completely covered with long white patches of foam lying along the direction of the wind. Everywhere the edges of the wave crests are blown into froth. Visibility is affected.	56–63	103–117
12.	Hurricane	...	The air is filled with foam and spray. Sea completely white with driving spray. Visibility very seriously affected.	64 and over	118 and over

26.0 National Map Policy, 2005

Preamble

All socio-economic developmental activities, conservation of natural resources, planning for disaster mitigation and infrastructure development require high quality spatial data. The advancements in digital technologies have now made it possible to use diverse spatial databases in an integrated manner. The responsibility for producing, maintaining and disseminating the topographic map database of the whole country, which is the foundation of all spatial data vests with the Survey of India (SOI). Recently, SOI has been mandated to take a leadership role in liberalizing access of spatial data to user groups without jeopardizing national security. To perform this role, the policy on dissemination of maps and spatial data needs to be clearly stated.

Objectives

- To provide, maintain and allow access and make available the National Topographic Database (NTDB) of the SOI conforming to national standards.
- To promote the use of geospatial knowledge and intelligence through partnerships and other mechanisms by all sections of the society and work towards a knowledge-based society.

Two Series of Maps

To ensure that in the furtherance of this policy, national security objectives are fully safeguarded, it has been decided that there will be two series of maps, namely:

a. Defence Series Maps (DSMs)—These will be the topographical maps (on Everest/WGS-84 Datum and Polyconic/UTM Projection) on various scales (with heights, contours and full content without dilution of accuracy). These will mainly cater for defence and national security requirements. This series of maps (in analogue or digital forms) for the entire country will be classified, as appropriate, and the guidelines regarding their use will be formulated by the Ministry of Defence.

b. Open Series Maps (OSMs)—OSMs will be brought out exclusively by SOI, primarily for supporting development activities in the country. OSMs shall bear different map sheet numbers and will be in UTM Projection on WGS-84 datum. Each of these OSMs (in both hard copy and digital form) will become "Unrestricted" after obtaining a one-time clearance of the Ministry of Defence. SOI will ensure that no civil and military Vulnerable Areas and Vulnerable Points (VAs/VPs) are shown on OSMs.

The SOI will issue from time to time detailed guidelines regarding all aspects of the OSMs like procedure for access by user agencies, further dissemination/sharing of OSMs amongst user agencies with or without value additions, ways and means of protecting business and commercial interests of SOI in the data and other incidental matters. Users will be allowed to publish maps on hard copy and web with or without GIS database. In addition, the SOI is currently preparing City Maps. These City Maps will be on large scales in WGS-84 datum and in public domain.

National Topographical Data Base (NTDB): SOI will continue to create, develop and maintain the NTDB in analogue and digital forms consisting of following data sets: (a) National Spatial Reference Frame, (b) National Digital Elevation Model, (c) National Topographical Template, (d) Administrative Boundaries, and (e) Toponomy (place names). Both the DSMs and OSMs will be derived from the NTDB.

Fig. A.1 Layout of the Open Series Map, 2005

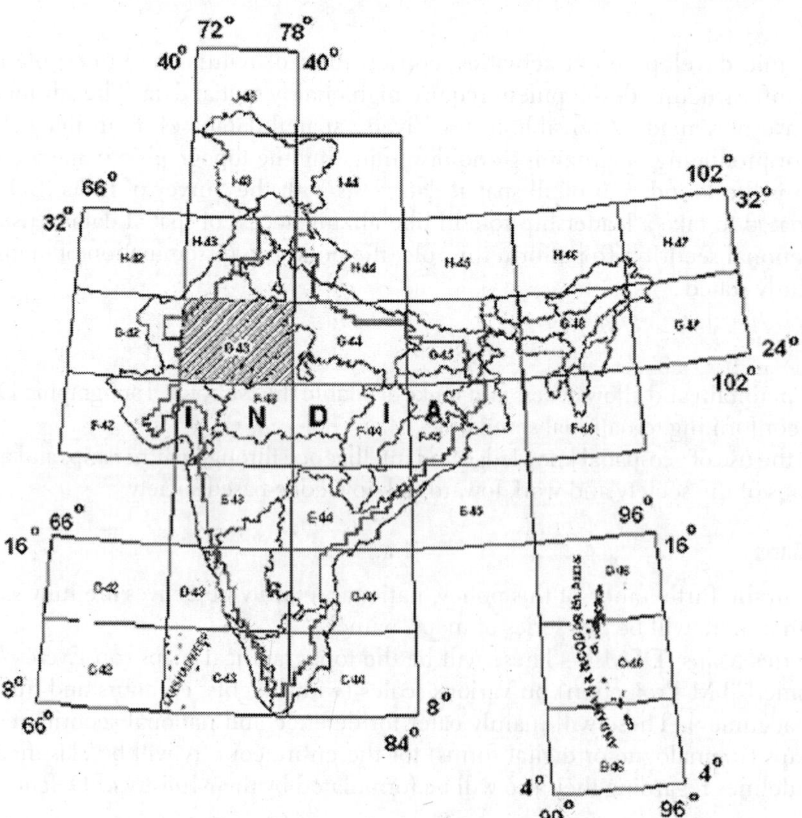

Dissemination

Open Series Maps of scales larger than 1:1 million either in analogue or digital formats can be disseminated by SOI by sale or through an agreement to any agency for specific end use. This transaction will be registered in the Registration database with details of the receiving agency, end use etc.

- Through the agreement, SOI will allow a user to add value to the maps obtained (either in analogue or digital formats) and prepare his own value-added maps.
- The user should be able to share these maps with others—the information of all such sharing will also require to be logged in the Map Transaction Registry.

27.0 Land Remote Sensing Systems

Country	Programme	Sensor	No. of Bands	IFOV
USA	NOAA 12–14	AVHRR	5 : 0.58–12.5	1.1 KM
	NASA Landsat 5	TM	7 0.45–12.5	28.5 m VNIR, SWIR 120 m TIR
	NASA Landsat 7	ETM+	7 0.45–12.5 1 pan 0.52–0.90	30 m 28.5 m VNIR, SWIR 60 m TIR 15 m
	EOS-Terra	ASTER	14 0.5–12	15 m VNIR 30 m SWIR 90 m TIR
		MODIS	36 0.4–14.5	250 m 500 m, 1 km
		MISR	4 : 0.425–0.886	275 m, 550 m, 1.1 km
	Space Imaging	IKONOS 2	1 pan 4 MSS	1m pan 4 m MSS
	Earth Watch	Quickbird 2	1 pan 4 MSS	0.61m pan 2.5 m MSS
France	SPOT 3	HRV	1 pan 3 XS : 0.5–0.89	10 m pan 20 m XS
	SPOT 5	HRV	1 pan 4 XS 10 m VNIR	5 m pan 20 m SWIR
ESA and Canada	ERS 1, 2	SAR	C-band	25 m
	Envisat-1	SAR	C-band	5.7 m
Japan	JERS 1	OPS SAR	7 0.52–2.40 L-band	18 m
	ADEOS	AVNIR	4 0.42–0.89	8 m pan 16 m VNIR

India	IRS 1A, 1B	LISS I, II	4	72 m, 36 m
	IRS 1C, D	LISS III	4 0.52–1.7	5.8 m pan 23.5 m VNIR 70.5 m SWIR
	IRS P3 !RS P4 (Oceansat)	WiFS, MOS OCM, MSMR		
	IRS P5 (Cartosat 1) IRS P6 (Resourcesat 1)			
Canada	RADARSAT	SAR	C–band HH	10–50 m
Germany	MOMS-02 / D2	Pan, M/S	7 0.44–0.81	4.5 m HR pan 13.5 m VNIR
	MOMS-02 / P	Pan, M/S	7 0.44–0.81	6.0 m pan 18.0 km VNIR

27.0(A) Landsat TM and ETM+ Spectral Bands

Band	*Name*	*Band Width (l, μm)*	*Spatial Resolution (m)*
1	Blue	0.450–0.515	30
2	Green	0.525–0.605	30
3	Red	0.63–0.69	30
4	Near Infrared	0.75–0.90	30
5	Shortwave IR-1	10.40–12.50	60 / 120*
6	Thermal IR-2	2.09–2.35	30
7	Shortwave IR-2	2.09–2.35	30
8*	Panchromatic	0.52–0.90	15

Note:

1. TM and ETM+ sensors have 8 bit quantization.
2. The values above are the specifically the spectral ranges for the ETM+ sensor; the TM sensor spectral ranges vary just slightly for some bands.
3. The thermal band on Landsat 7 (ETM+ sensor) has a spatial resolution of 60 m; it is 120 m on Landsats 5 and 4 (TM sensors).
4. Only Landsat 7 has a 15 m panchromatic "sharpening band".
5. Landsats 1–3 had a four band MultiSpectral Scanner (MSS) sensor.

True Colour: For the true colour rendition, band 1 is displayed in the blue colour, band 2 is displayed in the green colour, and band 3 is displayed in the red colour. The resulting image is fairly close to realistic—as though you took the picture with your camera and were riding in the satellite. But it is also pretty dull—there is little contrast and features In the image are hard to distinguish.

False-Colour, also called Near Infrared or NIR: In this image, band 2 is displayed in blue, band 3 is displayed in green, and band 4 is displayed in red. This rendition looks rather strange—vegetation jumps out as a bright red because green vegetation readily reflects infrared fight energy! It is similar to pictures taken from aircraft when using infrared film; very useful for studying vegetation.

Short-Wavelength Infrared (SWIR) or "Pseudo Natural Colour": In this SWIR image, band 2 is displayed in blue, band 4 is displayed in green, and band 7 (or 5) is displayed in red. This rendition looks like a jazzed up true colour rendition—one with more striking colours.

Land Cover Type	*In Natural Colour (3,2,1), Appears:*	*In False Colour: (4,3,2), Appears:*	*In Pseudo Natural Colour (7,4,2), Appears:*
Trees and Bushes	Olive Green	Red	Shades of Green
Crops	Medium – Light Green	Pink – Red	Shades of Green
Wetland Vegetation	Dark Green – Black	Dark Red	Shades of Green
Water	Shades of Blue and Green	Shades of Blue	Black — Dark Blue
Urban Areas	White – Light Blue	Blue – Gray	Lavender
Bare Soil	White – Light Gray	Blue – Gray	Magenta, Lavender, or Pale Pink

27.0(B) Band, EMS and Characteristics

Band	*EMS*	*Characteristics*
1	Blue Light	scattered by the atmosphere and illuminates material in shadows better than longer wavelengths; penetrates clear water better than other colours; absorbed by chlorophyll, so plants don't show up very brightly in this band; useful for soil/vegetation discrimination, forest type mapping, and identifying man-made features
2	Green Light	penetrates clear water fairly well, gives excellent contrast between clear and turbid (muddy) water; helps find oil on the surface of water, and vegetation (plant life); reflects more green light than any other visible colour; man-made features are still visible
3	Red Light	limited water penetration; reflects well from dead foliage, but not well from live foliage with chlorophyll; useful for identifying vegetation types, soils, and urban (city and town) features
4	Near IR (NIR)	good for mapping shorelines and biomass content; very good at detecting and analyzing vegetation
5	Shortwave IR (SWIR)	limited cloud penetration; provides good contrast between different types of vegetation; useful for measuring the moisture content of soil and vegetation; helps differentiate between snow and clouds
6	Thermal IR (TIR or LWIR)	useful to observe temperature and its effects, such as daily and seasonal variations; useful to identify some vegetation density, moisture, and cover type; ETM+ TIR has 60-metre pixels; TIR pixels on Landsat-5 are 120 metres
7	Another SWIR	limited cloud penetration; provides good contrast between different types of vegetation; useful for measuring the moisture content of soil and vegetation; helps differentiate between snow and clouds
8	Panchromatic ("pan")	on Landsat 7 only, has 15 m resolution, used to "sharpen" images

27.0(C) Band Combination and Information Content

R, G, B	*Potential Information Content*
4,3,2	The standard "false colour" composite. Vegetation appears in shades of red, urban areas are cyan blue, and soils vary from dark to light browns. Ice, snow and clouds are white or light cyan. Coniferous trees will appear darker red than hardwoods, This is a very popular band combination and is useful for vegetation studies, monitoring drainage and soil patterns and various stages of crop growth. Generally, deep red hues indicate broad leaf and/or healthier vegetation while lighter reds signify grasslands or sparsely vegetated areas. Densely populated urban areas are shown in light blue. This TM band combination gives results similar to traditional colour infrared aerial photography.
3,2,1	The "natural colour" band combination. Because the visible bands are used in this combination, ground features appear in colours similar to their appearance to the human visual system, healthy vegetation is green, recently cleared fields are very light, unhealthy vegetation is brown and yellow, roads are gray, and shorelines are white. This band combination provides the most water penetration and superior sediment and bathymetric information. It is also used for urban studies. Cleared and sparsely vegetated areas are not as easily detected here as in the 4 5 1 or 4 3 2 combination. Clouds and snow appear white and are difficult to distinguish. Also note that vegetation types are not as easily distinguished as the 4 5 1 combination. The 3 2 1 combination does not distinguish shallow water from soil as well as the 7 5 3 combination does,
7,4,2	This combination provides a "natural-like" rendition, while also penetrating atmospheric particles and smoke. Healthy vegetation will be a bright green and can saturate in seasons of heavy growth, grasslands will appear green, pink areas represent barren soil, oranges and browns represent sparsely vegetated areas. Dry vegetation will be orange and water will be blue. Sands, soils and minerals are highlighted in a multitude of colours. This band combination provides striking imagery for desert regions. It is useful for geological, agricultural and wetland studies. If there were any fires in this image they would appear red. This combination is used in the fire management applications for post-fire analysis of burned and non burned forested areas. Urban areas appear in varying shades of magenta. Grasslands appear as light green. The light-green spots inside the city indicate grassy land cover - parks, cemeteries, golf courses. Olive-green to bright-green hues normally indicate forested areas with coniferous forest being darker green than deciduous.
4,5,1	Healthy vegetation appears in shades of reds, browns, oranges and yellows. Soils may be in greens and browns, urban features are white, cyan and gray, bright blue areas represent recently clearcut areas and reddish areas show new vegetation growth, probably sparse grasslands. Clear, deep water will be very dark in this combination, if the water is shallow or contains sediments It would appear as shades of lighter blue, For vegetation studies, the addition of the Mid-IR band increases sensitivity of detecting various stages of plant growth or stress; however care must be taken in interpretation if acquisition closely follows precipitation. Use of TM 4 and TM 5 shows high reflectance in healthy vegetated areas. It is helpful to compare flooded areas and red vegetated areas with the corresponding colours in the 3 2 1 combination to assure correct interpretation. This is not a good band combination for studying cultural features such as roads and runways.
4,5,3	This combination of near-IR (Band 4), mid-IR (Band 5) and red (Band 3) offers added definition of land-water boundaries and highlights subtle details not readily apparent in the visible bands alone. Inland lakes and streams can be located with greater precision when more infrared bands are used. With this band combination, vegetation type and condition show as variations of hues (browns, greens and oranges), as well as in tone. The 4,5,3 combination demonstrates moisture differences and is useful for analysis of soil and vegetation conditions. Generally, the wetter the soil, the darker it appears, because of the infrared absorption capabilities of water.

R, G, B	*Potential Information Content*
7,5,3	This band combination also provides a "natural-like" rendition while also penetrating atmospheric particles, smoke and haze. Vegetation appears in shades of dark and light green during the growing season, urban features are white, gray, cyan or purple, sands, soils and minerals appear in a variety of colours. The almost complete absorption of Mid-IR bands in water, ice and snow provides well defined coast lines and highlighted sources of water within the image. Snow and ice appear as dark blue, water is black or dark blue. Hot surfaces such as forest fires and volcano calderas saturate the Mid-IR bands and appear in shades of red or yellow. One particular application for this combination is monitoring forest fires. During seasons of little vegetation growth the 7 4 2 combination should be substituted. Flooded areas should look very dark blue or black, compared with the 3 2 1 combination in which shallow flooded regions appear gray and are difficult to distinguish.
5,4,3	This combination provides the user with a great amount of information and colour contrast. Healthy vegetation is bright green and soils are mauve. While the 7 4 2 combination includes TM 7, which has the geological Information, the 5 4 3 combination uses TM 5 which has the most agricultural Information. This combination is useful for vegetation studies, and is widely used in the areas of timber management and pest infestation.
5,4,1	This will look similar to the 7 4 2 combination in that healthy vegetation will be bright green, except the 5 4 1 combination is better for agricultural studies.
7,5,4	This combination involves no visible bands. It provides the best atmospheric penetration. Coast fines and shores are well defined. It may be used to find textural and moisture characteristics of soils. Vegetation appears blue. If the user prefers green vegetation, a 7 4 5 combination should be substituted. This band combination can be useful for geological studies.
5,3,1	This combination display topographic;. textures while 7 3 1 may display differences in rock types.

28.0 Thematic Mapping with GRAM ++

Functionality

- Input/Output; Edit; Vector Analysis; Raster Analysis; Terrain; Image Processing; Statistics; Geo-Sql; Gramnet; Layout

Operational Steps

- Scanning the Base Map; Registering the Base Map; Creating Layers; Digitising a Layer; Adding Labels and Permanent Attachment; Attaching Data to a Layer; Performing Vector Analysis; Thematic Mapping

Scanning the Base Map

- Before scanning draw a rectangular boundary around the Map, convert Map Scale into Ratio Form and calculate the rectangular coordinates of the 4 cardinal points
- Then scan the Base Map, save in a *desired folder* for the Raster Map Image and always keep a hard copy

Registering the Raster Image

- Open Gram++, click Map Edit and maximise the screen
- Click File, click New: File Wizard opens, put File Parameters and save
- Give map name and map description, click Next, keep Projection Parameter as it is, enter the value of ratio scale

- Click Next: Original Parameter Box opens up
- Substitute the value of false Easting and false Northing by the metre value of the denominator of the R.F., substitute all other values by zero, click Finish and save
- A Vec-File is created
- Click File, click Open Image, click Name of the Raster Map Image File in the folder and double click
- Image File opens, Click Edit, go to Add, click Tick Marks and the Tick Mark Box opens up
- Enter the coordinates of 4 cardinal points *one-by-one* with the help of arrow keys by adding Tic Marks on 4 occasions
- Click Calculate Errors and check if errors are within 2%, click OK and save. Otherwise, click View, click Tick Marks, and re-register carefully

Note

One may change Tick Marks any number of times, but only four will have RIGHT signs.
Map registration must be carried out only once for each map with a different name.

Creating a Thematic Layer

- Step I: Open Grampp, click Map Edit, open Vec File, and open Image
- Step II: Click Layer, click New, enter Name for a desired Layer (Point, Segment or Polygon)
- Step III: Click File and click Save

Note

If no layer is formed, Registration is to be done all over again.

Digitising a Layer

- Open GRAM ++, go to Map Edit, open the VEC File, open Image File, activate the Thematic Layer to be digitised
- Click Edit, go to Add and click Point (for Point Layer) <u>or</u> Segment (for Segment <u>or</u> Polygon)

Note

Cursor changes to a + sign. Begin with a Right Click. Keep moving by clicking the left button of the mouse along a line or on a point and stop with a Right Click.

For Polygon Layer

- Complete digitising
- Ensure there are no incomplete Polygons and that the junction points of adjacent Polygons are clicked
- Click Operation, click Clean, and click OK three times
- When Polygons are complete, Form Polygons opens up, click and save
- When it is complete, Label Polygons is activated, click it
- A small Map Edit Box appears: Click NO
- A small Map Edit Box with Finished Attaching Labels appears if there are no errors
- Click OK and save

Attaching Data to a Polygon Layer

- Follow all the previous steps, click Operations, click Create Table
- Enter mdb File and Table Name Box appears, enter a data source File (*.mdb) and enter Table Name, click Open

- Enter mdb File and Table Name Box reappears. Enter Table Name and Click OK. A small Map Edit Box with Table Creation Complete appears. Click OK
- Map Edit Window reappears. Click File and Save

Open MS-Access

- Open File, Insert Column, Add Field Name and then DATA as required
- Manage Data through Addition, Deletion and Mathematical Operations
- Click File, Click Save and Exit

ector Analysis and Thematic Mapping

- Open GRAM ++ and Click Vector Analysis
- Click File, Click Open, Double Click the Vec File and activate the Layer
- Click Operations, open Thematic Map Generation, choose the WAY
- Select Table, Click Fields and Other Particulars, Click OK

Go To Query, Build ONE as required, verify and save

- Go to Display, Map OUTPUT
- Go to Layout, Enter Range, Colour, and Set Attribute
- Click on the Screen, Right Click and Set Title, North Arrow, Scale, Legend and their Attributes

Click File and save or print the Output Map.

29.0 Thematic Mapping with Map Info

Functionality of Map Info

- Scanning the Base Map
- Registration of the Map (i.e., Making it Geo-Referenced)
- Creating the Layers (i.e., Cosmetic Layer, and Other Layers for Point Feature, Line Feature and Polygon Feature)
- Table creation for different aspects separately (Creation of Internal Table or Attribute Data in the software itself.
- With appropriate colour and in accordance with Data, Digitisation for different features like, cities or places (Point Layer), Drainage Network or Transport Network (Segment or Line Layer) and An Area or Administrative Boundary (Polygon or Region Layer)
- Attaching Data to the respective aspects separately for Point, Line and Polygon Topologies
- Creation of Thematic Map with Graphical Scale, Legend, North Arrow, Title, etc.
- Creation of Layout

The MapInfo Software Opens with a Quick Start Window, with 4 Start Up Options—

- Restore Previous Session
- Open Last Used Workspace
- Open a Workspace
- Open a Table

1. **Click on** "Open a Table" **and Click on Open. The** "Open Table" **Window opensup.**
2. **Choose** New Mapper as the Preffered View
3. **Choose** Raster Image Option
4. **Give** File Name **specifying** Drive **and** Folder **(i.e., Locating the Raster File), and**
5. **Click on** Open

A New Box Window pops up with 2 Options

- Display (the Image)
- Register (the image)

The Register Option is CLICKED and The Registration Window Pops-up

Registering the Image

- Choose Projection (Latitude/Longitude OR Planar Co-ordinates)
- Choose Unit (degrees OR metre)
- Click on a point (with known global coordinates); zooming may be done, if necessary for accurately locating the cursor
- A small Coordinate Box appears. Put Map-X and Map-Y and Click on OK.
- Repeat this for all the Map Points; Observe the magnitude of error in the upper part of the window; Edit the location of a point, if necessary
- Registration is complete, click OK
- Image Window appears with the Raster Image

A *New Mapper* Window opens-up with the *File Name* mentioned on its top. Now, before proceeding any further, right clicked on the *New Mapper* to find the *Layer Control Option*, where a *Cosmetic Layer* is found along with another Layer that shows the Base Map which is Open in the *New Mapper*.

Map Info Table Creation

This step is very important because without The MapInfo Table, Digitisation of the Raster Image cannot be performed

- Select the specific layer from the Layer Control Menu for Point, Line OR Area Features
- Go To FILE and click on New Table
- The "Create New Table" Window opens with 3 Options
  a) "Open New Browser" b) "Open New Mapper" c) "Add To Current Mapper"
- Select The "Add To Current Mapper" Option
- Click on Create
- Give Field Information—Name, Type, Width, Projection and Click on Create
- Give File Name (*. Tab Extension) and specify with desired Folder and Drive
- Click on SAVE

The Process is repeated to create Tables for all the Layers of user's need. Each time the "New Table Structure" Window automatically opens up, where necessary "Field Information" is Input, by adding as many fields as necessary, followed by create and save

Note

As the Table Structure is saved, the Digitisation Tools are Enabled

File Structure in Map Info

MapInfo organises all its information, whether textual or graphic, in the form of Tables. Each Table is a Group of MapInfo Files that constitutes either a Map File or a Database File. In MapInfo, when a Data File is opened, it creates a Table, that consists of at least 2 separate Files

- The first one contains the Structure of the Data, and
- The second File contains the Raw Data

All MapInfo Tables contains the following Files

- .tab: This describes the Structure of a Table. It is a small Text File describing the Format of the File containing Data
- .dat: This File contains all the raw Data
- .map: This File describes the Graphic Objects
- .id: It is a Cross-reference File that Links the Data with the Objects
- .ind: It is the Index File, that allows one to search for Map Objects using the Find Command. To locate an address, a city or a state using the Find Command, these fields must be indexed in the Table.

Digitisation in Map Info

- Now the Map is ready for digitisation. From the Layer Control menu, choose the Layer by selecting the desired Table, and making it editable
- Zoom the Area. Select appropriate Tool (i.e., Symbol Tool for Points, Polyline Tool for Lines and Polygon Tool for Boundary), and digitise it by clicking the mouse
- Digitisation can be started from any Node or Vertex. Make sure that SNAP is ON; if not, simply press S-button
- The common boundary is digitised by pressing Shift and by dragging the mouse from one common end to another and then by simply clicking on it
- When digitisation is complete, double click on each of the digitised areas to get a Window called "Region Object", that gives the details of the area digitised
- Click on "Style", Customize the "Region Style" with Fill Colour and Boundary Colour
- Click OK. This way, the Digitisation and other related jobs are DONE
- The whole work is saved as an Workspace in a desired folder and drive.

Creation of Thematic Maps

The Thematic Maps are created based on the Table Structure created

- Go to Map Option (in The Mapper) and click on "Create Thematic Map" Option
- Choose a desired Option and go on stepwise with the MAP WIZARD WINDOW
- Customize the SETTING, STYLE and LEGEND
- Insert North Arrow, Scale and other Contextual Map elements and click on OK
- The Thematic Map is now ready for the Layout

Creation of Map Layout

The LAYOUT refers to the DESIGN of the SPATIAL ARRANGEMENT of the MAP ELEMENTS on the DRAWING SHEET for enhancing the GET-UP and EFFICIENCY of CARTOGRAPHIC COMMUNICATION, maintaining all cartographic conventions and Norms/Rules

- After a Thematic Map is prepared, click on "New Layout Window" or press F5

- A New Layout is CREATED, that allows one to arrange and annotate the contents of one or several windows for printing
- Choose "One frame for Window", click OK, maximise the Layout Window, view entire Layout. Click the Frame Tool. The Frame Object Dialog Box appears. Fill it with Coordinates (Map Corners and Centroid, Fill Frame with Contents, Scale on Paper) and click OK.
- Write Map Title, Fill the Layout with all Contextual Map Elements.
- The Map is ready for a PRINTOUT: choose File, print setup, verify the information. Click OK and click Print.
- Save the Layout once again as an WORKSPACE.
- Also save the Layout as an Image File in either *.BMP, *.JPEG OR *.TIFF extensions from the "Save as Type" Option.

30.A Synoptic Chart

Each element of the observation, with the exception of wind, is plotted in a fixed position around the station circle so that individual elements can be easily identified.

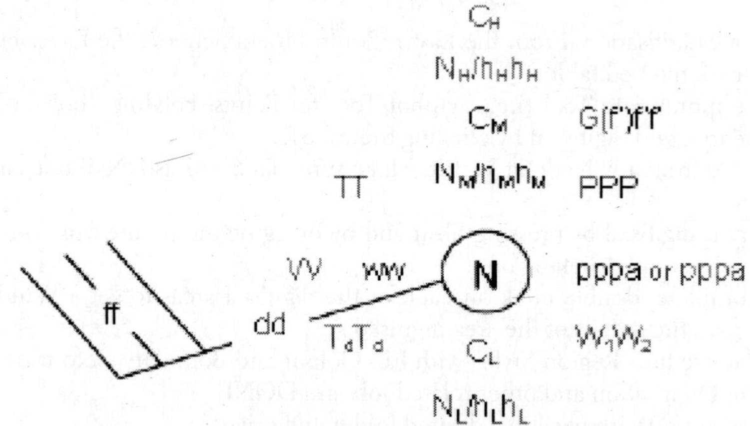

Identifier	Description
N	Total amount of cloud (in oktas)
C_L	Type of low cloud
N_L	Amount of low cloud (in oktas)
h_Lh_L	Height of low cloud (in feet)
C_M	Type of medium cloud
N_M	Amount of medium cloud (in oktas)
h_Mh_M	Height of medium cloud (in feet)
C_H	Type of high cloud
N_H	Amount of high cloud (in oktas)
h_Hh_H	Height of high cloud (in feet)
TT	Dry-bulb air temperature (in degrees Celsius)

Identifier	Description
ww	Present weather
dd	Wind direction (in degrees)
ff	Wind speed (in knots)
W	Visibility (in metres or kilometres)
T_dT_d	Dew point temperature (in degrees Celsius)
W_1W_2	Past weather
pppa or pppa	Pressure tendency and trend (bold rising, normal: falling) (in millibars)
PPP	Atmospheric pressure (in millibars)
G(f')f'f'	Wind gust (in knots)

30.B Synoptic Chart

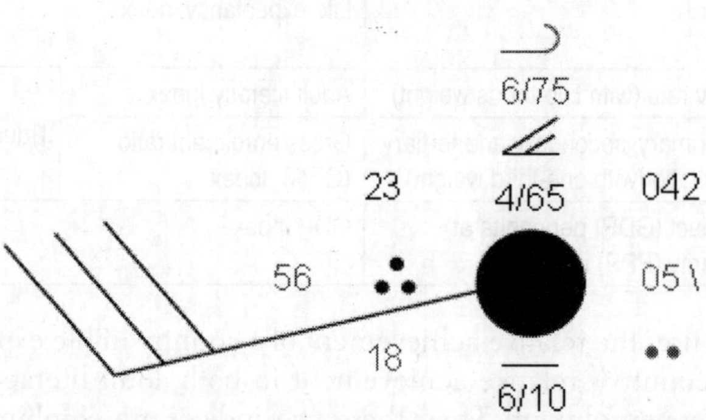

Weather as observed	Code group	Description
8 oktas	**N**	Total amount of cloud (in oktas)
23°C	**TT**	Dry-bulb air temperature (in degrees Celsius)
Continuous moderate rain	**ww**	Present weather
260°	**dd**	Wind direction (in degrees)
30 knots	**ff**	Wind speed (in knots)
6 km	**W**	Visibility (in metres or kilometres)
18°C	T_dT_d	Dew-point temperature (in degrees Celsius)
Stratus (6 oktas at 1000 feet)	C_1 **or C**	Type of low cloud
Rain	W_1W_2	Past weather
Falling 0.5mb in last 3 hours	**pppa** or pppa	Pressure tendency and trend (bold rising, normal: falling) in millibars)
1004.2mb	**PPP**	Atmospheric pressure (in millibars)
Dense alto stratus (4 oktas at 15000 feet)	C_m **or C**	Type of medium cloud
Cirrus (6 oktas at 25000 feet)	C_H **or C**	Type of high cloud

31.0 Human Development Index

The UN Human Development Index (HDI) is a comparative measure of poverty, literacy, education, life expectancy, and other factors for countries worldwide. It is a standard means of measuring human well being, developed in 1990 by the Pakistani economist *Mahbub ul Haq,* and has been used since 1993 by the UNDP in its annual reports. The HDI measures the average achievements in a country in three basic dimensions of human development life expectancy, adult literacy and gross enrollment ratio, and gross domestic product (Table 1). For each of these dimensions, an *index* is created based on the minimum and maximum values for each underlying indicator. Performance in each dimension is expressed as a value between 0 and 1 by applying the following general formula:

Dimension Index = (Actual Value – Minimum Value) / (Maximum Value – Minimum value)

Table-1: Dimension of HDI with Indicators and Indices

Dimension	Indicator Used	Dimension Index	
A long and healthy life	Life expectancy at birth	Life expectancy index	
Knowledge	1) The adult literacy rate (with two-thirds weight)	Adult literacy Index	Education Index
	2) The combined primary, secondary and tertiary gross enrolment ratio (with one-third weight)	Gross enrolment ratio (GER) Index	
A decent standard of living	Gross domestic product (GDP) per capita at purchasing power parity (PPP) in USD	GDP Index	

The life expectancy index measures the relative achievement of a country in life expectancy at birth. The education index measures a country's relative achievement in both adult literacy and combined primary, secondary and tertiary gross enrolment. Then these two indices are combined to create the education index, with two-thirds weight given to adult literacy and one-third weight to combined gross enrolment. Gross Domestic Product (GDP) is the total value of final goods and services produced within a country's borders in a year. It is one of the measures of national income and output. It may be used as an indicator of the standard of living in a country. It represents the total dollar value of all goods and services produced over a specific time period. The GDP index is calculated using adjusted GDP per capita (PPP US$). In the HDI income serves as a surrogate for all the dimensions of human development not reflected in a long and healthy life and in knowledge. Income is adjusted because achieving a respectable level of human development does not require unlimited income. Accordingly, the logarithm of income is used. The HDI is then calculated as a simple average of the dimension indices. For the data (Table 2), the indices are as follows:

1. Life expectancy index = (73.4 – 25) / (85 – 25) = 0.807
2. Adult Literacy Index = (85.3 – 0) / (100 – 0) = 0.853
3. Gross enrolment Index = (69 – 0) / (100 – 0) = 0.690
4. Education Index = (2/3) x 0.253 + (1/3) x 0.690 = 0.798
5. GDP Index = {log (3680) – log (100)} / {log (40000) – log (100)} = 0.602

Once the dimension indices have been calculated, determining the HDI is straightforward. It is a simple average of the three dimension indices, as follows:

HDI of Country X, 2001 = (0.807 + 0.798 + 0.602) / 3 = 0.735

Table 2: HDI Database and Goalposts for Country X, 2001

Indicator	Actual Value	Maximum value	Minimum value
Life expectancy at birth (years)	73.4 years	85	25
Adult literacy rate (%)	85.3%	100	0
Combined gross enrolment ratio (%)	69%	100	0
GDP per capita (PPP US$)	3,680(PPP US$)	40000	100

32.0 Gender Development Index

The GDI is a measure that adjusts the Human Development Index by gender inequalities. It is computed based on gender inequalities concerning the life expectancy at birth, adult literacy rate, combined gross enrolment ratio for primary, secondary, and tertiary schools, and estimated earned income. Basically, the GDI shows average national level of human development - but with additional focus on gender. Gender empowerment does not depend on income and women's empowerment does not depend on the level of national income. Gender empowerment varies between countries and even within countries, it may vary widely across regions. A comparison of the countries ranked by HDI and GDI shows the level of gender discrepancies. The GDI is computed in the following three steps :

First, female and male indices in each dimension are calculated as -
Dimension Index = (actual value - minimum value) / (maximum value - minimum value)

Second, the female and male indices in each dimension are combined in a way that penalizes differences in achievement between men and women. The resulting index is simply the harmonic mean of the female and male indices. It is called the equally distributed index and is calculated as, -
Equally Distributed Index
= [{female population share (female index)$^{-1}$}+[{male population share (male index)$^{-1}$}]$^{-1}$.

Third, the GDI is calculated by combining the three equally distributed indices in an unweighted average as,
GDI = (ED Life Expectancy Index + ED Education Index + ED Income Index) / 3

GDI Database of Country Y, 2001

Population Share: Female = 0.508 Male = 0.492

Indicator	Maximum Value	Minimum Value	Actual Value
Female Life Expectancy at Birth (years)	87.5	27.5	73.2
Male Life Expectancy at Birth (years)	82.5	22.5	64.9
Adult Literacy Rate (%)	100	0	Female = 94.1 Male = 97.3
Combined Gross Enrollment Ratio (%)	100	0	Female = 69.3 Male = 74.6
Estimated earned Income (PPP $)	40,000	100	Female = 4875 Male = 7975

Step-1: Computation of Equally Distributed Life Expectancy Index

Life Expectancy Index for Female = (73.2 – 27.5) / (87.5 – 27.5) = 0.762

Life Expectancy Index for Male = (64.9 – 22.5) / (82.5 – 22.5) = 0.707

Therefore, Equally Distributed Life Expectancy index = 1 / {(0.508 / 0.762) + (0.492 / 0.707)} = 0.734

Step-2: Computation of Equally Distributed Education Index

Adult Literacy Index for Female = 0.941 GE Index for Female = 0.693

Adult Literacy Index for Male = 0.973 GE Index for Male = 0.746

Female Education Index = 2 / 3 (0.941) + 1 / 3 (0.693) $\Rightarrow$ 0.858

Male Education Index = 2 / 3 (0.973) + 1 / 3 (0.746) $\Rightarrow$ 0.897

Therefore, Equally Distributed Education Index = 1 / {(0.508 / 0.858) + (0.492 / 0.897)} $\Rightarrow$ 0.877

Step-3: Computation of Equally Distributed Income Index

Income Index for Female = {log (4875) – log (100)}/{log (40000) – log (100)} = 0.649

Income Index for Male = {log (7975) – log (100)}/{log (40000) – log (100)} = 0.731

Therefore, Equally Distributed Education Index = 1/{(0.508 / 0.649) + (0.492 / 0.731)} $\Rightarrow$ 0.687

Step-4: Computation of GDI

Therefore, GDI of Country Y, 2001 = (0.734+0.877+0.687) / 3 $\Rightarrow$ 0.766

Bibliography

Abler R, J S Adams and P Gould (1971): Spatial Organization: Geographer's View of the World, Prentice-Hall, NJ
Alexander J W (1958) : Location of Manufacturing: Methods of Measurement, *AAAG*, 48(1), 20–26
Allan A L (2004): Maths for Map Makers, Whittles Publishing
Anderssen R S and M R Osborne (ed. 1970): Data Representation, Queensland University Press
Anderson J M and E M Mikhail (1985): Introduction to Surveying, McGraw Hill Book Co., NY
Andrews H C (1970): Computer Techniques in Image Processing, Academic Press, NY
Anson R W and F J Ormling (1993-96): Basic Cartography, Vols 1-3, Elsevier AS Pub., NY
Aronoff S (1989): GIS—A Management Perspective, WDL, Ottawa
Avery T E and G L Berlin (1992): Fundamentals of Remote Sensing and Air Photo Interpretation, MacMilan, NY

Bachi R (1963): Standard Distance Measures and related Measures for Spatial Analysis, *Reg. Sci. Assoc.*, 14(1), 83–132
Bagrow L (1964): History of Cartography, Watts & Co. Ltd., London
Bailey T C and A C Gatrell (1995): Interactive Spatial Data Analysis, Harlow: Longman
Balchin W G V and W V Lewis (1945): The Construction of Distribution Maps, *Geography*, 30(3), 91
Barber G M (1988): Elementary Statistics for Geographers, The Guilford Press, London
Barnes J A and A H Robinson (1940): A New Method for the Representation of Dispersed Rural Settlements, *Geog. Rev.*, 30(2), 134–137
Bernhardsen T (1999): Geographic Information Systems—An Introduction, John Wiley & Sons, NY
Berry B J L and D F Marble (ed. 1968): Spatial Analysis—A Reader in Statistical Geography, Englewood Cliff, NJ
Bhattacharyya D S and T C Bagchi (1973): Elements of Geological Map Reading and Interpretation, Orient Longman, Kolkata
Bondy J A and U S R Murty (1976): Graph Theory with Applications, North Holland, NY
Boots B and A Getis (1988): Point Pattern Analysis, London: Sage
Bugayevskiy L M and P Synder (1995): Map Projections—A Reference Manual, London : Taylor and Francis
Burrough P A and R A McDonnell (1998): Principles of GIS, OUP, NY

Cairns S (1961): Introductory Topology, New York: Ronald Press Ltd
Calkins H W and R F Tomlinson (1977): GIS—Methods and Equipments for Landuse Planning, RALI PROGRAM, USGS, Reston, Virginia
Campbell J P (1996): Introduction to Remote Sensing, Guilford, NY

Chang K T (2002): Introduction to GIS, Tata McGraw-Hill, New Delhi
Chorley R J and P Haggett (1965): Trend-Surface Mapping in Geographical Research, *Transactions of the Institute of British Geographers*, 37, 47–65
Chorley R J and P Haggett (ed. 1967): Models in Geography, Methuen, London
Chrisman N R (2002): Exploring GIS, John Wiley & Sons, NY
Christensen D E (1960): A Simplified Traffic Flow Map, *Prof. Geogr.*, 12(2), 21–22
Clark D (1963): Plane and Geodetic Surveying for Engineers, Vols. I & II, Constable, London
Clark G and T Wareham (2003): Geography @ University, Sage, London
Clarke J I (1959): Statistical Map Reading, *Geography*, 44(2), 96–104
Clarke J I (1966): Morphometry from Maps, in Dury G H (ed.) Essays in Geomorphology, Heinemann Education Books, London
Clark P G and C F Evans (1954): Distance to Nearest Neighbour as a Measure of Spatial Relationship in Population, *Ecology*, 35(1), 45–53
Clark W A V and P Hosking (1986): Statistical Methods for Geographers, New York: John Wiley
Clarke K C (2001): Getting Started with GIS, Upper Saddle River, NJ: Prentice Hall
Clarke K C (1995): Analytical and Computer Cartography, Prentice-Hall, NJ
Cleveland W S (1993): Visualising Data, Summit, NJ: Hobert Press
Cliff A D and J Ord (1981): Spatial Processes—Models and Applications, London: Pion
Clifford N and G Valentine (2004): Key Methods in Geography, Sage Pub., London
Cohew L and M Holliday (1982): Statistics for Social Sciences, Harper & Row Pub., NY
Cole J P (1969): Mathematics and Geography, *Geography*, 54(2), pp 152–164
Cole J P (1975): Situations in Human Geography—A Practical Approach, Blackwell, Oxford
Cole J P and C A M King (1970): Quantitative Geography, John Wiley & Sons, NY
Colwell R N (ed. 1983): The Manual of Remote Sensing, Falls Church, Virginia
Corbett J (1979): Topological Principles in Cartography, Washington DC: Census Bureau
Cracknell A P and L W B Hayes (1991): Introduction to Remote Sensing, Taylor & Francis, Washington
Cressie N A C (1993): Statistics for Spatial Data, John Wiley & Sons, NY
Cromley R G (1999): Digital Cartography, Englewood Cliffs, NJ: Prentice Hall
Croxton F E, D J Cowden and S Klein (1968): Applied General Statistics, Prentice-Hall, NY
Cuff D J and M T Mattson (1982): Thematic Maps—Their Design and Production, Methuen, London
Curran P J (1985): Principles of Remote Sensing, Longman, London

Dale P (2005): Introduction to Mathematical Techniques used in GIS, CRC Press, London
Dalton R et al (1975): Sampling Techniques in Geography, George Phillip & Sons, London
Davis J C (1986): Statistics and Data Analysis in Geology, John Wiley & Sons, NY
Davis J C and M McCullagh (1975): Display and Analysis of Spatial Data, John Wiley & Sons, NY
Debenham F (1937): Exercises in Cartography, Blackie & Sons Ltd., London
Deetz C H and O S Adams (1934): Elements of Map Projection, *USCGS Spl. Bull.* 8, 117–136
DeMers M N (2000): Fundamentals of GIS, John Wiley & Sons, NY
Dent B D (1990): Cartography—Thematic Map Design, William Brown Pub., Dubuque, IA
Dickinson G C (1963): Statistical Mapping and Presentation of Statistics, Edward Arnold, London
Diggle P (1983): The Statistical Analysis of Point Patterns, London: Academic Press
Doornkamp J C and King C A M (1971): Numerical Analysis in Geomorphology, Edward Arnold Pub., London
Drury S A (1993): Image Interpretation in Geology, Allen-Unwin, London
Duda R O, P E Hart, and Stork D G (2001): Pattern Classification, John Wiley & Sons, NY

Dury G H (ed.1967): Essays in Geomorphology, Heinemann Education Books, London
Dury G H (1972): Map Interpretation, Pitman Pub., London

Ebdon D (1977): Statistics in Geography—A Practical Approach, Basil Blackwell, Oxford
Esson C C and G S Philip (1943): Map Reading Made Easy, George Philip and Sons, London

Finch J K (1920): Topographic Maps and Sketch Mapping, John Wiley & Sons, NY
Fordham H G (1927): Maps—Their History, Characteristics and Uses, CUP
Fotheringham A S, C Brunsdon and M Charlton (2000): Quantitative Geography, Sage, London
Frew J (1993): Advanced Geography Fieldwork, Nelson

Garnier B J (1963): Practical Works in Geography, Edward Arnold Pub., London
Geddes A and A G Ogilvie (1938): The Technique of Regional Geography, *Jour. of MGS*, 13(2), 121–132
Gehrke W T (1944): The Wind Flow Diagram, AAAG, Vol. 34, No. 1, 63–66
Getis A (1964): Temporal Landuse Pattern Analysis with the Use of Nearest Neighbour and Quadrat Methods, AAAG, 54(4), 391–399
Ghosh A and G Rushton (1987): Spatial Analysis and Location—Allocation Models, Nostrand Reinhold, NY
Ghosh S K (1979): Analytical Photogrammetry, Pergamon, NY
Gibbs J P (ed.1961): Urban Research Methods, D Van Nostrand Co. Inc., NJ
Glendinning R M (1937): The Slope and Slope-Direction Maps, *Michigan Papers in Geography*, 7, Ann Arbor
Gonzalez R C and R E Woods (2003): Digital Image Processing, Pearson Education, Singapore
Goudie A et al (1990): Geomorphological Techniques, Routledge
Greasley B (1987): Project Fieldwork, Bell & Hymen, London
Green F W (1948): Motor-Bus Centers in S W England, *Trans. Instt. Brit. Geogr.*, 14, 58–60
Gregory S (1963): Statistical Methods and the Geographer, Longman, London
Griffith D A (1987): Advanced Spatial Statistics, Dordrecht: Kluwer Academic
Guest P G (1961): Numerical Methods in Curve Fitting, Cambridge University Press
Gupta R P (2003): Remote Sensing Geology, Springer Verlag, Berlin

Haggett P (1965): Locational Analysis in Human Geography, Edward Arnold, London
Haggett P, A W Cliff and A Frey (1977): Locational Methods, Vols. I & II, Edward Arnold, London
Haggett P and R J Chorley (1969): Network Analysis in Geography, Edward Arnold, London
Hammond R and McCullagh P S (1974): Quantitative Techniques in Geography, Claredon Press, Oxford
Harary F (1969): Graph Theory, Addison-Wesley, Reading
Haring L L and J F Lounsbury (1983): Introduction to Scientific Geographical Research, Dibuque, IA
Harley J B (2001): The New Nature of Mapping, MD: Johns Hopkins University Press
Harmon J E and S J Anderson (2003): The Design and Implementation of GIS, Wiley, NJ
Hart J F (1954): Central Tendency in Areal Distribution, *Economic Geography*, 30(1), 48–59
Harvey D H (1969): Explanation in Geography, Edward Arnold Pub., London
Heywood I, S Cornelius and S Carver (2002): An Introduction to GIS, Prentice-Hall, Harlow
Hinks A R (1921): Map Projections, Cambridge University Press
Hogg W H (1948): Rainfall Dispersion Diagrams, *Geography*, 33(1), 31–37
Hoggart K et al (2002): Researching Human Geography, Arnold, London
Hope K (1968): Methods of Multivariate Analysis, University of London Press
Hunt A J and H A Moisley (1960): Population Mapping in Urban Areas, *Geography*, 45(1), 78–79

Iliffe J C (2002): Datums and Map Projections, Whittles Publishing
Inkpen R (2005): Science, Philosophy and Physical Geography, Routledge, London
Issaks E H and R M Srivastava (1989): An Introduction to Geostatistics, OUP, Oxford

Jahne B (1997): Digital Image Processing—Concepts, Algorithms and Scientific Applications, Springer-Verlag, NY
Jensen J R (2005): Introductory Digital Image Processing, Upper Saddle River, NJ: Prentice Hall
Jensen J R (2007) : Remote Sensing of the Environment: An Earth Resource Perspective, Upper Saddle River, NJ: Prentice Hall
Johnston R J (1978): Multivariate Statistical Analysis in Geography, New York: London
Jones C (1997): Geographic Information Systems and Computer Cartography, Longman, Reading, MA
Joseph G (2003): Fundamentals of Remote Sensing, Universities Press, Hyderabad

Kadmon N (1982): Cartograms and Topology, *Cartographica*, 19(1), pp1-17
Kanetkar T P and S V Kulkarni (1965): Surveying and Levelling, Parts I & II, AVG Prakashan, Pune
Kaplan E D (1996): Understanding GPS—Principles and Applications, Artech House, Boston
Kansky K J (1963): Structure of Transportation Networks, Chicago University Press
Keates J S (1973): Cartographic Design and Production, Longman, London
Keates J S (1982): Understanding Maps, John Wiley & Sons, London
Kendall M G (1965): A Course in Multivariate Analysis, Griffin, London
Kennedy M (1996): The Global Positioning System and GIS, Ann Arbor, MI
King C A M (1966): Techniques in Geomorphology, Edward Arnold Pub., London
King L J (1969): Statistical Analysis in Geography, Englewood Cliffs, NJ: Prentice Hall
Kitanidis P K (1997): Introduction to Geostatistics, Cambridge University Press
Kitchin R and Tate N J (2000): Conducting Research into Human Geography—Theory, Methodology and Practice, Prentice Hall, Harlow
Konecny G (2003): Geoinformation—Remote Sensing, Photogrammetry and GIS, Taylor and Francis, London
Kraak M J and F Ormerling (1996): Cartography—Visualisation of Spatial Data, Longman, Harlow
Kraak M J and A Brown (2000): Web Cartography, Taylor & Francis, London
Kundu A (1975): Construction of Indices for Regionalisation, *Geog. Rev. of India*, 37(1), 21–29

Lackey E E (1937): Annual Variability Rainfall Maps of the Great Plains, *Geog. Rev.*, 27, 665–670
Lawrence G R P (1971): Cartographic Methods, Methuen, London
Lenon J and P J Cleaves (2001): Techniques and Fieldwork in Geography, Collins, London
Lillesand T M and Kiefer R W (1994): Remote Sensing and Image Interpretation, New York: Wiley
Limb M and C Dwyer (ed. 2001): Qualitative Methodologies for Geographers, London: Arnold
Lo C P (1976): Geographical Applications of Aerial Photographs, David & Charles, CRC Inc., NY
Lo C P (1986): Applied Remote Sensing, Longman, London
Lo C P and A K W Yeung (2002): Concepts and Techniques of GIS, Prentice Hall of India, ND
Lobeck A K (1924): Block Diagrams, John Wiley & Sons Ltd., NY
Longley P A et al (2001): Geographical Information—Systems and Science, New York: Wiley

MacEachern A M and D R F Taylor (1994): Visualisation in Modern Cartography, Pergamon Press, London
Marchington T (1975): Reading Maps, McDonald Educational Holywell House, London
Mainwaring J (1960): An Introduction to the Study of Map Projections, McMillan & Co., London

Maling D H (1973): Co-ordinate Systems and Map Projections, George Phillip & Sons Ltd., London
Mandal R B (1982): Statistics for Geographers and Social Scientists, Concepts Pub. House, ND
Maquire D J et al (ed. 1991): GIS—Principles and Applications, Harlow: Longman Scietific
Mardia K V and P E Jupp (2000): Directional Statistics, Wiley, NY
Mather P M (1976): Computational Methods of Multivariate Analysis in Physical Geography, John Wiley & Sons, NY
Mather P M (1991): Computer Applications in Geography, Wiley: New York
Mather P M (2004): Computer Processing of Remotely-Sensed Images, John Wiley & Sons, NY
Matthews M H and I D L Foster (1989): Geographical Data—Sources, Presentation and Analysis, OUP
McConnell D (1966): Quadrat Methods in Map Analysis, *Discussion Paper-13*, Geography, Iowa University
Medvedkov Y V (1966): Concept of Entropy in Settlement Pattern Analysis, *Reg. Sci. Assoc*, 18(2), 165–8
Meux A H (1962): Reading Topographical Maps, University of London Press Ltd, London
Mikhail E, J Bethel and J C McGlone (2001): Introduction to Modern Photogrammetry, NY: John Wiley & Sons
Miles M and Huberman A (1994): Qualitative Data Analysis, Sage, London
Miller O M and C H Summerson (1960): Slope-Zone Maps, *Geog. Rev.*, 50, 196
Moffitt F H and J D Bossler (1998): Surveying, MenloPark, Calif: Addison-Wesley
Moffitt F H and E M Mikhail (1980): Photogrammetry, New York: Harper and Row
Money D C (1969): An introduction to Mapwork and Practical Geography, UT Press Ltd, London
Monkhouse F J and H R Wilkinson (1967): Maps and Diagrams—Their Compilation and Construction, Methuen, London
Monmonier M S (1982): Computer-assisted Cartography—Principles and Prospects, Englewood Cliffs, NJ: Prentice Hall
Monmonier M and S A Schnell (1986): Map Appreciation, Prentice Hall, NJ
Morley F (1979): Advanced Geological Map Interpretation, Edward Arnold, London
Morrison D F (1967): Multivariate Statistical Methods, McGraw Hill, NY
Muehrcke P C (1981): Maps in Cartography, *Cartographica*, 18, 1–41
Munns E N (1922): The Climatograph—A New Form of Chart for Climatic Phenomena, *Monthly Weather Review*, 50(1), 67–68

Neft D S (1966): Statistical Analysis for Spatial Distributions, *Reg. Sci. Res. Inst.*, Monograph No.2, Philadelphia

Ormsby H (1950): France, Methuen, London
O'Sullivan D and D J Unwin (2003): Geographic Information Analysis, Wiley, NJ

Paine D P and J D Kiser (2003): Aerial Photography and Image Interpretation, Wiley, NJ
Pearl R M (1955): How to Know Rocks and Minerals, McGraw Hill, London
Pearson F (1990): Map Projection—Theory and Application, Boca Raton, Florida
Peterson M P (1995): Interactive and Animated Cartography, Prentice-Hall, NJ
Philipson W (ed 1996): Photointerpretation, *Am Soc for Photo & RS*, Bethesda, MD
Porter P W (1958): Putting the Isopleths in its Place, *Proceed. Minnesota Acad. Sci.*, 25, 372–384
Pratt W K (2001): Digital Image Processing, John Wiley & Sons, NY

Quattrochi D A and M F Goodchild (ed 1996): Scale in RS and GIS, Lewes Pub, Florida

Raisz E (1934): The Rectangular Statistical Cartogram, *Geog. Rev.*, 24(3), 292–296
Raisz E and J Henry (1937): An Average Slope Map of Southern New England, *Geog. Rev.* 27, 467–472
Raisz E (1938): General Cartography, McGraw Hill Book Co., NY
Raisz E (1946): Landform, Landscape, Land-Use and Land-Type Maps, *Surveying and Levelling*, 6, 220–03
Raisz E (1957): The Use of Air-Photos for Landform Maps, AAAG, 41, 324 –30
Ramachandran H (1975): Analysis of Network Geometry, *Geog. Rev. of India*, 37(1), 81–86
Rees W G (1990): Physical Principles of Remote Sensing, Cambridge University Press
Reeves E A (1910): Maps and Map Making, Royal Geographical Society, London
Rhind D W (ed. 1991): GIS—Principles and Applications, Longman, London
Richards J A and X Jia (1999): Remote Sensing Digital Image Analysis, Springer Verlag, Heidelberg
Richardus P and R K Adler (1972): Map Projections, North Holland Pub. Co., Amsterdam
Ripley B D (1981): Spatial Statistics, Wiley, NY
Robinson A H (1952): The Look of Maps, Wisconsin Univ. Press, Madison
Robinson A H et al (1995): Elements of Cartography, John Wiley & Sons, NY
Robinson G (1998): Methods and Techniques in Human Geography, Wiley, NY
Rogerson P A (2001): Statistical Methods for Geography, Sage, London
Ross Mackay J (1949): Dotting the Dot Maps, *Surveying & Mapping*, 9, 102–112
Ross Mackay J (1952): Some Problems and Techniques in Isopleth Mapping, *Surveying & Mapping*, 12(1), 32–38
Ross Mackay J (1953): Percentage Dot Maps, *Economic Geography*, 29(3), 263–266
Ross Mackay J (1955): An Analysis of Isopleth and Choropleth Intervals, *Economic Geography*, 31(1), 71–81

Sabins F F (1997): Remote Sensing—Principles and Interpretation, Freeman, NY
Sandover J A (1961): Plane Surveying, Edward Arnold Pub., London
Sarkar, A (1991): Tendency of the Urban Settlement Patterns in West Bengal, 1901–1981, *Indian Journal of Landscape Systems and Ecological Studies* 14(2), 50–55
Sarkar, A (1995): Bibhinna Prakar Manchitra—O—Tather Byabahar, *Bhugolika (Bengali)* 3(1), 23–29
Sarkar, A (1995): Field Work in Geography—The Basic Principles, *Indian Journal of Landscape Systems and Ecological Studies* 18(2), 62–68
Sarkar, A (1999): Geographical Data Analysis—Concepts and Issues, *Traverse*, September Issue, Geographical Institute, Presidency College, Calcutta, 10–15
Sarkar, A (2000): Fundamentals of Digital Cartography—An Appraisal, *Traverse*, November Issue, Geographical Institute, Presidency College, Calcutta, 10–15
Sarkar, A (2000): Manchitrabijnan—Bikash o Sadharan Dharana, in Roy R (ed: 2000): *Prateeti* (Bengali), October, 41–51
Sarkar, A (2002): Manchitrabidyar Pracheen Itihas, *Bhugolika (Bengali)*, BBM, 8 (1&2), 3–7.
Sarkar, A (2003): Maps—An Interface to Geographic Information, *Traverse*, September Issue, Geographical Institute, Presidency College, Calcutta, 1–5
Sarkar, A (2006): Principles of Mapping and Analysis of Terrain—A Modern Concept, *Traverse*, January, vol. 38, Geographical Institute, Presidency College, Kolkata
Sarkar, A (2007): GIS and Cartographic Visualization, *Autumn Annual*, PCAA, vol. 36, 149–57
Sarkar, A (2007): Statistical Analysis of Drainage Networks—A Study in the BMB—CGC Complex, *Indian Journal of Landscape Systems and Ecological Studies* 30(1), 5–16
Sarkar, A (2009): David Harvey's Explanation in Geography and TFL, *Autumn Annual*, PCAA, vol. 38, 110–26
Sarkar, A (2011): Maps—Yesterday, Today and Tomorrow, *Bhugolika* 17(2), January, 10–18

Sarkar, A (2012): Analysis of Human Settlement Patterns Using RS and GIS in the Plains of West Bengal, in Mandal, D K (ed. 2012): Applications of Remote Sensing and GIS in Resource Management, University of North Bengal, 246–62
Sarkar, A (2013): Quantitative Geography—Techniques and Presentations, Orient BlackSwan, New Delhi
Sarkar, A and U Mukhopadhyay (1990): Working Out a Geological Map on Geographic Perspective, *Research Bulletin*, 1990, No. 2, CAAGR, 53–63
Sarkar, A and P P Patel (2008): Hypsometric Analysis of the Dulung N. Basin and Its Sub-basins, *Geographical Review of India*, December 2008, 36(4), 409–22
Sarkar, A and P P Patel (2010): Terrain Characterization using SRTM Data, *Journal of the Indian Society of Remote Sensing* 38(1), March, 11–24
Sarkar, A and P B Roy (1981): Some Selected Map Projections for India—Their Relative Efficiencies, *Geographical Review of India*, 43(2), 182–195
Sarkar, A and I Saha (2013): Regional Disparities in Development—A Village Level Study of Binpur-I Block, West Bengal, in Majumder. S (ed.): Contemporary Issues on Environment and Development of India and Adjacent Countries, 460–68, 2013. Sandhya Prakashani, Kolkata
Shaw N and J G Garbett (1933): A New Sort of Wind Roses, *Quarterly. Jour. RMS.*, 59(1), 39–44
Shaw R L and D Wheeler (1985): Statistical Techniques in Geographical Analysis, John Wiley & Sons, NY
Sheskin I M (1985): Survey Research for Geographers, *Assoc. Am. Geogr.*, Washington
Singh R L (1979): Elements of Practical Geography, Kalyani Pub., New Delhi
Short N M and R W Blair (1986): Geomorphology from Space, *NASA SP*-486, US GOV PO, Washington
Slama C C (ed. 1981): Manual of Photogrammetry, Am Soc of Photogm, Falls Church, Virginia
Slocum T A (1999): Thematic Cartography and Visualization, Prentice-Hall Inc, NJ
Smith G H (1935): The Relative Relief of Ohio, *Geog. Rev.*, 25, 272–284
Snyder J P (1997): Flattening the Earth—2000 Years of Map Projections, Chicago University Press
Snyder J P (1987): Map Projections—A Working Manual, USGS, USG PO, Washington DC
Soh Z A (1980): Post-Certificate Practical Geography, Oxford University Press
Star J and J Estes (1990): GIS—An Introduction, Prentice-Hall, NJ
Starck J L, F Murtagh and A Bijaou (1998): Image Processing and Data Analysis, CUP, Cambridge
Starling T D and S V Pollack (1968): Introduction to Statistical Data Processing, Prentice-Hall, NJ
Stoddard R H (1982): Field Techniques and Research Methods in Geography, Kendall Hunt Pub. Co., Iowa
Strahler A N (1954): Statistical Analysis in Geomorphic Research, *Jour. Geol.*, 62(1), 1–25
Strahler A N (1956): Quantitative Slope Analysis, *Bulletin of the* GSA, 67, 571–96
Streich T A (1986): Geographic Data Processing—An Overview, California Univ. Press, Santa Barbara
Sviatlosky E E and W C Eells (1937): The Centrographical Method and Regional Analysis, *Geog. Rev.*, 27, 240–254
Swan A R H et al (1996): Introduction to Geological Data Analysis, Blackwell, Austria

Taylor P J (1977): Quantitative Methods in Geography—An Introduction to Spatial Analysis, Houghton Mifflin, Boston
Tobler W R (1959): Automation and Cartography, *Geog. Rev.*, 49(4), 526–534
Tomlin C D (1990): GIS and Cartographic Modeling, Prentice-Hall, NJ
Townsend J R G (ed. 1981): Terrain Analysis and Remote Sensing, George Allen & Unwin, London
Tufte E R (2001): The Visual Display of Quantitative Information, Graphics Press, Chesire
Tukey J W (1977): Exploratory Data Analysis, Addison-Wesley, Mass

Unwin D (1981): Introductory Spatial Analysis, New York: Methuen

Walford N (2002): Geographical Data—Characteristics and Sources, Wiley, NJ
Walsh J (ed.1966): Numerical Analysis—An Introduction, Academic Press, London
Waters R S (1958): Morphological Mapping, *Geography*, 43, 10–17
Watson G S (1971): Trend Surface Analysis, *Mathematical Geology*, 3, 215–226
Weaver J C (1954): Crop Combination Regions in the Middle West, *Geog. Rev.*, 44, 175-200
Wentworth C K (1930): A Simplified Method of Determining the Average Slope of Land Surfaces, *American Journal of Science*, Series 5, Vol. 20
White W A (1943): Topographic Sketches from Contour Maps, *Surveying & Mapping*, 3(4)
Winterbotham H J L (1934): Dots and Distribution, *Geography*, 19(3), 211–213
Winterbotham H J L (1936): A Key to Maps, Blackie & Sons Ltd., London
Wolf P R (1983): Elements of Photogrammetry, McGraw-Hill, NY
Wolf P R and R C Brinker (1989): Elementary Surveying, Harper & Row Pub., NY
Wolf P R and B A Dewitt (2000): Elements of Photogrammetry with applications in GIS, McGraw-Hill, NY
Wood D (1993): The Power of Maps, Routledge, London
Wood C H and C P Keller (1996): Cartographic Design, Wiley, Chichester
Worboys M (2004): GIS—A Computing Perspective, London: Taylor and Francis
Worthington B D R and R Gont (1975): Techniques in Map Analysis, McMillan Ltd, London
Wright J K (1936): A Method of Mapping Densities of Population, *Geog. Rev.*, 26(1), 78–89
Wright J K (1937): Some Measures of Distributions, AAAG, 27(2), 177–211
Wrigley N and R J Bennett (ed.1981): Quantitative Geography, Methuen, London

Zipf G K (1949): Human Behaviour and the Principle of Least Effort, CUP
Zobler L (1957): Statistical Testing of Regional Boundaries, AAAG., 47(1), 83–95

Index